Techniques and Technologies in Electrical Stimulation for Neuromuscular Rehabilitation

Other related titles:

You may also like

- PBHE016 | Wai Yie Leong | EEG Signal Processing: Feature extraction, selection and classification methods | 2019
- PBCE114 | Toshihisa Tanaka and Mahnaz Arvaneh | Signal Processing and Machine Learning for Brain–Machine Interfaces | 2018
- PBCE108 | Shaoping Bai, Gurvinder Singh Virk and Thomas Sugar | Wearable Exoskeleton Systems: Design, control and applications | 2019
- PBHE031 | Nachiappan Chockalingam | Technologies and Techniques in Gait Analysis: Past, present and future | 2022

We also publish a wide range of books on the following topics:
Computing and Networks
Control, Robotics and Sensors
Electrical Regulations
Electromagnetics and Radar
Energy Engineering
Healthcare Technologies
History and Management of Technology
IET Codes and Guidance
Materials, Circuits and Devices
Model Forms
Nanomaterials and Nanotechnologies
Optics, Photonics and Lasers
Production, Design and Manufacturing
Security
Telecommunications
Transportation

All books are available in print via https://shop.theiet.org or as eBooks via our Digital Library https://digital-library.theiet.org.

IET HEALTHCARE SERIES 62

Techniques and Technologies in Electrical Stimulation for Neuromuscular Rehabilitation

Edited by
Ian Swain, Jane Burridge and Tamsyn Street

The Institution of Engineering and Technology

About the IET

This book is published by the Institution of Engineering and Technology (The IET).

We inspire, inform and influence the global engineering community to engineer a better world. As a diverse home across engineering and technology, we share knowledge that helps make better sense of the world, to accelerate innovation and solve the global challenges that matter.

The IET is a not-for-profit organisation. The surplus we make from our books is used to support activities and products for the engineering community and promote the positive role of science, engineering and technology in the world. This includes education resources and outreach, scholarships and awards, events and courses, publications, professional development and mentoring, and advocacy to governments.

To discover more about the IET please visit https://www.theiet.org/.

About IET books

The IET publishes books across many engineering and technology disciplines. Our authors and editors offer fresh perspectives from universities and industry. Within our subject areas, we have several book series steered by editorial boards made up of leading subject experts.

We peer review each book at the proposal stage to ensure the quality and relevance of our publications.

Get involved

If you are interested in becoming an author, editor, series advisor, or peer reviewer please visit https://www.theiet.org/publishing/publishing-with-iet-books/ or contact author_support@theiet.org.

Discovering our electronic content

All of our books are available online via the IET's Digital Library. Our Digital Library is the home of technical documents, eBooks, conference publications, real-life case studies and journal articles. To find out more, please visit https://digital-library.theiet.org.

In collaboration with the United Nations and the International Publishers Association, the IET is a Signatory member of the SDG Publishers Compact. The Compact aims to accelerate progress to achieve the Sustainable Development Goals (SDGs) by 2030. Signatories aspire to develop sustainable practices and act as champions of the SDGs during the Decade of Action (2020–30), publishing books and journals that will help inform, develop, and inspire action in that direction.

In line with our sustainable goals, our UK printing partner has FSC accreditation, which is reducing our environmental impact to the planet. We use a print-on-demand model to further reduce our carbon footprint.

British Library Cataloguing in Publication Data
A catalogue record for this product is available from the British Library

ISBN 978-1-83953-876-6 (hardback)
ISBN 978-1-83953-877-3 (PDF)

Typeset in India by MPS Limited

Cover image: Tamsyn Street

Contents

16 Paediatric applications   485
Kirsten Hart, Matt White, Chris Smith and Kristin Girshin

Abbreviations

AB	Able Bodied
AC	Alternating Current
ACh	Acetylcholine
ACL	Anterior Cruciate Ligament
ACPIN	Association of Chartered Physiotherapist in Neurology (UK)
AD	Autonomic Dysreflexia
AFO	Ankle Foot Orthosis
AI	Artificial Intelligence
AIS	(ASIA) Impairment Scale
ANPT	Academy of Neurological Physical Therapists (US)
AP	Action Potential
APCP	Association of Chartered Physiotherapists in Paediatrics
APTA	American Physical Therapy Association
ARAT	Action Research Arm Test
ASIA	American Spinal Injuries Association
ATP	Adenosine Triphosphate
BCI	Brain–Computer Interface
BMAC	Burst Mode Alternating Current
BMD	Bone Mineral Density
BMR	Basal Metabolic Rate
BoNTA	Botulinum Toxin A
BWS	Body Weight Support
CA	Competent Authority
CE	Conformite Europenne
CG	Control Group
CMIT	Constraint Induced Movement Therapy
CMT	Charcot Marie Tooth
CNS	Central Nervous System
COPD	Chronic Obstructive Pulmonary Disease
CPD	Continual Professional Development
CPG	Clinical Practice Guideline

CST	Corticospinal Tract
CT	Computerised Tomography
CVA	Cerebral Vascular Accident
CP	Cerebral Palsy
CSA	Cross-Sectional Area
CVD	Cardiovascular Disease
DC	Direct Current
dEMG	Diaphragm EMG
DGNS	Dorsal Genital Nerve Stimulation
DM	Diabetes Mellitus
DP	Diaphragm Pacing
DSD	Detrusor Sphincter Dys-synergia
DTI	Deep Tissue Injury
DVT	Deep Vein Thrombosis
DXA	Dual Energy X-Ray Absorptiometry
ECMO	Extracorporeal Membrane Oxygenation (Heart-Lung Machine)
ED	Erectile Dysfunction
EDSS	Expanded Disability Scale Score
EEG	Electro Encephalograph
EMG	Electro Myograph
EMI	Electromagnetic Interference
Eq-5D-5L	EuroQoL Health Outcome Score
ERS	Electrical Rectal Stimulation
ES	Electrical Stimulation
EU	European Union
FDA	Food and Drugs Administration (US)
FDP	Flexor Digitorum Profundus
FDS	Flexor Digitorum Superficialis
FES	Functional Electrical Stimulation
FTW	Failure to Wean
FSP	Familial Spastic Paraplegia
GHTF	Global Harmonisation Task Force
GMFM	Gross Motor Function Measure
GMP	Good Manufacturing Practice
GSPR	General Safety and Performance Requirement
GSW	Gun Shot Wound
HAD	Hospital Anxiety and Depression Index
HIIT	High Intensity Interval Training

HOOS	Hip Disability and Osteoarthritis Outcome Score
HUI3	Health Utility Index v3
HVMPC	High Voltage Monophasic Pulsed Current
HVPC	High Voltage Pulsed Current
ICD	Implanted Cardiac Defibrillator
ICF	International Classification of Function Disability and Health
ICU	Intensive Care Unit
ICUAW	ICU Acquired Weakness
IFT	Interferential Therapy
IL-x	Interleukin-x
IMU	Inertial Measurement Unit
IRB	Institutional Review Board (US)
IPG	Implantable Pulse Generator
ISNCSCI	International Standards for Neurological Classification of SCI
ISO	International Standards Organisation
IT	Ischial Tuberosity
KOOS	Knee Injury and Osteoarthritis Outcome Score
LCE	Leg Cycling Exercise
LDF	Laser Doppler Flowmetry
LL	Lower Limb
LMN	Lower Motor Neuron
LOS	Length of Stay
LSCI	Laser Speckle Contrast Imaging
LVAD	Left Ventricular Assistive Device
LVEF	Left Ventricular Ejection Fraction
MAS	Modified Ashworth Score
MDR	Medical Devices Regulations
MEP	Motor Evoked Potential
MP	Motor Point
MRC	Medical Research Council (UK)
MRI	Magnetic Resonance Imaging
MS	Multiple Sclerosis
MSK	Musculoskeletal
MTS	Modified Tardieu Score
MUAP	Motor Unit Action Potential
MV	Mechanical Ventilation
MVA	Motor Vehicle Accident
MVIC	Maximum Voluntary Isometric Contraction

xMWT	xMetre Walking Test
NB	Notified Body
NBD	Neurogenic Bowel Dysfunction
NDO	Neurogenic Detrusor Overactivity
NDT	Neuro Development Therapy
NED	Neuromuscular Electrophysiological Disorder
NESS	Neuromuscular Electrical Stimulation System
NHS	National Health Service (UK)
NICE	National Institute for Health and Care Excellence (UK)
NIRS	Near Infrared Spectroscopy
NIV	Non-Invasive Ventilation
NLUTD	Neurogenic Lower Urinary Tract Dysfunction
NMES	Neuromuscular Electrical Stimulation
NMJ	Neuro Muscular Junction
NR	Neuromuscular Rehabilitation
OAB	Over Active Bladder
QALY	Quality Added Life Years
QMS	Quality Management System
QoL	Quality of Life
PAS	Paired Associative Stimulation
PCI	Physiological Cost Index
PD	Parkinson's Disease
PIADS	Psychological Impact of Assistive Devices Score
PICU	Paediatric ICU
PMS	Post Market Surveillance
PNS	Peripheral Nervous System
PPG	Photoplethysmography
PROM	Patient Reported Outcome Measure
PRISM	Patient Reported Impact of Spasticity Measure
PRP	Posterior Root Reflex
PRT	Progressive Resistance Training
PSP	Perisynaptic Schwann Cell
PTNS	Percutaneous Nerve Stimulation
PU	Pressure Ulcer
PVS	Penile Vibratory Stimulation
PW	Pulse Width
RCT	Randomised Controlled Trial
RFM	Radio Frequency Microstimulator

RM	Repetition Maximum
ROM	Range of Movement
RST	Reticulospinal Tract
RTA	Road Traffic Accident
SARS	Sacral Anterior Root Stimulator
SCI	Spinal Cord Injury
	C – Cervical
	T – Thoracic
	L – Lumbar
SCS	Spinal Cord Stimulation
SD	Strength Duration
SDR	Selective Dorsal Rhizotomy
SDT	Strength Duration Test
SET	Stimulus Electrodiagnosis Test
SMA	Spinal Muscular Atrophy
SNM	Sacral Neuromodulation
SOC	Standard of Care
SPG	Stereo Photogrammetry
TBI	Traumatic Brain Injury
TcPO2	Transcutaneous Oxygen Monitoring
tDCS	Transcutaneous Direct Current Stimulation
TENS	Transcutaneous Electrical Nerve Stimulation
TF	Technical File
TMS	Transcutaneous Magnetic Stimulation
TNF	Tumour Necrosis Factor
TTNS	Transcutaneous Nerve Stimulation
UDI	Unique Device Identifier
UL	Upper Limb
UMN	Upper Motor Neuron
VIDD	Ventilator Induced Diaphragm Dysfunction
WHO	World Health Organisation

Glossary

Term	Definition
Acetylcholine	A neurotransmitter that plays a role in memory, learning, attention, arousal and muscle contraction.
Action Potential	A rapid rise in the membrane potential of a nerve cell leading to a nerve impulse travelling along the axon from the cell. It also causes depolarisation of adjacent cells.
Active Electrode	The negative electrode.
Adenosine Triphosphate	A nucleoside triphosphate that provides energy to drive and support many processes in living cells, such as muscle contraction and nerve impulse propagation.
Afferent Nerve Fibre	A (sensory) nerve fibre leading from the body to the central nervous system.
Autonomic Dysreflexia	An over-reaction of the autonomic nervous system that causes sudden and severe rise in blood pressure (and other symptoms). People with a high-level spinal cord injury are most at risk.
Axon	The nerve fibre carrying (efferent) signals away from the cell body. While a nerve has many dendrites it only has one axon.
Cauda Equina	Literally 'horses tail'. The bundle of peripheral nerves that extend within the spinal column from T12.
Cell Soma	The body of the cell.
Central Nervous System	The part of the nervous system comprising the brain and spinal cord. Below the level of T12 the spinal cord becomes the corda equina, a large bundle of peripheral nerve fibres – i.e. not part of the central nervous system.
Chronaxie	Length of time a current at twice the rheobase must be applied to produce a measurable response.
Circumneutral Electrodes	Implanted electrodes that wrap around a nerve. i.e. a bundle of nerve fibres often including both sensory and motor fibres.
Corticospinal Tract	Nerve fibres leading from the cortex of the brain to the spinal cord.
Dendrite	A short, branched extension of the nerve cell that transmits signals towards the cell body. Each nerve cell has numerous dendrites.
Denervated Muscle	Muscle that lacks a peripheral nerve connecting it to the central nervous system.

(Continues)

(Continued)

Term	Definition
Efferent Nerve Fibre	A nerve transmitting signals from the central nervous system to the body.
Electrical Stimulation in Medicine	Any form of electrical stimulation used in the treatment of patients.
Epimysial Electrodes	Surgically implanted electrodes sutured to the surface of a muscle near the motor point i.e. where a nerve enters the muscle.
Epineural Electrodes	Neural interface devices that are placed on or around peripheral nerves for recording or stimulating. Typically, they are longitudinal strips with multiple contact sites and are sutured to the epineurium.
Frequency	Measured in Hz. It is the number of pulses/second.
Functional Electrical Stimulation	Application of electrical stimulation that, at the time of application, results in a change in physical function or behaviour.
Implanted Electrode	Any type of electrode that is surgically implanted within the body.
Indifferent Electrode	The positive electrode.
Innervated Muscle	Muscle that has peripheral nerves connecting it to the central nervous system.
Interferential Therapy	Two different frequencies of electrical current that interfere with one another to produce a medium frequency current within the tissues. It is commonly used to reduce oedema, relieve pain and stimulate muscle, without causing a contraction.
Lower Motor Neuron	Nerve pathways leading from or to the central nervous system.
Motor Unit	An anterior horn cell, its axon and nerve branches, their neuromuscular junctions, and all of their innervated muscle fibres.
Motor Unit Action Potential	The sum of the extracellular potentials of muscle fibre action potentials of a given motor unit. The waveform is determined by the inherent properties of the motor unit.
Myelin Sheath	An insulating layer around nerves, including those in the brain and spinal cord. It is made up of protein and fatty substances. The myelin sheath allows electrical impulses to transmit quickly and efficiently along the nerve cells.
Nerve Conduction	The process by which nerve signals are transmitted along nerve fibres.
Neuromodulation	Electrical stimulation that modifies nervous system activity for a therapeutic effect.
Neuromuscular (Electrical) Stimulation	Any form of electrical stimulation applied to the muscles via the nerves.
Neuromuscular Junction	The region where a motor nerve synapses with a muscle. It includes the motor nerve terminal, the synaptic space or cleft, and the muscle endplate region. The neurotransmitter ACh is synthesised in the motor nerve terminal and stored in vesicles.

(Continues)

(*Continued*)

Term	Definition
Neuron	A nerve fibre (motor and sensory).
Node of Ranvier	A gap in the myelin sheath that allows ions to flow in and out of the neuron. Nodes of Ranvier speed-up transmission of nerve signals by allowing them to 'leap' from one node to the next.
Notified Body	An organisation designated by an EU Member State (or other countries under specific agreements) to assess the conformity of medical devices and products.
Percutaneous Electrodes	A thin wire that is inserted through the skin to provide an electrical interface with the nervous system. They can be used to either stimulate and record.
Peripheral Nerve	Nerve connecting the central nervous system with the muscles or other innervated organs.
Post Market Surveillance	The process of monitoring the safety and performance of medical devices after they have been released to the market. This includes collecting data from users, analysing that data, and acting if necessary.
Pulse Width	The duration of a single electrical pulse.
Quality Management System	Business processes and procedures which aim to ensure that the quality of products or services meets (or exceeds) customer expectations. Such systems are typically repeatable and measurable and rely on continuous improvement.
Resting Potential	The electrical charge difference between the inside and outside of a neuron when it is not actively sending or receiving signals.
Rheobase	Magnitude of current just sufficient to excite the muscle.
Skeletal Muscle	Muscles that are under voluntary control and are able to generate movement. At least one end of the muscle will be attached to the skeleton.
Surface Electrodes	Electrodes placed on the skin surface. They can be used to both stimulate (i.e. transmit electrical impulses to the body) or record signals from the neuromuscular junction.
Technical File	A set of documents that describes a product and can prove that the product was designed in accordance with the requirements of a quality management system.
Transcutaneous Electrical Nerve Stimulation	Low-voltage electrical currents to relieve pain. A TENS unit is a small device that delivers the current at or near nerves to block or change perception of pain. It is thought to utilise the 'gate theory' of habituation.
Upper Motor Neuron	A nerve within the central nervous system (i.e. the brain and spinal cord above the level of T12).
Waveform	The shape of a stimulation pulse.

About the editors

Ian Swain is a professor of clinical engineering at Bournemouth University, UK. Ian helped established the National Clinical FES (Functional Electrical Stimulation) Centre in Salisbury and took several products they designed to market. Ian was clinical director of OML until 2017 and personally assessed over 3500 patients. Ian has served as a board member of the International Functional Electrical Stimulation Society (2002–05). He has 7 patents and has written over 200 papers in rehabilitation engineering.

Jane Burridge is a professor of restorative neuroscience at the University of Southampton where she leads the Neurorehabilitation Research Group. Drawing from her research, she designs and evaluates rehabilitation technologies that will aid recovery following central nervous system lesions such as stroke or spinal cord injury. She has written over 350 journal articles and conference papers. She is a former President of IFESS (2010).

Tamsyn Street is an interdisciplinary research fellow at Salisbury NHS Foundation Trust with a background in motor neurorehabilitation. She is currently an executive board member of the IFESS. Her research interests include the use of electrical stimulation technologies for improving motor neurorehabilitation, neurogenic bowel, bladder, sexual functioning and respiratory functioning and exploring the underlying mechanisms of electrical stimulation to optimise outcomes for patients.

Introduction

This chapter explains the scope and aims of the book, introduces the concept of electrical stimulation (ES), defines terms used throughout the book, explains the structure and provides brief synopses of each chapter.

Who is this book for?

It is for everyone interested in ES. Engineers, designing ES systems, will learn about clinical research and ES applications and about the underlying physiology and neuroscience. The non-technical style of the book will ensure that it is accessible to therapists who do not have a technical or engineering background. It will extend their understanding of the neurophysiology and neuroplasticity that explains the effects of ES. Anyone conducting research, whether they be from a clinical, scientific or engineering background, will find that the book fills in the gaps in their knowledge and understanding.

What is electrical stimulation?

It is, in its simplest definition, the elicitation of muscle contraction using electric impulses. It involves the application of electrical impulses to the nerves that supply muscles, by means of surface electrodes placed over the nerves or muscle belly or via electrodes implanted within the body. The aim is to evoke a muscle contraction that may or may not complement the patient's own voluntary effort. People may not be able to activate their own muscles sufficiently to execute effective movement, either due to damage to the central nervous system preventing the signals from the brain getting to the muscles due to a stroke, spinal cord injury, multiple sclerosis, etc., and/or due to disuse as a result of pain, immobility, etc. In such cases, ES can be used to restore or improve impaired function by initiating or complementing muscle activity. Stimulation can be used either to provide exercise and hence build up strength and endurance or timed to a physical activity such as walking to improve quality of movement and function.

ES is not a new technique and Chapter 1 provides a brief entertaining history, from torpedo fish used by the ancient Egyptians and Greeks to emerging control of stimulation using artificial intelligence. Recent advances in microelectronics, computer control systems and smart fabrics, alongside a better understanding of neuroscience and neuroplasticity, have enabled great strides towards the effective use of ES, but, although there has been an increase in the number of patients treated, range of clinical applications, number of therapists trained to use it and evidence for its effectiveness, clinical use still lags scientific research. An important aim of this book is therefore to provide practical guidance, built on sound evidence, to clinicians so that they can use ES with greater confidence.

What topics does the book cover, and which are excluded?

There is a very large number of clinical applications of ES and this book does not encompass them all. Our focus is on applications that stimulate muscle contraction that results in movement; but we have also included chapters on bladder, bowel and sexual function, applications to improve respiratory function, afferent stimulation and wound healing and prevention. The book does not include widely used applications such as cardiac pacemakers, stimulation for pain management such as transcutaneous nerve stimulation (TNS), cochlear implants or deep brain stimulation. The book also includes chapters on safety and regulatory issues and on providing a clinical service.

Stimulation terms and parameters

Electrical stimulation (ES)

This is the overarching term and covers all forms of therapeutic ES and all types of ES listed below.

Surface electrical stimulation (SES)

Surface electrical stimulation encompasses all types of ES but, as its name implies, only when electrodes are placed on the skin to stimulate neural structures transcutaneously.

Neuromuscular electrical stimulation (NMES)

Is ES that causes a muscle contraction with the aim of increasing strength or endurance.

Functional electrical stimulation (FES)

Is ES used functionally to provide or assist function. An example is a drop foot stimulator that provides dorsiflexion of the foot during walking or stimulation to the forearm to open and close the hand to provide a functional grasp.

Transcutaneous electrical nerve stimulation (TENS)

TENS is typically used for pain relief by stimulation of the sensory or afferent fibres. Pulses are typically of smaller pulse width than NMES/FES and are not intended to cause a muscular contraction.

Afferent stimulation

Is ES of afferent nerves to modify the neural network with aim of improving motor behaviour?

How is the book structured?

Except for Chapters 1–3: Introduction, Physiology and Safety and Regulatory issues, which do not relate to the clinical application of functional electrical stimulation (FES), chapters follow a common format, although emphasis will vary, for example applications that are currently widely used clinically will have more detail about treatment and those that are currently mainly used in research will focus more on theoretical concepts, evidence and development. Each chapter will:

- Give an overview of the topic
- Present a brief survey of the background literature and, where appropriate evidence and uncertainty; presenting arguments for and against and discuss current challenges
- Discuss the current state of the art and where relevant, perceived and objective data of efficacy and the views of stakeholders
- Provide practical clinical considerations
- Make suggestions for future directions of engineering and scientific research and development, clinical applications and use.

Chapter synopses

Chapter 1. History and introduction to electrical stimulation (ES) provides a brief historical background to the clinical use of electrical stimulation and explains, from a physical perspective, the key principles and terminology associated with electrical stimulation.

Chapter 2. Neuromuscular physiology provides a comprehensive explanation of muscle and nerve physiology which underpins the use of ES and will enable the reader to understand why and how ES is used clinically. It explains the difference between normal nerve conduction and nerve conduction elicited by ES and the effect of ES on muscle and neuroplasticity.

Chapter 3. Overview of regulatory matters and electrical safety is essential reading for anyone proposing to undertake clinical research or for anyone involved in the development of devices intended for medical use.

Chapter 4. Afferent and therapeutic stimulation treatment after spinal cord injury distinguishes between central afferent and peripheral afferent stimulation. Three goals are explained: peripheral (neuropathic) pain relief, modification of spasticity or muscle stiffness and augmentation of impaired motor control or sensory perception. Evidence is reviewed and methods of application using surface, transcutaneous or epidural stimulation are explained

Chapter 5. Upper extremity rehabilitation describes the anatomical and neurophysiological bases of movement, the impact of upper extremity impairment on quality of life and the objectives of using ES with a range of conditions. It provides practical guidance on how to use ES, when and with whom. It discusses therapeutic and orthotic applications and the use of ES in combination with other therapies

Chapter 6. Electrical stimulation in musculoskeletal populations defines and describes musculoskeletal (MSK) conditions and the impact they have on patients. It explains the principles of using ES, especially to overcome problems of muscle atrophy, weakness and pain. It presents the evidence for effectiveness in a range of conditions, including pre- and post-operative lower limb joint replacement, shoulder surgery and osteoarthritis. The chapter also provides clinical guidance for the therapeutic use of ES with MSK patients.

Chapter 7. FES to support walking in people with upper motor neuron lesions acknowledges that this is the most common application of ES and has translated into routine clinical practice. The chapter defines normal gait and the commonly encountered abnormalities, explaining how ES can be used to improve walking, both as an orthosis and as a therapeutic tool. The mechanisms which are thought to account for improvements are described and evidence is presented. Different ES approaches – implanted vs. surface stimulation, electrode placement and stimulation control and timing, for example, are described. Recommendations are made for outcome measures and two case studies are presented.

Chapter 8. Functional electrical stimulation of denervated muscles describes the physiological effects of denervation, the clinical consequences, the opportunities for ES and potential benefits. The issue of clinical scepticism, due to the need for long pulse duration and the risk of inhibiting the reinnervation process is explained and supported by research evidence. Practical guidance is given for clinical application, illustrated by outcomes in a range of cases.

Chapter 9. Prevention and treatment of pressure ulcers using electrical stimulation describes the problem that ES addresses and its use both to prevent and treat pressure ulcers (PUs) The mechanisms underpinning its effect are explained and the research in this emerging field is reviewed. Practical recommendations are made including types of stimulators, stimulation parameters, muscle selection, electrodes, placement and therapy protocols.

Chapter 10. Diaphragm stimulation and respiratory function outlines the development of intramuscular diaphragm pacing and reviews clinical indications in spinal cord injury (SCI) and phrenic nerve/diaphragm dysfunction including unilateral abnormalities. ES of the diaphragm aims to reduce the need for positive pressure ventilation with tracheostomy mechanical ventilation to improve quality of life and decrease mortality.

Chapter 11. Practical considerations for running a clinical service is essential reading for clinicians. It provides guidance on building a business case – suggesting potential funding, selection of equipment, training of staff, support for patients and protocols for use. The importance of defining patient goals and using appropriate outcome measures and providing long-term support is explained. It also recommends audit and feedback to funders to ensure that the service is sustainable.

Chapter 12. Neurogenic bowel, bladder and sexual function management outlines the developments and practical applications of ES that have led to techniques to treat the underlying neurological deficit rather than simply manage the symptoms of neurological damage that cause loss of bladder bowel and sexual function. Implanted and wearable devices utilising neurostimulation and neuromodulation are described and evaluated.

Chapter 13. Use of electrical stimulation in the intensive care unit acknowledges that it is rarely used clinically but identifies potential benefits. The chapter explains the rationale for its use, especially to improve mobility and minimise muscle loss. Evidence from research is presented and ways it could be implemented are discussed.

Chapter 14. Health, fitness, and physiological adaptations after functional electrical stimulation Exercise describes the effect of ES exercise on musculoskeletal, cardiovascular and metabolic health. Evidence for benefits, including muscle hypertrophy, physical fitness and decreased risk of inactivity-based morbidities, is presented. Recent advances in hybrid and high-intensity interval training are discussed and suggestions are made for advancing ES as a therapeutic modality in clinical practice.

Chapter 15. Spasticity management describes the pathophysiology of spasticity within the context of upper motor neuron lesions and describes how ES can be used to address these complex problems, as a stand-alone therapy and in conjunction with other modalities. Outcome measures are discussed and evidence for clinical benefit is presented. Three case studies illustrate how ES can be used to manage spasticity in clinical practice.

Chapter 16. Paediatric application describes the practical application of surface ES in the management of children with congenital and acquired neurological conditions. It emphasises the differences in neuromuscular physiology between child and adult. Evidence is reviewed and the rationale for and aims of a wide range of applications are presented, including treatment for both upper limb and gait impairment. Clear guidelines are given for clinical application.

Chapter 1

History and introduction to electrical stimulation

Ian Swain[1], Tim Watson[2] and Jane Burridge[3]

1.1 History of electrical stimulation up until 1970s

The history of the use of electrical stimulation in medicine follows the developments of both the understanding of the neuromuscular system and, equally importantly, the developments in electrical engineering, for until there were reliable
sources of electrical power, studies were limited to forms of electricity occurring in
the natural world. These included static electricity, such as occurring by rubbing
amber, or that present in certain animals such as electric eels or torpedo fish which
can produce electric shocks of the order of 100–150 V (Figure 1.1 Torpedo fish).

Both torpedo fish and rubbed amber were used in the ancient world to treat a
wide variety of conditions. Amber, a fossilised resin, was found to produce static
electricity when rubbed and small pieces of electrically charged amber were given
as pills to cure haemorrhages, nausea and catarrh [1]. In fact, a derivative of
'electrica', from the Latin word for amber, was used in the 17th century to describe
the force which activated the sensory and motor nerves when applied to the body
[2]. Torpedo fish were the first example of the use of electricity as a cure. Anthero,
a freed slave of the Emperor Tiberius (AD14–37), stepped on one when walking on
a beach. After the initial numbing effect of the stimulation subsided, he found that
his gout was cured [1]. Following this accidental cure, Torpedo fish became widely
used to treat a variety of conditions such as pain relief in arthritis. When boiled and
eaten they were thought at the time to cure asthma and when strapped to the head to
ease headaches. They were also apparently an effective treatment for haemorrhoids,
but exactly how they were applied is not known [2,3].

These examples of electrical stimulation remained the 'state of the art' for the
next 1500 years until in 1658, Jan Swammerdam amused his patron, the Duke of
Tuscany, by causing twitches of an isolated muscle of a frog's leg by cutting and
pinching the nerve. In 1700, Duverney, a French anatomist, showed that it was
possible to electrically stimulate muscle [4] and the first reported use of static

[1]Orthopaedic Research Institute, Faculties of Science and Engineering and, Health and Social Care,
Bournemouth University, UK
[2]Department of Allied Health Professions, Midwifery and Social Work, School of Health and Social
Work, University of Hertfordshire, UK
[3]School of Health Sciences, Southampton University, UK

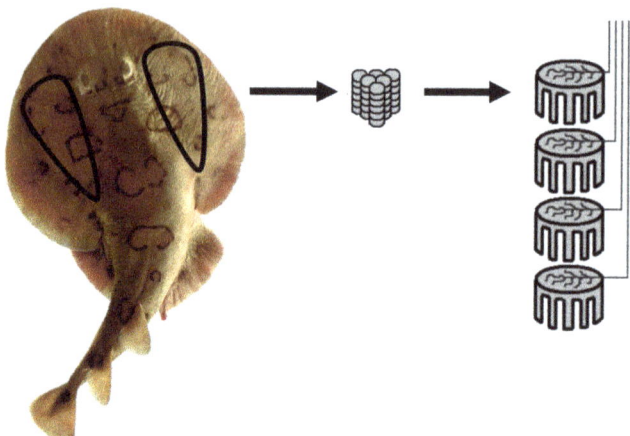

Figure 1.1   Torpedo fish showing locations of electric organs and electrocytes stacked within them. © Donald Orth, 2023. Fish, Fishing, and Conservation by Donald Orth is licensed under a Creative Commons Attribution 4.0 International License, except where otherwise noted.

electricity as a treatment in this period is attributed to Kratzenstein, a German physician, who wrote in March 1744 of '*a woman who lost the paralysis in her small finger within one quarter of an hour by electrification*'. He later wrote '*I also enabled a learned man, after a single electrification, to play the piano again with his two paralysed fingers*'. Possibly unsurprisingly, others were unable to repeat Kratzenstein's success [5].

A major change occurred with the invention of the Leyden jar in 1745, a forerunner of the modern capacitor, which enabled charge to be reliably stored [6]. A glass jar, lined inside and out with tin foil, is filled with water, enabling it to store electrical charge until required (Figure 1.2).

This development led to new treatments and in 1748 Jallerbert reported a new successful cure.

> *The wrist was flexed towards the inner side of the forearm and hung limp and motionless, in addition, he limped on the right side and could only walk with a cane. I connected his paralysed hand to the Leyden jar and drew a spark from his other hand. He felt a violent blow to his right shoulder followed by tingling throughout the arm. By January 15, 1748, he began to flex the last phalanx of his thumb voluntarily and by the 12th of March had begun to resume work* (McNeal)

Then in 1753 Samuel Quelmalz reported a successful cure for hemiplegia.

> *A young man of 18 with hemiplegia of two years duration, was unable to stand or walk and had lost his speech. His fingers were held in involuntary flexion so that he was unable to put his shoes on by himself. His arm was*

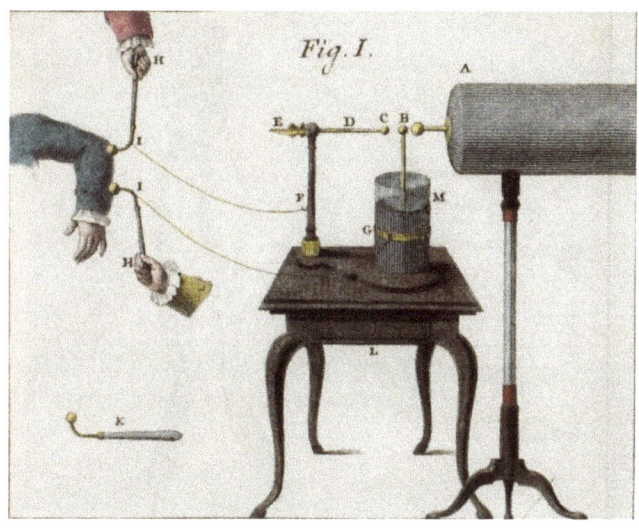

Figure 1.2   Leyden jar. Courtesy of MIT libraries [6]

motionless and his hand cold. His mother urged me to administer electricity because she had heard so much about its marvellous effects. I applied some shocks to his hand in the morning and again in the afternoon. After a few days he returned and was able to move the arm more freely and also to speak with greater ease. Electric shocks were given once or twice a week. Soon he recovered so much function that he no longer complained of inability to finger the violin as he had previously [5].

Such successes attracted the interest of people from other fields, notably John Wesley, founder of the Methodist movement, and Benjamin Franklin [1,5], one of the founding fathers of the United States. It is unfortunate that over 250 years later, and with vastly improved technical facilities, we are not always that successful today.

Despite such convincing reports, it was not until 1791 that a link was made between muscle contraction and electrical stimulation of the nerve, when Galvani observed that the application of dissimilar metals applied to the nerve of a frog's leg muscle caused that muscle to contract (Figure 1.3) [8].

However, Galvani assumed that these contractions were due to inherent 'animal electricity', generated by the nerves and stored in the muscles. Opposing this hypothesis, Volta, a physicist at the university of Pavia, put forward the case that, although frogs could react to the connection of the metals, they themselves were devoid of any inherent electricity, and that the electricity originated from the junctions of the dissimilar metals used. This controversy between Galvani and Volta was considered by some, e.g., [8] to be one of the most important in the history of science as it led both competitors to carry out fundamental experiments to prove their own hypotheses. Volta's, in 1799 to the creation of the voltaic pile, a forerunner of the battery and Galvani's to the basis of electrophysiology.

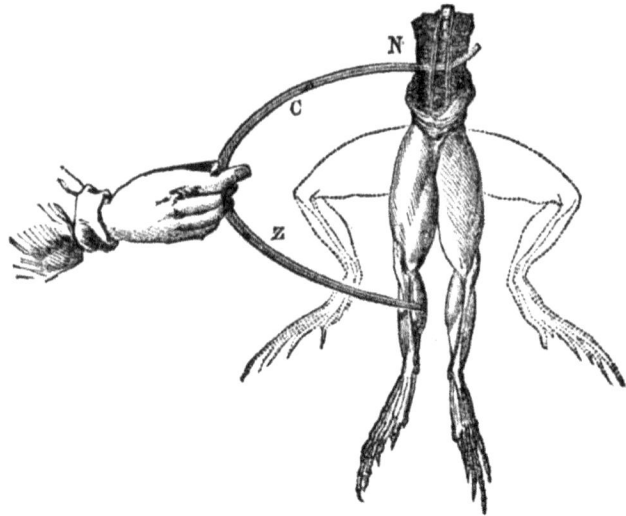

Figure 1.3   Galvani stimulating the leg muscle of a frog. Courtesy Wikipedia [7]

The voltaic pile consisted of alternating silver coins and zinc discs separated by water-soaked card and could deliver a significant electric shock. The steady output of the pile was named 'Galvanic current' (McNeal) and Volta noticed that a muscle contraction only takes place at the first flow of electricity and sometimes at the end [2]. In 1801, Ritter connected a tapped voltaic pile of a hundred cells in series to a muscle by adding one cell at a time until the full strength was achieved. He noticed that if the current was not applied 'briskly', the muscle would not contract [2].

In the second century, Galen had already divided nerves into motor and sensory types, but it was not until Magendie that the final distinction between them was made and led to the first experimentation with electropuncture in 1822. In this, electrodes were inserted into muscles along the lines of Japanese acupuncture and muscles stimulated. However, it was found to be too painful to be used as a treatment [2].

What it did do, however, was to attract the attention of Duchenne who became intrigued by the technique and continued to explore it throughout the 1830s and indeed throughout the rest of his life. He soon found that he could stimulate muscles electrically without piercing the skin by using cloth covered percutaneous electrodes over specific locations on the body. These locations were later found by Remak to be where the nerve entered the muscle, or the 'motor points' as they are known today. He was also one of the first to use the new alternating electric current which was produced by Faraday's discovery of electrical induction in 1831 and he suggested the word 'Faradic' to describe that current (Figure 1.4).

Until Duchenne introduced his percutaneous induction, the method for 'general Faradisation' required the patient to remove most of their clothing and sit on a

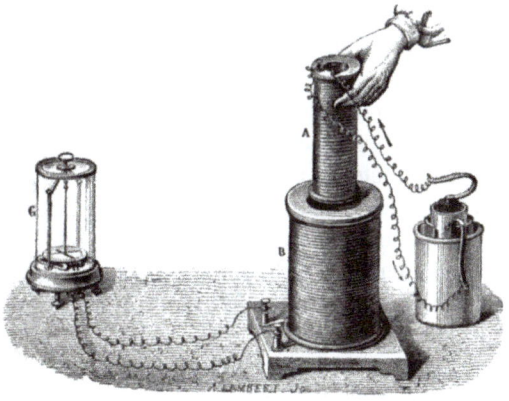

Figure 1.4   Faraday induction experiment. Courtesy of Wikipedia [9]

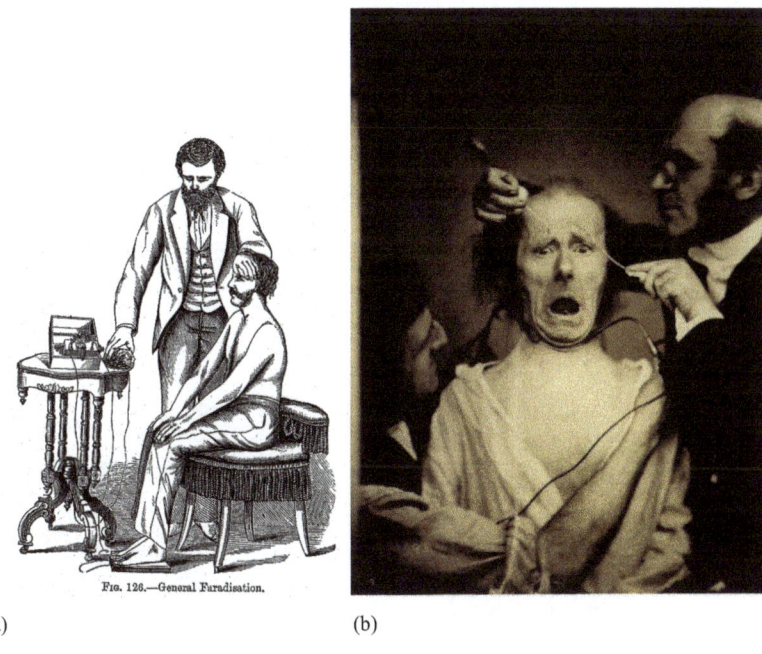

(a)   (b)

*Figure 1.5   (a) General Faradisation courtesy of Wikipedia [10]. (b) Applied to
the frontalis muscle, courtesy of Wikipedia [11].*

seat with their feet on a copper plate. The copper plate was attached to one pole of
the generator and the other attached to a wet sponge on a pole which the treating
clinician applied to the appropriate part of the body (Figure 1.5) [3].

In the 1840s, the introduction of electricity into muscles was proposed as a
diagnostic tool, with several experiments observing that paralysed muscles, those

without a nerve supply, responded only to 'Galvanic' stimulation and not to 'Faradic', although it was not until the 1860s that the reason why was determined [3]. It was also noted that if a continuous or Galvanic current was rapidly interrupted when applied to a paralysed muscle, no contraction was produced by the muscle if those interruptions exceeded a certain rate. Or rather, that the duration of the current was a deciding factor in causing that contraction. It was not until 1909 that Madame and Louis Lapicque showed the relationship between the length of the stimulation pulse and the strength of stimulation. In 1916, Adrian continued to investigate the effect of stimulation on both innervated and denervated muscles and enabled him to produce 'Strength–Duration Curves' which can be used as a diagnostic tool and to chart recovery [2,3].

With the increased understanding of the mechanism of electrical stimulation and the development of improved equipment, such as the battery and the alternating current generator, the latter half of the 19th century became the 'Golden Age of Medical Electricity' [5] with most physicians in the United States possessing one or more electrical stimulation device. These devices were used to treat a wide range of conditions such as rheumatism, gout, cuts and bruises, fractures, circulatory problems and impotence. Stimulation was even used in exorcism. Of course, although there is no scientific data to support their effectiveness, there are plenty of documented claims. Below is just one example taken from McNeal's article on the history of electrical stimulation [5] which describes the 'Pulvermacher Bi-Polar Electric Belt and Suspensory Appliance (Latest Improvement)' as featured in the Pulvermacher Galvanic Company's 1889 catalogue, promising a 'cure of weak, nervous and debilitated conditions of the generative organs' (Figure 1.6). Here is the description:

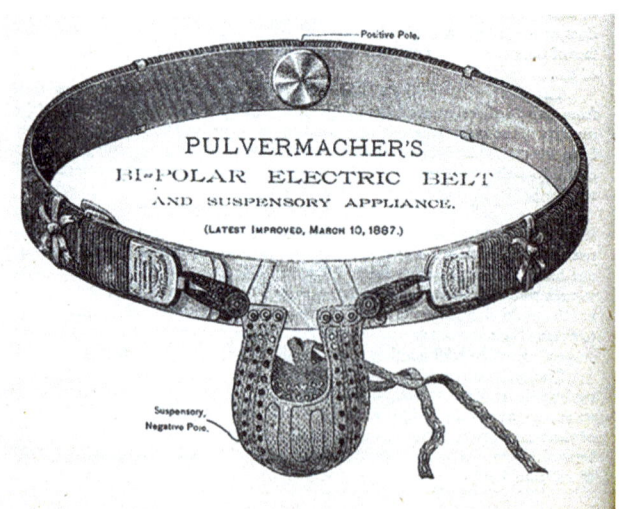

Figure 1.6   Pulvermacher bi-polar electric belt, produced with permission from the Spark Museum for electrical invention [12]

'Continuous electric currents are constantly travelling through the generative organs, the small of the back and every adjacent part, at once healing, strengthening and invigorating the organs and speedily removing every symptom of waste and decay...'

'Happily for the sufferer, these disorders, in all their various forms, yield readily to electricity... This treatment is now offered to every sufferer, and may be relied on for a complete cure, and restoration of health and manly vigor (stet)'.

Despite all these strange and probably ineffective uses, progress in the use of electrical stimulation to treat diseases began to produce meaningful results when in 1931 Albert Hyman successfully resuscitated animals following induced cardiac arrest, and although he reportedly used the same equipment to resuscitate his patients. Even though he never published the results [5], news leaked out and caused public outcry and even lawsuits accusing him of playing God. Despite this, Hyman and others continued their research.

The modern era of cardiac pacing began in 1952 when Paul Zoll used an artificial cardiac pacemaker with electrodes placed on the chest to maintain heartbeats for 20 minutes in a patient with repeated Stokes-Adams attacks of several hours' duration. Although this patient died, subsequent attempts were successful and in 1958 Furman and Schwedel undertook the first long-term use of stimulation in cardiac pacing in a 76-year-old man with Morgagni–Adams–Stokes seizures; he was successfully paced for 96 days. The equipment needed to be placed on a trolley, but it did enable him to move around the hospital [5].

As before, developments in electrical and electronic engineering enabled significant progress in the application of electrical stimulation, such as the development of the transistor by Shockley, Bardeen and Brattain of Bell Labs in 1947 which enabled electronic devices to be made significantly smaller and led to the first fully implanted pacemaker being fitted by Senning in Sweden in October 1958. The energy source was two nickel-cadmium batteries which were inductively charged through the skin. About the same time, Glenn *et al.* developed a radio frequency coupled device that was first implanted in January 1959. The first patient died after three weeks due to a broken lead, but other subsequent cases were successful. In April 1960, a man who had been in complete heart block for years was given a fully implanted system that required no charging, being powered by ten mercury cells. This can be considered the forerunner of the many millions of cardiac implanted devices that have been fitted since [5]. It is now such a common, successful treatment that a detailed investigation of it is considered beyond the scope of this book.

Another area of medicine in which electricity is commonly used is pain management [13,14]. As with cardiac pacing, a detailed description of the techniques and equipment is beyond the scope of this book, but we are reminded that this began with those ancient torpedo fish. Melzack and Wall proposed the 'gate control' theory of pain in the 1960s, the notion being that large nerve fibres could carry non-nociceptive information to excite cells that inhibit the transmission of nociceptive stimuli [15] (i.e. if large touch fibres are activated, the gate would close, and pain would lessen). Therefore, this led to the concept of using repetitive low level electric stimuli to stimulate these large nerve fibres as a method of pain relief. Wall and Sweet were the first to be credited with verifying this 'gate control' theory when they demonstrated that pain perception in a peripheral nerve could be reduced by the application of a non-painful electrical stimulus [16]. Initially, the technique was limited by lack of suitable equipment, but miniaturisation of electronics has now made such devices cheap and widely available.

The 1960s also saw significant progress in the use of electrical stimulation to improve function [17]. A technique that is now known as 'Functional Electrical Stimulation' or FES. Like pain relief and cardiac pacing, its development was a combination of engineering, neuroscience and clinical science. The first 'Functional' use of electrical stimulation is widely accredited to Liberson and colleagues in 1961 [18] when they electrically stimulated the common peroneal nerve of people with hemiplegia to provide dorsiflexion of the ankle to overcome foot drop during the swing phase of the gait cycle. The stimulus was timed to the gait cycle by a switch in the shoe so that as the person lifted their foot, the stimulation was applied to the common peroneal nerve to initiate dorsiflexion.

Liberson's work encouraged other groups who extended the clinical applications of FES [19]. In 1963, the first FES hand splint was proposed by Long and Masciarelli [20] and enabled the user to achieve both hand opening and wrist extension. The promising results of this work encouraged others, particularly a group in Ljubljana, at that time in the former Yugoslavia and now in Slovenia [21,22]. Many of the group such as Vodovnik, Dimitrijevic, Krajl and Gracanin led electrical stimulation research for the next 20 years. They collaborated widely with other centres around the world including Case Western Reserve University in Cleveland Ohio, Rancho Los Amigos Hospital in California, The Texas Institute for Rehabilitation Research in Houston and Salisbury Hospital in England. They also set up a company called Gorenje in their institute in Ljubljana to manufacture equipment which was sold internationally. However, the company is no longer active in the medical field.

The interests of this group were not just limited to electrical stimulation but were applied to the wider area of the use of engineering in rehabilitation, in particular robotics. This led to a series of meetings being organised, held every three years starting in 1967 in Dubrovik, entitled 'Advances in External Control of Human Extremities.' These were the forerunner of the current 'Rehab Week' conferences which are now held worldwide every other year. Luckily for anyone wanting to know more about this early work, all the conference proceedings from these meetings 1967–1990 are available on the IFESS (International Functional Electrical Stimulation Society) website and are free for members to download. ECHE – ADVANCES IN EXTERNAL CONTROL OF HUMAN EXTREMITIES – IFESS. Readers of these papers will find it interesting to see how many of the ideas and concepts proposed at these early meetings are still being considered today, but at that time were severely limited by the state of electronic devices at the time, particularly the lack of computing power.

In their 1977 paper, Krajl and Vodovnik [22] give a list of potential uses of and areas of research for FES. These are:

1) Control problems associated with FES, particularly in the upper extremities
2) One channel stimulation of the hand to provide hand opening in people with hemiplegia and paraplegia
3) The learning or 'carry over' effect of FES
4) Multichannel FES gait assistance
5) Optimal stimulation sequence determination
6) Gait studies and gait evaluation to determine the needs of FES systems
7) FES in pain suppression
8) FES-induced bone growth and healing
9) Prevention of atrophy following peripheral nerve lesions
10) Rehabilitation procedures for children with cerebral palsy
11) FES-induced suppression of spasticity
12) Idiopathic scoliosis prevention and correction
13) FES of bladder and sphincter mechanisms for retention and continence control.

This list was produced over 45 years ago, but looking at it today, it is frustrating that despite many years of research and clinical evidence to support the efficacy of the devices developed, there are very few devices used routinely today in clinical care. The reasons for this poor uptake are unclear and are a subject that we will return to at the end of this chapter.

1.2 Basic physiological effect of electrical stimulation

Irrespective of the exact stimulation parameters used, the aim of electrical stimulation is to have a physiological effect usually on muscles and nerves, although it often has a wider systemic effect. Although the physiological effects will be dealt with in more detail in the following chapters and Chapter 2, the basics are summarised here.

For most clinical situations covered in this book, the aim of electrical stimulation is to cause a muscle contraction, which in most cases is achieved by stimulating the nerve, either proximally or at the motor end plate (Figure 1.7) [23]. To achieve this, the stimulation must be of sufficient intensity to generate an action potential in the relevant nerve. The exceptions are the applications of electrical stimulation in wound healing, Chapter 7, and in the stimulation of denervated muscle in Chapter 8.

Therefore, as the amplitude/current or the pulse width is increased, the number of motor neurons that are activated is also increased, with those nearest the electrodes being stimulated first (Figure 1.8). Once the initial threshold to generate an action potential is achieved, there will be a twitch response in the muscle. As the number of motor neurons activated increase, more muscle force will be produced

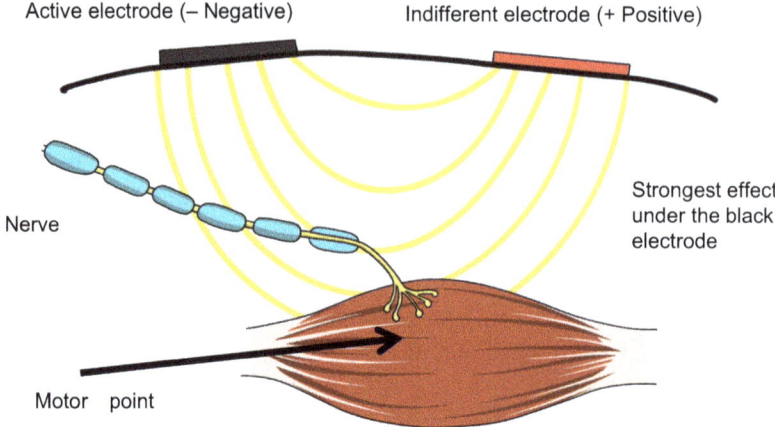

Figure 1.7   Stimulating a muscle via the motor point (Courtesy of Odstock Medical Limited – Lower Limb FES course)

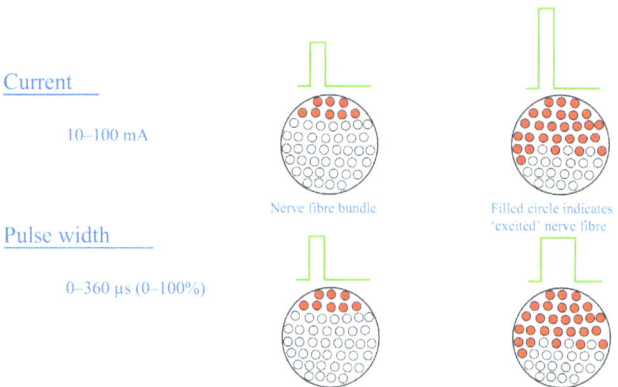

Figure 1.8   The effect of increasing current and pulse width on the number of nerve fibres activated (Courtesy of Odstock Medical Limited – Lower Limb FES course)

until all the muscle fibres are contracting. The force profile of the muscle in such a situation is shown in Figure 1.9 [3].

As can be seen in Figure 1.10 [3], a single pulse will cause an increase in muscle tension and hence force which will then gradually decay. If a second impulse arrives before the muscle tension has totally decayed, there will be a summation of the forces developed. As the pulse repetition rate increases, the curve

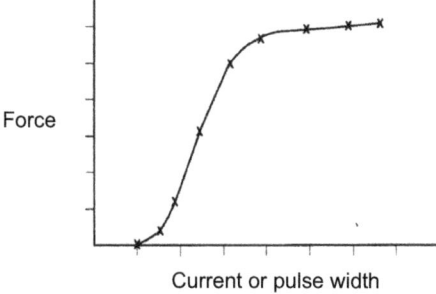

Figure 1.9   Force generated when increasing current or pulse width.

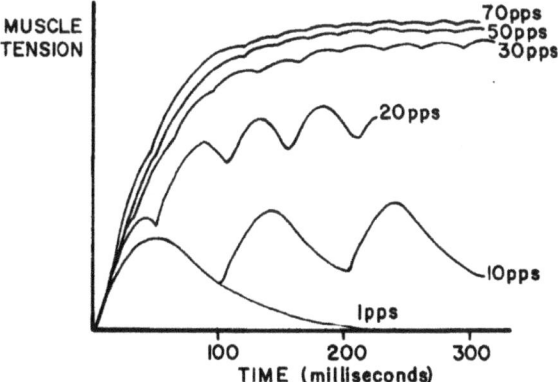

Figure 1.10   The effect of changing the number of pulses per second on the resulting force generated

of force production gets smoother until a 'fused' contraction is produced. The pulse repetition rate at which this occurs depends upon the specific properties of the muscle. This will be covered in greater depth in the next chapter.

Therefore, in general, the faster the pulse repletion rate, the more force that is produced and the smoother the contraction. However, stimulating at a higher pulse repetition rate does induce greater muscle fatigue and hence a reduction of force over time. The effect of this is shown in Figure 1.11 [3].

Therefore, the chosen pulse repetition rate is a compromise between force generation and fatigue. The optimal frequency will depend on the type of muscle being stimulated and the aim of applying stimulation, whether that be maximising force such as push of when walking or maintaining stimulation for a longer time such as to maintain standing.

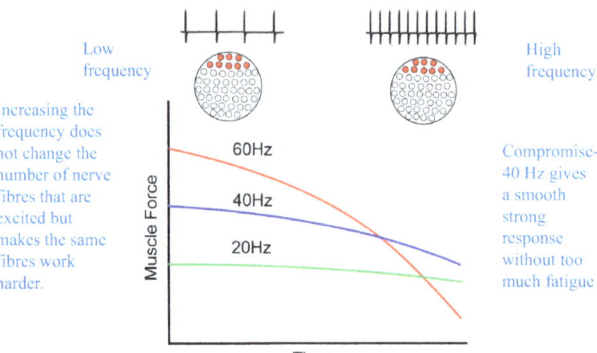

Figure 1.11   The effect of pulse repetition rate on muscle fatigue (Courtesy of Odstock Medical Limited – Lower Limb FES course)

1.3   The difference between stimulating innervated and denervated muscle

There is a great difference between stimulating innervated and denervated muscle. An innervated muscle is one in which there is a connection to the central nervous system. It might be that the appropriate signal is not getting to the muscle through damage to the appropriate part of the central nervous system such as after a stroke, multiple sclerosis or damage to the spinal cord above T11/12. However, the nervous system below the level of the injury is still working and can be easily stimulated via the nerve. If, however, there is damage to a the lower motor neuron system such as to a peripheral nerve or the spinal cord below T11/12, in the cauda equina where the nature of the spinal cord changes, then it is not possible to stimulate the nerve and the muscle needs to be stimulated directly (Figure 1.12). This requires far more energy as shown in Figure 1.13 where much longer stimulation pulses are required. However, it should be noted that even where the injury is above T11/12, there might be damage to the cauda equina or the peripheral nerves as they leave the spinal column, which could lead to a lower motor nerve injury.

Rheobase is defined as the magnitude of current just sufficient to excite the muscle. Chronaxie is the length of time a current at twice the rheobase must be applied to produce a response [3].

This means that electrical stimulation using parameters to cause a muscle contraction via stimulating the nerve can be used to assist people with some conditions, but not with others. The conditions that are suitable for stimulation with these parameters include stroke, multiple sclerosis (MS), spinal cord injury (SCI) above T12, Parkinson's disease (PD), hereditary spastic paraplegia, traumatic brain injury (TBI) and cerebral palsy (CP). Conditions which do not response to such stimulation parameters and therefore require longer pulse widths include; peripheral nerve injury, SCI below T12, motor neuron disease, poliomyelitis and Guillain–Barre disease.

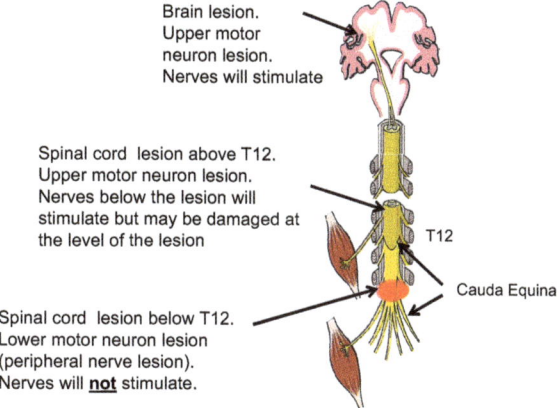

Brain lesion.
Upper motor
neuron lesion.
Nerves will stimulate

Spinal cord lesion above T12.
Upper motor neuron lesion.
Nerves below the lesion will
stimulate but may be damaged at
the level of the lesion

T12

Cauda Equina

Spinal cord lesion below T12.
Lower motor neuron lesion
(peripheral nerve lesion).
Nerves will **not** stimulate.

Figure 1.12   The difference between an upper and a lower motor neuron injury (Courtesy of Odstock Medical Limited – Lower Limb FES course)

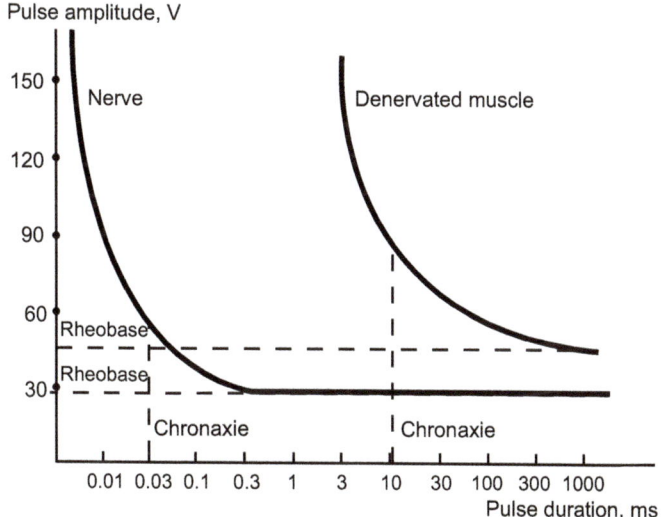

Figure 1.13   Strength duration curves showing the difference between stimulating innervated and denervated muscles with pulse duration (MS) being plotted against the required stimulating voltage

1.4   Principles of electrical stimulation

This section explains, from a physical perspective, the key principles and terminology associated with electrical stimulation. It does not explain the detailed physiological responses to stimulation, as this is covered in Chapter 2 and subsequent application chapters.

The essential principle of electrical stimulation of nerve or muscle is straightforward: an electrical stimulus initiates a membrane depolarisation creating an action potential. Once the action potential is instigated, nerve transmission continues as a normal physiological response. The pattern of the resulting muscle activity is not identical to a 'normal' physiological contraction, but each action potential is physiologically the same. The electrical stimulus may constitute a single pulse, a package of pulses or a train of stimuli from an alternating current.

If the magnitude (amplitude) of the stimulus depolarises the membrane potential beyond its threshold, the 'all or none' principle means that the action potential remains the same. Action potentials are not larger or smaller. Both the frequency of firing and number of nerve fibres activated affect the strength of muscle contraction and intensity of sensory response and are both influenced by the electrical stimulation.

The nerve membrane will repolarise following the stimulus. This restores the membrane to its resting potential – a process which takes a finite time – typically 1 ms in an average peripheral nerve fibre. During this time frame (the absolute refractory period), it is not considered possible to depolarise the membrane again – and thus, there is a finite limit as to how many times per second a nerve fibre can be stimulated. For a short time after the absolute refractory period, the nerve membrane has a raised threshold – a greater stimulus is required to initiate the action potential. This is usually referred to as the relative refractory period and would typically be or 10–15 ms duration (but varies with nerve type).

This is a simple explanation of the process. Other variables and factors influence outcomes; some are related to the parameters of the electrical stimulus and others to the physiological characteristics of the nerve fibre (not all nerve fibres behave in the same way). For a more detailed explanation and consideration of these variables, the reader is referred to Chapter 2 of this text or to other sources such as Selkowitz [24] or Robinson [25].

1.5   Characteristics, classification and terminology

There is no single globally accepted classification of electrical stimulation, although there are similarities. Classifications vary between countries and to some extent, between professional groups (for example, terminology employed by a research electrical engineer may differ from that employed by a clinical therapist) and can lead to confusion and ambiguity. Many manufacturers compound this by 'inventing' new names for their stimulation mode to make it sound 'new, different, and interesting'. In this section, the most commonly employed and technically accurate terms will be employed, and synonymous terms will be provided where relevant.

Neuromodulation, although not specifically covered in this text as a standalone topic, has received considerable attention in recent years. Neuromodulation is somewhat different to electrical stimulation in that it involves interventional technologies (most commonly electrical or chemical) with the intention of modifying nervous system activity for a therapeutic effect. It can involve the brain [26] or peripheral nervous system [27] and is employed in disorders of the central nervous system [28], pain syndromes [29,30], continence issues [31]; post stroke [32]; fibromyalgia [33] and many others. Chapters 4 looks in more detail at afferent stimulation.

1.5.1 Basic divisions

Classification of stimulating waveforms, also known as stimulating currents, allocates them to three core groups: direct currents (DC); alternating currents (AC); pulsed currents.

1.5.2 Direct currents

A DC flows in one direction. In clinical applications, there is often a 'proviso' in that it flows in one direction for at least a second. Historically, DC has also been termed Galvanic current. DC has some clinical utility, for example stimulating a denervated muscle (see Chapter 8).

1.5.3 Alternating currents

An AC changes the direction of flow at least once per second.

Pure alternating currents are rarely used in therapy, but interrupting the current in one way or another can instigate useful therapy outcomes. Burst mode alternating currents (BMAC) employ AC (in the low kHz range) with an interrupting (burst) at a low frequency (say 50 Hz). The nerve will not respond to the underlying 2500 Hz AC signal but will respond to the 50 Hz 'burst'. Russian Current/Russian Stimulation and the more recently named 'Aussie Currents' are examples of this type of clinical stimulation (see Figure 1.14).

Interferential Therapy (IFT) – or Interferential Current Therapy (IFC) – employs a pair of kHz AC currents which 'interfere' as they pass through the tissue. The resulting interference is at a low frequency, and it is this interference pattern that the nerve responds to – not the underlying kHz signal.

1.5.4 Pulsed currents

A pulsed current can have a myriad of formats, and the terminology generates abundant confusion. It is widely advocated that any pulsed current is best described by unambiguous descriptors (e.g. biphasic symmetric balanced square wave) rather than by machine or trade names.

Technically, a pulsed current consists of a pulse (more usually a series of pulses) of short duration – typically in the millisecond (ms) or microsecond (μs) duration range.

The first division of pulsed currents is into **MONOPHASIC** or **BIPHASIC** groups. In the past, these have also been termed **UNIDIRECTIONAL** and **BIDIRECTIONAL** – though these terms are rarely used in modern devices.

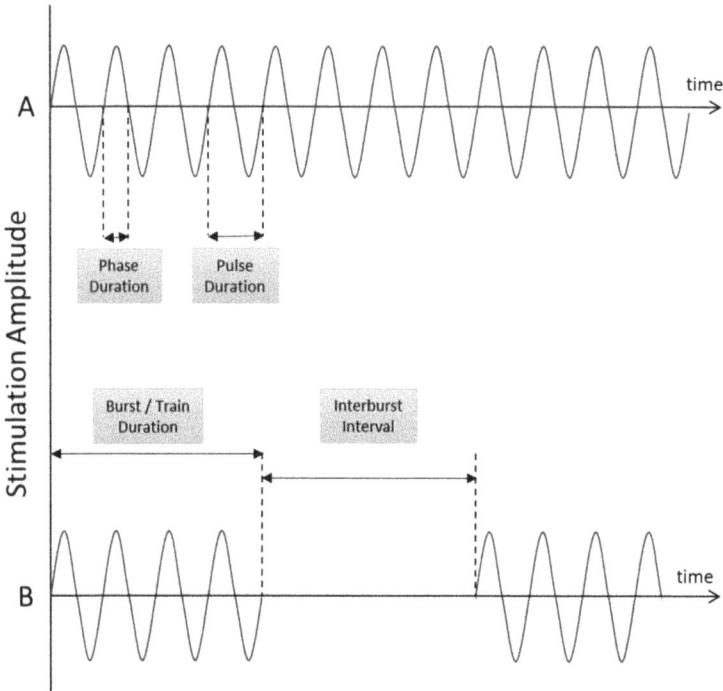

Figure 1.14   Alternating current (A) and an example of a clinical application –
burst mode alternating current (BMAC) (B) in which the alternating
current is 'chopped' into separate bursts

A **MONOPHASIC** pulse has only one phase. The shape of the pulse provides a supplementary descriptor (such as square, rectangular, triangular). During a MONOPHASIC pulse, the current will only flow in one direction, one electrode always being the anode and one the cathode.

A **BIPHASIC** current has two opposing phases in each pulse. These can also be of several different 'shapes' (square, triangular, trapezoidal). The two phases can be **SYMMETRICAL** or **ASYMMETRICAL** and furthermore the **CHARGE** delivered during each phase can be **BALANCED** (equal and opposite) or **UNBALANCED** (opposite but not equal). If a pulsed current is BIPHASIC and BALANCED, then the term **ZERO NET DC** is often invoked which means that after a number of pulses, there is no more positive than negative charge delivered (examples of pulsed current variations are shown in Figure 1.15).

When describing a pulsed current, the following sequence can be usefully adopted for consistency: monophasic/biphasic; symmetrical/asymmetrical; balanced/unbalanced; then pulse shape as the final descriptor (Figure 1.15). Some authors prefer an alternative sequence, but the principle is that the key terms should be included.

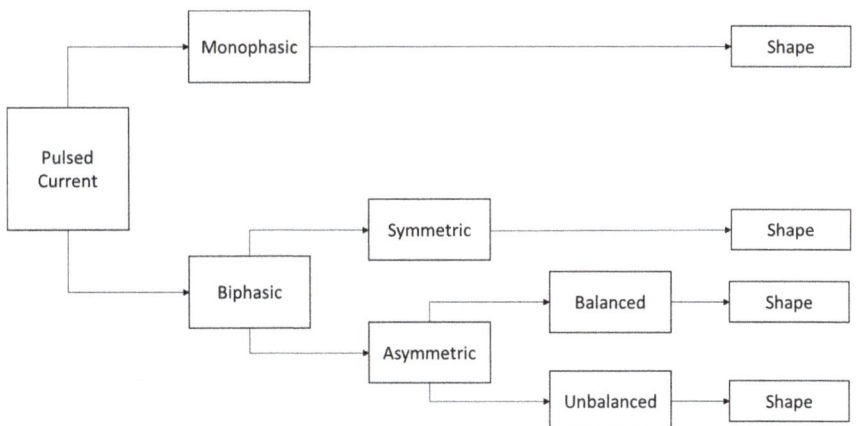

Figure 1.15 Pulsed current descriptor sequence

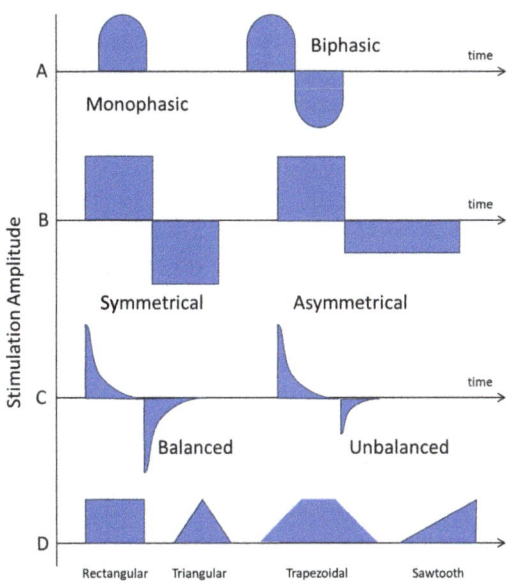

Figure 1.16 Essential pulse characteristics: (A) number of phases, (B) symmetry, (C) balancing, and (D) example pulse shapes

A stimulation current could therefore be described as monophasic triangular wave or biphasic, symmetric, balanced square wave.

The basic pulsed current pulses are illustrated in Figure 1.16.

Using these limited number of terms enables a stimulation current to be identified using factual terminology rather than trade or proprietary names.

1.6   Stimulation parameters

Given that the aim of stimulation is to achieve depolarisation of the nerve membrane and thereby an action potential (whether in the motor, sensory or other nerve fibre), stimulation parameters will influence the effect. The parameters which have been historically regarded as being 'vital', 'critical' or 'core' have been challenged to some extent in some of the more recent literature [24,34]. In published papers, especially clinical research, stimulation parameters are often incompletely defined.

1.6.1   Waveform

As identified in the previous section, stimulation is best described by a consistent and limited set of descriptors – such as 'Biphasic, symmetric, balanced square wave' or '2500Hz AC with 50Hz Burst'. See Figure 1.15.

1.6.2   Amplitude (current/voltage)

This often represents the 'strength' of the applied current. The operator can adjust this (whether the patient or the therapist). On some devices, turning up the intensity will increase the applied voltage (on a constant voltage (CV) controlled device) or the applied current (constant current (CC) controlled device). The maximum voltage of a clinical device will commonly range from 100 to 150 V and an applied current from 80 to 100 mA. There are exceptions; for example, a high voltage pulsed current (HVPC) device may deliver pulses up to 500 V, but for a very short duration. Conversely, microcurrent devices are limited to a 1 mA current maximum.

Whilst practitioners (and patients) often fret over the current that 'should' be used for a particular application, in practice it varies between patients to achieve the same effect. For example, using TENS for pain relief, the 'target' may be to achieve a 'strong but comfortable' sensation. For patient A, this might be achieved with a very low current – maybe 10 mA, whilst in Patient B, with apparently the same clinical problem, they may not reach this sensation level until the machine is delivering 50 mA or more. The same variability is found when delivering motor stimulation. Prescribing a treatment by identifying a specific current (or voltage) which should be employed effectively negates this substantial difference between individuals and is rarely appropriate; using descriptors has become more normal in recent literature. Not only does the required current (voltage) intensity vary between patients, it will also vary for the same patient on different occasions and with different electrode sizes and positions. Recording the applied current is appropriate; proscribing what it 'should be' is not.

1.6.3   Pulse/phase duration

Referred to as the pulse width in some clinical handbooks and literature, pulse duration is a more accurate term – it represents the **DURATION** of the pulse or its phases. In the clinical domain, pulses (phases) will typically be in the millisecond (ms) – thousandths of a second, or microsecond (μs) (millionths of a second) range.

If the current is monophasic, it will have a single phase. A biphasic current will have two phases in each pulse, which may be separated by an interphase interval. It is possible to deliver triphasic or polyphasic pulses, though this is less commonly encountered. By way of clarification, a **PHASE** related to a period of unidirectional current flow (flow from baseline which returns to baseline). A **PULSE** may consist of one or more PHASES.

Pulse and phase timings and their terminology are illustrated in Figure 1.17. In clinical stimulation, the interval between the individual pulses (the inter-pulse interval) will be considerably longer than the pulse duration. In a typical NMES type application in which stimulation is delivered at 50 pps with a pulse duration of 300 μs (=0.3 ms), the inter-pulse interval will be 197.7 ms – which is more than 650 × longer than the pulse itself – the pulses are SHORT and the inter-pulse intervals are LONG.

The nerve membrane depolarisation does not require long duration pulses – it is a fast event. If a muscle is denervated (i.e. does not have an active nerve supply) and the aim of the stimulation is to achieve depolarisation of the muscle membrane itself (rather than the nerve membrane), then pulses of long duration – typically in

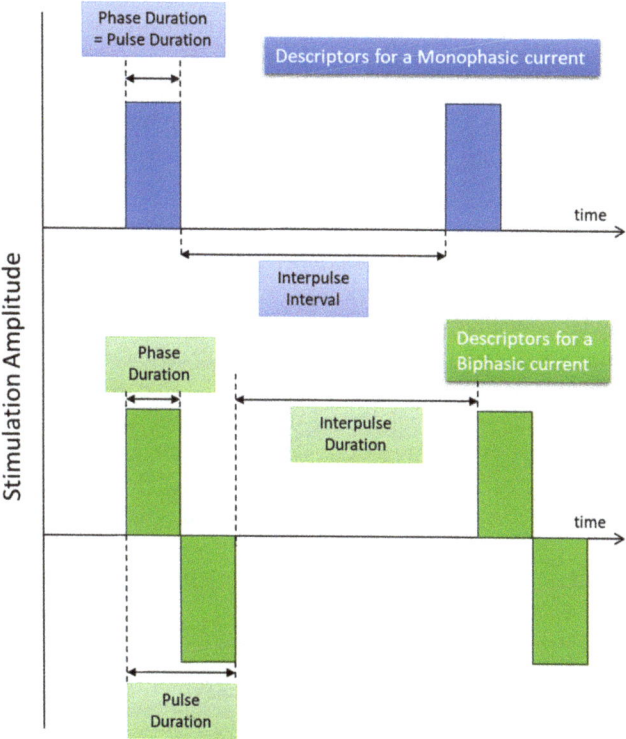

Figure 1.17   Diagrammatic representation of key terms applied to monophasic and biphasic stimulating currents (not to scale)

the tens or hundreds of milliseconds – may be needed as the muscle membrane will require substantially greater charge in order to achieve a depolarisation threshold (see Chapters 2 and 8 for further details).

If a stimulating pulse is BIPHASIC, each PHASE will have a known duration (see Figure 1.17).

The combination of the **PULSE (PHASE) AMPLITUDE** and the **PULSE (PHASE) DURATION** will determine the **PULSE (PHASE) CHARGE** (vide infra) which is currently thought to be a critical parameter. The pulse (phase) charge needs to be sufficient to achieve a threshold nerve membrane activation. Short pulses at high amplitude can carry the same charge as a longer pulse at lower amplitude.

1.6.4   Frequency/pulse rate

Technically, the frequency of a stimulating current, being described as so many Hertz(Hz) or cycles per second, should only when it is used to describe an alternating, not a pulsed current be done. However, in the clinical domain, this rule is a widely broken and a pulsed current is frequently described as being delivered at so many Hz. When describing a pulsed current, a frequency identified in pulses per second (pps) is probably more accurate.

The pulse rate or stimulating frequency will typically be in the lower range when nerve stimulation is required – commonly 1–250 pps. There are exceptions, but the majority of clinically employed currents will operate somewhere in this range given the potential responsiveness of nerve membrane depolarisation/repolarisation rates (see Chapter 2).

1.6.5   Modulation

Modulation involves the automatic variability of frequency, duration and amplitude either alone or in combination. The primary reason for delivering stimulation in a modulated mode is to limit the adaptation of the nerve to an unvarying stimulus (accommodation). It is frequently employed in TENS stimulation for pain relief as the user may use the device for long periods of time (10's of minutes up to several hours) and it thus reduces the number of times the user needs to increase the stimulation amplitude during the treatment session to maintain an effective response level. Whilst there is evidence to support the use of modulated output, especially for long-term users, there is no unambiguous evidence of superiority of one method over another.

Examples of the basic modulation methods are illustrated in Figure 1.18:

1.6.6   Phase/pulse charge

The **PHASE CHARGE** is the time integral of current for a single phase – the charge delivered by a single phase – diagrammatically, this is the area under the curve. The **PULSE CHARGE** is the sum of the phase charges during the delivered pulse. If the pulsed current is monophasic in nature, the **PHASE CHARGE** and the **PULSE CHARGE** will be equal.

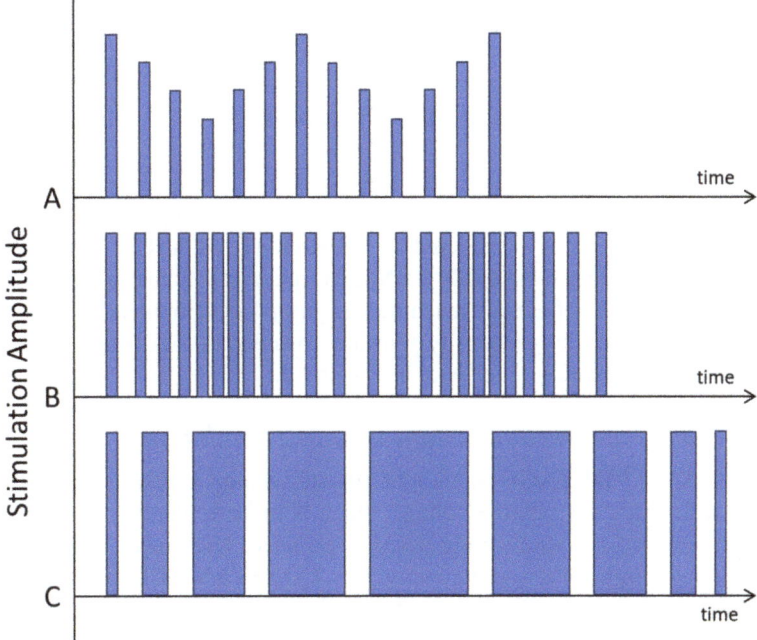

Figure 1.18   Representation of modulated stimulation: (A) pulse intensity (amplitude) modulation, (B) pulse frequency modulation, (C) pulse duration modulation

The PHASE and PULSE CHARGES are measured in Coulombs (C), though in clinical applications, microcoulombs (μC) are the typical unit of measurement. (By definition, the Coulomb is a unit of electrical charge which represents the quantity of electricity conveyed in 1 sec by a current of one ampere.)

Phase/pulse charge is a critical parameter in generating an action potential. The phase/pulse charge is more important than the pulse shape, amplitude or duration. If the phase/pulse charge is insufficient to generate an action potential, amplitude, duration or shape the pulse are irrelevant. Higher amplitude, shorter duration pulses can carry the same charge as longer duration, lower amplitude pulses. Akhtar [34] and Selkowitz [24] amongst others discuss the importance of pulse/phase charge. Additionally, phase charge is identified as a primary concern with regard to potential (neural) tissue damage [35,36].

1.6.7   Pulse interval(s)

In a monophasic pulsed current, the duration of the pulse combined with the time interval between pulses will determine the frequency at which the stimulating current is delivered. If, for example, the pulse is of 1 ms duration and the inter-pulse interval is 9 ms, then it will take 10 ms for a combination of the pulse and the gap – thereby delivering 100 pulses per second. In a biphasic pulsed current, the pulse duration will be the summed duration of the phases plus any interphase and inter-pulse intervals.

1.7   Electrodes + issues of stimulation density

Whilst a detailed consideration of electrodes is not included in this chapter, the size of the electrode does have an important effect on the 'density' of the applied stimulus. Sometimes this is considered as **CURRENT DENSITY** (mA/cm^2) but some prefer to consider **CHARGE DENSITY** (µC/cm^2). Some safety legislation and guidance (which influences machine design and development) cites maxima for both current and charge density. Further consideration is found in the later section on safe stimulation levels and in Chapter 3.

1.7.1   Constant current/constant voltage stimulation

A **CONSTANT VOLTAGE** stimulator (or stimulation control system) simply means that the operator sets the voltage which is to be applied between the electrodes. If there is a variation in the tissue impedance (resistance), then the actual current flow through the tissue may vary. In a **CONSTANT CURRENT** stimulation system, the operator will 'set' the current that will flow between the electrodes and the voltage difference will be affected by the impedance (resistance) of the tissues through which current passes. There are advantages/disadvantages with each. In a CONSTANT VOLTAGE system, if the electrode contact pressure is increased and the impedance lowered, the current flowing through the tissue will be increased and discomfort/tissue damage may ensue. In a CONSTANT CURRENT system, reduction of the effective electrode contact area (e.g. a self-adhesive electrode becomes partly detached), the same current will flow through a smaller area (greater current density) which may similarly result in discomfort/pain/ damage. The CONSTANT CURRENT system will deliver a more consistent level of stimulation and is often employed in clinical stimulators. The CONSTANT VOLTAGE system will be clinically advantageous in that the current will be reduced if the electrodes become partially detached. There is no 'best' approach. Some machines offer the operator a choice, whilst others are fixed.

1.8   Safe stimulation levels

A current density (as above) of up to 10 mA/cm^2 is 'safe' through some advocate lower levels such as 1.5–4.0 mA/cm^2. There is no one set of safety standards that can be easily cited in this regard. The 'safe' upper limits are different for implanted electrodes and surface electrodes and differ between regulatory bodies. This material together with its implications for electrode design and utilisation is usefully reviewed in numerous publications – for example, Cogan [35] considering electrical stimulation and tissue damage; Gunter *et al.* [36] and Paggi *et al.* [37] Larson *et al.* [38] consider the complex field of electrode-tissue interface, whilst

Zheng [39] consider these issues with a cellular level emphasis. Safety issues are considered further in Chapter 3 of this text and additionally in Marjenin [40]. Regulatory authorities each publish their specific standards and regulations for medical devices, e.g. (European Medical Device Regulation (MDR 2017/745 in Europe and various FDA standards in the United States (https://www.fda.gov/medical-devices). These regulations are updated at frequent intervals to reflect changes in evidence and legislation.

Whilst clinically the lowest current required to achieve the intended physiological response is often recommended, there are some applications, both sensory and motor, in which the highest tolerable current is employed and is described in subsequent application chapters.

1.9   Stimulation ramping

In various electrical stimulation applications, predominantly, but not exclusively muscle stimulation, it is common practice to employ a stimulation train **RAMPING**. If a series of pulses at a given frequency are being used to bring about a muscle contraction, it is considered advantageous to gradually increase the amplitude of the stimulation, then hold at a constant level for a period of time, then progressively decrease the stimulus strength back to a zero baseline. The approach is a closer mimic of a physiological contraction and is more comfortable (less irritating) for the user.

The increasing intensity at the start of the pulse train is referred to as the **RAMP UP** and conversely, at the end of the pulse train, the reduction is termed the **RAMP DOWN**. The device operator usually has some control over this function, though it does vary between machines. It is typical to employ a longer ramp up (2–4 sec) and a shorter ramp down (1–2 sec) but this will vary with application.

Figure 1.19 illustrates a 10 sec stimulation train which incorporates a 2 sec RAMP UP, a 7 sec PLATEAU (or SUSTAIN) phase and 1 sec RAMP DOWN packaging of the stimulation.

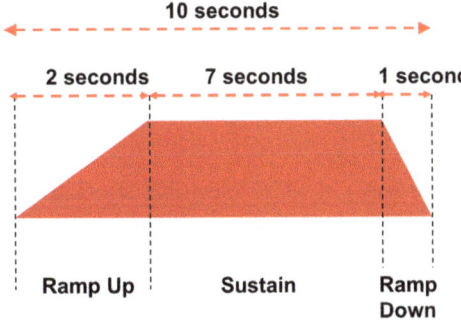

Figure 1.19   Stimulation package with both ramp up and ramp down applied waveform packaging + stimulation trains

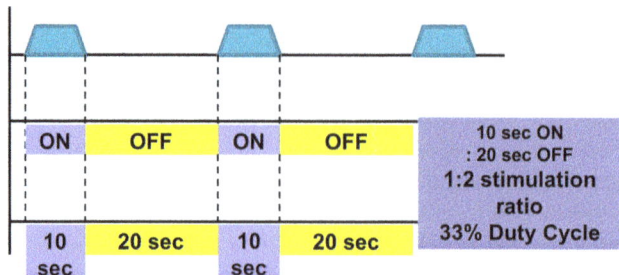

Figure 1.20   Stimulation packaging using a 33% duty cycle (1:2 ON:OFF ratio)

In some circumstances, it is not appropriate to deliver a continuous mode of nerve stimulation. This is especially true in motor stimulation applications. Ideally, there will be periods when the stimulator is **ON** followed by a period of time when it delivers no stimulation (**OFF**). When setting the parameters on the stimulator, the user will have at least some control over the timing of each ON:OFF phase. When taken as a proportion, this is referred to as the **DUTY CYCLE**. When using proportions, it is often called the **ON:OFF Ratio**.

For example, when delivering motor level stimulation to a specific group of muscles, a 10 sec period of stimulation (ON) followed by a 20 sec rest period (OFF) may be used. The ON:OFF ratio will be 1:2 and expressed as a percentage, this is a 33% DUTY CYCLE and a 5 sec stimulation followed by a 10 sec rest would deliver the same DUTY CYCLE. Whilst the DUTY CYCLE is a useful parameter to cite, as can be seen from the example above that, it is insufficient to explain all the parameters utilised. Many research publications detail either the ON:OFF times and/or the duty cycle, though unfortunately, it is sometimes information which is absent, making parameter replication impossible. Many guidelines [41] include ON:OFF ratios and duty cycle details to facilitate practical application of research based evidence.

The duty cycle will typically be varied based on the assessment findings, stage of recovery and rate of progress – which is considered as appropriate in subsequent chapters. Figure 1.20 illustrates a stimulation schedule for motor (muscle) stimulation employing the settings described in the earlier example.

1.10   Electrodes

There are basically three ways in which electrical stimulation can be provided by using surface/transcutaneous, percutaneous or implanted electrodes. Surface or transcutaneous electrodes are placed on the surface of the skin and therefore the stimulation must be of sufficient intensity to activate the underlying neural structures. Percutaneous electrodes actually pass through the skin to the underlying nerves and muscles and therefore require lower levels of stimulation, but due to the increased risk of infection are seldom used for long-term applications. However, they can be very useful in the short term to determine the likelihood of success before proceeding to a fully implanted system. Implanted electrodes are attached directly to the specific muscle and nerves and are entirely under the skin and therefore control signals and/or power need to be provided via a radio link.

1.10.1   Surface electrodes

Surface electrodes are most commonly used in clinical practice. They come in a wide range of shapes and sizes and are simple to apply on the skin over the underlying motor point or nerve that is required to be stimulated. They consist of a conductive backing, usually of metal coated fabric or conductive rubber with a coupling medium placed between the backing and the skin to ensure good electrical connection and uniform current distribution. This could simply be water, a hydrogel or a conductive adhesive to hold the electrode in place. If water is used as the conductive medium, then straps or an electrode garment will be required to hold the electrodes in place. Hydrogel electrodes are the most commonly used clinically with hypoallergenic versions being available. With care they can be reusable, lasting three weeks or so if in daily use.

1.10.2   Percutaneous electrodes

Percutaneous electrodes are implanted via a cannula and located next to the nerve or muscle that is required to be activated. They have been used in some long-term applications [42], but as far as we are aware, none are in use today as infection did become a major problem leading to many systems having to be removed. However, they are used to determine if an implanted system is likely to be beneficial, particularly to improve bladder NICE IPG536 [43] and bowel function NICE MIB 259 [44].

1.10.3   Implantable electrodes

The practical design of implantable electrodes for ES strongly depends on the specific application and the degree of invasiveness. Often, electrodes are designed based on the anatomical morphology of a single application. Although the following list is far from exhaustive, here we highlight two principal designs of implantable extraneural[1] electrodes for peripheral nerve stimulation.

1.10.4   Epineural electrodes

Electrodes that are placed near a nerve or are sewn onto the epineurium that serves as an external covering to the nerve are referred to as epineural. Epineural electrodes are the least invasive group of extraneural electrodes [45] and examples of their application include implanted peroneal nerve stimulators to treat drop foot [46] and phrenic nerve stimulation for diaphragm pacing [47]. They are advantageous in applications where whole peripheral nerves are to be stimulated, e.g. to activate synergistic muscle groups, like the dorsiflexors in the example of drop foot stimulators, rather than single muscles. The benefits of relatively low invasiveness, however, come at the expense of limited selectivity of the induced electrical field.

This presents challenges when only a subpopulation of a nerve is to be stimulated and increases the potential for unwanted coactivation of nearby nervous structures.

1.10.5   Epimysial electrodes

Epimysial electrodes comprise a polyester weave reinforced silicone rubber backing with a platinumiridium – 10% disc approximately 3.5 mm in diameter [48]. The backing is sutured to the lateral surface of the target muscle near to the motor point. They have been used in a number of applications including improving upper limb function in people with tetraplegia [49] and in the lower limb to assist with walking [50].

1.10.6   Circumneural electrodes

Circumneural, or nerve cuff, electrodes encircle a peripheral nerve and thus create an intimate interface between that nerve and multiple electrode contacts incorporated into the inner surface of the cuff. The cuffs are made of an insulating material, typically silicone, which confines the electrical field within them. Selective stimulation with cuff electrodes has been demonstrated in various clinical applications including standing systems in paraplegic subjects [51] and control of the upper extremity in human subjects [52]. The increased invasiveness of cuff electrodes over other extraneural electrodes is associated with greater risks. A major challenge of circumneural electrode interfaces is the risk of squeezing the nerve and thus restricting blood flow. In order not to exceed safe levels of intraneural pressure [53] self-sizing spiral [54] and helix [55], electrode cuffs have been developed. These cuffs can accommodate some swelling of the nerve without dangerously increasing intraneural pressure. Such swelling might occur during post-operative inflammation, but such a cuff could also maintain a close fit after the swelling is resolved.

Electrodes must be biocompatible but also electrochemically stable under pulsed conditions. Most commercially available electrodes are manufactured from electrochemically stable noble metals such as platinum, titanium or iridium.

1.11   Usage – mismatch between evidence and practice

As we have seen, electrical stimulation is not a new technique and has been in use, in some form or another, for over 2000 years, with its development led by a combination of advances in physical, as well as medical science. Despite that, electrical stimulation is still not widely used in routine clinical practice, with a few exceptions such as heart pacemakers and the use of TENS for pain relief. It is however often the subject of 'miracle cures' featuring on news items, enabling people with spinal injuries to walk, etc. The reasons for this are multifactorial and complex.

In general, electrical stimulation is usually used as a form of rehabilitation and as such shares many of the problems inherent in rehabilitation research where the gold standard of providing research evidence by RCTs is difficult. It is particularly difficult in electrical stimulation where blinding of the subject is virtually impossible. It is also difficult to get a large sample size as 'hands-on' treatment time is often significant which makes research grants expensive. As a result, many trail are small, often less than 50 people, and given the heterogeneity of the subjects, results are not always conclusive. Combined with this, there is no standard treatment protocol in many cases and hence meta-analysis of these small trials is also difficult.

The upshot of all this is that when healthcare funders look at the available evidence, they often conclude that there is insufficient evidence to commission services.

This book obviously cannot provide an answer to all of these questions, but what it does attempt to do is to consider each area of electrical stimulation in turn and to provide evidence of the current state of the art and to provide basic treatment protocols to provide a basis for future work. In addition to helping the practicing clinician, this book also aims to inform the research engineer of what is known clinically, what equipment and techniques exist and what still needs to be developed. It is hoped that, like the whole history of electrical stimulation, that this will lead to close cooperation between clinicians and researchers and result in systems that are easy to use, provide a significant clinical benefit and that are affordable by healthcare systems.

References

[1]   Cambridge NA. Electrical apparatus used in medicine before 1900. *Proc R Soc Med.* 1977;70:635–641.

[2]   Licht S. History of electrodiagnosis. In: Licht S, ed., *Electrodiagnosis and Electromyography. 3rd edn.* Baltimore: Waverly; 1971. pp. 1–23.

[3]   Benton LA, Baker LL, Bowman BR, and Waters RL. *Functional Electrical Stimulation – A Practical Clinical Guide. 2nd edn.* Downey, California: Rancho Los Amigos Rehabilitation Engineering Centre; 1981.

[4]   Licht S. The history of electrodiagnosis. *Bull Hist Med.* 1944;16(5):450–467.

[5]   McNeal DR. 2000 years of electrical stimulation. In: Hambrecht FT, and Reswick JB, eds, *Functional Electrical Stimulation; Applications in Neural Prostheses.* New York: Marcel Dekker; 1977. pp 3–35.

[6]   https://en.wikipedia.org/wiki/Leyden_jar Accessed 26 March 2024.

[7]   *Galvani-frogs-legs-electricity – Galvanism* – Wikipedia (Accessed 26 March 2024).

[8]   Piccolino M. Animal electricity and the birth of electrophysiology; the legacy of Luigi Galvani. *Brain Res Bull.* 1998;46(5):381–407.

[9]   https://commons.wikimedia.org/wiki/File:Induction_experiment.png (Accessed 26 March 2024).

[10] https://commons.wikimedia.org/wiki/File:Julius_Althaus,_General_Faradisation,_ 1873_Wellcome_L0002393.jpg (Accessed 26 March 2024).

[11] https://commons.wikimedia.org/wiki/File:Guillaume_Duchenne_de_Boulogne_ performing_facial_electrostimulus_experiments.jpg (Accessed 14 August 2024).

[12]   https://ieeexplore.ieee.org/document/9113370/figures#figures (Accessed 26 March 2024).

[13]   Ottenstat E, and Orvovich DS. History of peripheral nerve stimulation – update for the 21st century. *Pain Med.* 2020;21(S1):S3–5. doi:10,1093/pm/pnaa165.

[14]   Melzack R, and Wall PD. Acupuncture and transcutaneous electrical nerve stimulation. *Postgrad Med J.* 1984;60(710):893–896.

[15]   Melzack R, and Wall PD. Pain mechanisms: a new theory. *Science.* 1965;150 (3699):971–979.

[16]   Wall PD, and Sweet WH. Temporary relief of pain. *Science.* 1967;155 (3785):108–109.

[17]   Reswick JB. A brief history of functional electrical stimulation In: Fields WS, and Leavitt LA, (eds.) *Neural Organisation and its Relevance to Prosthetics.* New York: International Medical Book Corp.; 1973. pp 3–13.

[18]   Liberson WT, Holmquest HJ, Scott D, and Dow M. Functional electro-therapy: stimulation of the peroneal nerve synchronised with the swing phase of gait of hemiplegic patients. *Arch Phys Med Rehabil.* 1961;42:101–105.

[19]   Dimitijevic MR. Clinical practice of functional electrical stimulation from 'yesterday' to 'today'. *Artif Organs.* 2008;32(8):577–580.

[20]   Long C, and Masciarelli V. An electrophysiological splint for the hand. *Arch Phys Med Rehabil.* 1963;44:499–503.

[21]   Krajl A, and Vodovnik L. Functional electrical stimulation of the extremities part 1. *J Med Eng Technol.* 1977;1(1):12–15. doi:10.3109/03091907709161582.

[22]   Krajl A, and Vodovnik L. Functional electrical stimulation of the extremities part 2. *J Med Eng Technol.* 1977;1(2):75–80. doi:10.3109/03091907709160608.

[23]   Perotto AO. *Anatomic Guide for the Electrode Myographer. 5th edn.* Springfield, USA: Pub Charles C Thomas; 2011. ISBN 987-0-398-08649-7.

[24]   Selkowitz DM. Foundations of electrical stimulation. In: Watson T, and Nussbaum E, eds, *Electro Physical Agents: Evidence-Based Practice.* 13th edn. Amsterdam: Elsevier Health Sciences; 2020. p. 432.

[25]   Robinson AJ. *Clinical Electrophysiology: Electrotherapy and Electro-physiologic Testing.* Philadelphia, PA: Lippincott Williams & Wilkins; 2008.

[26]   Oriuwa C, Mollica A, Feinstein A, *et al.* Neuromodulation for the treatment of functional neurological disorder and somatic symptom disorder: a systematic review. *J Neurol Neurosurg Psychiatry.* 2022;93(3):280–290.

[27]   Goroszeniuk T, and Pang D. Peripheral neuromodulation: a review. *Curr Pain Headache Rep.* 2014;18(5):412.

[28]   Johnson MD, Lim HH, Netoff TI, *et al.* Neuromodulation for brain disorders: challenges and opportunities. *IEEE Trans Biomed Eng.* 2013;60(3):610–624.

[29]   Greig J, Mak Q, Furrer MA, Sahai A, and Raison N. Sacral neuromodulation in the management of chronic pelvic pain: a systematic review and meta-analysis. *Neurourol Urodyn.* 2023;42(4):822–836.

[30]   Patel CB, Patel AA, and Diwan S. The role of neuromodulation in chronic pelvic pain: a review article. *Pain Physician.* 2022;25(4):e531–e542.

[31] Tilborghs S, and De Wachter S. Sacral neuromodulation for the treatment of overactive bladder: systematic review and future prospects. *Expert Rev Med Devices*. 2022;19(2):161–187.

[32] Li L, and Huang H. Noninvasive neuromodulation for unilateral neglect after stroke: a systematic review and network meta-analysis. *Neurol Sci*. 2022;43 (10):5861–5874.

[33] Cheng YC, Hsiao CY, Su MI, Chiu CC, Huang YC, and Huang WL. Treating fibromyalgia with electrical neuromodulation: a systematic review and meta-analysis. *Clin Neurophysiol*. 2023;148:17–28.

[34] Akhtar A, Boyce B, and Bretl T. *The relationship between energy, phase charge, impedance, and perceived sensation in electrotactile stimulation.* Paper presented at the 2014 IEEE Haptics Symposium (HAPTICS), 2014.

[35] Cogan SF, Ludwig KA, Welle CG, and Takmakov P. Tissue damage thresholds during therapeutic electrical stimulation. *J Neural Eng*. 2016;13 (2):021001.

[36] Günter C, Delbeke J, and Ortiz-Catalan M. Safety of long-term electrical peripheral nerve stimulation: review of the state of the art. *J Neuroeng Rehabil*. 2019;16(1):13.

[37] Paggi V, Akouissi O, Micera S, and Lacour SP. Compliant peripheral nerve interfaces. *J Neural Eng*. 2021;18(3):031001.

[38] Larson CE, and Meng E. A review for the peripheral nerve interface designer. *J Neurosci Methods*. 2021;332:108523.

[39] Zheng XS, Tan C, Castagnola E, and Cui XT. Electrode materials for chronic electrical microstimulation. *Adv Healthc Mater* 2021;10(12):2100119.

[40] Marjenin T, Scott P, Bajaj A, *et al.* FDA perspectives on the regulation of neuromodulation devices. *Neuromodulation: Technol Neural Interface*. 2020;23(1):3–9.

[41] Nussbaum EL, Houghton P, Anthony J, Rennie S, Shay BL, and Hoens AM. Neuromuscular electrical stimulation for treatment of muscle impairment: critical review and recommendations for clinical practice. *Physiother Can*. 2017;69(5):1–76.

[42] Kobetic R, Triolo RJ, and Marsolais EB. Muscle selection and walking performance of multichannel FES systems for ambulation in paraplegia. *IEEE Trans Rehabil Eng* 1997;5(1):23–9. doi:10.1109/86.559346.PMID: 9086382 Clinical Trial

[43] https://www.nice.org.uk/guidance/ipg536/chapter/3-The-procedure Accessed 24 January 2024.

[44] https://www.nice.org.uk/advice/mib259/chapter/The-technology Accessed 24 January 2024.

[45] Tyler DJ. Electrodes for the neural interface. In: ES Krames, PH Peckham, and AR Rezai (eds), *Neuromodulation*. 2nd edn. London: Academic Press; 2018. pp. 239– 274.

[46] Strojnik P, Acimovic R, Vavken E, Simic V, and Stanic U. Treatment of drop foot using an implantable peroneal under knee stimulator. *Scand J Rehabil Med* 1987;19(1):37–43. PMID: 3495033.

[47] Glenn WW, and Phelps ML. Diaphragm pacing by electrical stimulation of the phrenic nerve. *Neurosurgery.* 1985;17(6):974–84. doi: 10.1227/00006123-198512000-00021. PMID: 3908966.

[48] Grandjean PA, and Mortimer JT. Recruitment properties of monopolar and bipolar epimysial electrodes. *Ann Biomed Eng.* 1986;14(1):53–66. doi:10.1007/BF02364648. PMID: 3706855.

[49] Kilgore KL, Peckham PH, Keith MW, and Thrope GB. Electrode characterization for functional application to upper extremity FNS. *IEEE Trans Biomed Eng.* 1990;37(1):12–21. doi:10.1109/10.43606. PMID: 2154398.

[50] Uhlir JP, Triolo RJ, Davis JA Jr, and Bieri C. Performance of epimysial stimulating electrodes in the lower extremities of individuals with spinal cord injury. *IEEE Trans Neural Syst Rehabil Eng* 2004;12(2):279–87. doi: 10.1109/TNSRE.2004.827224. PMID: 15218941.

[51] Fisher LE, Miller ME, Nogan SJ, *et al.* Preliminary evaluation of a neural prosthesis for standing after spinal cord injury with four contact nerve-cuff electrodes for quadriceps stimulation. *Conf Proc IEEE Eng Med Biol Soc* 2006;2006:3592–5. doi:10.1109/IEMBS.2006.260833. PMID: 17947042.

[52] Polasek KH, Hoyen HA, Keith MW, Kirsch RF, and Tyler DJ. Spiral nerve cuff electrodes for an upper extremity neuroprosthesis. *Conf Proc IEEE Eng Med Biol Soc* 2006;2006:3584–7. doi:10.1109/IEMBS.2006.259625. PMID: 17947040.

[53] Rydevik B, Lundborg G, and Bagge UJ. Effects of graded compression on intraneural blood blow. An in vivo study on rabbit tibial nerve. *Hand Surg Am* 1981;6(1):3–12. doi:10.1016/s0363-5023(81)80003-2. PMID: 7204915.

[54] Naples GG, Mortimer JT, Scheiner A, and Sweeney JD. A spiral nerve cuff electrode for peripheral nerve stimulation. *IEEE Trans Biomed Eng* 1988;35 (11):905–16. doi:10.1109/10.8670. PMID: 3198136.

[55] Agnew WF, McCreery DB, Yuen TG, and Bullara LA. Histologic and physiologic evaluation of electrically stimulated peripheral nerve: considerations for the selection of parameters. *Ann Biomed Eng* 1989;17(1):39–60. doi:10.1007/BF02364272.

[56] https://bio.libretexts.org/Bookshelves/Ecology/Fish_Fishing_and_Conservation/01%3A_Chapters/1.03%3A_Sensory_Capabilities_of_Fish (Accessed 14 August 2024).

Chapter 2

Neuromuscular physiology

Jonathan Jarvis[1] and Steffen Eickhoff[1]

2.1 Overview of neuromuscular physiology in relation to electrical stimulation

Skeletal muscle is a form of linear motor that generates force on skeletal attachments and delivers mechanical work by shortening or lengthening against internal and external loads. It is made up of cells known as muscle fibres, which are highly specialised for their motile function. They exist in various forms adapted for endurance or sprint activity, and often mixed together in a single anatomical muscle.

Muscle is normally activated by electrical signals from the central nervous system (CNS) that travel in motor nerves to the muscles. Artificial stimulation of those nerves, and therefore muscle activity, can also be achieved by electrical stimulation (ES). Functional electrical simulation (FES) attempts to restore the missing function of paralysed muscles, and NMES (neuromuscular electrical stimulation) is used to achieve therapeutic but not necessarily biomechanically functional outcomes in compromised muscles, such as maintenance of muscle mass, or improvement of blood flow.

A continuous power output of about 10 watts per kg can be produced by muscle, and in a typical human of about 80 kg with 30 kg of muscles, about 300 watts or up to 1 kW during brief bursts of activity.

Skeletal muscles provide motive power for breathing, eating, standing and walking and all movements of the limbs. The direct energy source is adenosine triphosphate (ATP) whose chemical energy is converted to mechanical work by a multifunctional enzyme (the myosin cross bridge) that also provides a nanometric change in conformation while bound to filamentous actin. Summation of billions of such cyclic molecular movements causes sliding of actin filaments over myosin filaments to provide macroscopic shortening of the order of centimetres in typical human muscle.

ATP is synthesised within muscle cells using energy derived from food and the oxygen delivered in the blood supply to the muscle. ATP can also be supplied for

[1]School of Sport and Exercise Science, Liverpool John Moores University, UK

short periods by non-oxidative pathways, especially in fibres adapted for bursts of high-power output.

Muscle fibres are activated as they receive input from the motor neurons of the peripheral nervous system (PNS). The temporal patterns of action potentials in those motor neurons determine the recruitment of muscle fibres within a muscle and the force that the fibres will generate. The force produced on the muscle attachment is determined by the proportion of fibres activated at any one time, and by the firing frequency of the action potentials they receive.

The power produced by muscle is determined by the degree of activation and the load against which the muscle shortens. Efficient movement can be assisted by external apparatus (such as lever systems) that modify the muscle load.

Muscle fibres respond to action potentials received via their motor neurons whether those action potentials are generated by synaptic transmission from neurons of the CNS, or by electrical or magnetic depolarisation of their excitable membranes. This is the principle of FES in which external stimulation is used to substitute for voluntary activation that has been lost through damage to the nervous system. It is difficult to reproduce the complexity of normal motor patterns with external stimulation because it is difficult to steer currents within the neural structures that can reproduce the spatial and temporal complexity on which normal function is based.

Muscle is a highly plastic tissue. A population of resident stem cells called satellite cells can repair muscles that suffer damage through, for example, mechanical overload. Muscles also lose mass quickly when disused and increase in mass when challenged by an increase in average loading. They can adapt to an increase in average daily activity by increasing their resistance to fatigue. In most applications of FES, the principles of muscle training will apply to ensure that the remaining muscle is prepared and used appropriately for the desired task.

Practical systems for FES must take account of the anatomical disposition of motor nerves and nerve endings. Control of muscle action can be by control of proportional recruitment via amplitude or pulse width, but also by control of the temporal pattern of activating pulses. The choice of activating pulses will depend on the balance between innervated and denervated or partially denervated muscles.

2.2   Nerve conduction and the effect of injury

Before we consider the properties of the skeletal muscle motors in human physiology, we will first discuss how their mechanical function to generate force and power is controlled by signals arising in the motor nervous system. Some of this discussion is based on part of the Doctoral Thesis submitted to Liverpool John Moores University by Steffen Eickhoff.

The nervous system provides the infrastructure for electrical communication and is made up of the CNS, which is the organism's central processing unit, and the PNS, which includes a system of intricate connections between the CNS and the muscles. The CNS comprises the brain and spinal cord, suspended in a shock absorbing fluid inside the skull and the spinal canal. The PNS is a network of conductive pathways, the nerve fibres or axons, that connect the CNS to every other part of the organism. While the PNS contains some peripheral processing units (ganglia), its two main tasks are providing sensory input to the CNS, via afferent fibres, and transmitting commands from the CNS to the periphery via efferent fibres. The structural units fulfilling these tasks are the individual neurons.

Neurons have a cell body, the soma, from which a network of fine processes extends to contact neighbouring or remote cells (Figure 2.1). A neuron can be seen as a miniature nervous system itself: it receives input via synapses on the cell body and the nearby dendritic processes, integrates these input data and, according to that input, may generate an output signal called an action potential, which it sends along its efferent pathway to its axonal terminals. At the dendritic/somatic inputs and the terminal outputs, the neuron is in close contact with its neighbouring cells, but for the most part, nerve cells are enveloped by satellite cells, such as the Schwann cells that form an insulating myelin sheath around many mammalian nerve axons.

In the neuromuscular systems of vertebrates, functional complexes are formed by neurons interacting with adjacent cells. For example:

- *Motor units* consist of a motor neuron and several skeletal muscle fibres that are innervated by terminal branches of that neuron [2].

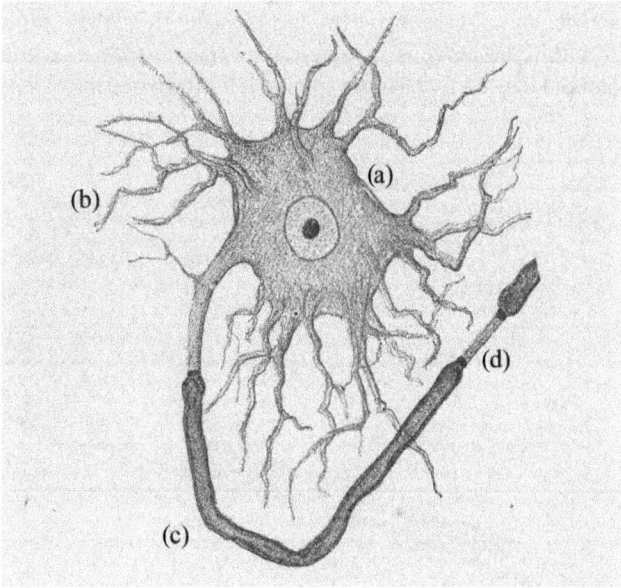

Figure 2.1 *The neuron is the structural unit of the nervous system. Dendritic processes (b) and the myelinated axon (c) originate from the cell soma (a). Myelination is interrupted at nodes of Ranvier (d), where the axonal membrane is exposed [1]. Copyright by Wellcome Collection. CC BY. Adapted with permission.*

- *Neuromuscular junctions (NMJ)* are tripartite synapse complexes consisting of a pre-synaptic terminal nerve process of a motor nerve, a post-synaptic muscle fibre and a perisynaptic Schwann cell (PSC).
- *Muscle spindles* and *Golgi tendon organs* are receptors that sense changes in muscle length and tension and provide feedback to the CNS. They comprise afferent sensory fibres interacting with muscle and tendon tissue.

2.3   The motor neuron

Axons leaving the CNS form bundles of fibres that run to the region they innervate. These peripheral nerve trunks are enclosed by the epineurium, a dense connective tissue layer and typically comprise multiple bundles of axons (the fascicles), small arterioles and venules, fibroblasts, mast cells and lymphocytes [3]. The fascicles and axons are enclosed in thinner connective tissue layers called the perineurium and endoneurium, respectively (Figure 2.2).

2.4   The nerve membrane

Axons are not just passive electrical conductors like copper wires in a cable. Rather they are active impulse generators by virtue of ionic gradients maintained across their cell membranes by active transport, and the presence within that cell membrane of channels that are sensitive to the electric field they experience. The quantal signal within the nervous system is the action potential, a brief change of membrane potential caused by changes in channel properties that propagates along

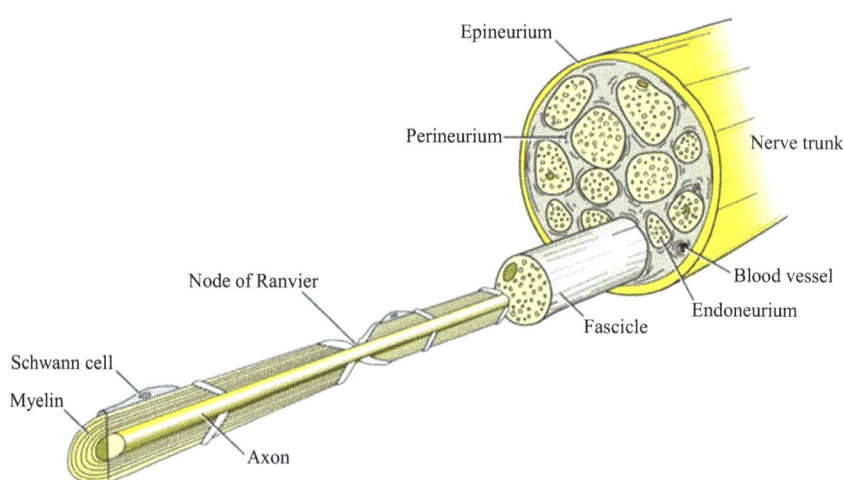

Figure 2.2   The peripheral nerve and its components [3]. Copyright by Elsevier. Adapted with permission.

axons because it triggers similar channel openings and closings in adjacent parts of the membrane.

Special transmembrane channel proteins that enable selective passage of electrical charge-carrying ions allow generation and propagation of action potentials in nervous tissues [4]. The so-called leak channels are open when the nerve membrane is at rest and let specific ions cross the membrane following the electrical and/or concentration gradient. Other ligand-gated channels close or open when a specific neurotransmitter binds. For example, the nicotinic acetylcholine receptor (nAChR) is of central importance for excitation–contraction coupling at the NMJ. Voltage-gated channels change their conformation in response to local changes of transmembrane potential. Examples include the voltage-gated sodium ion channels ($Na_{V)}$ or potassium channels (K_V). Active transport mechanisms use energy from ATP to move ions against their concentration gradient across the cell membrane. The sodium–potassium pump (Na-K ATPase pump) is an important example of an active transmembrane transport system that allows nerve cells to maintain their resting potential [3].

2.5   The membrane resting potential

In excitable cells, the concentration and electrical gradients of osmotically active particles create a resting potential across the cell membrane. The cell interior is between 60 and 90 mV more negative than the extracellular space.

The cylindrical fibres of nerve and muscle cells can be considered electrical conductors in the sense that an electrolytic, conductive interior is surrounded by the insulating cell membrane. However, their passive cable properties are very poor compared with an insulated metal conductor. But experimental observations with varying separation between stimulation source and site of recording show that APs can travel long distances without any distortion [5]. This discrepancy between the limited cable properties of nerve and muscle fibres and observations of action potentials being conveyed over long distances without attenuation suggested that other mechanisms must be involved in action potential propagation.

The search for these mechanisms was the driving force behind research into electrophysiology during the early twentieth century. In 1952, Hodgkin, Huxley and Katz published five papers summarising a series of experimental findings that represent a breakthrough in our understanding of the action potential [6–10]. The large diameter of the unmyelinated squid giant axon (up to 1 mm) allowed the authors to apply the voltage clamp technique, placing an electrode inside the axon and measuring the voltage across the membrane while injecting charge into the nerve cell. They measured the voltage-dependent transmembrane permeabilities to sodium and potassium ions. Further, they developed a mathematical model that described these nonlinear ionic conductances as functions of voltage and time and correlated them to discrete hypothetical ion channels with different gating properties. Numerical solution of that model enabled Hodgkin and Huxley to make a quantitative simulation of the action potential [7].

2.6   Generation of action potentials by voltage sensitive channels

Voltage-gated potassium K_V channels are modelled as a pore with an activation gate, and so can either be 'open' or 'closed'. The voltage-gated sodium Na_V channel has an additional deactivation gate, and therefore can adopt a third, 'inactivated' state. As the transmembrane potential changes, the probability for individual ion channels to transition to another gating state changes.

At rest, most Na_V and K_V activation gates are closed and most Na_V deactivation gates are open. Upon membrane depolarisation, that is a reduction of the negative transmembrane resting potential, the Na_V activation gates open rapidly and Na^+ ions, following their electrical and concentration gradient, enter the intracellular space. This influx of positive charged Na^+ ions further depolarises the membrane. However, the reduced negative transmembrane potential also leads to a steady opening of K_V activation gates, causing a gradual efflux of K^+ ions. If the net influx of Na^+ exceeds the efflux of K^+ to depolarise the membrane to the threshold potential of approximately 55mV, then a rapid opening of Na_V activation gates causes a strong Na^+ influx and membrane depolarisation gives rise to an action potential. The Na_V deactivation gates then close, stopping further influx of Na^+, and the continuing efflux of K^+ leads to a repolarization of the membrane. During this period of repolarisation, the Na_V channels gradually return to their resting state, closing their activation gates and opening their deactivation gates.

Physiologically, the initial stimulus of gradual membrane depolarisation occurs at the neuron's cell body. This could be the cell body of a motor neuron in the anterior horn of the spinal cord, or a neuron within the complex cortical structures within the brain. At the cell body, many subthreshold input signals, arriving via synapses either at the dendrites or at the cell soma membrane, are integrated. If the integrated input signal reaches the threshold for depolarisation, an action potential is triggered. From there the AP propagates along the axon, leaving the membrane that it passed in a state of temporary refractoriness.

2.7   The neuromuscular junction

At the NMJ, also called the motor endplate, the presynaptic action potential of a motor nerve terminal is conveyed, via release of the neurotransmitter acetylcholine ACh, to a skeletal muscle fibre. ACh receptors respond to generate a postsynaptic action potential which then propagates along the excitable muscle membrane and initiates muscle contraction. In general, each skeletal muscle fibre is innervated by a single terminal branch of a motor neuron and the NMJ is typically located in the middle of that muscle fibre (Figure 2.3.a).

Although action potentials are physiologically initiated at motor neuron cell bodies, the nerve cell is excitable along its whole length and can therefore be activated by an external field and this is the principle of FES. Electrical charge is delivered to a target excitable structure in order to influence voltage-gated channels,

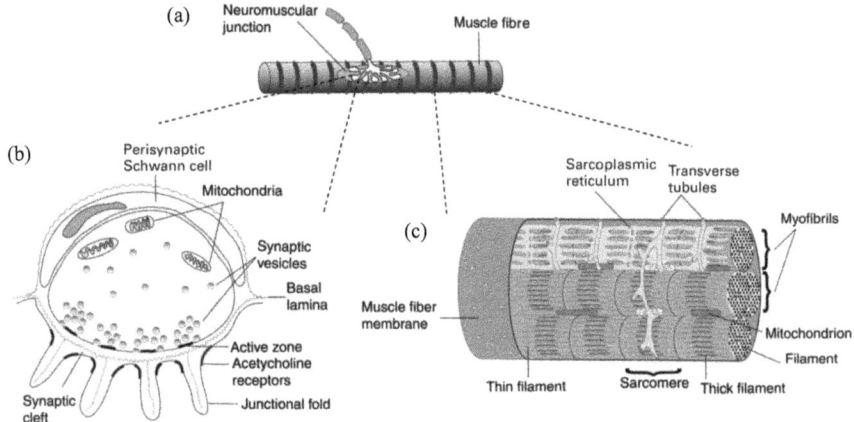

Figure 2.3 *Schematic representation of the neuromuscular system. (a) The neuromuscular junction where the terminal branch of a motor nerve is in synaptic contact with a single muscle fibre. (b) Cross-sectional representation of a synaptic bouton at the neuromuscular junction. (c) An electron microscopic representation of a longitudinal section of a single muscle fibre segment. [11] Copyright by Elsevier. Reprint with permission.*

usually to generate action potentials, and thereby to control the release of neurotransmitters.

This injected charge creates an artificial potential field between the poles of the stimulation source, altering the state of voltage-gated ion channels affected by that field. This modulation effect can either be inhibitory or excitatory, and can in the latter case, if sufficiently intense, evoke self-propagating action potentials leading to neurotransmitter release at the axonal terminals. While some ES therapies are claimed to be 'subthreshold', the targeted control of neurotransmitter release provides the basis for the therapeutic effects of most well-studied ES applications. Increased neurotransmitter release may, for example, be used to activate muscle fibres and thus generate movement, or to provide visual or auditory perception. A decreased neurotransmitter release can be achieved by blocking action potential conduction pathways. Such a block can be achieved both with kilohertz frequency alternating current or with direct current and may be used to inactivate or downregulate the nervous system [12], e.g. to reduce spasticity or chronic pain.

2.8 Recruitment and pulse duration relationships

2.8.1 Recruitment behaviour

For an ES pulse of any given waveform (e.g. monophasic, cathodic rectangular) to evoke an action potential, it needs to be of sufficient intensity. The stimulation intensity of a single stimulus depends on the stimulation amplitude and duration (or phase width). For a fixed stimulation phase width one can thus define the stimulation amplitude required to evoke an action potential (in the target excitable fibre) as a threshold [13]. An amplitude below the threshold is insufficient to elicit excitation and is termed subthreshold, whereas an amplitude above this minimum required intensity for activation is called suprathreshold. While reaching the threshold for activation, and thus action potential generation, is an all-or-nothing event for a single fibre, activation happens gradually, i.e. over a range of threshold amplitudes, in a stimulated population of fibres. The two main determinants for activation (via the magnitude of the extracellular potential) are the distance of the nearest Node of Ranvier from the electrode and the internodal distance, which is correlated to axon diameter. In a typical motor nerve both these parameters are normally distributed and recruitment is thus typically described by a sigmoidal curve (Figure 2.4). For stimulation of a whole nerve one can thus express relative activation as a percentage threshold, e.g. the 50% activation threshold $I_{th50\%}$ is the amplitude required to activate 50% of the individual nerve fibres in that nerve. Further, dynamic ranges, e.g. the range of amplitudes between 5% and 95% relative activation, which are of interest for many ES applications that seek to stimulate specific subpopulations selectively, can be determined (Figure 2.4(b)).

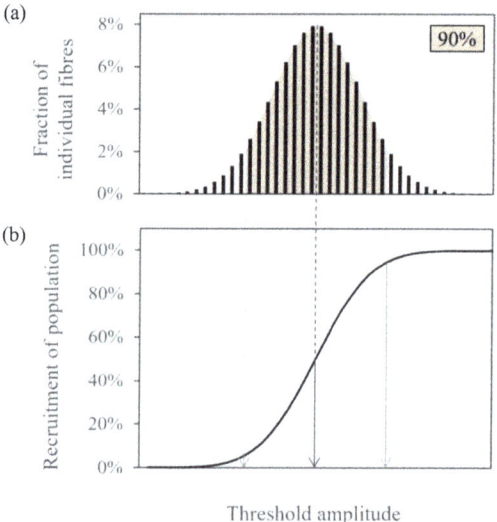

Figure 2.4   Recruitment behaviour of a homogeneous fibre population. (a) The normally distributed thresholds for individual fibres of a homogeneous population produce. (b) A characteristic sigmoidal recruitment curve. Percentage threshold for activation of the population $I_{th5\%}$, $I_{th50\%}$ and $I_{th95\%}$ are indicated.

2.8.2 Pulse duration relationships

The relationship of the pulse width PW and threshold amplitude I_{th} is called the strength-duration (SD) relationship. For monophasic rectangular pulses and a single excitable fibre, this relationship can mathematically be described by the Lapicque equation

$$I_{th}(PW) = \frac{I_r}{1 - e^{-\left(\frac{PW}{\ln(2) \cdot T_c}\right)}},$$

where I_r is the rheobase, that is the minimum amplitude required to activate the fibre with an infinitely long pulse and T_c is the chronaxie [13]. The chronaxie T_c is defined as the pulse duration at which the target fibre is activated with an amplitude of twice the rheobase I_r (Figure 2.5(a)).

Although this relationship was originally developed for stimulation of a single excitable fibre with monophasic rectangular pulses, the SD curve is often used to determine rheobase and chronaxie values for stimulation of whole nerves and muscles [14,15], e.g. at a specific percentage activation threshold, and also for different waveforms [16].

The linear charge-duration curve can be attained by integrating the threshold currents over the pulse duration (Figure 2.5(b)). The amount of charge injection Q_{th} required to reach threshold, and thus, when extrapolated to the whole stimulated excitable structure, to achieve the desired neuromodulation effect, is a critical parameter. Since charge injection is a main determinant of stimulation safety and scales proportionally with pulse duration, many ES applications typically use the shortest effective pulses.

$$Q_{th}(PW) = \int_0^{PW} I_{th}(PW) dt.$$

Integration of the power at threshold, which is proportional to I_{th} squared for a given load Z, over the pulse duration gives us the electrical energy E_{th} which is 'consumed' to reach the threshold for activation:

$$E_{th}(PW) \propto \int_0^{PW} I_{th}(PW)^2 dt.$$

An important characteristic of the energy–duration relationship (Figure 2.5(c)) is that it has a local minimum at a pulse duration equal to the chronaxie T_c. Using the least amount of electrical energy to achieve neural activation is desired in many battery powered ES implants, as the energy consumption per pulse directly impacts the battery life for a given battery technology and volume. More energy efficient stimulation would therefore enable increased battery life and/or reduced battery volume (and thus overall implant volume).

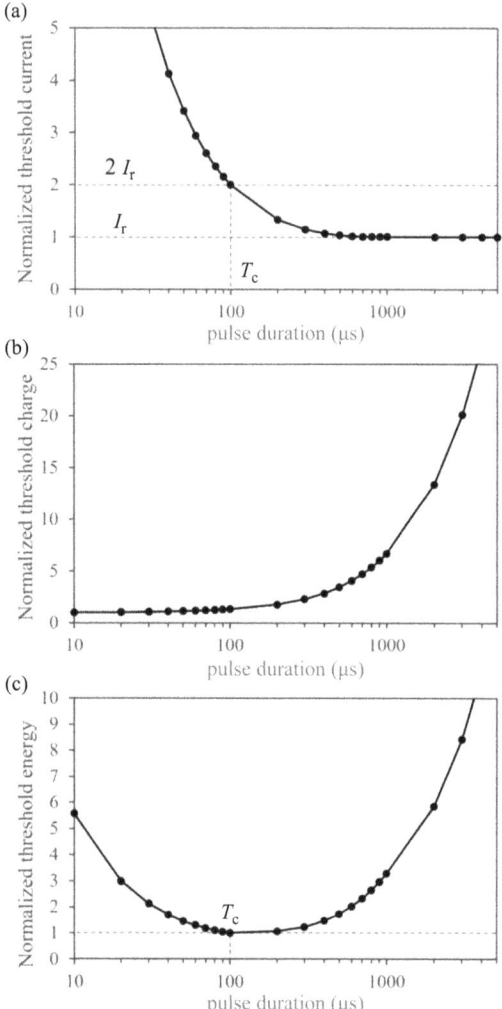

Figure 2.5   Pulse duration relationships. (a) Strength-duration curve as described by the Lapicque equation, normalised to multiples of the rheobase I_r. The pulse duration at which twice I_r is required to reach threshold is termed chronaxie T_c. (b) Normalised charge-duration curve and (c) normalised energy-duration curve with local minimum at T_c are based on the threshold current I_{th}.

2.9   Summary of neuromuscular physiology in relation to ES

Anyone contemplating the use of ES in humans should have a basic understanding of neurophysiology. The safety aspects are too complex to treat in this chapter, but because the normal function of the human system relies on secure electrical communication, then any external influence on this mechanism must be considered carefully, especially if current is anticipated to flow across parts of the CNS or the heart. Furthermore, though we have not discussed it here, the sensory system, just as much as the effector systems relies also on electrical signalling. Therefore, application of electrical current can cause disturbance to sensation, including direct activation of peripheral sensors, including pain sensors. Discomfort during artificial activation is a risk and a further complication in terms of appropriate application of external activation.

The use of direct activation of muscles in cases where the normal neural structures and mechanisms are damaged, for example, when peripheral nerves are damaged to the extent that a muscle receives little or no activation via the normal route is covered in Chapter 8. In this case, nerve regeneration may take place, or steps may be taken to encourage reinnervation. In the meantime, it is possible to bypass neural activation by direct muscle activation with high current, especially if skin sensation is also compromised.

Even in the case in which some muscles and not others are denervated it is possible to activate muscle directly with pulses that block the generation of action potential in undamaged nerves. This is the principle behind selective stimulation of the denervated larynx for example, without unwanted activation of innervated neck muscles.

2.10   Muscle contraction

Skeletal muscles are linear motors whose primary purpose is to convert energy derived from food into mechanical work. That is, they produce force on the skeleton, and via the skeleton to the outside world, to produce movement against a load.

They have other roles within the body, such as the maintenance of glucose homeostasis, and temperature control. They are highly plastic, existing in various specialised forms that allow matching of properties to various important movement tasks. For example, the work of breathing needs low level but unrelenting generation of power. By contrast. the muscles that allow us to accelerate our body mass rapidly upwards (that is to jump) produce large bursts of power but in most of us, only rarely.

This specialisation forces certain compromises in structure and function (Table 2.1). A muscle that must generate contractile function near continuously (the best example is the heart) has a large proportion of its volume committed to generation of ATP, the molecule that delivers energy directly to the contractile material, from oxygen and hexose units. This process happens in mitochondria which occupy volume, but do not themselves generate mechanical action. Sprint muscle on the other hand which has a much lower average requirement for ATP, has more of its cellular volume taken up by the apparatus for force generation known as sarcomeres, and less taken up by mitochondria. The sarcomere is the fundamental force generating unit. It is 2–3 microns long and about 1 micron in diameter. It is bounded by end plates known as Z disks containing structural proteins that can transfer force from one sarcomere to its neighbours, and ultimately to the tendons that link the active muscles to the skeleton. The Z disks are drawn together when myosin and actin within the sarcomere interact. A typical single muscle cell (also known as a muscle fibre) may contain tens of millions of sarcomeres, and a typical human upper limb muscle contains tens of thousands of fibres.

2.11   Muscle activation

Since muscle is a linear motor, it must deliver continuous power by cycles of shortening and re-lengthening. We are aware that most muscular work is cyclical (walking/running/chewing/breathing) and thus comprises short periods of activity (generation of force and shortening) with periods of relaxation in between. The control of the active and relaxed states of muscle is achieved by adjusting the calcium concentration experienced by the sarcomeres. The sarcoplasmic reticulum is a specialised system of membrane bound spaces that store, release and recycle calcium ions to achieve activation and relaxation of the sarcomeres.

The release of calcium from the sarcoplasmic reticulum is controlled by regenerative electrical pulses known as action potentials that travel in excitable membranes such as those bounding nerve and muscle cells. During purposeful voluntary movement, signals represented by bursts of action potentials are normally generated in the motor cortex of the brain, relayed through various spinal interchanges, and travel to the muscles of the body along the motor neurons of the PNS. Action potentials may also be generated in motor neurons without conscious intention by spinal reflex circuits, or by depolarisation of those neurons by external stimulation if that external stimulation causes sufficient current to flow across the nerve. As the action potentials reach the muscle fibres, they cause release of a chemical neurotransmitter, acetyl choline (ACh), into the synapses that form between motor nerve and muscle fibres: these specialised synapses are called motor end plates. ACh binding in turn generates new action potentials in the muscle fibres. These actional potentials travel into the body of the fibres via transverse or t-tubules, and cause calcium release from the sarcoplasmic reticulum and thus contraction of the sarcomeres. The calcium that triggers contraction is rapidly recycled back into the storage compartment, so that the muscle relaxes again unless new action potentials are received.

2.12   Muscle mechanics

A single action potential causes a brief activation known as a twitch contraction. In human muscle this lasts less than a tenth of a second (100 ms).

If we consider the required duration of typical contractions some approximate calculations are instructive. For example, human beings run with a gait cycle that lasts about 1 s from one footfall to the next. If we assume that muscles are active for half the time and relaxed for half the time, then typical muscle contractions might need to last about 500 ms. So, these contractions need to last longer than the response caused by a single action potential (twitch). As might be expected, such contractions are generated not by a single action potential but by a short sequence or burst of action potentials. If one activation is followed by another before the first has had time to dissipate, then muscle force builds up or summates.

Although we think of identifiable anatomical muscles, the functional units of muscular movement are at a smaller scale. The motor system controls muscular force progressively by recruiting single motor units. A single motor unit is the collection of muscle fibres that is innervated by the branches of a single motor neuron. This might be a collection of up to 1000 fibres in a large muscle, but only as few as 10 fibres in muscles adapted for fine motor control. A collection of 1000 muscle fibres produces a maximum force on the order of 20 g whereas a motor unit of only 10 fibres produces a force on the order of 0.2 g. The vastus lateralis (one of the thigh muscles) contains about 150 motor units, and the biceps brachii about 300 in human.

This maximum force is further modulated moment by moment by the degree of activation, which in turn is determined by the frequency at which action potentials arrive at the motor end plate. A single isolated action potential produces a brief force response known as a twitch, that lasts for less than 1 s. The production of sustained forced requires repetitive activation during which the force sums according to the typically sigmoidal force-frequency curve such as those shown in Figure 2.7. In human motor units, an average activation frequency of about 40 Hz is required to reach the force plateau for activation. Such a contraction is known as fused or tetanic because the individual twitch responses fuse together to produce a smooth force response. In movements controlled by the intact motor system, motor units within individual muscles, and within functional muscle groups, are recruited progressively to produce gross movement of a limb, or stabilisation of one joint to allow movement of another.

Furthermore, motor units can be recruited asynchronously to allow temporal summation of force. This is a strategy that may increase endurance, since the blood supply required for ongoing production of ATP is reduced by sustained tetanic contraction. Conversely, the recruitment of motor units synchronously and without the selection and rotation of motor units achieved by the normal motor system, tends to lead to premature fatigue with FES. More 'natural' recruitment can be attempted with artificial activation by recruiting parts of a muscle group in turn by sequential stimulation of several active electrodes arranged around a motor nerve or across a muscle. These aspects of muscle as a motor are usefully summarised in the following figures. They also illustrate the difference in behaviour of fast and slow motor units whose responses were recorded from a single muscle in the experiments illustrated in Figures 2.6 and 2.7. The fast motor units have a faster rise and a faster decline of force when activated by a single impulse (twitch). As bursts of impulses are delivered at progressively higher constant frequencies (pulses per second or Hz), the force of the individual responses summates because the processes of activation and relaxation are not completed before the next impulse is delivered. Summation occurs at lower frequencies for slow than for fast muscle. As

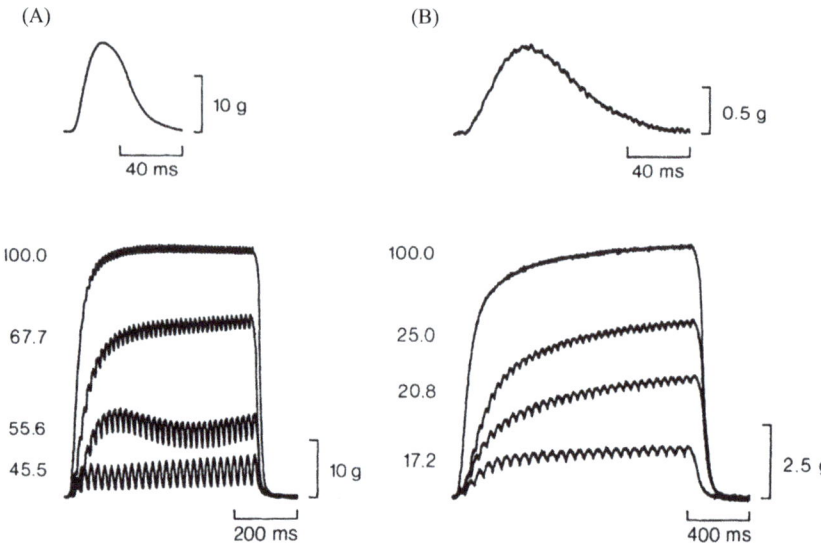

Figure 2.6   Isometric tension produced by fast (type IIX) motor unit (A) and slow (type I) motor unit (B) to stimulation at various frequencies, indicated to the left of the superimposed tetanic records (in pps). Force recordings from functionally isolated muscle units from cat flexor carpi radialis (FCR) muscle. Stimulation at 100 pps elicited max. tetanic tension for slow unit but only 94% of max. tension for fast unit. Upper record in A and B is max. potentiated twitch (avg. of 4). Used from [17] with permission.

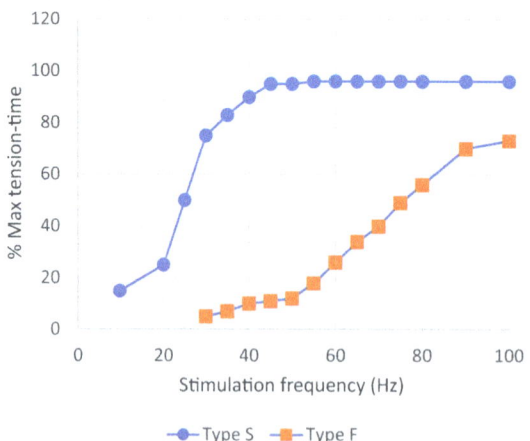

Figure 2.7   Frequency–tension curves for fast (type F, squares) and slow (type S, circles) motor units from cat muscle. Summarised from [17].

Table 2.1   Muscle fibre types

	Slow twitch – resistant to fatigue	Fast twitch – resistant to fatigue	Fast twitch – fatigable
Type	I	IIa	IIb
Speed of contraction	Slow	Fast	Fast
Myosin—ATPase activity	Low	High	High
Preliminary source of ATP production	Oxidative phosphorylation	Oxidative phosphorylation	Anaerobic glycolysis
Glycolytic enzyme activity	Low	Intermediate	High
Number of mitochondria	Many	Many	Few
Myoglobin content	High	High	Low
Glycogen content	Low	Intermediate	High
Fibre diameter	Small	Intermediate	Large
Rate of fatigue	Slow	Intermediate	Fast

the frequency increases further the force reaches a maximum in which all the sarcomeres are fully activated and generating force. This maximum or tetanic force is typically about ten times the twitch force for fast motor units, and three times the twitch force for slow motor units (Table 2.1).

2.13   Force–frequency relation

The behaviour described above can be illustrated in the force–frequency curves, which are shifted leftwards (towards lower frequencies) for slower muscles.

Although bursts of pulses at a constant frequency are useful to characterise muscle responses, the motor system rarely generates constant firing frequencies. At the start of a movement, rapid firing can achieve a brisk onset of contraction which may then be sustained by a lower following frequency. Such optimised patterns of doublets or triplets or n-lets may help to achieve the maximum force output for lower activation energy. It is counter-productive to use sustained frequencies above the fusion frequency for a particular muscle.

2.14   Force–velocity relationship

The discussion so far has considered only the production of force rather than work, which is defined as force multiplied by the distance over which that force is applied. Some contractions are near isometric (named because the muscles stay at approximately the same length), but many also involve substantial shortening of the muscles to move the skeleton. In such movements the external force is found to decline with increasing velocity, and because power is force times velocity there is an optimum combination of speed and force that produces the maximum power (Figure 2.8). This is one of the basic principles of ergonomics. If one wishes to move a quantity of sand

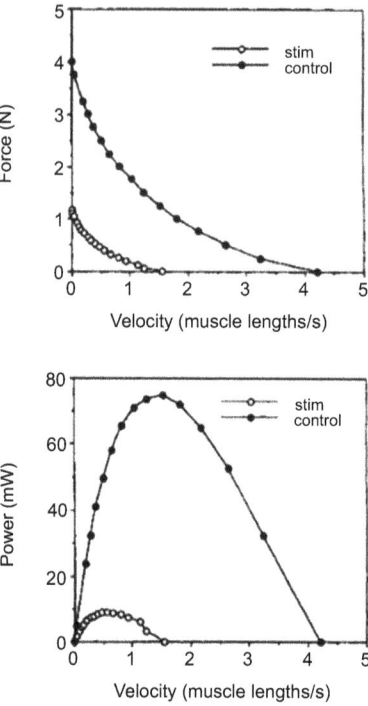

Figure 2.8   The force–velocity and power–velocity relationships for the rat extensor digitorum muscle during general anaesthesia measured in the Jarvis lab. The open circles represent the mechanical performance of muscle that has received continuous training for 8 weeks and has become much slower and less powerful, but (not shown here) more resistant to fatigue.

from one place to another, there is an ideal size of shovel that will load our muscles optimally to produce maximum power and minimise the time taken. Similarly, the gears on a bicycle help us to adjust the load on our muscles near to the optimum value whether the external load is great (climbing a hill) or small (downhill cycling). The power–velocity curves in Figure 2.8 also illustrate how different the power characteristic of fast and slow muscles can be, and thus how it is necessary to bear in mind the biomechanical arrangement and the intrinsic properties of the muscles when those muscles are used as motors. We need to be sure that the muscles of an FES participant can produce the power required to perform a particular task such as lifting the foot, or accelerating the body mass upwards for standing.

2.15   Sustained power and muscle fatigue

The maximum sustainable power is often known as the critical power in human physiology. Critical power is the asymptote of the hyperbolic relationship between the required power output and the duration before exhaustion in human exercise testing. The duration before exhaustion becomes longer and longer as required power is reduced until a level of around 250 W for a typical fit person. Assuming this power is produced from about 30 kg of muscle, that represents a sustainable muscle power output of about 8 W/kg. For comparison, the muscle of the human heart produces about 3 W/kg at rest and not much below that level continuously for the entire lifetime. The experience of muscle fatigue as a decline in the desired force or power is a complex phenomenon and can be a consequence of changes in cellular metabolism, neuromuscular transmission or central fatigue in which the drive from the nervous system to the muscles is disturbed [18].

2.16   Differences with electrically stimulated contraction

FES in persons that no longer have control of muscle activation because of damage to their motor systems is most commonly used in the case of upper motor neuron injury. The muscle fibres remain connected to a surviving motor neuron, but the activity of that neuron is no longer under normal voluntary control. Muscle recruitment can therefore be achieved by activation of the surviving lower motor neuron and its branches, usually by application of electrical current though electrodes on the skin surface. In the case of upper motor neuron injury some involuntary reflex activity may take place, and indeed such spastic activity may be troublesome because of unwanted or inconvenient uncontrolled movements. Since the muscles are at least occasionally active there is less risk of severe disuse atrophy.

If the lower motor neuron is damaged, then fibres are silenced completely and usually suffer substantial atrophy and then degeneration. However even in this case, especially within the first weeks or months after the injury, it is possible to achieve direct activation by generating action potentials in the muscle membrane directly. Such direct stimulation needs a much greater current than stimulation via the nerve. In situations in which innervated and denervated muscles or motor units are close together, it is possible to activate the denervated muscles selectively by using a ramped current that suppresses activation of motor nerves [19].

Practically, when one tries to activate muscles specifically by careful motor mapping, then the degree of denervation can be determined by using a typical nerve pulse and a typical muscle pulse. A typical nerve pulse is approximately 200 microseconds and about 5 milliamps whereas a typical muscle pulse would be more like 20 ms and 20 milliamps.

Denervated muscles are slower than normal and lack endurance. This is to be expected since they have suffered from lack of use or detraining for some period of time. But it seems to be more than that, because even after periods of external stimulation to restore mass and resistance to fatigue in denervated muscle, denervated muscles still lack endurance [20]. In this sense the presence of effective neuro muscular connection appears to have an effect greater than the delivery of contractile activity.

The aspects of the natural system of motor recruitment discussed in the muscle mechanics sections and in the paragraphs above illustrate how difficult it is to reproduce naturalistic movement patterns by external stimulation. In particular, it is difficult to activate motor units one by one. This is because the absolute threshold for activation by current delivered from the surrounding tissues is very similar for each of the axons contained in a motor nerve. As an illustration of this, consider the technique known as motor unit number estimation in which the activating current to a motor nerve is gradually increased while observing the force produced by the activated muscle. As each new axon is recruited, the additional force attributed to that motor unit will be noted as a step change in the total force. The typical increment for the experimental current to add one motor unit at a time is about 20 µA when the electrodes are close to a nerve bundle. And the typical window from the activation of the lowest threshold axons to activation of all the axons, let's say in a nerve of a few millimetres in diameter, is generally between 1 and 2 mA. Since the thresholds for activation of individual motor neurons are so similar, the degree of recruitment is also vulnerable to changes in the relative position of the nerve to the activating electrodes, or to the position of the nerve relative to other anatomical structures such as muscles, bones and tendons that will also affect the current field across an individual motor axons.

Furthermore, the phenomenon of reverse recruitment may be problematic. In progressive recruitment by the undamaged Motor System, the smaller motor units are generally recruited at lower required forces than the larger motor units. The smaller motor units tend to include the more oxidative fibre types that are adapted for endurance rather than for maximum force or acute power output. If the larger motor units have the lowest threshold for activation by external stimulation, then sub maximal contractions will tend to be subject to fatigue. Premature fatigue has been the subject of many studies and discussions in the FES field. No doubt the reverse recruitment phenomenon has contributed to this, but there are additional factors that must be considered such as detraining and retraining.

2.17   Growth and wasting

Spinal cord injury generally causes paralysis, that is the lack of voluntary control of skeletal muscle. This in turn dramatically affects the average activity of individual muscles. All biological tissues are in a constant state of turnover and skeletal muscle has a remarkably effective apparatus for deconstruction of disused muscle. If the daily activation of a muscle is reduced near to zero by mechanical immobilisation, silencing of its motor nerve, spaceflight, or external provision of its mechanical function such as in mechanical ventilation, the mechanisms of synthesis and degradation are shifted dramatically towards degradation. If silenced altogether, unloaded muscle can lose as much as three per cent of its mass per day. To increase muscle mass, or to prevent wasting, current recommendations are to exercise against resistance, typically with sets of repetitive contractions two or three times a week [21]. The precise prescription of such activity, the cellular mechanisms involved and the triggers for growth and maintenance are the subject of ongoing research. The effect of age and of prior disuse will also be influential. Many clinical centres have standard procedures for the preparation of muscles for use with FES. There is a danger of training too frequently, because damage and inflammation can be caused especially if the muscle is fixed or heavily loaded during activation, or for too long at a time especially when time in a clinical facility may be limited. It seems sensible to keep within the recommended guidelines for persons with intact neuromuscular systems, especially because those with paralysed limbs are likely to have disturbed or zero perception of muscle soreness. When muscles have been disused for some time, a progressive build up to the recommended training load is advisable.

References

[1] Ramón y Cajal S. *Histologie du systeme nerveux de l'homme et des verte-bres.* Paris: Maloine; 1911.

[2] Liddell EGT, and Sherrington CS. Recruitment and some other features of reflex inhibition. *Proceedings of the Royal Society of London Series B, Containing Papers of a Biological Character*, 1925;97:488–518.

[3] Bertorini TE. Neuromuscular anatomy and function. In: Bertorini TE (ed), *Neuromuscular Case Studies.* Philadelphia, PA: Butterworth-Heinemann; 2008. pp. 1–25.

[4] Koester J. Voltage gated channels and the generation of the action potential. In: Kandel ER, and Schwartz JH (eds), *Principles of Neural Science.* 2nd edn. New York: Elsevier; 1985. pp. 75–86.

[5] Katz B. *Nerve, Muscle, and Synapse.* New York: McGraw-Hill; 1966. pp. 1–193.

[6] Hodgkin AL, Huxley AF, and Katz B. Measurement of current-voltage relations in the membrane of the giant axon of *Loligo. J Physiol.* 1952;116(4): 424–448.

[7] Hodgkin AL, and Huxley AF. A quantitative description of membrane current and its application to conduction and excitation in nerve. *J Physiol.* 1952;117(4):500–544.

[8] Hodgkin AL, and Huxley AF. The components of membrane conductance in the giant axon of *Loligo. J Physiol.* 1952;116(4): 473–496.

[9] Hodgkin AL, and Huxley AF. Currents carried by sodium and potassium ions through the membrane of the giant axon of *Loligo. J Physiol.* 1952;116(4): 449–472.

[10] Hodgkin AL, and Huxley AF. The dual effect of membrane potential on sodium conductance in the giant axon of *Loligo. J Physiol.* 1952;116(4): 497–506.

[11] Ko CP. Neuromuscular system. In: NJ Smelser and PB Baltes (eds), *International Encyclopedia of the Social & Behavioral Sciences.* Oxford: Pergamon; 2001. pp. 10595–10600.

[12] Bhadra N, and Kilgore KL. Fundamentals of kilohertz frequency alternating current nerve conduction block of the peripheral nervous system. In: ES Krames, PH Peckham, and AR Rezai (eds), *Neuromodulation*. London: Academic Press; 2018 pp. 111–120.

[13] Lapicque L. Recherches quantitatives sur l'excitation électrique des nerfs traitée comme une polarisation. *J Physiol Pathol Gen*. 1907;9:620–635.

[14] Geddes LA. Chronaxie. *Australas Phys Eng Sci Med*. 1999;22:13–17.

[15] Ashley Z, Sutherland H, Lanmuller H, *et al.* Determination of the chronaxie and rheobase of denervated limb muscles in conscious rabbits. *Artif Organs*. 2005;29(3):212–215.

[16] Sahin M, and Tie Y. Non-rectangular waveforms for neural stimulation with practical electrodes. *J Neural Eng* 2007;4(3):227–233.

[17] Botterman BR, Iwamoto GA, and Gonyea WJ. Gradation of isometric tension by different activation rates in motor units of cat flexor carpi radialis muscle. *J Neurophysiol [Internet]*. 1986;56(2):494–506.

[18] Wan JJ, Qin Z, Wang PY, Sun Y, and Liu X. Muscle fatigue: general understanding and treatment. *Exp Mol Med*. 2017;49(10):e384. doi:10.1038/emm.2017.194. PMID: 28983090; PMCID: PMC5668469.

[19] Kurz A, Volk GF, Arnold D, Schneider-Stickler B, Mayr W, and Guntinas-Lichius O. Selective electrical surface stimulation to support functional recovery in the early phase after unilateral acute facial nerve or vocal fold paralysis. *Front Neurol*. 2022;13:869900. doi:103389/fneur.2022.869900. PMID: 35444611; PMCID: PMC9013944.

[20] Ashley Z, Sutherland H, Russold MF, Lanmüller H, Mayr W, Jarvis JC, and Salmons S. Therapeutic stimulation of denervated muscles: the influence of pattern. *Muscle Nerve*. 2008;38(1):875–86. doi:10.1002/mus.21020. PMID: 18563723.

[21] American College of Sports Medicine. American College of Sports Medicine position stand. Progression models in resistance training for healthy adults. *Med Sci Sports Exerc*. 2009;41(3):687–708. doi: 10.1249/MSS.0b013e3181915670 PMID: 19204579.

Chapter 3

Medical device safety and regulatory compliance – a guide for beginners

Steve Crook[1] and Anne Vanhoestenberghe[2]

3.1 Introduction

Research and development in electrical stimulation is desirable to provide new devices and clinical techniques. Although much of the field is mature, relying on technologies and parameters established for many years, there is a need for novel applications. This is particularly true for implanted devices. As our understanding of physiological mechanisms continues to improve, this unlocks the potential for more applications.

Medical device regulation scrutinises the entire product lifecycle. Of particular importance for researchers and early stage designers, there is a requirement to capture and document the user needs, design choices and decisions, and who approved them [1]. The eventual legal manufacturer will be subject to external auditing from the regulators and documented evidence of the early and subsequent stages of the design and development will be required. Where a project is seeking government or other 'seed' funding, there will probably be an expectation of subsequent commercial development of the medical device that is the outcome of the project. This will be much more straightforward if the documentation needed to address the requirements of one or more regulatory regimes (as applied in different territories) has been accumulated as part of the development cycle. In view of this, the generation, retention and organisation of this evidence must be considered as part of any medical device development project.

The intention of this chapter is to raise awareness of the regulatory requirements associated with medical devices, with a clear focus on electrical stimulation applications. Some of these requirements will be encountered at an early stage, especially if research is intended to lead to a finished commercial product. From the first use with a human, local ethics or governance should be insisting on a risk-benefit approach, hence looking for reference to appropriate product safety risk management documentation. For devices that are planned to be more widely used,

[1]Odstock Medical Ltd, National Clinical FES Centre, Salisbury District Hospital Salisbury, United Kingdom
[2]School of Biomedical Engineering and Imaging Sciences, King's College London, United Kingdom

then other questions arise. Will the development team become the legal manu-facturer? If not, at what stage will a collaboration or intercompany agreement be established? The contents of this chapter should make engineers, scientists and health professionals aware of the various aspects of medical device regulatory compliance. Our aim is not to provide a comprehensive review of the regulations, but to highlight the broad steps needed and point to some resources.

3.2   Brief historical outline

Prior to the 1990s, medical devices were subject only to national regulations – if they were regulated at all. In 1992, a combined group of industry and regulatory repre-sentatives from the European Union (EU), US, Canada and Japan met to discuss the establishment of a global partnership with the aim of harmonising the regulations applicable to medical devices [2]. The Global Harmonisation Task Force (GHTF) was born of these discussions and in 1993 Australia joined the group along with repre-sentatives from standards organisations and the World Health Organization (WHO). This later evolved into the International Medical Devices Regulators Forum (IMDRF).

A common agreement on the requirements for medical devices was attractive to manufacturers. This promised to reduce the burden of registering devices or seeking approval for use in individual countries by not having to apply to the regulator in each new country. In the EU, medical devices were to bear the CE mark once regulatory compliance was established. The CE, or Conformité Européenne, mark signifies that the manufacturer has followed the applicable regulations under one or more EU directives. This 'harmonisation' of requirements via the Medical Device Directive 93/42/EEC (and Directive 90/385/EEC for active implantable medical devices) enabled medical devices to be freely traded across borders of the EU member states. In addition, the CE mark provided a short cut to mutual recognition for markets such as Australia and Canada. This was due to the commonality of requirements established in the partnership discussions.

Both Medical Device Directives were later repealed and replaced by the Medical Device Regulations (EU) 2017/745 (MDR) [3].

The vote by the UK to leave the EU (BREXIT) has complicated matters for UK medical device exporters since they are required to follow the MDR and engage an Authorised Representative in the EU and an importer. This is the same as any other 'rest of world' manufacturer. Due to a lack of regulatory resources, compounded by the Covid pandemic, there are extensions in place which permit existing devices to be sold. The situation is beyond the scope of this chapter and the available guidance should be investigated.

3.3   Starting out – is your device regulated as a medical device?

The initial step must be to determine if the device you are developing is a medical device under the meaning of the legislation in your intended market. In this chapter, unless specifically stated, we refer to the EU Medical Devices Regulations 2017 (MDR) [3]. These are most directly aligned with the GHTF text and they are similar to the Canadian [4] and Australian [5] wording because of this common origin. Article 2 states that a medical device is:

any instrument, apparatus, appliance, software, implant, reagent, material or other article intended by the manufacturer to be used, alone or in combination, for human beings for one or more of the following specific medical purposes:

diagnosis, prevention, monitoring, prediction, prognosis, treatment or alleviation of disease,

diagnosis, monitoring, treatment, alleviation of, or compensation for, an injury or disability,

investigation, replacement or modification of the anatomy or of a physiological or pathological process or state,

and which does not achieve its principal intended action by pharmacological, immunological or metabolic means, in or on the human body, but which may be assisted in its function by such means.

The US has a similar criteria [6], *a medical device is:*

intended for use in the diagnosis of disease or other conditions, or in the cure, mitigation, treatment, or prevention of disease, in man or other animals, or

intended to affect the structure or any function of the body of man or other animals, and which does not achieve its primary intended purposes through chemical action

There are additional clauses and definitions, such as for the use of software as a medical device (SAMD), either stand alone or incorporated in a device. In alignment with this book's topic, our focus here is on electrical stimulation devices. It is likely that in the EU and US, as well as in other countries whose medical device regulations adhere to the spirit of the GHTF text, an electrical stimulation device intended to provide a diagnosis, deliver a therapy or a treatment will be regulated as a medical device.

The regulations are complex and require expert knowledge to interpret and ensure compliance. This chapter is only intended as a high-level introduction to raise the reader's awareness of the most important requirements, and the need to consider them from the earliest stage of design.

In the EU, each nation has a competent authority (CA) which is part of the government. They oversee compliance with the MDR but the audits and certification are carried out by notified bodies (NBs) who are registered with the Competent Authorities.

3.4 Medical device requirements stipulated in legislation and guidance

The purpose of this chapter is not to provide a comprehensive review of the regulations, but to highlight a few of the most salient requirements. Within the EU

MDR there is a requirement to have a quality management system (QMS). There is a similar requirement under the FDA rules. A QMS is a structured way of defining a manufacturers activity which encompasses the planning and implementation of the processes to do with the medical device. This includes the infrastructure and environment where these can have an influence. The FDA specifies 'Good Manufacturing Practice' (GMP) under 21 CFR Part 820 although this is now outdated (January 2024) and both FDA and EU specify ISO 13485:2016.

3.5   The general safety and performance requirements (GSPR)

Both the EU MDR and the FDA code 21-CFR Parts 800-1050 regulations take a risk-benefit approach to regulate devices and ensure user safety. Here, we start with the MDR, followed by a short paragraph on the FDA code 21-CFR.

Annex I of the MDR describes a set of 23 General Safety and Performance Requirements (GSPR) that must be met for regulatory compliance. At the top of the GSPRs is the need to look at the risks associated with a medical device and balance this against the benefits. Therefore, risk management is treated as its own section in this chapter.

In general, not every requirement is applicable to every device. For each requirement that is applicable to the device, an auditor will request to see an assessment or evidence showing that the requirement was met. If a requirement is not applicable, an unequivocal argument must be presented. For requirements that are covered by an international standard it is considered best practice to have a report from an accredited test house with an assessment of the compliance. A 100% compliance to a standard is not necessarily required, but the deviations or non-compliance must be documented, alongside a rationale for why this is acceptable.

All relevant requirements must be satisfied before a device is used by, or with, a human subject, even when it is in the prototype stage. Across the requirements there is a focus on maintaining safety throughout the lifetime of a devices, including after a device is made available to the public (post-market surveillance – PMS). The legal manufacturer's responsibilities are engaged not only as long as the devices are actively made available to the public, but also for years after discontinuation.

A useful document has been published by BSI [7] which lists and elaborates the GSPRs and cross references international standards.

3.6   FDA requirements for devices

The FDA medical device regulations [8] are within the Code of Federal Regulations (CFR). Title 21 of the CFR covers Food & Drugs with the medical device section being Chapter I, subchapter H. General requirements are given and there are Parts and Subparts up to Part 861 and then the Parts become more specific to groups of devices according to their technology and application. This differs from the EU MDR which remains more generic.

Relevant in this field is 21 CFR 890.5850, powered muscle stimulator, which is classified as a Class II Physical Medicine device and covered by FDA performance standards [9]. The external 'walking' stimulator is under Neurological devices 21 CFR 882.5810 and an implanted neuromuscular stimulator is at 21 CFR 882.5860. The description makes it clear that this is a dropped foot device, Class III. The FDA also reference the ISO standards [10,11].

3.7   Medical device risk classification and its implications

There are four bands of risk in the EU MDR (Annex VIII, Chapter 3). These are Class I, low risk, Class IIa low to moderate risk, Class IIb moderate risk, Class III, high risk. Dealing with electrical stimulation, a product like a skin surface electrode is considered a Class I device. A neuromuscular stimulator, dropped foot stimulator and TENS device with transcutaneous electrodes are Class IIa. This is because they are powered (active device) and deliver energy to the body.

A Class IIb stimulator may deliver higher levels of energy (e.g. for denervated muscle) such that there is potential for harm and/or it may have a measuring or diagnostic purpose. An active implanted device is automatically a Class III device. Also, Class III devices are active therapeutic devices which have a diagnostic function that has a significant impact on the patient condition such as closed loop treatment systems or automatic external defibrillators.

The risk Class will determine certain compliance requirements. The main one is that, while all medical devices shall be manufactured under a QMS (see below), Class IIa and above require that this QMS be approved by an external body. In the EU, this body is called a NB.

In other countries the naming of the four risk Classes is I, II, III, and IV. The US has three classes of risk [12] and these correspond to devices where only general controls are required (Class I). Where special controls are required then these are Class II and Class III is where premarket approval (PMA) is required.

Comprehensive guidance is available on https://www.fda.gov/medical-devices/overview-device-regulation/classify-your-medical-device

A search of the FDA Product Classification database for 'muscle stimulator' gives a list from which a subset is presented I the table below.
https://www.accessdata.fda.gov/scripts/cdrh/cfdocs/cfpcd/classification.cfm

Product code	Device	Device class	Regulation number	Regulation description	Life sustaining	Implant
HII	Stimulator, Vaginal, Muscle, Powered, For Therapeutic Use	3	884.5940	Powered vaginal muscle stimulator for therapeutic use.	N	N
IPF	Stimulator, Muscle, Powered	2	890.5850	Powered muscle stimulator.	N	N
ISB	Stimulator, Muscle, Diagnostic	2	890.1850	Diagnostic muscle stimulator	N	N

(Continues)

(*Continued*)

Product code	Device	Device class	Regulation number	Regulation description	Life sustaining	Implant
NGX	Stimulator, Muscle, Powered, For Muscle Conditioning	2	890.5850	Powered muscle stimulator	N	N
NUW	Stimulator, Muscle, Powered, Dental	2	890.5850	Powered muscle stimulator	N	N
NYY	Stimulator, Muscle, Powered, With Limited Output, For Rehabilitation	2	890.5850	Powered muscle stimulator	N	N
NYZ	Stimulator, Muscle, Powered, With Limited Output, For Muscle Conditioning	2	890.5850	Powered muscle stimulator	N	N
QLK	Stimulator, Neuromuscular, Lower Back Muscles, Totally Implanted For Pain Relief	3	[blank result]	[blank result]	N	Y
QNO	Neuromuscular Tongue Muscle Stimulator For The Reduction Of Snoring And Obstructive Sleep Apnea	2	872.5575	[blank result]	N	N

The actual search produces links back to additional data which may include consensus standards, e.g. *IEC 60601-2-10 Edition 2.1 2016-04 Medical electrical equipment – Part 2-10: Particular requirements for the basic safety and essential performance of nerve and muscle stimulators* applying to powered muscle stimulators (Code IPF above). Also useful from this additional data is the route to market. Either a 510(k) application for the majority of the Class 2 or PMA for the devices in Class 3.

3.8   US routes to market – substantially equivalent device, 510(K) and pre-market approval

A 510(k) is an application to FDA in which the device manufacturer (domestic or foreign) submits evidence to demonstrate that the device they wish to market is 'substantially equivalent' to a device or devices already marketed. This can be a device from any manufacturer including the one making the application. The 510(k) is reviewed by FDA and if equivalence is shown to their satisfaction, then the device is *cleared* for marketing. Note that this is not approval.

PMA is for Class III (high risk) devices where a clinical trial is required or the available evidence has to be examined by the FDA. This is an expensive route to market and takes time for the scientific and regulatory review to be processed.

3.9   Product safety and risk management

One aim common to medical device legislation in many territories is to ensure that devices are safe for all users (patients, carers, clinicians, service engineers) and third parties and that any risk is outweighed by the potential benefits. Risk management is therefore a crucial aspect of the design of the product from the earliest, 'this is just an idea' stage, through to clinical evaluation, production and PMS of a medical device.

Quoting from the GSPR (EU MDR Annex 1):

Devices shall achieve the performance intended by their manufacturer and shall be designed and manufactured in such a way that, during normal conditions of use, they are suitable for their intended purpose. They shall be safe and effective and shall not compromise the clinical condition or the safety of patients, or the safety and health of users or, where applicable, other persons, provided that any risks which may be associated with their use constitute **acceptable risks when weighed against the benefits** to the patient and are **compatible with a high level of protection of health and safety, taking into account the generally acknowledged state of the art**.

3.10   Legal manufacturer

The design and development team does not necessarily become the legal manufacturer. Knowhow, expertise and patents may be transferred from the early inventors to a manufacturer. Such a manufacturer would be chosen for their experience and resources in getting devices through regulatory compliance and on to the market. In this case, the design information, including the risk management data, must also be transferred; hence it is essential that this shall be documented from the start. There is an international standard that gives guidance for risk management in medical devices: ISO 14971 [13]. The basic principles are also embedded in the EU MDR and these require a risk management plan and risk management report.

Throughout a project and in the post-market phase, existing risks are continuously considered. Clinical evaluation studies provide evidence as to the effectiveness of risk controls and identification of additional risks. Of specific concern with electrical stimulators is the potential for skin/tissue irritation and burns. This could be from the stimulator energy source and circuitry. Faults with components intended to provide safe operation must be considered in the risk assessment. Accidental or malicious connection to the mains may produce a shock hazard which can cause significant harm. Assessment of the current paths in this instance may suggest circuit or component changes to prevent this occurring.

3.11   Basic risk management (GSPR 2, 3, 4, 5, 6)

Establishing an acceptable risk-benefit ratio for the user is a primary requirement. This starts at the design concept where choices over techniques and materials will take account of what has gone before. In this way, risks can be reduced through design and engineering decisions. Selecting known or common methods will ensure reliable data as to any possible negative effects. Novel or untested materials will require biocompatibility assessments to an appropriate level and post-market clinical follow-up (PMCF).

For those risks which cannot be 'designed out' then the user must be alerted when the risk is present or imminent, e.g. a battery falling below a safe voltage or an output approaching a potentially hazardous level may require an alarm or indicator lamp/display. It would be advisable to consult electrical safety standards which apply to all medical devices [10] as well as the electrical stimulator specific ones [11].

The final influence that the manufacturer can have is the labelling and documentation which accompanies a medical device. This must include explanation of residual risks – i.e. those risks that remain regardless of the technical measures that have been taken to make a device as safe as it can be.

Once the risks have been identified and addressed, any remaining risk from use of the device shall be compared to the magnitude of the benefit resulting from its use. Bear in mind that 'use' covers the fitting of the device and its effect on third parties. This judgement assumes a level of clinical experience and objective data which may only be available after a clinical trial.

ISO 14971 has informative reference sections with a flowchart and descriptions of the fundamental risk concepts such as risk-benefit. So it is advised that the standard be obtained and any risk management tools (e.g. software) are validated against ISO 14971.

In general, a risk is composed of a hazard and a probability of that hazard occurring. A hazard will have a consequence which would cause harm so it can be rated for the magnitude of that harm. A probability will range according to data e.g. on component lifetime, or human action such as reversing a plug or battery. Giving a numerical value individually to the hazard and the probability enables them to be multiplied to give a numerical risk value. If a risk value is considered to be too high, then a risk control must be applied – either reducing the hazard/harm or the probability.

3.12   Technical file

The files made as part of the design and development of a medical device and evidence of compliance with the applicable regulations need to be readily accessible. The term for this assembly of documents, drawings, test reports, spreadsheets and so on is called the 'Technical File' (TF).

The contents of the TF are specified within the MDR. A compliance matrix is the recognised way of checking for completeness, it is also useful for internal or external auditors to follow the evidence trail. In addition, the entries in the Technical File can reference the particular MDR requirements. In order to assure third parties, such as representatives and NBs, a summary document is sufficient rather than access to the fine detail.

3.13   Quality management systems

Within medical device regulations is the requirement for a medical device manufacturer to institute and maintain a QMS. This links back to the initial stress on starting early with decisions over who might end up as the manufacturer. The development documents are required to be managed in a QMS to track the process and associated decisions. The proper creation and maintenance of documents is a core requirement of a QMS. Unsurprisingly there is an international standard and since that essentially establishes 'state of the art', that is the model that shall be followed. The Standard is **ISO 13485:2016 Medical devices — Quality management systems — Requirements for regulatory purposes**

ISO 13485 is a development of the general-purpose quality management standard, ISO 9001 (**ISO 9001:2015 Quality management systems — Requirements**). ISO9001 can be achieved by large organisations and small traders and would provide assurance of proper structure and customer responsiveness. The responsiveness aspect was one area that makes ISO9001 unsuitable for medical devices since a request for a particular feature or change could upset rigorous and well verified processes. So, ISO 13485 was based on ISO9001 but tailored to the medical devices industry. As such it will bring in considerations of document control, management responsibility, device and possibly component traceability, planning, design, production, PMS, dealing with adverse incidents, continuous improvement and so on.

3.14   Choices on how to construct the QMS

The days of running a medical devices QMS by documenting processes on paper, computer folders and spreadsheets are very much numbered. At Odstock Medical Limited the QMS was put in and managed from 1996 by keeping documentation in distributed paper files and tracking updates with a self-written database. Also, this was largely one person's responsibility. After approximately ten years the device regulatory compliance became shared. With the development of new device, we introduced a commercial risk management database system alongside our pre-existing, internally constructed, document tracking database. The documents themselves were in identified network folders. This arrangement ran for another ten years or so, but two major factors were hard to manage, firstly, the requirement to link changes throughout the design and after launch with identified risks and secondly, capturing all customer feedback and complaints along with non-conformance reports and managing trending, corrections and corrective and preventive action.

The people with the appropriate skills to follow all these processes through correctly also had other roles so targets were being missed. If this is a situation that could arise in your organisation, then it may be time to consider one of the eQMS solutions – an electronic QMS. An eQMS will hold and control the QMS related documentation and force correct management of the other requirements. Running an eQMS still requires people with QMS and regulatory knowledge but omissions from missing paper or overrunning responses to complaints and audits are more visible.

The new product development lead could see that the additional burden of MDR compliance over the MDD was not going to be manageable with our existing resources so eQMS systems were investigated and one was selected. The advantages are great but there is still input needed from skilled regulatory resources to make sure all compliance aspects are captured. With an eQMS it is easier to have internal and external audits since no searching across systems is required. The document sign-off is clear and multiple sign-offs can be managed within the eQMS.

3.15   Post-market surveillance

Under the EU MDR (article 10), the legal manufacturer of a medical device must implement PMS activities (for details see articles 83–86 and annexes 3 and 14). In essence, knowledge of the risks, and benefits, of using a device must continue to be gathered after a device has been approved to be placed on the market. This requirement aims to ensure that any risk not identified during the testing phase is identified, and addressed, in a timely manner to avoid, or minimise, harm to users. Manufacturers must implement a system to gather use information both reactively and proactively. For devices of Class II, IIb and III, article 86 requires the publication of Periodic Safety Update Reports (at least annually for Class III devices including implantable devices), to demonstrate timely analysis of the PMS data to identify any safety or performance concerns. Where applicable, this may lead to re-evaluation of the risk-benefit profile, or more broadly of the state of the art of the medical device.

3.16   Unique device identifier system and EUDAMED

A unique device identifier (UDI) is a globally recognised identifier that enables regulatory authorities to identify unequivocally each and every medical device on the market. It is the responsibility of the manufacturer to comply with the UDI requirements.

In the EU MDR (article 27), a UDI system must be in place for all medical devices, regardless of their class. This is a recent change as part of the increased responsibilities for PMS, as the UDIs are expected to improve traceability throughout the device's lifecycle.

UDIs have also been introduced in the US and lodged within a Global Unique Device Identification Database (GUDID) [14].

UDIs are designed to improve medical device safety by assisting the regulatory authority's monitoring duties, improving the reporting of adverse events and enhancing the effectiveness of PMS. Healthcare professionals and individual patients should also benefit from the UDI system as they will be able to seek information pertaining to a device they are using or wish to use without doubt about knowing the exact model or version.

UDIs will be collected in a database (article 28) that itself will be integrated in EUDAMED (articles 33 and 34), defined as

> 'a European database on medical devices that should integrate different electronic systems to collate and process information regarding devices on the market and the relevant economic operators, certain aspects of conformity assessment, notified bodies, certificates, clinical investigations, vigilance and market surveillance'.

The core purpose of EUDAMED is traceability of devices and tracking of adverse incidents. Certification can be checked and a publicly accessible module is planned for Class III devices, including implantable devices, but excluding custom-made devices and investigational devices. A summary of safety and clinical performance must be made available to the public via EUDAMED. so that consumers can determine if there are any problems with a device they may be considering using. Custom made device are devices that are made for a named patient to the specification of a medical professional. They may not have a CE mark, but they still must comply with MDR.

3.17   Becoming a medical device manufacturer

If we consider a product that isn't a medical device then the view of their operation can be relatively easily described as a flow chart (Figure 3.1).

There will be industry standards, but the control of the manufacturer will be assumed to be adequate unless there are some obvious shortcomings in their safety or hygiene. They are free to change their product with little or no reference to anyone else. It is in in their commercial interest to achieve good sales and they may make customer specific changes in order to satisfy particular requests (Figure 3.2).

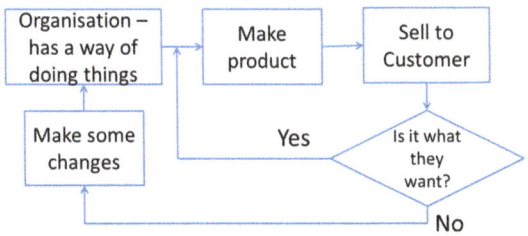

Figure 3.1   Simple manufacturer chart

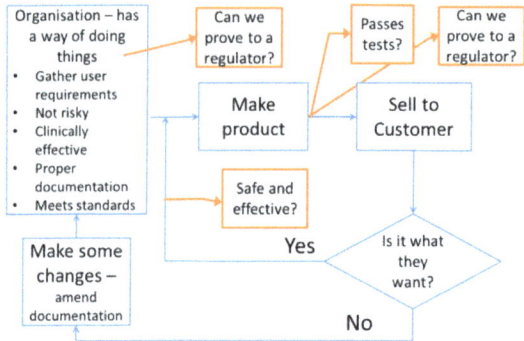

What if things are more demanding?

Figure 3.2   Adding to the simple manufacturer chart

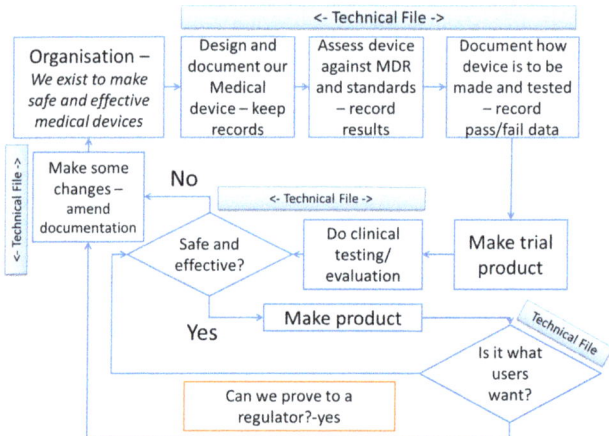

Figure 3.3   Additional role of the technical file for medical device production

Introducing what we already know about the requirements around a medical device we can see how the complexity begins to increase. During the design of the device and the processes required to manufacture it, reference must be made to the regulatory requirements. The documentation must be of a sufficient standard that a regulator can check that this was done properly so records have to be kept. The document control that ensures the current versions and availability of records is a core function of any QMS.

Within Figure 3.3 we see a combination of requirements from the QMS and MDR and this is how it should be. The inspection aspects of Figure 3.2 diagram have been omitted for clarity but they are still necessary. The QMS is a shell or support structure to place the MDR requirements within. In this way they can be recorded, monitored/analysed and retrieved. In the 'Organisation' box is the 'policy' of the medical device manufacturer. In the following section we will see how the other boxes are covered.

The ISO 13485 QMS standard [15] places an emphasis on having descriptions of processes rather than procedures and a way of showing the linkage of these processes. The process of making the device itself is called 'product realisation' and in a QMS audit this will extensively checked.

3.18   Implementing a quality management system – QMS

As introduced in the 'Manufacturer' section, the task of organising and making available documents and records, as well as all the other compliance requirements, requires some sort of systematic approach. This approach should also be common between manufacturers and auditors so that they understand the framework on which the QMS is built. Using the ISO 13485 standard ensures that manufacturers and regulators are working to the same reference.

At this point it would be simple to write several hundred pages on how to implement a QMS. That is not the intention – the aim is to expose researchers to the ideas that underpin the QMS and regulatory requirements. The structure of the 13485 standard will be described. If the approach presented here is too simple or too complicated then the internet has many alternatives, some of which will be referenced.

The sections of the standard are briefly outlined. The sub-sections are condensed and some explanation given. The numbering from the standard is retained (0–3 are omitted since they contain a technical preamble and definitions of terms that assist in aligning this standard with others).

3.19   Quality management system

Section 4 contains the *General requirements, Documentation requirements* (Document Control) and the Quality Manual. The Quality Manual is the document that contains the purpose of the organisation such as the scope of devices manufactured and the regulatory territories. It is also a useful place to locate the company structure and commitment to risk management, monitoring, measurement and analysis of processes leading to continuous improvement.

3.20   Management responsibility

It is necessary to show that the top management of the organisation is involved and committed to the medical device processes. A representative from top management must sign the *Quality policy* which encapsulates the purpose and function of the company/manufacturer. There must be objectives with measurable parameters along with planning visible for the processes and implementation of the QMS. When the stage of seeking external audit and certification is reached a Management Representative and Person with Responsibility for Regulatory Compliance (PRRC) will be required. Evidence of management action is captured in a Management Review which takes place at least annually.

3.21   Resource management

Commitment to the manufacturing and regulatory resources along with their planning is covered. The necessary infrastructure plus Work environment and contamination control should be addressed. This is not onerous for external portable devices but may involve substantial investment and maintenance when dealing with implanted and sterile devices. Even with portable devices – they may be returned in unknown states of contamination.

3.22   Product realisation

This section covers the planning, design and development and Control of the design changes. Risk management also requires planning and changes must be risk assessed. Product storage and preservation shall be considered, and equipment used for monitoring and testing has to be appropriately controlled. In practice, test equipment used for product release shall be calibrated against national standards.

Product realisation requires support functions from across the organisation, for instance IT infrastructure is necessary for the ordering and recording of device related components. Traceability is necessary for all devices but has to be taken to component and process level for implanted/Class III devices.

3.23   Measurement, analysis and improvement

A very general heading that applies to the QMS processes and the device(s) through the manufacture and into the clinical usage. Monitoring and measurement of the devices addresses quality assurance (repeatability and safety) prior to release. It also applies via customer feedback and complaints related to devices and the accompanying instructions and labelling. A complaint or report alleging that harm has resulted from use of the device may require reporting to one or more national authorities and EUDAMED. Even if harm has not resulted, complaints and feedback require analysis and trend reporting.

Checking up on the QMS and regulatory compliance requires regular internal audit. The portions covered shall be prioritised according to risk but the audit plan shall be produced in advance and adhered to.

The findings from audit, analysis of trends and other sources should be used for improvements in the QMS. Errors or an episode where something happened that shouldn't have, is termed a 'non-conformance'. Addressing non-conformances may require corrective and/or preventive action. Preventive action is taken where an occurrence or trend may lead to a non-conformance. The action is designed to avoid or head off the situation so that an actual non-conformance does not occur.

Corrective action is taken to prevent recurrence of a non-conformance that has occurred. Risk shall be considered so that any action does not raise or introduce risk, this shall be documented.

3.24   Conclusions

In conclusion, the stated aim has been to introduce the wide-ranging complexity associated with medical device regulatory compliance. It is not possible to deliver a working model to suit all manufacturers since there are many territories and many means of achieving the objective of a finished medical device to be placed on the market.

With the complexity comes increased expense for the necessary human resources and possibly licensing of an eQMS. There will be fees to NBs, auditors and other consultants. So, an appreciation of this and the knowledge to seek out the appropriate information at an early stage will assist the determined researcher/ engineer in achieving their goal of a safe and beneficial medical device being available to the patients that need it.

Resources

Emergo – territory requirements – steps to compliance for different territories. Charts available on registration, may become fee-based in the life of this chapter. https://www.emergobyul.com/resources/process-charts

- BSI – Medical Device resources & standards https://knowledge.bsigroup.com/categories/medical-devices
- Risk Management guide from a regulatory services company.
- https://www.orielstat.com/blog/iso-14971-risk-management-basics/
- FDA Code of Federal Regulation for Medical Devices – CFR
- https://www.ecfr.gov/current/title-21/chapter-I/subchapter-H

Glossary

CAPA	**C**orrective **A**nd/or **P**reventive **A**ction, planned activity to correct or prevent the occurrence of a non-conformance or complaint.
Clinical evaluation	putting a device into clinical use so that the safety and effectiveness can be verified prior to general launch. May be conducted under a formal trial, overseen by a Competent Authority (or equivalent)
Clinical trial	putting a medical device into use in order to assess its safety and efficacy. For a manufacturer to use clinical trial data to justify a novel application it will have to be conducted under the auspices of a Competent Authority or the equivalent in the applicable territory.

Competent authority	A national government department or agency which has the responsibility for making sure that Medical Devices are properly regulated. Also, the body that any adverse incidents or field corrective actions shall be notified to.
Corrective action	action to put right a situation that has led to a non-conformance.
EUDAMED	European medical devices database for recording manufacturers, notified bodies and reporting of adverse incidents. Intended to have a publicly accessible module for self-research on device safety and performance.
GSPR	General Safety and Performance Requirements, the section of the EU MDR where specific device requirements are described.
GUDID	US/FDA Universal device identifier and the name of the associated database.
Lifecycle	the making, use and 'end of life' stages of a medical device
MDD	EU Medical Device Directive
MDR	EU Medical Device Regulations 2017 (see references)
Non-conformance	an identified lapse in the product, quality system or regulatory compliance.
Notified body	An organisation established in the EU which has responsibility for checking compliance with Medical Device requirements on behalf of an EU government agency.
Post market clinical follow up (PMCF)	Gathering clinical data, including proactive surveys of users, to regularly re-evaluate the risks and benefits after device launch.
Preventive action	action to avoid a non-conformance
QMS	Quality Management System
UDI	Universal Device identifier/Identification

References

[1] FDA Design control guidance for medical device manufacturers, relates to FDA 21 CFR 820.30. https://www.fda.gov/media/116573/download.
[2] Global Harmonisation Task Force. https://www.imdrf.org/ghtf/history.
[3] EU Medical Device Regulation (MDR) *"Regulation (EU) 2017/745 of the European Parliament and of the Council of 5 April 2017 on medical devices"*. http://data.europa.eu/eli/reg/2017/745/oj.
[4] Canadian Medical Devices Regulations (SOR/98-282). https://laws-lois.justice.gc.ca/eng/regulations/SOR-98-282/page-1.html.

[5] Australian Government Therapeutic Goods Act 1989. https://www.legislation.gov.au/C2004A03952/latest/versions.

[6] FDA medical device determination. https://www.fda.gov/medical-devices/classify-your-medical-device/how-determine-if-your-product-medical-device.

[7] https://www.bsigroup.com/LocalFiles/es-MX/dispositivos-medicos/General_Safety_and_Performance.pdf.

[8] An Introduction to FDA's Regulation of Medical Devices, Mallis, E. https://www.fda.gov/media/123602/download.

[9] https://www.fda.gov/regulatory-information/search-fda-guidance-documents/guidance-document-powered-muscle-stimulator-510ks-guidance-industry-fda-reviewersstaff-and.

[10] BS EN 60601-1:2006+A13:2024 Medical electrical equipment – General requirements for safety.

[11] IEC 60601-2-10:2012/AMD2:2023 Amendment 2. Medical electrical equipment – Part 2-10: Particular requirements for the basic safety and essential performance of nerve and muscle stimulators.

[12] US FDA device classification. https://www.accessdata.fda.gov/scripts/cdrh/cfdocs/cfcfr/CFRSearch.cfm?fr=860.3.

[13] BS EN ISO 14971:2019+A11:2021 Medical devices. Application of risk management to medical devices.

[14] FDA GUDID. https://www.federalregister.gov/documents/2013/09/24/2013-23059/unique-device-identification-system.

[15] ISO 13485:2016 Medical devices – Quality management systems – Requirements for regulatory purposes.

Chapter 4

Afferent and therapeutic stimulation treatment after spinal cord injury

Nick Donaldson[1], Lynsey Duffell[1] and Winfried Mayr[2]

4.1 Introduction

4.1.1 History

Electrical therapy has a much longer history than functional electrical stimulation (FES). Liberson *et al.* is usually regarded as the seminal paper about FES; they reported the first portable stimulator for immediately dorsiflexing a dropped foot and thereby enabling *function* [1]. The portable stimulator was made possible by transistors becoming available during the 1950s.

Electrical therapy goes back much further. In 46 AD, Scribonius Largus recommended the application of a live black torpedo fish for treating headache or gout, until numbness was achieved. During the eighteenth century, electrostatic generators became available for shock treatment, which became practicable after the invention of the Leyden jar (capacitor). In 1748, Jallabert reported a case study of a patient whose paralysed hand recovered after three months of shock treatment sufficiently for him to return to work [2]. By the 1780s, several major hospitals in Britain had electrical machines which were used to treat intractable cases. At St Thomas's in London, a respected surgeon called John Birch, whose patients included the royal family, ran the electrical department for 20 years. In 1793, the London Electric Dispensary opened on City Road and treated about 300 patients per year for at least the following two decades, claiming a cure for at least half of them [3]. In the nineteenth century, better sources of current were invented by Volta (Galvanic) and by Faraday (Faradic) and there were great advances in electrophysiology. Particularly notable was Duchenne, who found the motor points for muscle activation. Claims were made for the treatment of many diseases, and, towards the end of the century, many doctors had electrical machines and their use was widely accepted [4]. Without electronics, they could produce stimulation currents which would be familiar to us: the *Leduc Interrupter* was a motor-driven

[1]Department of Medical Physics and Biomedical Engineering, University College London, UK
[2]Center for Medical Physics and Biomedical Engineering, and Department of Physical Medicine, Rehabilitation and Occupational Medicine, Medical University of Vienna, Austria

commutator with adjustable brushes that could produce pulse frequencies up to 100 Hz and pulse lengths down to 1 ms.

In 1905, Cumberbatch published a textbook on medical electricity [5]. He was the Medical Officer in Charge at St Bartholomew's Hospital in London and former President of the Section of Electrotherapeutics at the Royal Society of Medicine. On the treatment of paralysis, he says the following.

Paralysed muscles can be made to contract by electrical stimulation. In this way they can be artificially exercised. Rhythmic contraction and relaxation also increase the circulation of the muscles. The temperature of a limb is raised if a number of its muscles are exercised in this way....

The use of electric currents for the treatment of paralysis does not end with the production of artificial exercise. Muscles can be reached not only by voluntary impulses from the brain, but also by others which come from the skin. The sensory nerve endings in the skin, the motor nuclei in the spinal cord, and the motor nerves which supply the muscles form a reflex arc. The efferent side of the arc also includes the nuclei in the lateral horns of the spinal cord and the nerves which pass from them to the sympathetic nervous system and thence to the blood vessels and other parts....

But, apart from the explanation of biological phenomena on physiological grounds, there is abundant clinical evidence that the stimulation of skin by electrical methods aids recovery from paralysis, even if there is no actual contraction of the muscles.

After explaining the difference between Upper and Lower Motor Neuron injuries, he continues:

When paralysis is due to a lesion affecting the upper motor neuron .... if the lesion is not progressive .... experience shows that in cases where a certain amount of voluntary power has returned, it is often possible to effect further improvement by electrical stimulation of the affected limbs. The reason why such treatment can procure improvement is not easy to give, but clinical evidence of the fact is seen when cases of hemiplegia due to cerebral haemorrhage, thrombosis or embolism are treated...

He opens the chapter with the following warning.

Electricity is a valuable agent in the treatment of paralysis and will, in suitable cases, give results that cannot be obtained by other methods. It is, however, of the first importance that correct methods of application should be employed, and it is the failure to recognise and adopt them that is generally responsible for the inability to obtain good results, and for the scepticism that still prevails in some quarters regarding the value of the treatment.

There is not much in these quotations which we would disagree with today. He distinguishes direct motor effects producing artificial exercise (FES) from the therapeutic effect achieved by sub-threshold stimulation. We would say that the sub-threshold stimulation is activating the proprioceptors as well as the cutaneous afferents, but it is probably the reflex arc from the former to the motor neurons that is more important, compensating for lost input and raising excitability.

Most intriguing is why this knowledge seems to have been forgotten. Perhaps the last paragraph quoted is a clue, that too many practitioners were not using the correct method, results were disappointing, and that was bringing electrotherapy into disrepute. Once discredited, presumably patients would not pay for treatment and the hospitals would close the department and the know-how lost as the practitioners retired.

For us in the twenty-first century who are working towards the re-establishment of electrotherapy, the message from Cumberbatch seems to be: treat suitable cases and use the right method.

4.2   Why stimulate afferents after spinal cord injury?

The spinal cord is a neural network of interneurons with inputs from the afference peripheral (sensory) nerve fibres, and outputs being the motor neurons that activate motor units in muscles. Some interneurons cross between segmental levels, for example, coupling arm and trunk movement with leg movement. The functions of the network include reflexes that produce stereotype movements and responses. There are also inputs from the brain via axons in corticospinal and reticulospinal tracts (RST) of the spinal cord that either cause direct responses in muscle or modulate the reflexes. The spinal network also has outputs that carry sensory information to the brain through afferent fibres in the tracts. The sensory-motor behaviour of the network depends on the strengths of the synapses (number of synapses or amount of neurotransmitter) between all these neurons. These strengths can change, allowing neuroplastic adaptation, which in intact people enables improvement in motor performance by practice.

Traumatic spinal cord injury may break some or all of the afferent and efferent fibres in the tracts, thus immediately paralysing some muscles below the lesion, by depriving them of input from the cortex or the reticular formation and causing loss of sensation. In the longer term, if the muscles are no longer used, for example because the person stops using their legs because of the paralysis, the strength of the synapses may alter to reduce excitability, to cause spasticity or to cause spasmodic responses such as clonus.

Afferent stimulation is used either for paraesthesia or to modify the neural network to improve motor behaviour. We will only be considering the latter in this chapter. The treatment may be used acutely for the alleviation of symptoms, or, applied over a longer period, as a therapy.

In this chapter, we distinguish peripheral from central afferent stimulation, giving three examples of the former: hands, feet and peroneal nerve. Central afferent stimulation is so-called *spinal cord stimulation* (SCS) which actually targets the proprioceptive neurons which enter spinal segments through the dorsal

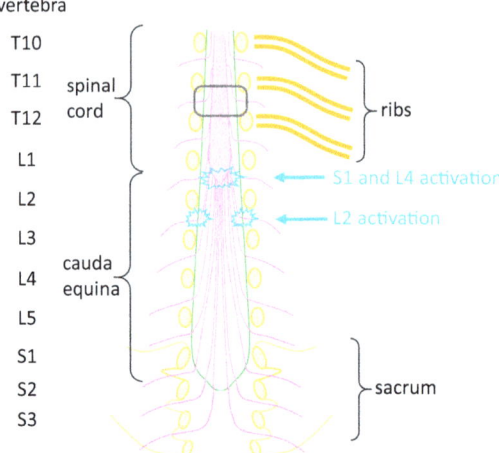

*Figure 4.1   Nerve activation due to transcutaneous SCS. This diagram
representing results from Ladenbauer et al. [137]. The coronal section
through the spinal cord and cauda equina shows the posterior nerve
roots. The position of the stimulating electrode on the back is shown at
T11-12 but activation is caudal to the electrode, either at the conus
(S1 and L4) or where the root passes through the dura (L2). The dura
is shown green. The positions where activation occurs is strongly
influenced by curvature along the nerve due to changing direction.*

nerve roots. These are the neurons that convey information on the body position for
motor control. Central afferent stimulation is divided into *transcutaneous*, in which
the electrodes are on the skin, and *epidural,* where they are implanted into the
vertebral foramen of the spine, adjacent to the roots where they emerge from the
cord (Figure 4.1). These various methods are considered separately in the following
sections.

4.3   General principles of afferent nerve stimulation

Electrical stimulation provides a unique tool for eliciting action potentials in afferent
and efferent neurons and – only in the case that the nerve supply is missing – also in
denervated muscle fibres. Nerve fibre recruitment follows a size principle, starting
with the largest neurons, reacting at the lowest intensity, followed by smaller diameter neurons, so that there is co-recruitment at higher stimulation intensities.
Threshold intensities, at which axons first respond, are inversely proportional to their
size. As a second influence, the distance from electrode to nerve fibre matters; as a
rough estimate, the threshold grows with square of the distance, which can make a
difference in recruitment order, when, for example, small cutaneous fibres are closer
to the electrode than larger proprioceptive fibres in an underlying mixed nerve
branch.

The largest neurons in a mixed peripheral nerve or in the purely afferent dorsal roots, having the lowest threshold, are proprioceptive 1a afferents, primarily, though not solely, providing feedback from muscle spindles sensing actual muscle length. They connect via posterior nerve roots to segmental interneuron networks, which react with a fast, monosynaptic reflex activation of the same muscle, whilst blocking activation of the antagonist via reciprocal inhibition. This input can come from peripheral stimulation sites near distal mixed nerves or central activation of purely afferent dorsal roots. The latter modality is what is commonly meant with 'spinal cord stimulation'.

Goals of afferent nerve stimulation can be

- Peripheral (neuropathic) pain relief
- Modification of spasticity or muscle stiffness
- Augmentation of impaired motor control or sensory perception

These effects are produced by stimulus trains with intensities between subsensory to moderate suprasensory intensity, depending on the problem to be solved.

Intensity determines the fibres recruited, beginning with proprioceptors, which is usually sufficient for reduction of spasticity and stiffness. If mild paraesthesia is perceived, larger cutaneous afferents are also being recruited, eventually influencing perceived neuropathic pain. Still higher intensity can show augmentation effects of voluntary muscle contractions and movement. This is only true for patients with *discomplete* paralysis, a clinical profile between 'complete' and 'incomplete'. They appear to be completely paralysed (i.e. AIS A or B) on conventional clinical tests, but retain intact neural structures and have reduced suprasegmental input and low excitation of segmental interneuron structures ('central state of excitability'); consequently, there are neither voluntary movements nor, in AIS A, perceived peripheral sensations. In a very limited number of patients, we can observe afferent stimulation, elevating the central state, which enables some voluntary movements and sensory perception despite the deficits in the neural pathways.

There are pros and cons speaking for central posterior root versus the peripheral nerve approaches.

Central stimulation with surface electrodes (transcutaneous SCS) activates afferent fibres with low selectivity. These fibres pass behind the relatively-insulating bone of the vertebral bodies, separated by better-conducting intervertebral discs. The neurons in the roots are densely packed and therefore their activation profile depends mainly on their diameters. As relatively high stimulation intensity is necessary for eliciting action potentials in spinal roots with skin-attached electrodes, stimulation also causes co-contraction of the trunk muscles and sensations from the cutaneous afferents in the corresponding dermatome. This may lead to sensory discomfort, but, more seriously, in persons with impaired respiratory capacity, such as after cervical spinal cord injury, the continuous contraction of trunk muscles is an additional obstruction of breathing.

In comparison, peripheral stimulation of afferent neurons can take advantage of selective access to peripheral structures, e.g. flexor and extensor regions in the proximal or distal extremity, or locations with extra high density of sensory nerves like in hand or foot. On the other hand, at higher stimulation intensity with co-activation of efferent neurons, there are limitations by collision block phenomena [6], when directly elicited antidromic action potentials in motor-fibres meet orthodromic ones, produced as the central response in the faster acting reflex loops.

Generally, tailored individual setups need to be compiled along with meticulous assessment of functional profile and practical considerations. For example, the choice can be limited by the presence of metal implants, as after spinal stabilisation surgery, and skin condition, such as pressure sores caused by the upper edge of a wheelchair backrest, or hypersensitivity at the intended stimulation sites. Also, there are usability limitations, for example due to impaired hand function, or a lack of carers to give assistance. These may be insurmountable obstacles to otherwise very effective treatments.

Electrode positions and the intensity-dependent recruitment of neurons are essential to achieve specific physiological responses, but equally relevant is adjustment of stimulation frequency which strongly influences interneuron processing in the spinal networks. The choice of frequency can lead to inhibition or augmentation of neural activity, but specific values vary from one person to another. In an uninjured organism, a continuous train of stimuli tends to cause progressive inhibition in reflex responses: relative to a single stimulus, the response amplitude declines with growing frequency, beginning with 2 Hz towards complete suppression at approximately 100 Hz. This characteristic can be strongly altered after spinal cord injury: augmenting or inhibitory effects can then occur at vastly different person-specific frequency values, which need to be identified for intervention planning and can change under various therapeutic modalities over time.

So, to summarize:

- Amplitude determines neuron recruitment, starting with largest proprioceptive fibres and co-recruiting smaller ones with growing stimulation intensity. For neuromodulation slightly subsensory (slightly below sensory threshold for patients with intact sensory perception) to slightly suprasensory (perceiving mild paraesthesia) stimulation strength is most effective. If sensory perception is impaired, an intensity about 10% below motor threshold can be taken as starting point for fine tuning.
- The frequency of tonic stimulation influences augmenting and inhibitory effects in spinal interneuron processing, which can be substantially altered by a neural injury to an individual functional profile (i.e. voluntary and involuntary motor control features, reaction to external stimulus parameters) and requires personal optimization.
- Stimulation sites can be near the spinal posterior nerve roots or at distal nerve sites.

We want now to come to useful practical application with requirements for instrumentation, basic protocols for setups and assessment guided parameter adjustments.

4.4   Peripheral afferent stimulation

4.4.1   Principles of instrumentation and method

Afferent nerve stimulation is a powerful and easy protocol for modifying spasticity and rigidity, or more generally, influence excitability of spinal interneuron networks. Stimulation intensity for such applications is relatively low, in the range slightly below sensory threshold to slightly above motor threshold, tonic (continuous) stimulation is delivered and the pulse frequency is a particularly important parameter, with strong influence on spinal interneuron processing, and thus to be carefully adjusted to patient's needs and therapeutic goals.

4.4.1.1   Instrumentation

For afferent nerve stimulation monophasic as well as biphasic stimulators are suitable, so long as they can provide continuous (tonic) stimulation. Many low-cost TENS devices provide the necessary features, those with symmetrical biphasic pulses should be preferred due to their better neuron recruitment characteristic. Frequency should be adjustable in a range from 1 to about 100 Hz. 1 Hz is a useful test pattern to find motor threshold intensity, based on vision or palpation. For the therapeutic intervention, the optimal frequency for the patient should be identified for augmentation or inhibition of motor or sensory functions, that can vary between approximately 20 and 100 Hz, depending on individual patient's symptoms and therapeutic goals, but also generally change over the course of treatment (Table 4.1).

Appropriate electrodes vary with the application setup and the respective anatomical situation. Standard hydrogel electrodes are available in various sizes and brands, more or less all similar in quality and ease of use. It is difficult to give general recommendations for optimal electrode placement, in all applications an active electrode should be placed close to a peripheral nerve or terminal afferent nerve branches and counter-electrode, usually with a bigger surface area, somewhere along the course of the same nerve. Polarity of electrodes matters: for monophasic pulses nerves near the active electrode react at a lower threshold if it is

Table 4.1   Essential features of a stimulator for afferent nerve stimulation

Pulse form	Symmetric biphasic rectangular (Asymmetric, monophasic rectangular/exponential decay recharge phase)	*phase < 0.5 ms* *period* τ *frequency* $f = 1/\tau$
Amplitude range	$0 - \geq 50$ mA, continuous adjustment or increments <0.5 mA	
Pulse width per phase	100 μs – 1 ms	
Frequency range	1 Hz – ≥ 100 Hz	
Pulse train	Tonic (continuous)	

connected as cathode, for biphasic pulses this effect is less pronounced, but the active electrode should be the anode with reference to the first pulse phase.

Below in Section 4.5, useful exemplary setups are explained in detail: mesh glove, mesh sock, and peroneal nerve stimulation, but many more clinical problems can be addressed just using general purpose electrodes at different anatomical locations.

4.4.1.2   Application protocol

After preliminary placement of electrodes, it can get necessary later to modify their position for better responses, correct connection polarity to the stimulator is important. As explained above, for biphasic pulses the anode of the first pulse phase (usually red terminal) is connected to the active electrode, for monophasic (sometimes named asymmetric biphasic) the active electrode is connected as cathode.

Now we are ready to assess physiological responses and determine stimulation parameters:

1) The stimulator should be adjusted to the longest pulse duration in the available range, in some devices 1 ms, more often just 400 or 500 μs.
2a) If sensory perception is preserved, first a continuous (tonic) pattern with 30 Hz pulse frequency is chosen. Then amplitude should be slowly increased and the patient asked to report the first tingling sensation. Now intensity is slightly reduced to subsensory level
2b) If sensory perception appears altered or absent, the appropriate intensity can be estimated by searching for the motor threshold via single twitch reaction (actually, a tonic 1 Hz pulse train), visually observed or palpated, and then reduction of the amplitude level by about 10%. This level is then used for the further applied 30 Hz stimulation. Eventually this initial amplitude estimate needs to be slightly modified to find the individual optimum.
3) If reducing muscle tone is the therapeutic goal, a pronounced effect should get visible more or less immediately or within a few minutes after the start of stimulation. At the first therapy session, tonic stimulation should be applied for 30 minutes and, in many cases, a carry-over effect prolongs the muscle relaxation for an extended period of time. This carry-over varies a lot between individuals and gets longer with repeated applications
4) If the results are not sufficient, it may help to slightly increase or decrease amplitude; if that does not work, next try higher, more inhibitory, frequencies. As changes in response appear quickly, a relative optimum can usually soon be found.
5) An important application of afferent nerve stimulation is modification of neural excitability. Disuse often leads to significantly higher activation thresholds for sensory and motor neurons, which can be assessed with the above-mentioned test patterns: thresholds are reached with much higher amplitude values than usual (<10 mA). Regular application – once or twice 30 minutes per day – of the 30 Hz stimulation, as described above, can lead to normalisation of excitability

within weeks or months. Developments can be monitored via regular threshold testing that is also useful for keeping intensity optimal.

6) Reduction of muscle tonus, e.g. in the finger flexors, can significantly support impaired grasp movements by itself, but stimulation of proprioceptors can additionally augment residual voluntary movements, though not in all cases. 30 Hz is a useful compromise to begin with, higher frequencies tend to have more inhibitory, lower frequencies more excitatory influence on central inter-neuron processing. Responses to stimulation are highly individual, and so, to find a personal optimum, an interactive search for appropriate stimulus ampli-tude and frequency is necessary.

7) Last but not least, similar stimulation setups can also help in chronic peripheral pain, just with slightly higher, supra sensory threshold amplitude, inducing frequency-dependent paraesthesia. The immediate changes in pain perception are the indicators for individually optimal strength and frequency of stimulation. Adjustments can only rely on interaction with the patient: how an alteration to the intensity and frequency affect the perceived pain, not just additional 'tin-gling'. As in other modalities of pain therapy by stimulation, relief can only be achieved in some cases, but due to its easy application, without side effects, it is always worth a try.

4.5 Three methods for peripheral afferent stimulation

4.5.1 Mesh glove stimulation

This is a powerful and easy method for reducing tone in the spastic hand or, more generally, of neural excitability in the upper extremity [7].

The classical mesh glove application relied on stainless steel gloves used for protecting the hand and fingers during dangerous tasks with sharp hand tools. As a whole-hand electrode, the stainless-steel mesh is reliable, but not very comfortable, as it has no elasticity, and is quite difficult to get over a hand with paralyzed hypertonic finger flexors. Also, the necessary filling of the mesh with electrode gel plus coverage of the whole arrangement with an elastic rubber glove is incon-venient. Though the electrical properties of the wire mesh gloves are excellent, conductive textile gloves are a better choice, and are available at moderate cost as medical products. It is important to make sure that the gloves have a socket for the electrical connection, as some are offered that are just conductive, but lack any means for connecting a stimulator. They should fit tightly (different sizes are available) and be slightly wetted with just water or even used in the dry condition. The choice depends on skin impedance and sensitivity, which can lead to uncomfortable perception. It is worth trying dry, because glove handling is easier: the practical criterion is whether there are local burning sensations, due to low-impedance spots, where current density gets too high.

As a counter-electrode, two interconnected hydrogel electrodes are positioned at the dermatomes of the flexor- and extensor muscle groups. Interconnection requires a 'y-cable' with 1 female and 2 male 2 mm-connectors, or alternatively a

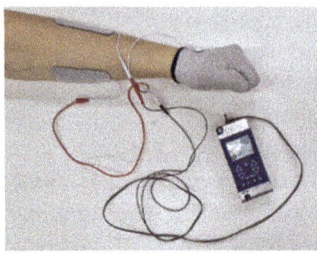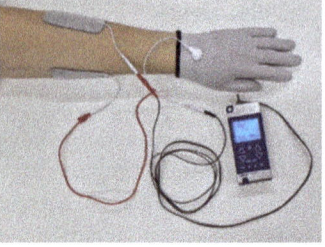

Figure 4.2   Mesh glove application with conductive textile glove and biphasic stimuli, red connector indicates anode of first pulse phase. Ready-made laboratory cables with cascadable 2 mm connectors can be used to interconnect electrodes.
NOTE: It is important to have the terminal of conductive textiles insulated against skin contact (lower right) – in some products this detail is not considered.

short laboratory cable with 2 cascadable 2 mm-connectors: male with integrated female connector on both sides. Polarity of stimuli matters, for monophasic pulses the anode is to be connected to the glove, the cathode to the interconnected hydrogel electrodes, for biphasic stimuli polarity is reversed in relation to the first pulse phase, anode generally labelled by a red connector (Figure 4.2).

4.5.2   Meshsock stimulation

This is a powerful and easy method for reducing tone in the foot and calf muscles, or, more generally, modify spasticity and neural excitability in the lower extremity.

The approach is very similar to the mesh glove application. Steel mesh socks do not exist but conductive textile socks are available at moderate cost as medical products. Again, it is important to make sure that the sock provides a socket for the electrode cable. The same conditions apply as for the glove: the sock should fit tightly (different sizes are available) and be slightly wetted with water or used dry. As counter electrode we configure two 5 × 9 (or10) cm hydrogel electrodes over ankle plantar and dorsi flexors, interconnected via Y cable or laboratory cable (Figure 4.3) as described in detail in Section 4.5.1.

4.5.3   Peroneal nerve afferent stimulation

Peroneal nerve stimulation has been used since early times of neuroprostheses, as the nerve is relatively easy to access via surface electrodes and can be stimulated to elicit the withdrawal reflex with flexion in ankle, knee and hip joints simultaneously. This effect can help correction of dropped foot syndrome and more generally weakness in control of lower extremity flexor control. In these still widely used applications the underlying mechanism is a combination of activating proprioceptive afferents for reflex activation and efferent motor fibres for direct activation of the tibialis anterior. Stimulation is phasic, controlled by ground reaction force or inertia sensors.

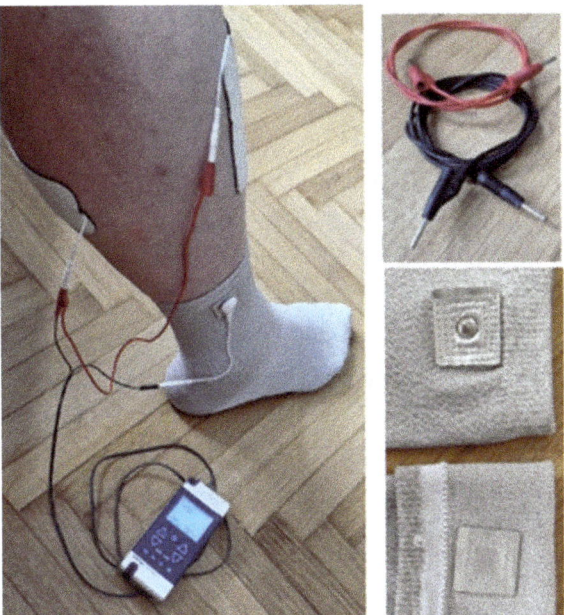

Figure 4.3 *Instrumentation for mesh sock application with conductive textile sock, two 5 × 9 (or 5 × 10) cm hydrogel electrodes, one placed longitudinally over ankle dorsi and one horizontally over the plantar flexors, both interconnected via Y cable or laboratory cable and, stimulator delivering a continuous train of biphasic stimuli, red connector indicates anode of first pulse phase.*
NOTE: It is important to have the terminal of conductive textiles insulated against skin contact (lower right) – not all products have this detail considered!

The same electrode setup can also be used very effectively for modification of spasticity and stiffness in the lower extremity, applying the same protocol as described above (Sections 4.4.1–4.4.2), and for mesh glove and mesh sock arrangements (Sections 4.5.1–4.5.2).

The positioning of the active electrode is critical but, at the same time, also relatively easy to access for wheelchair bound persons with intact hand function. The best choice is a round hydrogel electrode with a diameter of approximately 3 cm, but square electrodes (4 × 4 cm) can also be used. The best position is just above the easily palpable fibula head and over the biceps femoris tendon. This electrode is to be connected to the cathode of the stimulator for monophasic stimuli, or to the anode (red connector) of the first phase of biphasic ones. The counter electrode is larger in size (5 × 9 or 5 × 10 cm), placed longitudinally over the tibialis anterior muscle with its edge about 1 cm lateral to the edge of the shin bone (Figure 4.4).

Again continuous 30 Hz stimulation is a reasonable default, but searching for an individual optimum may be necessary. Intensity should be adjusted to just below

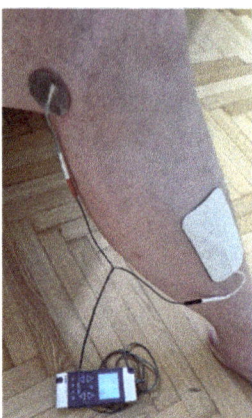

Figure 4.4 *Instrumentation for peroneus nerve stimulation with a 3 cm round active electrode over the superficial branch of the peroneal nerve and a 5 × 9 (or 5 × 10) cm counter electrode over the tibialis anterior muscle, longitudinally along the shin edge, and stimulator delivering a continuous train of biphasic stimuli, red connector indicates anode of first pulse phase.*

sensory perception (if intact), otherwise the described assessment of motor threshold minus approx. 10% can be performed via 1 Hz twitches. Ideally, the exact reflex threshold intensity should be identified, but in practice this is only clearly detectable in a minority of cases.

This setup can modify spasticity and stiffness efficiently, but also provide augmenting effects in impaired gait control, so it can be applied in supine or sitting, but also for gait support in cases of partly preserved ambulation.

4.6   Central afferent stimulation (spinal cord stimulation)

SCS may be applied by electrodes placed over the surface of the skin (*transcutaneous*) or by electrodes implanted in the epidural space (*epidural*). Epidural SCS is a clinically established technique used to treat chronic pain in the non-SCI population. It has also been used for many years, and in various species, to provide evidence for the existence of neuronal networks that produce rhythmic activation of limb muscles (i.e. Central Pattern Generators) in the lumbosacral cord [8]. In the 1990s, Dimitrijevic and colleagues in Vienna were the first group to observe the emergence of rhythmic, locomotor-like electromyographic (EMG) activity with flexion/extension movements in the lower limbs when epidural SCS was applied tonically in humans with motor complete SCI [9–15]. SCS is thought to excite afferent nerve fibres where they enter spinal segments through dorsal nerve roots, returning sensory information from proprioceptive and cutaneous receptors in the limbs and trunk to the central nervous system. Activating afferent nerve roots at low stimulation intensity causes paraesthesia (tingling sensations) and, at higher intensity, motor pathways and inter-neuronal spinal networks are trans-synaptically activated [12]. When applied at low frequency (<5 Hz), and at an intensity high enough to activate motor pathways, SCS elicits motor unit action potentials (MUAP) in lower limb muscles (analogous to the H-Reflex) termed *posterior root reflexes* (PRR). At higher frequencies (5–15Hz), sustained activation of lower limb muscles occurs [11,12,16], similar to that observed during FES. At 15–50 Hz, rhythmic, locomotor-like EMG activity emerges, indicative of Central Pattern Generators [11–13,15]. At >50 Hz muscle activity becomes diminished.

The majority of therapeutic applications of SCS occur when it is applied at an intensity below motor threshold (i.e. below the intensity that causes motor activity). For example, sub-motor threshold epidural SCS has been found to be an effective modality to treat spasticity [17,18] and a clinical service has been established at the former Maria-Theresien-Schloessel Neurological Hospital in Vienna, Austria [19]. Under certain conditions, SCS has also been found effective for augmentation of impaired movement control [20–22]. As the underlying mechanisms for spasticity and/or lack of voluntary motor control is the loss of suprasegmental input to the interneuron network, the basic idea of the intervention is to provide additional afferent input to segmental interneuron networks to compensate. The altered neural structures after spinal injury or other central lesions are highly individual and hardly ever similar between different persons. Consequently, the electrode arrangement and stimulation parameters must be optimised for each person to achieve better control over spasticity and movement. Both transcutaneous and epidural application methods will be described.

4.6.1 *Transcutaneous spinal cord stimulation*

4.6.1.1 **Instrumentation**

As with peripheral afferent stimulation, charge-balanced monophasic and biphasic stimulators are suitable. Typically, only one stimulation channel is required, and the device may be either constant current or constant voltage but should allow modification of stimulation intensity. Constant current devices are preferable to define the intensity despite the inevitable variations in the electrode impedance. Stimulation currents are typically within the range 10–200 mA. Low-cost constant voltage stimulators may be used but beware of variation in the current causing corresponding changes in neurological response. Pulse width adjustment is desirable but not essential and, as with peripheral afferent stimulation, frequency should be adjustable in a range from 1 to about 100 Hz.

For transcutaneous SCS, one or two electrodes are applied over the trunk or neck to depolarise primary afferent nerve fibres within the posterior roots or rootlets. Self-adhesive stainless steel/hydrogel or silver-silver chloride hydrogel electrodes are typically used to apply current, similar to those used for FES or NMES. Keeping in mind that stimulation with electrodes placed at the neck or trunk leads to strong neuromuscular and sensory activation in those anatomical regions before the deep-lying nerve roots are recruited. Therefore, a first principle is to use small electrodes: 5 cm or smaller, round or square, are suitable for cathodes and similar sized or larger oval or rectangular electrodes (5 × 9 cm) can be used as anodes, depending on the electrode configuration used.

4.6.1.2   Application protocol

Single-site SCS

Several different electrode configurations have so far been adopted dependent on whether upper or lower limbs are targeted and including bipolar, trans-spinal or unipolar configurations; these will be summarised below but a recent review article details all of the configurations that have been used to date [23]. In bipolar configurations, two adjacent electrodes of similar size are placed over the targeted nerve root(s). In trans-spinal and unipolar configurations, the cathode is placed over the targeted nerve root(s) on the midline, and one or two larger anode(s) are positioned on the anterior part of the body (close to the cathode) or over bony prominences at a distant location. In these configurations, larger anodes can be used to spread the current over a larger area, minimising discomfort.

For upper limb SCS, the nerve roots of the brachial plexus are targeted: innervation extends from the C5 to T1 vertebral levels (Figure 4.5 maps the peripheral nerves from the nerve roots at the spinal cord to upper limb muscles). Note that there are seven cervical vertebrae but eight cervical nerve roots; the eighth nerve root exits between the C7 and T1 vertebrae. Upper limb innervation separates

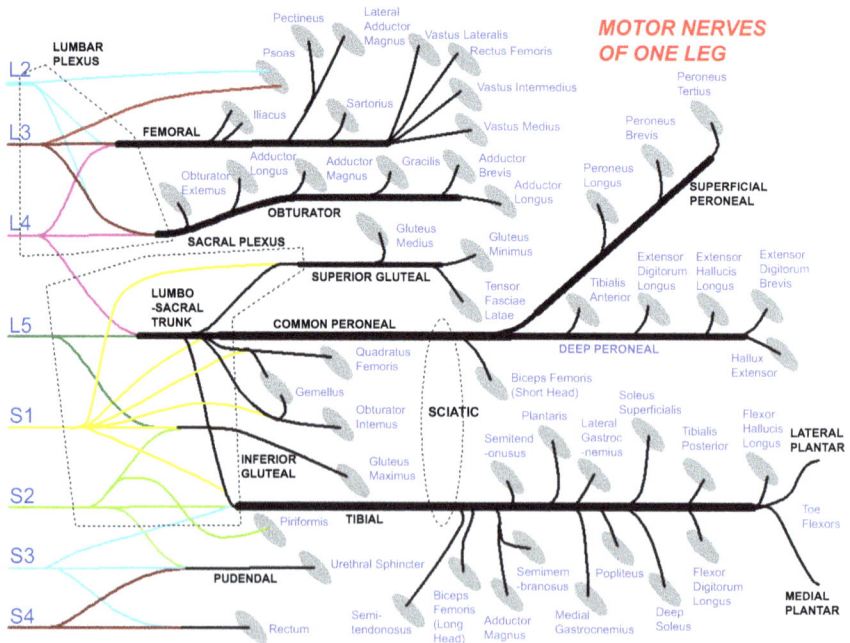

Figure 4.5   Peripheral neuro-anatomy for one leg

into median, ulnar and radial nerves. The median nerve originates at the C5-C7 roots; it is the principal nerve of the anterior forearm, supplying all of the flexor muscles (except flexor carpi ulnaris), the lateral part of the flexor digitorum profundus (flexion of the first and middle finger) and the abductor pollicis brevis and flexor pollicis brevis (flexion and abduction of the thumb). The ulnar nerve, originating at the more caudal C7-T1 roots, innervates the flexor carpi ulnaris and the medial part of the flexor digitorum profundus (flexion of the ring and little finger). The radial nerve, originating at the C5-T1 roots, is the principal nerve of the posterior forearm, supplying the extensor muscles and the abductor pollicis longus.

The brachial plexus posterior roots can be activated using a bipolar set up by positioning electrodes over the midline of the neck between C5-T1 vertebral levels. The vertebral levels may be determined by palpating for the most prominent vertebrae at C7: the caudal electrode positioned at the C4/C5 intervertebral space and the rostral one at the C7/T1 intervertebral space, to span the full brachial plexus innervation. The cathode may be positioned in the rostral or caudal location; it is recommended to test both configurations by switching polarity. Care should be taken when placing electrodes close to the C3-C5 nerve roots, corresponding to phrenic nerve innervation, which supplies the diaphragm muscle, particularly in individuals with impaired respiratory function. To increase selectivity of activated nerve roots, smaller electrodes may be used, however; there will be a trade-off with greater discomfort as the electrode size reduces at similar stimulation intensity. Recently, electrode arrays have been trialled, with the aim of enabling selective stimulation of multiple nerve roots. Arrays contain a high number of smaller electrodes ($\sim$10–20 mm diameter) with up to 40 electrodes arranged in, for example, a 5×8 array [24]. Specific electrode(s) from the array are systematically selected as cathode and anode pairs and reflex responses are monitored in targeted muscles to explore selectivity. To date, the application of non-invasive SCS using electrode arrays is limited and with mixed success.

In a unipolar configuration, the cathode is placed at the intervertebral space of the C5/C6, C6/C7 or C7/T1 vertebral levels, depending on the targeted upper limb muscles and two large, interconnected anodes are placed bilaterally near both iliac crests, scapulae or clavicles. In a trans-spinal setup, the cathode is positioned similarly to unipolar configurations and a single anode is placed anteriorly over the neck. However, this setup should be used cautiously in individuals with cervical injuries and impaired respiratory function, due to the risk of obstructing breathing or swallowing. Respiratory and/or speech and language therapists should be consulted prior to applying this setup.

For lower limb SCS, the lumbosacral plexus is targeted: innervation extends from the L2 to S2 nerve roots (Figure 4.6 maps the peripheral nerves from the nerve roots at the spinal cord to lower limb muscles). It should be noted that the levels at which the lumbar and sacral nerve roots exit the spinal cord is not aligned with their respective vertebral levels. The lumbar and sacral nerve roots exit the cord at approximately the T10-L1 vertebral levels (the lumbar enlargement). As with the upper limb, the lumbosacral plexus dorsal roots can be activated using either bipolar, trans-spinal or unipolar arrangements. In a unipolar or trans-spinal

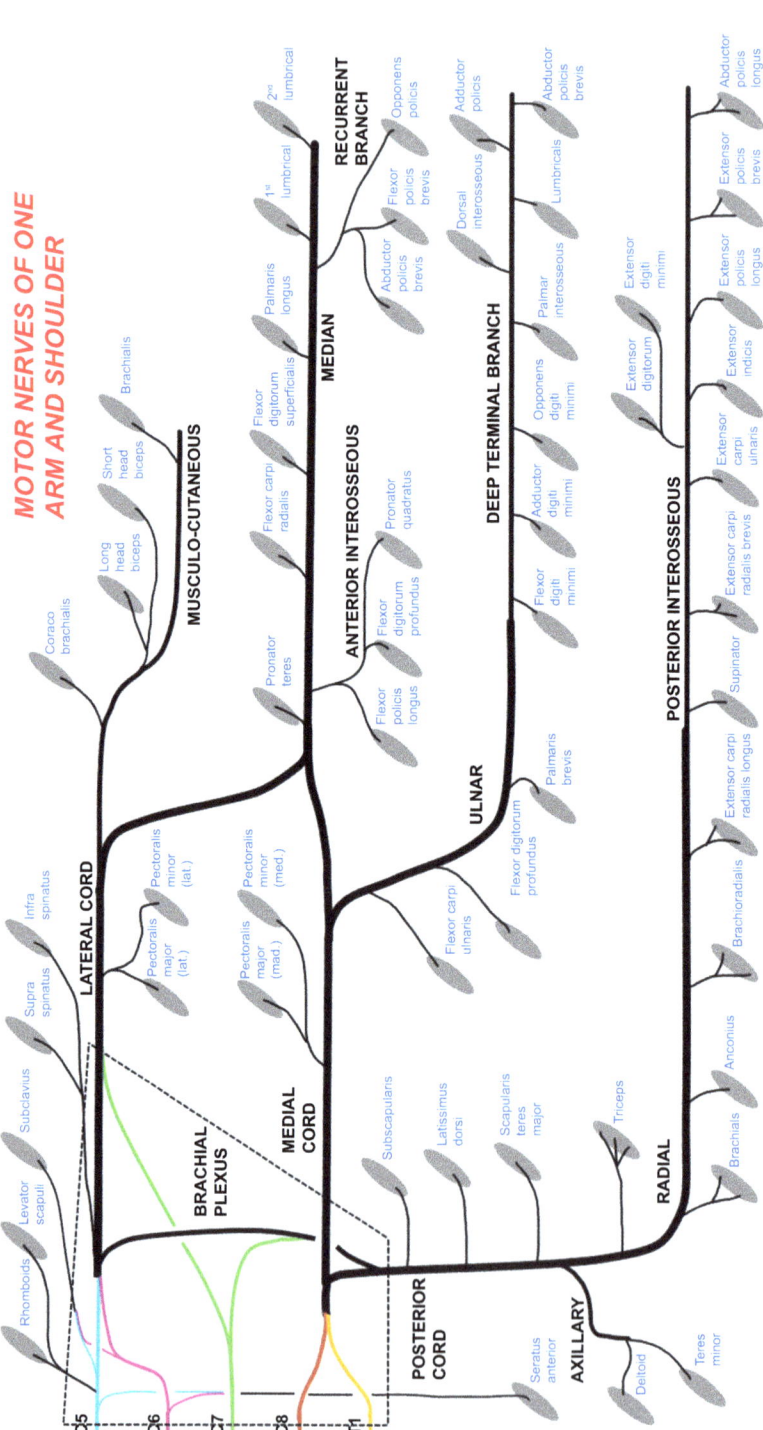

MOTOR NERVES OF ONE ARM AND SHOULDER

BRACHIAL PLEXUS

LATERAL CORD
MEDIAL CORD
POSTERIOR CORD

MUSCULO-CUTANEOUS
Coraco brachialis
Long head biceps
Short head biceps
Brachialis

MEDIAN
Pronator teres
Flexor carpi radialis
Flexor digitorum superficialis
Palmaris longus
1st lumbrical
2nd lumbrical

RECURRENT BRANCH
Abductor policis brevis
Flexor policis brevis
Opponens policis

ANTERIOR INTEROSSEOUS
Flexor policis longus
Flexor digitorum profundus
Pronator quadratus

ULNAR
Flexor carpi ulnaris
Flexor digitorum profundus
Palmaris brevis

DEEP TERMINAL BRANCH
Flexor digiti minimi
Abductor digiti minimi
Opponens digiti minimi
Dorsal interosseous
Palmar interosseous
Lumbricals
Adductor policis

RADIAL
Brachialis
Anconius
Brachioradialis
Extensor carpi radialis longus

POSTERIOR INTEROSSEOUS
Supinator
Extensor carpi radialis brevis
Extensor carpi ulnaris
Extensor digitorum
Extensor digiti minimi
Extensor indicis
Extensor policis longus
Extensor policis brevis
Abductor policis longus

AXILLARY
Serratus anterior
Deltoid
Teres minor

Triceps
Subscapularis
Latissimus dorsi
Scapularis teres major

Rhomboids
Levator scapuli
Subclavius
Supra spinatus
Infra spinatus
Pectoralis major (lat.)
Pectoralis minor (lat.)
Pectoralis major (mad.)
Pectoralis minor (med.)

C5
C6
C7
C8
T1

Figure 4.6   Peripheral neuro-anatomy for one arm and shoulder

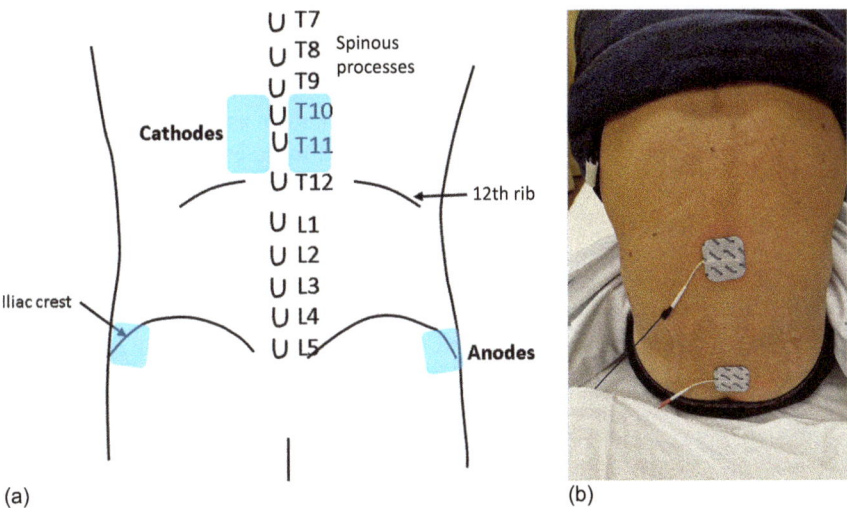

(a)   (b)

Figure 4.7   *(a) Anatomical Landmarks for placing transcutaneous SCS electrodes for lower limbs. (b) In this alternative arrangement, the cathode is placed level with the lower aspect of the rib cage, close to T10/T11 and the anode is placed caudal to the cathode, level with the iliac crests.*

configuration (see Figure 4.7), the cathode(s) are placed over the midline of the back at the T11/T12 intervertebral space [25–30] or slightly more caudal at the L1/L2 intervertebral space [31,32]. Others use a larger cathode to span the area both caudal and rostral to T11/T12, within the range T10 to L4/5 [33–36] or two cathodes are placed paraspinally [13,14,35,37]. The T11/T12 spinous processes may be identified by palpating for the iliac crests and identifying the L4 vertebrae, level with the iliac crests, then palpating and counting each spinal process towards L1 and T12. For anodes, two large interconnected electrodes are placed bilaterally over both iliac crests or anteriorly over the lower abdomen or umbilicus. Anodes placed over the abdomen should be used with caution as it can strongly co-activate abdominal muscles with potentially unpleasant resistance to respiratory movements and muscular metabolic overload by continuous strong contraction. In a bipolar configuration, the cathode and anode are placed over the midline of the back, covering the full lumbosacral innervation. The cathode is placed level with the lower aspect of the rib cage, close to T10/T11 and the anode is placed caudal to the cathode, either at T12/L1 [38] or level with the iliac crests, close to L4.

Electrode placements are recommended based on anatomical knowledge; however, anatomy can vary between individuals, nerve root anomalies are infrequent but possible, and the nerve roots may move with respect to other anatomical structures depending on body position [39]. Therefore, verifying that the stimulation is reaching the intended posterior root primary afferent fibres is a crucial step in performing SCS. This enables slight adjustment of the electrode placement to optimise activation of the intended nerve roots. If available, the optimal way to achieve this is through electromyography (EMG) recordings from targeted muscles.

Depolarisation of primary afferent roots by a single pulse of SCS evokes a short latency monosynaptic spinal reflex response, known as the PRR. PRRs can be measured by EMG recordings, taken from targeted muscles in the upper or lower limb. Self-adhesive EMG electrodes placed over relevant muscles are suitable to record PRRs, but users should be aware of the limitation to superficial muscles and susceptibility to cross talk in smaller muscles such as in the forearm. When SCS is applied non-invasively, EMG electrodes placed over limb muscles will detect artefacts from the applied stimulation, followed by a PRR at short latency. The PRR latency should be measured and verified against estimated or reference latencies, based on the distance between the stimulation and EMG electrodes, and estimations of nerve conduction velocity (note that nerve conduction velocity can be affected by age and neurological damage). Low limit of nerve conduction velocities for specific upper and lower limb motor nerves in the neurologically intact population is shown in Table 4.2. Activation of the anterior nerve roots by SCS would shorten these latencies by approximately 5 ms due to the slightly shorter distance travelled and because the response no longer crosses the synapse.

Single pulses of SCS should initially be applied at low frequency (≤ 1 Hz) while progressively increasing stimulation intensity. A *reflex threshold* intensity should be defined by measuring the peak-to-peak amplitude of recorded PRRs. The

Table 4.2   *Low limits of motor nerve conduction velocity in neurologically intact people (data taken from Chen et al. [139]).*

Motor nerve	Age/sex/height subgroup	Conduction velocity (m/s)
Median	All	49
	19–49 y men	49
	19–49 y women	53
	50–79 y men	47
	50–79 y women	51
Ulnar (below elbow)	All	52
Ulnar (across elbow)	All	43
Ulnar (above elbow)	All	50
Fibular peroneal (Ankle to	All	38
below fibular head)	19–39 y < 170 cm	43
	19–39 y >170 cm	37
	40–79 y <170 cm	39
	40–79y >170 cm	36
Tibial	All	39
	19–49 y <160 cm	44
	19–49 y 160–170 cm	42
	19–49 y $\geq$ 170 cm	37
	50–79 y <160 cm	40
	50–79 y 160–170 cm	37
	50–79 y $\geq$ 170 cm	34

lowest stimulation intensity that elicits PRRs with a peak-to-peak amplitude ≥ 0.05 mV can be considered as reflex threshold for that muscle. PRRs may not be evident in some or all muscles and reflex threshold may be different in each muscle. The position of the cathode can be slightly modified to optimise PRRs in targeted muscles. The aim should be to find the electrode position where threshold is lowest and, where possible, targeted muscles reach threshold at a similar stimulation intensity.

As stated above, PRRs are monosynaptic reflex responses; action potentials cross one synapse and, as such, the pathway is not hard-wired. Repetitive depolarisation of primary afferents is associated with altered neurotransmitter release at the synapse and modulation in the transmission of primary afferent activation to motoneurons. Both facilitatory and inhibitory pathways exist, with different recovery times, but inhibitory processes are dominant for longer duration (several seconds), known as *post activation depression* [40]. As a result, previous activation of the primary afferent pathway causes a substantial reduction in the amplitude of the subsequent PRR response, dependent on the frequency of repetitive stimulation. This physiological phenomenon can be measured with EMG to provide evidence that the afferent (as opposed to efferent) nerve roots have been activated [41].

This measurement can be made by applying pairs of pulses with an interstimulus interval of between 50 ms and 1 s at a stimulation intensity greater than reflex threshold (~ 1.1–$1.2 \times$ threshold). Evidence of posterior root activation comes from suppression of the PRR to the second applied stimulus relative to the first. If little or no suppression is observed, it is likely that the anterior root (efferent pathway) is being activated (at least in part). If the stimulator being used does not allow for the application of paired pulses, verification of posterior root activation can be done by applying continuous (tonic) trains of stimulation pulses at a progressively increasing stimulation frequency. Suppressed reflexes should be evident at frequencies of 2–5 Hz and 20–100 Hz, with greater suppression evident at increasing frequency until complete suppression is observed at 100 Hz.

Where EMG is not available, the person's sensation and palpable muscle responses may be monitored to verify electrode placement. Initial sensation of SCS occurs at the location of the electrodes, due to activation of sensory nerve fibres at that anatomical area. With increasing stimulation intensity, sensation may be felt in the targeted muscle groups as the sensory afferent roots are activated. For upper limb applications, sensation is typically either towards the medial part of the hand (ring and little fingers) or lateral part (first and middle fingers) indicative of activation of the ulnar or median pathways, respectively. In lower limb, sensation may be felt initially in the proximal (quadriceps) or distal muscle groups initially, indicative of activation of the femoral or peroneal pathways, respectively. As stimulation intensity is further increased, palpable and visible responses to the stimulus are evident, which can be used to estimate reflex threshold, as defined above.

Multisite SCS

Multisite SCS involves the application of SCS simultaneously at two cervical [42–45] or lumbosacral [31,33,37,46] levels, or at both one or two cervical and lumbosacral levels simultaneously [27,47–49]. A wide range of electrode configurations have been trialled in multisite SCS with single or pairs of interconnected cathodes being placed at vertebral levels ranging from C3-T2 and T10-L2 and pairs of interconnected anodes, typically placed around the iliac crests (see Rehman *et al.* [23] for full details). A multi-channel stimulator is required for this application, with the number of channels dependant on the number of vertebral levels simultaneously targeted.

In general, greater motor output has been reported with multisite than single-site SCS [23], however more research is needed to better understand the therapeutic and neurophysiological effects, particularly considering the variety of electrode configurations that have been tested. Combined cervical and lumbosacral SCS have been applied following reports that walking performance was improved in people with neurological injury following combined arm and leg cycling exercise compared with leg cycling alone [50–52], which was attributed (at least in part) to improvements in corticospinal drive [53] and modulation of cervico-lumbar connectivity [52].

SCS has also been combined with peripheral nerve stimulation or FES [49,54–56] with similar reasoning of facilitating motor output. Complex neural connections and the possibility of collision block due to antidromic firing should be a consideration when applying electrical stimulation simultaneously at different sites (either two central afferent sites or central afferent combined with peripheral stimulation).

Kilohertz-modulated transcutaneous SCS

While monophasic or biphasic waveforms are typically used, some groups deliver non-invasive SCS modulated with high frequencies ranging from 5 to 10 kHz [27,33,43,44,48,57–59]. Specifically, each applied pulse is replaced by a short (1ms) burst of 5–10 pulses, each with a 50 μs pulse width, applied at 5–10 kHz. These waveforms have been referred to as 'carrier frequencies', 'Russian stimulation' or 'painless cutaneous enabling motor control (pcEmc)' [27]. There are currently only a few commercially available stimulators with the capacity to apply kilohertz-modulated stimulation and those are much higher cost than constant-voltage TENS devices.

When applied for neuromuscular activation, kilohertz-modulated waveforms have been reported to produce maximum force with less discomfort [60]; it was therefore proposed that they could be applied to modulate the spinal cord with least discomfort [27]. Recent evidence however casts doubt on this notion as, while higher stimulation amplitudes could be tolerated when kilohertz-frequency modulated waveforms were applied, reflex thresholds were at least equally higher [37]. That is, when applied at similar intensity with respect to reflex threshold, similar discomfort levels were experienced with and without kilohertz-modulation [61]. It has long been known that continuous stimulation applied in the range 1–20 kHz results in neural block [62,63], with the largest fibres blocked first as stimulus intensity is increased [62]. When kilohertz-frequency stimulation is applied in 1 ms bursts, the first few pulses of each burst summate, causing the nerve fibre to fire repetitively, providing the stimulation intensity is sufficiently above threshold. This appears to be a less efficient way to recruit the afferent nerve roots however, since substantially greater charge was required to reach reflex threshold when kilohertz-modulated waveforms were applied compared with single 1ms biphasic pulses [61,64].

Current understanding of the neurophysiological effects of kilohertz-modulated SCS is limited but the effects are likely to be different from unmodulated waveforms [65]. For example, it has been reported that kilohertz-frequency modulated SCS may contribute to cortical [57] and spinal [66] inhibitory effects that were not present following unmodulated SCS. More research is needed to clarify the role of kilohertz-modulated waveforms in SCS.

4.6.2   Epidural spinal cord stimulation

4.6.2.1   Instrumentation

Epidural SCS involves implantation of electrode leads or arrays in the epidural space, connected via an extension lead to an implantable pulse generator (IPG) placed under the skin of the abdomen, buttocks or chest. The stimulation is controlled by an external hand-held controller. Some IPGs are rechargeable whereas others require battery replacement every few years (depending on the intensity and duration of use), which needs to be done under local anaesthetic. Implantation of a spinal cord stimulator is a reversible procedure.

There are several commercially available IPGs from medical device companies including Medtronic, Boston Scientific, Abbot, Nevro and Saluda. They are recommended by NICE in the clinical management of chronic pain, and thousands of people have received epidural stimulation implants over the past 40 years. Epidural spinal cord stimulators are not routinely applied for neuromodulation beyond their intended use for pain management; therefore, any other application is currently 'off-label'. Since early reports dating back to the 1970s [17,67–69], there has been growing interest in the use of epidural SCS in neurorehabilitation, targeting motor and autonomic functions, predominantly in people with SCI (for a recent review article see [70]) but also for other neurological conditions such as multiple sclerosis, stroke and Parkinson's disease. Epidural SCS after SCI was recently reported to have a similar safety profile as when it is applied to treat chronic pain, despite significant comorbidities in the SCI population [71].

Epidural SCS can be applied either using cables with electrodes on the cylindrical surface (known as *leads*) or flat electrode arrays (known as *paddles*). Leads can be inserted percutaneously via a needle (typically two leads are inserted, each containing four to eight electrodes). Paddles contain multiple electrodes [2–24], which are made from biocompatible and biostable materials, typically platinum or platinum/iridium encapsulated in silicone. Most paddles need to be implanted surgically; however recently paddles that can be implanted percutaneously have been made available.

Electrodes are implanted close to the targeted nerve roots in the epidural space, therefore, much lower stimulation intensities are required compared to transcutaneous SCS. Percutaneous leads are used most commonly in pain management, where two parallel leads are placed along the midline of the spinal cord with one lead more rostral or caudal to the other to ensure full coverage of the painful area. This arrangement is used to target activation of the dorsal column rather than the nerve roots. In the spinal injury population, the vast majority of clinical trials to date have used paddle electrodes [20–22,71–77] as they can be positioned more precisely and are believed to provide superior lateral coverage with greater selectivity. A few trials have used percutaneous leads in people with spinal cord injury [78–83] or after stroke [84]. Percutaneous leads placed more lateral to the midline may be superior to target nerve roots.

4.6.2.2 Application protocol

Prior to undergoing implantation of a spinal cord stimulator, it is recommended that multidisciplinary assessments, including a psychological evaluation, are carried out. As epidural SCS for this indication is not currently approved by NICE, patient expectations should be managed, ensuring the individual is fully informed of risks and benefits based on the currently available evidence. To inform the surgical procedure, imaging of the spinal cord should be done to assess any injury to the spinal cord and the implantation site, and for preplanning of electrode placement [85].

Implantation procedures are dependent on whether lead or paddle electrodes are used. Leads can be placed into the epidural space through a needle using x-ray guidance, while the patient is under local anaesthetic and awake with sedation. Paddle leads are usually inserted surgically by laminectomy or laminotomy, under sedation or general anaesthesia. An incision is made in the back and a small part of the lamina is removed to create space for insertion of the paddle. A laminotomy is preferred over laminectomy because it is a bone sparing procedure, leaving more intact ligaments and muscles attached to the spinous process. This is particularly true in people with spinal cord injury who may have weak or unstable vertebrae as a result their injury.

There is greater risk involved with epidural than transcutaneous SCS, due to the risks of surgery and living with an active implanted device. While most of the risks are minor, rarely, more serious problems can occur and the risks should always be weighed against benefits. The most common risks and precautions for epidural SCS are listed below.

•   As with any surgical procedure, there is a risk of infection. While most infections can be resolved by a course of antibiotics and do not generally result in serious problems, infections can be more difficult to eradicate in individuals with spinal cord injury. The risk of infection is higher when paddle electrodes are implanted surgically compared with leads inserted with a needle.

- The surgical procedure is conducted with the individual in a prone position on the theatre table. There are risks of anaesthesia in a prone position including loss of airway, aspiration, visual loss and peripheral nerve injury.
- As with any spinal cord surgery, there is a risk of nerve damage, which could be temporary or permanent.
- There is a risk of dural tear during surgery and subsequent cerebrospinal (CSF) fluid leak. This can result in headaches, which are generally treatable but are persistent in some cases.
- There is a risk of electrode migration or fracture after implantation – the risk of migration is higher with lead electrodes compared with paddles. To minimise the risk of migration/fracture, physical activity can be limited for up to two months after implantation. Electrode migration/fracture could result in loss or reduction of any therapeutic benefit from the device and a repeat surgery may be required to reposition the electrodes.
- There may be bleeding, bruising and discomfort around the implantation sites, mainly in the short-term during healing but, rarely, long-term discomfort may be experienced. Loss of superficial subcutaneous tissue at the IPG implantation site can also occur, requiring further surgery.
- The individual may not experience any benefit from the SCS. Experience with SCS is currently limited, therefore the chance of success is not currently predictable. When spinal cord stimulators are implanted for pain management, they have been helpful in five to seven out of every ten patients globally. Non-invasive SCS should be trialled prior to implantation of a spinal cord stimulator to ensure the targeted nerve roots respond to stimulation.
- It is generally recommended by manufacturers that individuals with implanted spinal cord stimulators should not have MRI scans (although x-rays are safe). This can limit diagnostics and monitoring on the individual, so it is an important consideration when deciding whether to have a spinal cord stimulator implanted. The MRI compatibility of IPGs varies–some do allow MRIs but only for certain body areas, others are MRI-conditional, meaning that specific sequences need to be used, so it is important that radiographers are aware of the IPG type and position. In people with SCI, MRI-compatible spinal cord stimulators are preferable.
- Non-invasive electrical or magnetic stimulation should not be applied close to an IPG as it may interfere with the device.
- Short-wave diathermy, a therapy sometimes offered by physiotherapists, can be dangerous for people with a spinal cord stimulator.
- It is important that Institutions performing implantation of spinal cord stimulators have a suitable infrastructure in place for long-term follow up of implanted patients. Stimulation parameters require regular review to maintain effectiveness and non-rechargeable IPGs will need to be replaced every few years.

Implantation of the electrodes and device should be performed according to the recommendations made by the manufacturer of the device. Positioning of the

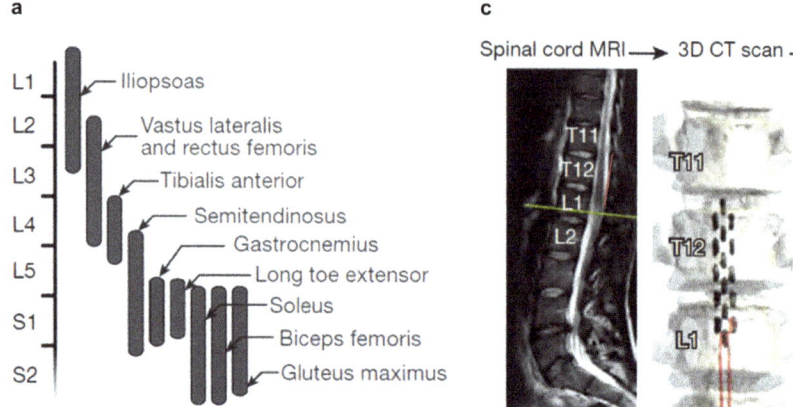

*Figure 4.8   The two pictures (originals in Wagner et al. 2018, Figure 4.2(a) and
(c)) show the placement of the electrode array in relation to posterior
afferent nerve roots providing neural feedback from the main lower
extremity muscle groups [76]. Reproduced under agreement with
Springer Nature.*

electrode array is based on the nerve roots being targeted (see Figure 4.8 for
electrode array placement targeting lower limb muscles) and intraoperative
mapping performed to confirm optimal positioning of the electrode array.
Epidural stimulators may be configured to apply stimulation with anode-cathode
configurations from any electrodes in the array. Low frequency (<2Hz) stimulation may be applied systematically to anode–cathode pairs across the length and
width of the array whilst recording EMG activity from relevant upper or lower
limb muscles. Threshold intensities (as described in Section 4.1) may be determined, and the position of electrode array adjusted to optimise activation of the
targeted nerve roots.

Following implantation, an x-ray of the spinal cord can be taken at the level
of electrode implantation to confirm exact positioning of the electrode array
relative to the vertebral bodies [85]. Following recovery from the surgical
procedures, more extensive mapping should be performed. This may include
investigations of optimal spatial anode–cathode configurations, similar to those
conducted intraoperatively. Detailed descriptions of mapping procedures for
epidural stimulation can be found in the published literature [85–88].
Investigations of optimal stimulation parameters including amplitude, pulse
width and frequency should also be performed, depending on the targeted therapeutic effect. Procedures for establishing parameters for spasticity management
and augmentation of voluntary control for postural control, standing and locomotion are provided in the following section.

4.6.3   SCS parameters for therapeutic applications

The duration and frequency of SCS sessions and the procedures to establish SCS stimulation parameters, in terms of intensity (relative to threshold), pulse width and frequency, vary depending on the intended therapeutic effect, spasticity modification or augmentation of residual movement. Procedures and suggested frequencies, pulse widths and intensity (relative to motor threshold) are similar across transcutaneous and epidural applications.

4.6.3.1   Spasticity modification

Trials by Dimitrijevic's group in Vienna, performed in the 1980s, demonstrated the capacity of SCS to modulate spinal excitability and treat spasticity in people living with SCI [17,18,89], which has since been supported by further reports that SCS can attenuate neural hyperexcitability after SCI [19,26,41,42,90]. A wide range of parameters have been found to be effective in spasticity modulation for each participant, including frequencies ranging from 0.2 to 130 Hz (most commonly 50–100 Hz), intensities between sensory and motor threshold or below sensory threshold and pulse widths between 210 and 1000 μs. Identifying an appropriate therapeutic window for each participant is an important aspect of SCS application [90] and may be performed using the procedure described below. It should also be acknowledged that an effective therapeutic window may change with time, so stimulation parameters should be regularly reviewed.

Biomechanical manoeuvres for spasticity assessment, such as slow and fast passive multi-joint movements, should be performed as a baseline to compare the influence of stimulation. These manoeuvres are to be repeated under continuous stimulation with reflex threshold intensity and different frequencies, usually beginning with wider ranging frequencies, such as 20 Hz, 50 Hz and 80 Hz. If any of those show effects towards relaxation of passive movements, fine tuning of the frequency and then the amplitude can further improve the result. If an epidural array is used, the spatial configuration should be guided by mapping procedures. Wide-ranging frequencies may also be trialled with different spatial electrode configurations to determine the optimal site for stimulation.

Once a suitable parameter set has been identified, it should be adapted to different postures: supine, sitting, standing, for useful daily application, and the participant should be able to adjust the stimulation amplitude for different body positions, when applying SCS at home. It may also be necessary to provide several programmes with different frequencies and/or pulse widths for home use. Ensure the participant is aware how different programmes should be used at home.

It is important to observe development of carry-over effects over time and plan daily timing on activities such as sleeping or calm early morning activity. Hydrogel electrodes cannot remain attached to the skin permanently, as biochemically induced skin irritation can occur, but two to three episodes of about an hour are usually unproblematic. In case of lacking carry-over effect, but permanent disturbing spasticity with good direct reaction to stimulation, implantation of an epidural implant can be considered.

4.6.3.2   Augmentation of residual movement control

Early trials in people with SCI demonstrated that tonic SCS applied at an intensity above motor threshold and at a frequency between 5 and 15 Hz, i.e. causing sustained muscle contraction [11,12,16], could facilitate standing in people with motor complete SCI [91]. Harkema and colleagues reported on one individual with complete SCI (AIS B) who was able to stand without manual facilitation but with 65% bodyweight support, on their first attempt with epidural SCS [20]. Full weight-bearing standing under SCS with minimal support for balance in people with motor complete injuries has since been demonstrated [13,33,73]. However, SCS applied above motor threshold at higher frequencies (15-50 Hz), where rhythmic, locomotor-like motor activity emerges, was found to impede functional walking as the bursting activity occurred out of phase with useful muscle activation.

In selected cases of spinal injury, brain injury or dystrophy diseases with a high degree of 'discompleteness' (intact neural motor control structures but reduced central state of excitability for performance of voluntary movement), there is an opportunity for motor output or gait improvements by tonic SCS applied at an intensity *below* motor threshold. Suitable individuals with discomplete SCI may be identified by neurophysiological testing, such as the Brain Motor Control Assessment (BMCA) [92] or TMS. The BMCA can be used to test the patient's capacity to inhibit intact reflex responses; demonstrating that non-functional preserved neural pathways can be used to influence spinal excitability caudal to the lesion [93]. In these cases, motor output or function may be acutely enhanced in the presence of stimulation that does not directly activate afferent or motor pathways. An example of enhanced EMG activity measured in the quadriceps muscle of an individual with SCI during voluntary knee extension in the presence of epidural stimulation is shown in Figure 4.9. Other Examples can be

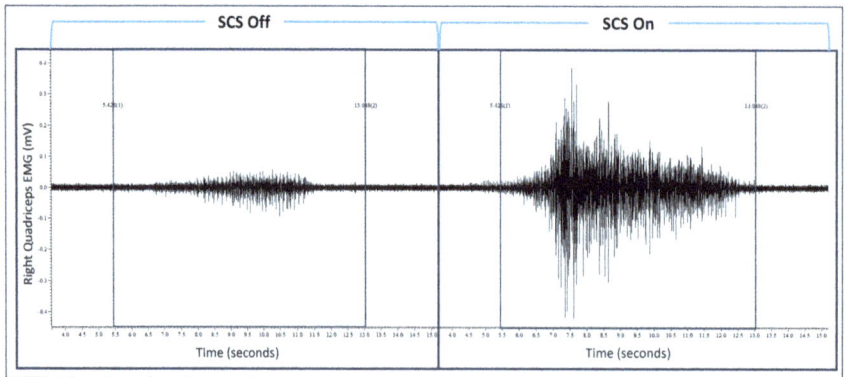

Figure 4.9   Right quadriceps muscle activity (EMG) from an individual with chronic SCI and an implanted epidural stimulator (paddle array over the conus). In the left plot the spinal cord stimulation (SCS) is switched off and in the right plot SCS is on throughout at an intensity below motor threshold. The blue shaded area indicates the time (7.5s) during which voluntary right knee extension was attempted. Voluntary muscle activity was substantially enhanced when SCS was switched on.

found in several case series involving people with chronic SCI in the published literature. Observations include enhanced range of motion, muscle activity, force and gait parameters in the presence of sub-motor threshold epidural [20–22,71–77] or transcutaneous [14,25,27,33,43,94,95] SCS. Harkema's group in Louisville have implanted 25 people with chronic SCI (AIS A or B) over the period 2009–2020. They report that 100% of participants achieved voluntary movement in the lower extremities after implantation with the stimulation on [77].

After several months of epidural [21,22,75,81,96,97] or transcutaneous [27,33,98,99] SCS combined with rehabilitation, progressive improvements in volitional movements, standing and stepping (in the presence of SCS) have been observed. In a few cases, recovered movement is maintained even when SCS is switched off. Changes in motor scores following long-term SCS (measured with SCS off) are reported in the next section. For both transcutaneous and epidural applications, long-term interaction with medication should be considered. Clinically controlled reduction of standard medication like Baclofen is reasonable in parallel to applying afferent stimulation, resulting in improved active spasticity control and often also improved voluntary motor control.

4.6.3.3   Epidural stimulation for augmentation of movement control with phasic patterns

Most published demonstrations of enhanced movement control after spinal cord injury rely on continuous (tonic) stimulus patterns with individually adjusted stimulus amplitude, to recruit a defined afferent neuron pool, and individually optimized frequency to optimise the influence on spinal interneuron processing. This strategy, in transcutaneous as well as in epidural application setups, can lead to positive outcomes in many but not all individuals with spinal cord injury, but in most with gradual improvements in spasticity or movement control.

An alternative strategy is pursued by Courtine and colleagues in Lausanne, Switzerland, who rely on delivering phasic stimulus patterns during the gait cycle and switch electrode combinations accordingly to directly influence flexor or extensor motor neuron pools, augmenting or inhibiting, when best fitting in the impaired gait cycle ('Spatiotemporal Epidural Electrical Stimulation') [76]. The basic mechanisms of neural interaction are like tonic method, but with selective influence on motor control through step cycles, and potentially other movement patterns, resulting in reduced muscles stiffness and augmented voluntary motor control as required to optimise execution of the movement pattern. Technically this requires a real-time movement feedback and closed loop stimulation control. Selected case studies have been presented with motor incomplete (AIS C and D) [76] as well as motor complete (AIS A and B) injury [100], where basic ambulation could be improved or enabled after a five-month intensive post-implantation rehabilitation program.

In a publication in 2019, this approach was first demonstrated in three individuals, one with an AIS D lesion and two with AIS C lesion, from which one improved to AIS D after the intervention. Figure 4.8 shows placement of a standard 16-electrode paddle array, designed for pain therapy (*Specify* 5-6-5,

Medtronic, Minneapolis, MN, USA) in the dorsal spinal canal of the vertebra T11 and T12, where the afferent nerve roots for the main lower extremity flexors and extensors are situated, and a map of root origins as provided in the paper [76]. The outcome was visibly improved motor control overall, to a comparable extent to the reports of Angeli *et al.* (2014 and 2018) [21,22], who had applied tonic epidural stimulation in individuals with AIS A and B lesions.

In a later publication [100], the group reports on three more individuals (two with AIS A lesions and one with an AIS B lesion) who also underwent an intensive five-month post-operative rehabilitation program, and accomplished some ambulation, though with less ability than the earlier presented AIS C and D cases. The main difference is a larger custom-made electrode array, that also covers stimulation sites below T11 and T12 vertebra, reaching more distal nerve branches, where afferent and efferent neurons are already united in a mixed nerve. This is evident in Figure 4.10, where responses to stimulus trains at 20 Hz appear as neuromuscular

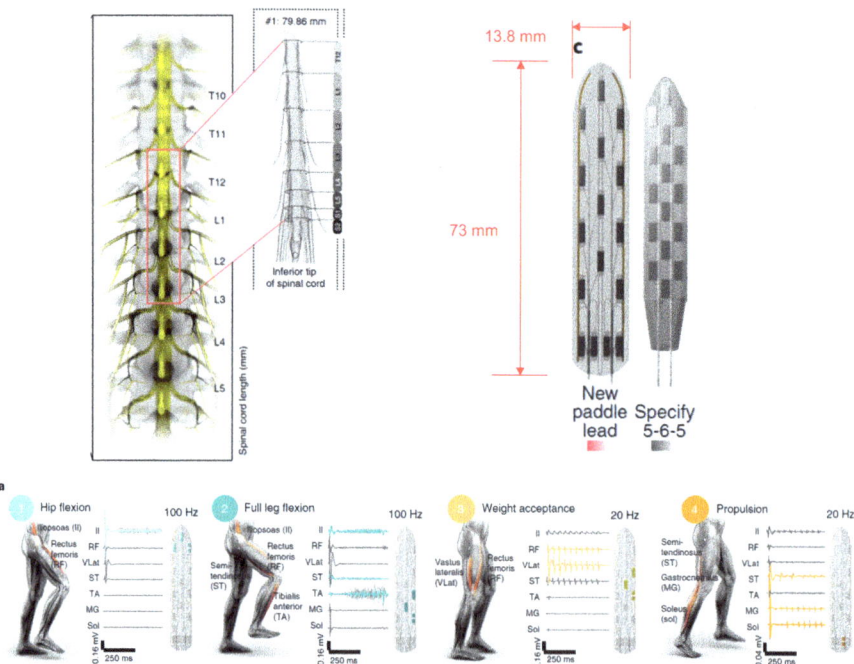

Figure 4.10   Illustration from Rowald et al. (2022) (part of Figure 4.1(a) and (b), Figure 4.2(f), part of Figure 4.3(c), and Figure 4.4(a)) shows the enlarged custom-made electrode array and its position between vertebra T11 and L3 (shown in comparison to the Medtronic Specify 5-6-5 paddle array). The lower sequence shows delivery of phasic stimulus trains with 20 Hz (motor-wave responses in extensors) and 100 Hz (flexion reflex response to afferent input) in the step cycle [100]. Reproduced under agreement with Springer Nature.

activation in the EMG recording, whereas 100 Hz responses appear as augmented flexion reflex responses.

In a recent report, the same group presents a case study [101], where one of the subjects of the Wagner 2019 [100] study (AIS D) received a pair of BCI-implants (Brain Computer Interface) three years after the first intervention. The BCI detects 'intention to move' patterns from the recorded EEG as an additional control input and provides further improved motor control after saturation of the first phase gain (Figure 4.11).

Overall, phasic SCS is a very interesting approach to gain more selectivity and has the potential to contribute to further progress in the development of neuro-prostheses for restoration of movement function. On the other hand, we need to

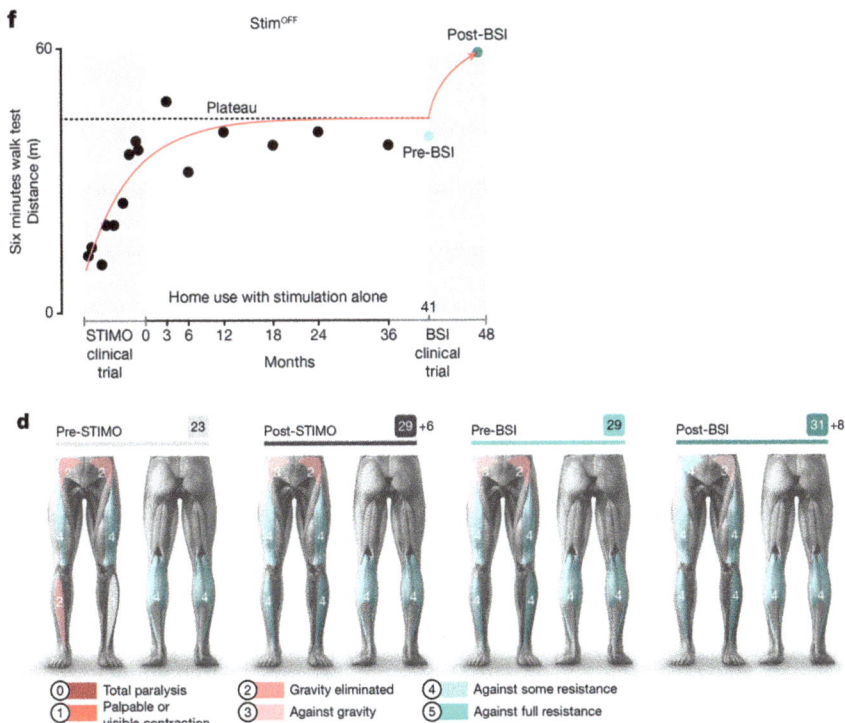

Figure 4.11   *Original Figure 4.4(d) and (f) from Lorach et al. [138] shows improvements in six-minute walk test, pre and post implantation and the postoperative intensive five months rehabilitation, then further improvements after BCI-implantation and subsequent rehabilitation period. The main improvements are in control of illiopsoas and tibialis anterior (person was AIS D motor and sensory incomplete spinal cord injury). Reproduced under terms of the Creative Commons, with acknowledgement to the authors and Springer Nature.*

respect limitations regarding applicability. First, it is essential that patients have a certain degree of 'discompleteness' in their post-injury neural control. By a long way, not all persons with spinal cord injury can benefit from the approach. Secondly, highly complex additional instrumentation is required for accurate bio-mechanical movement recordings, as feedback for closed loop control needs to be worn and kept calibrated. Last, but not least, costs for implants and long-lasting intensive rehabilitation programs need to be covered, considering that we already experience growing cost restrictions for much less costly interventions in our current health care systems.

4.7 Therapeutic electrical stimulation

4.7.1 Electrotherapeutic effects from FES

Therapeutic or *carry-over* effects from peroneal nerve stimulation to correct drop-foot were observed by Liberson *et al.* [1] in the seminal paper on FES, and by many subsequent FES practitioners. In his survey of 1981, Vodovnik [102] cited 15 papers that reported therapeutic effects of FES prior to that date (not all in SCI patients). Taylor *et al.* made similar observations in their clinical trial: some patients who had been treated for footdrop by peroneal nerve stimulation recovered sufficient voluntary function for them to stop using the stimulator [103]. The devices were primarily intended for artificial activation of the dorsiflexors (*orthotic effect*) but caused a useful neurological change (therapeutic effect). Two papers reported recovery due to FES cycling. In a case study, Donaldson *et al.* [104] reported that a man with incomplete T11/12 spinal cord injury, whose motor scores has declined slightly in the previous ten years, recovered voluntary knee extension after 15 months of FES cycling on a recumbent tricycle. A remarkable recovery occurred in an AIS Grade A patient, Christopher Reeve (the actor) who, 5 years post injury, started recovering after 16 months of FES-cycling, three times per week and had improved to AIS Grade C after three years of this regime [105]. Assuming that the FES caused the recovery in these cases, there was clearly the possibility of developing an *electrical stimulation therapy*. Since then, in the past 20 years, there have been many studies related to such therapies, and there are now three approaches which are described below. To put these electrical therapies into the context of rehabilitation after SCI, non-electrical therapies will first be briefly reviewed.

4.7.2 Non-electrical therapy

It has been known at least since Sherrington that if cats with transected spinal cords have their hind paws placed on a moving belt, the legs will walk, driven only by afferent input to the spinal cord. In the late 1980s, it was further found [106] that this walking improved with training, exhibiting neuroplastic learning. Previous human case reports had found no evidence of locomotor responses in men with complete lesions [107] but nevertheless, locomotor tests with spinal cord injured humans were reported in the 1990s. Wernig *et al.* reported in 1995 on

89 incomplete-lesion patient who had had *Laufband* (treadmill) *Therapy* with Body Weight Support (BWS), compared to 64 patients at the same hospital who had only conventional therapy [108]. The results were striking. For example, of the 45 chronic patients who were wheelchair-bound before the Laufband therapy, 92% learnt to manage without a wheelchair, whereas only 50% of the 24 who had conventional therapy. However, the therapy did not enable stepping in another seven patients with complete lesions so, although the trial was inspired by results from cats with complete transections, it found that humans are different, in that the spinal circuits (central pattern generator) cannot be so easily activated.

Barbeau *et al.* published a review in 1998 [109] of previous clinical trials, comparing several neurorehabilitation methods for walking. Four of the papers were studies of SCI patients treated with BWS and three of those showed significant improvements in one or more of the outcome measures (Function, Walking Endurance, Walking Speed). The systematic review by Wessels found that in AIS C and D patients, over-ground training with BWS was better than treadmill training, when possible, and they also found that assisting the leg motion with a *Locomat* (robot) during treadmill training was as effective as therapists (by Functional Independence Measure) [110]. Writing in *Physical Therapy* in 2006, Behrman *et al.* [111] argued that a paradigm shift was appropriate from assuming that SCI patients would not recover, to assuming that they might; changing from *compensation* for the injury to *therapy* to promote neuroplastic recovery.

In a spinalised cat study, Hodgson *et al.* had shown in 1994 that cats trained to walk could not stand, while cats trained to stand, could not walk [112]. Since one wants SCI patients to relearn all possible functions, not only walking, the idea of treadmill training was extended to *Activity-Based Therapy*, which added other training exercises [113,114]. Roy *et al.* [114] argued that the spinal cord without supraspinal input is capable for responding appropriately to changes on afferent input, and therefore that rehabilitation interventions should not be too restricted. This change of emphasis from walking to generalised neurological recovery means that other neurorehabilitation methods, including electrical stimulation methods can be compared to treadmill training and with each other. The obvious measure is the International Standard for Neurological Classification of Spinal Cord Injury (ISNCSCI).

Morrison [115] presented data from a Locomotor Training study with BWS from eight US rehabilitation centres. Treadmill training with body-weight support and manual assistance was provides for 120 sessions; the average time taken being 11 months. There were 69 participants, all initially classified AIS C or D. The primary outcome measures were 10-Meter Walk, 6-Minute Walk, Modified Functional Reach, and Berg Balance Scale. Autonomic function and ISNCSCI scores were also measured. By these measures, 77%, 80%, 58% and 62% surpassed the minimal detectable change, respectively. Only 3% did not improve by any measure. On average, detectable improvement was fastest at the beginning but continued throughout the therapy. Some people were late responders so, for example, stopping at 40 sessions would have prevented 30% from getting detectable change in gait speed, and 40% in endurance. The mean improvement in total motor score was 10 points (6 lower limb and 4 upper limb) and there were also sensory improvements.

These results vindicate the idea of treadmill therapy. Evidently, most incomplete SCI patients can make useful recovery but it may take a lot of therapy to achieve the effect. The 120 sessions costed six times more than the 20 sessions allowed by insurance cover at that time in the United States. The authors presented a reasonable argument that increasing the duration from the normal 20 up to 120 sessions is justifiable on the grounds of reduced future medical cost. This preventative treatment, an extra 100 sessions, would then have cost $18,000. But if the same improvement could be achieved more quickly using electrical stimulation, the high cost might be avoided. The following sections describe electrical stimulation methods for therapy.

4.7.3   Improvements on motor score due to FES

4.7.3.1   FES-walking

Barbeau [109] reviewed FES-walking as a therapy in 1998. In many FES-walking trials, the outcome measures relate to walking performance but without ISNCSCI scoring. One study that did include motor scores was Field-Fote [116] in which 19 chronic AIS C subjects used peroneal stimulation on their more affected side, with body weight support and treadmill training for 1.5 hours per day, three days/week for three months. The median improvement was three points in both legs. Tefertiller *et al.* trained AIS C and D chronic participants on an elliptical trainer with FES followed by non-FES over-ground walking for 12 weeks [117].

4.7.3.2   FES-cycling

Sadowsky [118] conducted a retrospective study of chronically injured AIS A-C, comparing a control group ($N = 20$) with a group who had been cycling ($N = 25$). The times that these patients had been cycling was very variable (3–168 months) at a rate of three sessions per week for 45–60 minutes. The controls were matched to the FES group and had only passive stretching. The mean change in the motor score was 8.1 in the FES group compared to 0.6 in the controls ($p = 0.004$). Pin-prick and light-touch improves by 5.8 and 6.0 points in the FES group but declined by 5.6 and 4.6 in the controls. The p values for the differences being 0.008 for both senses.

There are, on the market, several stationary ergometers designed for FES-cycling. They have motors to establish the cycling motion and a multi-channel stimulator that delivers stimulation usually at a fixed frequency over appropriate ranges of pedal position, usually to quadriceps, hamstrings and gluteal muscle groups. No thought is required from the users of such machines who often appeared bored, apparently making little voluntary effort. If neurological recovery is a Hebbian process, supraspinal drive should be essential so the users need motivation to push the pedals. Duffell *et al.* have developed an ergometer in which the torque at the crankshaft, which depends on voluntary effort, controls the speed of an avatar in virtual reality cycling. Her study [119] showed a mean increase in motor score of 4.7 at follow-up which can be compared to the study of Yasar [120] who used ergometers without biofeedback (i.e. virtual reality). Their subjects improved an equal amount (4.7 points) but three times more slowly.

4.7.4 Improvements on motor score due to SCS

Garcia *et al.* [121] reviewed tSCS studies, for upper and lower limb in 2020. Most were single-session so uninformative about long-term benefits. To date, there are few long-term studies in upper [43,44,99,122] or lower [34,98,123] limb, with recorded changes in motor score.

Inanici [44] is a case study of a tetraplegic, level C3, AIS D who had cervical tSCS during physical therapy for four weeks, then four weeks of only physical therapy, followed by one more week of both. The average time that tSCS was applied was 60 minutes per day. His upper extremity motor score improved by ten points during the first four weeks, and a further four points during the second four weeks, then remained the same at follow-up, three months later. Gad [43] reported an average 4.4-point improvement in motor scores following eight sessions of tSCS combined with training over four weeks in six people with chronic cervical SCI (AIS B or C). Moritz [99] is a safety and efficacy multi-centre trial of 60 people with chronic cervical SCI (AIS B, C or D). The participants undertook an intensive, standardized in-clinic rehabilitation program over four months with tSCS added to rehabilitation in the final two months. After four months, upper extremity motor scores had improved by 2.2 points on average.

For lower limb, Al'joboori [98] applied tSCS at sub-motor threshold level while subjects stood up, stood still and sat down. They repeated this sequence five times per sessions, three sessions per week, for eight weeks. Five people had the tSCS (3 AIS A, 1 AIS C, 1 AIS D) and two controls (no tSCS, 1 AIS A, 1 AIS C) followed the protocol. There was no change in the lower extremity motor score for the controls, nor for two AIS As in the treatment group. The others in the treatment group improved: AIS A +1, AIS C +5, AIS D +7. Bedi and Narkeesh [34] was a case study in which the AIS C subject undertook 12 hours of ABT and three 45-minute sessions of spinal stimulation per week for 24 weeks, gaining three points in LEMS. Alam [123] is a single-case study that reported a 12-point improvement in lower limb motor scores in an individual with chronic incomplete cervical SCI following 66 weeks of rehabilitation training with tSCS (standing, functional reaching, reclined sit-up, treadmill walking, and active biking).

With epidural SCS, only three trials have reported the long-term effects on motor scores in lower [22,76] or upper limbs [124]. Angeli reported a 1-point improvement in motor score in 1/4 participants. Wagner [76] implanted three people with motor incomplete chronic SCI (AIS C or D) and reported substantial improvements in motor scores in all three participants (+17, +13 and +6).

4.7.5 Improvements on motor score due to brain stimulation

Many forms of brain stimulation have been described as possible therapies following spinal cord injury: repetitive transcutaneous magnetic stimulation (rTMS), transcutaneous DC stimulation (tDCS), paired associative stimulation between motor cortex and muscles (PAS) and paired associative stimulation between loud click sounds and muscle stimulation (to activate the RST).

The justification given by Belci [125] for rTMS was that it reduces the cortical inhibition that occurs following spinal injury. Four AIS D cervical-injury out-patients were given one hour of sham stimulation on five consecutive days then a similar treatment of stimulation over the hand area of the motor cortex. Inhibition was reduced, while motor scores, sensory scores and a hand function test showed improvements which were maintained for three weeks.

The rather general idea suggested by Hebb for neural plasticity has been shown to be strengthening of synapses due to the near simultaneous firing of a neuron and the pre-synaptic neuron at the terminal. Bunday and Perez [126] were the first to show that this method can be used to strengthen the motor pathway by targeting the synapses between the corticospinal tract (upper motor neuron) and motor neurons, by arranging the timing of TMS over the cortex and peripheral motor nerve stimulation so that the descending volley arrived just prior to the antidromic volley from the peripheral nerve. This increased the motor evoked potentials from TMS, the voluntary strength in the hand, and the functional hand test. There were no follow-up measurements.

Jo and Perez [127] extended the method by applying the PAS for ten sessions with a follow-up after six months. To test whether exercise should accompany the PAS, they had three groups: PAS+exercise, sham PAS+exercise and only PAS. Stimulation intensities were above motor threshold to produce antidromic volleys in the motor neurons. While evoked potentials and maximum voluntary contraction increased after PAS, with or without exercise, but not sham PAS, there was only a long-term effect following PAS+ exercise. Motor scores were not measured.

Baker and colleagues have shown that motor neurons are normally driven from the reticulospinal tract (RST) as well as the cortico-spinal tract [128]. Since the RST is activated by sudden loud sounds, these can be used as an alternative to TMS of the cortex for paired stimulation which reinforces the synapses from the RST to the motor neurons, potentially providing an alternate pathway when the CST is injured [129].

Alder [130] published a critical review of PAS for strengthening the motor pathway in stroke patients: most studies were single-session and none investigated cumulative effect or long-term recovery. The authors made many suggestions for improving the quality of such experiments.

4.8   Discussion

Figure 4.12 is a comparison of the rates of neurological recovery for lower limb treatments, excluding implants and case studies. The numbers of participants are mostly small and the changes in LEMS are also small, but the results suggest: (i) that electrical stimulation treatment is faster than non-stimulation training (compare Field-Fote with Morrison); (ii) that the rate for FES-cycling is similar to FES-walking (compare Duffell with Field-Fote); (iii) that biofeedback by virtual reality can increases the rate of improvement (compare Duffell with Yasar); and that the

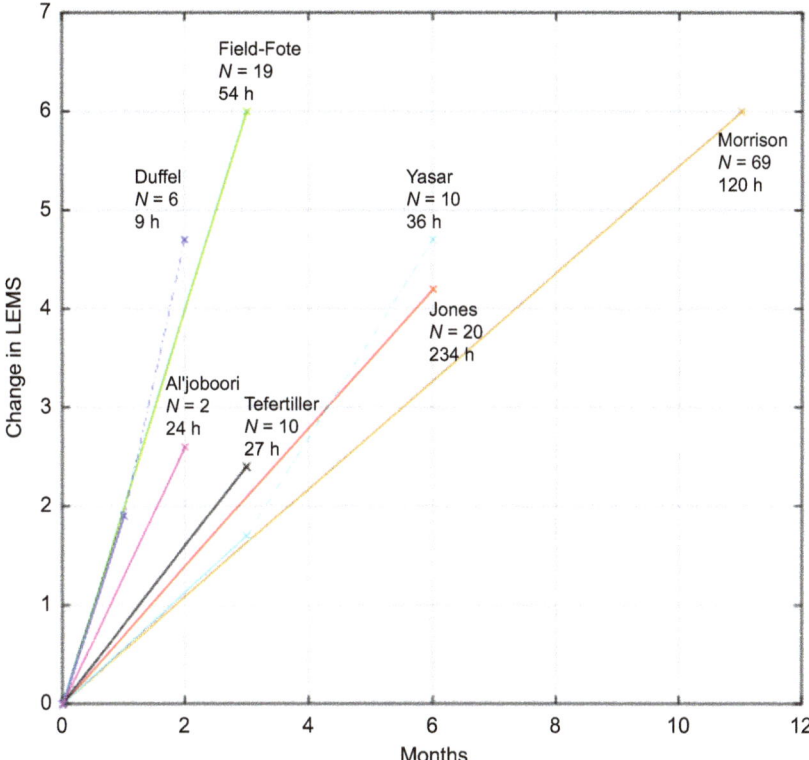

Figure 4.12 *Change in lower limb motor score (ISNCSCI) for seven surface stimulation studies with AIS C or D chronically-injured people. The labels show the first author, the number or participants treated, and the number of hours of treatment on a horizontal axis of the duration of the treatment in months. The therapies were: Morrison, treadmill with BWS [115]; Jones, Activity-Based Therapy [113]; Field-Fote, FES + treadmill+ BWS [116]; Yasar, FES-cycling without biofeedback [120]; Duffell, FES-cycling with biofeedback (virtual reality)[119]; Tefertiller (FES on elliptical trainer); Al'Joboori, tSCS for standing [98]. In the Yasar and Duffell studies, motor scores were measured again at follow up, after a period without treatment, and the LEMS had increased further, as shown by the dashed lines.*

amount of therapy time required may be much shorter than treadmill training with BWS (compare Duffell at 9 hours to Jones at 234 hours). The only SCS study with participants in this category is Al'joboori with $N = 2$.

Neurological recovery can occur due to changes in the strength of synapses. The three methods described above indicate that there are two, or possibly more, ways in which this can happen. The intensities used for SCS mean that only afferents are stimulated, raising excitability. This may enable the motor pathway, allowing muscle function during therapy or exercise and thereby natural strengthening at the junction between the upper and lower motor neurons. In contrast, PAS targets this junction directly, using precise timing of the brain and muscle stimulation to activate the lower motor neuron, which is the post-synaptic neuron, less than 2 ms after the presynaptic neuron. The precision required for this method is strong evidence that the change indeed occurs where the upper and lower motor neurons meet.

In FES, either or both these effects could occur. Because stimulation is above motor threshold, there will be antidromic action potentials in the lower motor neurons at the same time as, presumably, orthodromic action potentials in the upper motor neurons as the patient tries to use the muscles. In this case, the antidromic stimulation is not timed to coincide with the orthodromic, but because the muscle is being stimulated to assist the desired movement, there may be a PAS-like effect on average. There is also afferent stimulation raising excitability. What the relative importance of the two effects is in FES, is unknown.

Why is exercise necessary? The results in Figure 4.13 suggest that electrical methods are faster than ABT alone, but all three electrical methods are accompanied

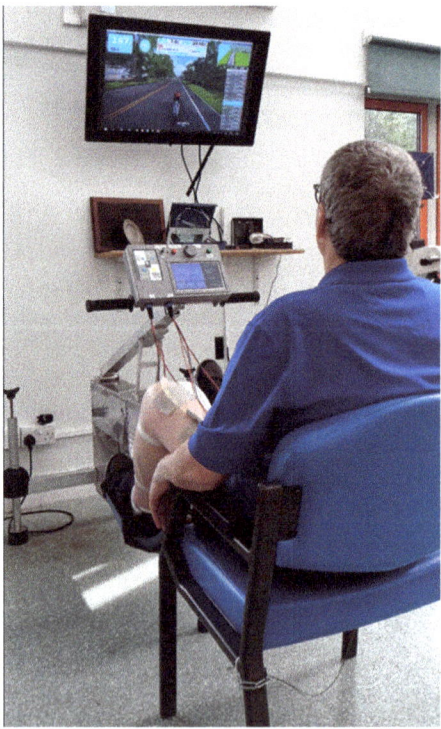

Figure 4.13   The iCycle ergometer for therapeutic cycling with stimulation of quadriceps, hamstrings and gluteal muscles, controlled through the dashboard. The user is motivated by participating in an online virtual reality cycle race, with choice of route and human competitors.

by exercise. Jo and Perez demonstrated that for PAS, increased motor-evoked potentials, maximal voluntary contraction and function tasks were only maintained in the group who also had exercise in addition to PAS. Without exercise, the raised excitability declines after about an hour so that the benefit is lost. This effect is familiar in synapse research where Long-term Potentiation is divided into early- and late-LTP [131]. What causes the latter, so that synaptic change to become permanent, is not fully understood. One hypothesis is the action of *exerkines* [132], molecules released into the bloodstream during exercise. The exerkine with perhaps the most recognisable acronym is BDNF, but there are many others [132]. The beneficial effect of exercise on cognitive function is generally accepted and this is thought to be due to it improving neuroplasticity by enhanced long-term potentiation. Most studies use animals but there have been indirect non-invasive demonstrations in humans. For example, visually-evoked EEG signals can show LTP responses to the rapid presentation of patterns, and Smallwood *et al.* [133] found that there was a significant difference in the late-LTP between a physically-active and a physically-inactive group. If this is true for the visual cortex, perhaps it is also true for the motor cortex and the spinal cord.

4.8.1   *Practical recommendations*

Two general themes emerge from all these studies. The first is that long-term improvement requires exercise to accompany the therapy; the patient must be trying to use the paretic muscles at the same time (FES) or immediately afterwards (PAS). Typically, sessions of electrical therapy last 30–60 minutes and are repeated about three times per week for 4–12 weeks. The second point, which is apparent when reading this literature, is that the response in individuals vary from none to useful reductions in disability, for reasons that have not yet been investigated.

For people with incomplete spinal cord injury, the easiest method for lower limb rehabilitation is probably FES-cycling because: (i) the patient is seated without risk of falling, (ii) cycling machines can be bought with stimulators that can apply appropriately-phased stimulation, (iii) little or no clinical supervision is required during the therapy. Cycling machines may be mobile (e.g. *Berkelbike*) or stationary ergometers (e.g. *RT300*). Stationary ergometers mostly have motors connected to the pedals, so that the cycling motion is established even if the user could not do so alone. Ideally the stationary machines should have biofeedback to motivate the user to push on the pedals, such as by virtual reality driven by mechanical power at the crankshaft [119]. An accompanying film, taken by another cyclist, whose speed is unrelated to the user's effort, will not do.

The stimulators usually have six channels, used bilaterally for quadriceps, hamstrings and gluteal muscle groups, but other combinations are possible and may be better for people with unsymmetrical paresis. Stimulation is applied a constant frequency, often 30Hz, and each channel switches on and off at set crankshaft angles. These angles are usually fixed parameters though for optimal stimulation, they should vary with cadence. EMG signals from able-bodied cyclists have been used to choose the switching angles, but then modified to get smoother motion [134].

4.9   Future research

Papers reporting neurological recovery are either case studies or results from groups who have had an experimental therapy. Selection is by inclusion and exclusion criteria, which do not include the results of tests that predict whether recovery is possible. It should not, therefore, be surprising that while some individuals improve, others do not. This range of outcome may have been regarded as showing that recovery is a matter of chance but with so many methods of imaging and electrophysiological testing, researchers should be looking for ways to predict what recovery is possible. An example that already exists is for AIS A patients who can be tested to see whether their lesion is discomplete. If it is, then some neurological recovery is possible: therapeutic stimulation or activity-based therapy should be considered. We do not yet have tests to show the potential for recovery for AIS B, C or D.

Recovery may occur because of synaptic changes between the corticospinal neurons and the motor neurons. Such pathways to specific muscles may be dormant but can be detected by the BMCA test [92] or by transcranial magnetic stimulation (TMS) while EMG electrodes detect the evoked potential at the muscle. Arora [135] advocates using TMS for more thorough investigation of individuals neurological condition for better prescription and prognostication.

The recent results from Baker's group, showing that there is an alternative pathway through the RST, that can also supply the motor neurons, gives an alternative that may help those with damage in the corticospinal tract. To test for dormant pathways in the RST, loud click sounds serve to activate the tract instead of TMS during investigations and, perhaps, for paired stimulation therapy.

Given this vision of more detailed neurological assessment, leading to customised therapy, we need to understand better the relative advantages the three types of therapy listed above. Tonic SCS is a not muscle-specific and may be the simplest or quickest way to initiate general recovery but perhaps deficits will remain, limiting functional improvement. In that case PAS, targeting the specific remaining paretic muscles might bring further improvement.

Trials comparing the different methods are needed but case studies where the customised treatment is applied according to a clinical protocol may lead to better understanding and eventually better outcomes.

4.10   Conclusions

The spinal cord is a very large neural network with inputs coming from the brain via the corticospinal tracts and the RST, with both excitatory fibres and inhibitory fibres. It also has inputs from afferent fibres coming from the periphery, including proprioceptors and cutaneous receptors. The network of inter neurons enables coordinated movements between joints in limbs, between left and right, and between upper and lower limbs. The network allows us to learn a large number of tasks by repetition, and this facility remains possible throughout life. This neural plasticity is possible because of the synapses linking the neurons, which change if both are active at the same time (Hebbian learning).

Spinal cord injury disrupts the network, first because of damage to the tracts, and second because the abnormal behaviour following injury – prolonged bed rest and then sitting in a wheelchair – causes neuroplastic changes which may be unfavourable. Patients present with paralysis or paresis and often spasticity. There is high redundancy in the central nervous system, so it is likely that dormant fibres could be recruited, but if there is no voluntary muscle activity or there is over-whelming spasticity, therapeutic exercises are impossible.

If the condition is mild, so that exercise therapy is possible. the patient can recover by the normal natural process. If spasticity is more severe, preventing the movements required for therapy, afferent stimulation by the mesh glove or mesh sock may relieve the hypertonia. If muscles are paralysed, SCS may sufficiently raise the excitability of the motor neurons for voluntary muscle contractions to occur. Both methods enable activity-based therapy and the natural process of recovery.

The problem with activity-based therapy is that it is slow. Patients may train with locomotion therapy, but it takes a large amount of their time, and therapist time, so is expensive and it is not widely available, but it does lead to improved walking, better motor score (Morrison) and the reflexes change towards normality [136].

Electrical therapy is faster, measured by changes in neurological motor score. But we do yet know whether one of the methods of electrotherapy is generally best, or, which may be a better question, how to choose the best method for each type of patient (injury level, completeness, spasticity, age, time available for therapy, treatment budget, etc.).

The mode of training also must be chosen. locomotion training, starting on a treadmill with partial body-weight support, may be the best way to re-learn how to walk, but, for example, is half an hour of locomotion training in a physiotherapy gym, once a week better than three hours of cycling at home? Or would it be better to use an elliptical trainer? Would SCS or FES or PAS be better?

As Electrical Therapy can reasonably claim to be the oldest medical application of electricity, going back over 250 years, it is, perhaps, surprising that we have so much still to find out. As mentioned in Section 1.1, this is partly the result from the collective amnesia that allowed knowledge to be forgotten around the middle of the twentieth century and electrotherapy having to be re-discovered in people using FES. The rea-lisation, about 25 years ago, that recovery should be possible and that it should be the aim of therapy was a milestone and the literature is now very large, with many clinical trials, new methods of measurements, new devices, and huge progress in relevant neuroscience.

In this chapter, we have only described methods with which we are familiar, and which in our experience, have been beneficial to appropriate SCI patients. However, the literature review has shown that there are many other possible methods and best practice may develop quickly in the next few years.

In this chapter, we have described four clinical uses of afferent stimulation for:

- Acute relief of spasticity (Sections 4.4 and 4.5),
- Demonstration of dormant motor pathways (Sections 4.4–4.6),

- Enabling voluntary movement during SCS (Sections 4.3 and 4.6.3.2), and
- As a course of therapy, with exercise and stimulation combined, to recover normal function (Section 4.7). This latter method may include efferent stimulation.

4.11   Appendix: Methods for Neurological Assessment

4.11.1   Brain motor control assessment (BMCA)

Recordings are made during movements, either voluntary or applied by the examiner. EMG recording from 12 muscles (quadriceps, adductors, hamstrings, tibialis anterior and triceps surae, bilaterally, plus midline abdominals and lumbar paraspinals). There are also two inclination sensors, a sensor for neck flexion or plantar force, and markers for application of vibration or tendon taps. The protocol is shown in Table 4.3 [92].

Table 4.3   Methods of assessment

		Name	Details
1	Relaxation	Lying down	≥ 5 minutes
2	Reinforcement manoeuvres	Valsalva Manoeuvre	Deep breath, held for 3 s, then forcefully exhaled. Repeat three times.
3		Eyes squeezed tightly closed	Repeat three times.
4		Clench jaw	Repeat three times
5		Strongly flex neck	Head against a force transducer Repeat three times
6		Modified Jendrassik Manoeuvre	Opposed palms pressed against force sensor Or variants for paralysed arms or hands
7	Voluntary movements	Bilateral voluntary hip and knee flexion	
8		Unilateral voluntary hip and knee flexion	Left and right
9		Unilateral plantarflexion/ dorsiflexion	Left and right
10	Passive movements	Unilateral hip and knee flexion	Left and right, repeated three times
11		Unilateral plantar and dorsiflexion	Left and right, repeated three times
12	Stretch testing	Patellar tendon taps	Repeated ten times
13		Achilles tendon taps	Repeated ten times
14	Clonus tests	Jerk on patellar tendon	Leg fully extended
15		Dorsiflexion jerks	Leg fully extended, then leg flexed
16	Tonic stretch reflexes	Vibrator applied to patellar tendon	Left and right
17		Vibrator applied to Achilles tendon	Left and right
18		Achilles vibrator with Jendrassik	Only if no response to previous test
19	Plantar reflex	Stroke sole of foot without warning	Left and right
20		Stroke sole of foot with warning	Left and right

Sherwood *et al.* summarise the responses from 40 SCI patients and show how the various responses were spread between the 4 AIS grades (A–D).

4.11.2   Electrophysiological tests of reflexes acting at the ankle joint

Responses are compared to normals. The following table is based on Smith and Knikou [136]. Tests 1–6 are variants for H-reflex measured in soleus while stimulating posterior tibial nerve (PTN).

Test		Method	Explanation
1	Soleus H-reflex excitability while standing	Stimulation of posterior tibial nerve while standing	
2	Soleus H-reflex phase-dependent modulation	Stimulation of posterior tibial nerve while walking	Should be modulated
3	Homosynaptic depression	Effect of frequency (habituation?) while seated	Before locomotor training, little habituation
4	Pre-synaptic inhibition of 1A afferents (detail in Knikou 2008) [140].	H-reflex with prior conditioning volley to common peroneal nerve	While seated and while stepping
5	Reciprocal 1A inhibition	Soleus H-reflex combined with conditioning pulse to ipsilateral TA large afferents, 2–4 ms earlier.	Seated and while walking
6	Nonreciprocal 1B inhibition	Soleus H-reflex combined with conditioning pulse to gastrocnemius medialis, 4–6 ms earlier.	
7	Flexor reflexes (detail in [141]	EMG of TA during walking. stimulation to ipsilateral sural nerve (300 Hz burst for 30 ms)	Both early and late reflexes change

4.11.3   Motor pathway through reticulospinal tract

Start-react test to see whether sudden loud sound reduces reaction time, indicating engagement of the RST. Baker and Perez [142] found that the faster response only occurs for power grip, not precision grips, which is consistent with recent results from Glover and Baker [128].

Measure gripping force between the proximal interphalangeal joint of the index finger and the thumb. Patient asked to respond as quickly as possible to a cue which is either an LED, flashed for 20 ms, or the same but accompanied by 20 ms of 500 Hz of quiet tone (80 dB) or startlingly loud tone (115 dB). For interpretation see [142].

4.11.4   Assessment by transcranial magnetic stimulation [135]

The magnetic stimulation coil is held over the motor cortex and stimulation is applied while observing EMG for muscles on the trunk or limbs.

The presence of an evoked response means that there is a corticospinal motor pathway. Thus, it may demonstrate that muscles are *discomplete* rather than completely paralysed. If there is no response at maximum stimulation intensity, while the patient is relaxed, a response may still be obtained if he tries to contract the muscle at the same time. Reflexes, such as the plantar reflex, may also be enabled by the TMS pulses.

The amplitude of the evoked response is diminished by SCI but a more satisfactory measure is the latency of the response which is increased.

Following the evoked response is the *contralateral silent period* due to inhibitory networks. The onset of this period can also be readily measured, and is also longer after SCI, but there is not agreement about its cause.

Several other methods use TMS, including mapping the muscles onto the cortex, and delivering paired pulses to investigate cortical excitability, but these seem remote from clinical application.

References

[1]   Liberson WT, Holmquest HJ, Scot D, and Dow M. Functional electrotherapy: stimulation of the peroneal nerve synchronized with the swing phase of the gait of hemiplegic patients. *Arch Phys Med Rehabil.* 1961;42:101–5.

[2]   Jallabert J. *Experience sur l'Electricite avec quelques conjectures sur la cause de ses effets.* Geneva: Barrillot & Fils; 1748.

[3]   Fara P. *An Entertainment for Angels.* Turney J, editor. Duxford: Icon Books; 2002.

[4]   McNeal DR. *Functional Electrical Stimulation.* Bekey GA, editor. New York: Marcel Dekker; 1977.

[5]   Cumberbatch EP. *Essentials of Medical Electricity.* 6th edn. London: Henry Kimpton; 1929. 443 p.

[6]   Formento E, Minassian K, Wagner F, *et al.* Electrical spinal cord stimulation must preserve proprioception to enable locomotion in humans with spinal cord injury. *Nat Neurosci.* 2018;21(12):1728–41.

[7]   Dimitrijević M. Mesh glove electrical stimulation. *Sci Med.* 1996;3:54.

[8]   Grillner S. Control of locomotion in bipeds, tetrapods and fish. *Handbook of Physiology. 1.* Maryland: Waverley Press; 1981. pp. 1179–236.

[9]   Rosenfeld J, McKay W, Halter J, and Polio F. Evidence of pattern generator in paralyzed subjects with spinal cord injury during spinal cord stimulation. *Neuroscience Meeting*; San Diego: Society for Neuroscience; 1995.

[10]   Gerasimenko Y, McKay B, Sherwood A, and Dimitrijevic MR. Stepping movements in paraplegic patients induced by spinal cord stimulation. *Soc Neurosci Abstr.* 1996;22:1372.

[11]   Dimitrijevic MR, Gerasimenko Y, and Pinter MM. Evidence for a spinal central pattern generator in humans. *Ann N Y Acad Sci.* 1998;860:360–76.

[12]   Minassian K, Jilge B, Rattay F, *et al.* Stepping-like movements in humans with complete spinal cord injury induced by epidural stimulation of the

lumbar cord: electromyographic study of compound muscle action potentials. *Spinal Cord.* 2004;42(7):401–16.

[13] Minassian K, Persy I, Rattay F, Pinter MM, Kern H, and Dimitrijevic MR. Human lumbar cord circuitries can be activated by extrinsic tonic input to generate locomotor-like activity. *Hum Mov Sci.* 2007;26(2):275–95.

[14] Hofstoetter US, Danner SM, Freundl B, *et al.* Periodic modulation of repetitively elicited monosynaptic reflexes of the human lumbosacral spinal cord. *J Neurophysiol.* 2015;114(1):400–10.

[15] Danner SM, Hofstoetter US, Freundl B, *et al.* Human spinal locomotor control is based on flexibly organized burst generators. *Brain.* 2015;138(Pt 3):577–88.

[16] Jilge B, Minassian K, Rattay F, *et al.* Initiating extension of the lower limbs in subjects with complete spinal cord injury by epidural lumbar cord stimulation. *Exp Brain Res.* 2004;154(3):308–26.

[17] Dimitrijevic MM, Dimitrijevic MR, Illis LS, Nakajima K, Sharkey PC, and Sherwood AM. Spinal cord stimulation for the control of spasticity in patients with chronic spinal cord injury: I. Clinical observations. *Cent Nerv Syst Trauma.* 1986;3(2):129–44.

[18] Dimitrijevic MR, Illis LS, Nakajima K, Sharkey PC, and Sherwood AM. Spinal cord stimulation for the control of spasticity in patients with chronic spinal cord injury: II. Neurophysiologic observations. *Cent Nerv Syst Trauma.* 1986;3(2):145–52.

[19] Pinter MM, Gerstenbrand F, and Dimitrijevic MR. Epidural electrical stimulation of posterior structures of the human lumbosacral cord: 3. Control of spasticity. *Spinal Cord.* 2000;38(9):524–31.

[20] Harkema S, Gerasimenko Y, Hodes J, *et al.* Effect of epidural stimulation of the lumbosacral spinal cord on voluntary movement, standing, and assisted stepping after motor complete paraplegia: a case study. *Lancet.* 2011;377 (9781):1938–47.

[21] Angeli CA, Edgerton VR, Gerasimenko YP, and Harkema SJ. Altering spinal cord excitability enables voluntary movements after chronic complete paralysis in humans. *Brain.* 2014;137(Pt 5):1394–409.

[22] Angeli CA, Boakye M, Morton RA, *et al.* Recovery of over-ground walking after chronic motor complete spinal cord injury. *N Engl J Med.* 2018;379 (13):1244–50.

[23] Rehman MU, Sneed D, Sutor TW, Hoenig H, and Gorgey AS. Optimization of transspinal stimulation applications for motor recovery after spinal cord injury: scoping review. *J Clin Med.* 2023;12(3):854.

[24] Chandrasekaran S, Bhagat NA, Ramdeo R, *et al.* Targeted transcutaneous spinal cord stimulation promotes persistent recovery of upper limb strength and tactile sensation in spinal cord injury: a pilot study. *Front Neurosci.* 2023;17:1210328.

[25] Hofstoetter US, Hofer C, Kern H, *et al.* Effects of transcutaneous spinal cord stimulation on voluntary locomotor activity in an incomplete spinal cord injured individual. *Biomed Tech (Berl).* 2013;58(Suppl 1):000010151520134014.

[26]   Hofstoetter US, McKay WB, Tansey KE, Mayr W, Kern H, and Minassian K. Modification of spasticity by transcutaneous spinal cord stimulation in individuals with incomplete spinal cord injury. *J Spinal Cord Med.* 2014;37 (2):202–11.

[27]   Gerasimenko YP, Lu DC, Modaber M, *et al.* Noninvasive reactivation of motor descending control after paralysis. *J Neurotrauma.* 2015;32(24):1968–80.

[28]   Minassian K, and Hofstoetter US. Spinal cord stimulation and augmentative control strategies for leg movement after spinal paralysis in humans. *CNS Neurosci Ther.* 2016;22(4):262–70.

[29]   Shapkova EY, Pismennaya EV, Emelyannikov DV, and Ivanenko Y. Exoskeleton walk training in paralyzed individuals benefits from transcutaneous lumbar cord tonic electrical stimulation. *Front Neurosci.* 2020;14:416.

[30]   McHugh LV, Miller AA, Leech KA, Salorio C, and Martin RH. Feasibility and utility of transcutaneous spinal cord stimulation combined with walking-based therapy for people with motor incomplete spinal cord injury. *Spinal Cord Ser Cases.* 2020;6(1):104.

[31]   Gerasimenko Y, Sayenko D, Gad P, *et al.* Electrical spinal stimulation, and imagining of lower limb movements to modulate brain-spinal connectomes that control locomotor-like behavior. *Front Physiol.* 2018;9:1196.

[32]   Bye EA, Heroux ME, Boswell-Ruys CL, *et al.* Transcutaneous spinal cord stimulation combined with locomotor training to improve walking ability in people with chronic spinal cord injury: study protocol for an international multi-centred double-blinded randomised sham-controlled trial (eWALK). *Spinal Cord.* 2022;60(6):491–7.

[33]   Sayenko DG, Rath M, Ferguson AR, *et al.* Self-assisted standing enabled by non-invasive spinal stimulation after spinal cord injury. *J Neurotrauma.* 2019;36(9):1435–50.

[34]   Bedi P, and Arumugam N. Activity based therapy and surface spinal stimulation for recovery of walking in individual with traumatic incomplete spinal cord injury: a case report. *Int J Recent Sci Res.* 2015;6(8):5581–3.

[35]   Bedi P, and Arumugam N. Tapping the neural circuitry: surface spinal stimulation in spinal cord injury: a case report. *J Exerc Sci Physiother.* 2016; 12(1):69–75.

[36]   Sutor TW, Ghatas MP, Goetz LL, Lavis TD, and Gorgey AS. Exoskeleton training and trans-spinal stimulation for physical activity enhancement after spinal cord injury (EXTra-SCI): an exploratory study. *Front Rehabil Sci.* 2022;2:789422.

[37]   Manson GA, Calvert JS, Ling J, Tychhon B, Ali A, and Sayenko DG. The relationship between maximum tolerance and motor activation during transcutaneous spinal stimulation is unaffected by the carrier frequency or vibration. *Physiol Rep.* 2020;8(5):e14397.

[38]   Al'joboori Y, Hannah R, Lenham F, *et al.* The immediate and short-term effects of transcutaneous spinal cord stimulation and peripheral nerve stimulation on corticospinal excitability. *Front Neurosci.* 2021;15:749042.

[39] Danner SM, Wilshin SD, Shevtsova NA, and Rybak IA. Central control of interlimb coordination and speed-dependent gait expression in quadrupeds. *J Physiol*. 2016;594(23):6947–67.

[40] Curtis DR, and Eccles JC. Synaptic action during and after repetitive stimulation. *J Physiol*. 1960;150(2):374–98.

[41] Hofstoetter US, Freundl B, Binder H, and Minassian K. Recovery cycles of posterior root-muscle reflexes evoked by transcutaneous spinal cord stimulation and of the H reflex in individuals with intact and injured spinal cord. *PLoS One*. 2019;14(12):e0227057.

[42] Murray LM, and Knikou M. Remodeling brain activity by repetitive cervicothoracic transspinal stimulation after human spinal cord injury. *Front Neurol*. 2017;8:50.

[43] Gad P, Lee S, Terrafranca N, *et al*. Non-invasive activation of cervical spinal networks after severe paralysis. *J Neurotrauma*. 2018;35(18):2145–58.

[44] Inanici F, Samejima S, Gad P, Edgerton VR, Hofstetter CP, and Moritz CT. Transcutaneous electrical spinal stimulation promotes long-term recovery of upper extremity function in chronic tetraplegia. *IEEE Trans Neural Syst Rehabil Eng*. 2018;26(6):1272–8.

[45] Inanici F, Brighton LN, Samejima S, Hofstetter CP, and Moritz CT. Transcutaneous spinal cord stimulation restores hand and arm function after spinal cord injury. *IEEE Trans Neural Syst Rehabil Eng*. 2021;29:310–9.

[46] Sayenko DG, Atkinson DA, Dy CJ, *et al*. Spinal segment-specific transcutaneous stimulation differentially shapes activation pattern among motor pools in humans. *J Appl Physiol (1985)*. 2015;118(11):1364–74.

[47] Barss TS, Parhizi B, and Mushahwar VK. Transcutaneous spinal cord stimulation of the cervical cord modulates lumbar networks. *J Neurophysiol*. 2020;123(1):158–66.

[48] Parhizi B, Barss TS, and Mushahwar VK. Simultaneous cervical and lumbar spinal cord stimulation induces facilitation of both spinal and corticospinal circuitry in humans. *Front Neurosci*. 2021;15:615103.

[49] Atkinson DA, Steele AG, Manson GA, *et al*. Characterization of interlimb interaction via transcutaneous spinal stimulation of cervical and lumbar spinal enlargements. *J Neurophysiol*. 2022;127(4):1075–85.

[50] Klarner T, Barss TS, Sun Y, Kaupp C, Loadman PM, and Zehr EP. Long-term plasticity in reflex excitability induced by five weeks of arm and leg cycling training after stroke. *Brain Sci*. 2016;6(4):54.

[51] Klarner T, Barss TS, Sun Y, Kaupp C, Loadman PM, and Zehr EP. Exploiting interlimb arm and leg connections for walking rehabilitation: a training intervention in stroke. *Neural Plast*. 2016;2016:1517968.

[52] Zhou R, Alvarado L, Ogilvie R, Chong SL, Shaw O, and Mushahwar VK. Non-gait-specific intervention for the rehabilitation of walking after SCI: role of the arms. *J Neurophysiol*. 2018;119(6):2194–211.

[53] Zhou R, Alvarado L, Kim S, Chong SL, and Mushahwar VK. Modulation of corticospinal input to the legs by arm and leg cycling in people with incomplete spinal cord injury. *J Neurophysiol*. 2017;118(4):2507–19.

[54]   Wiesener C, Spieker L, Axelgaard J, *et al.* Supporting front crawl swimming in paraplegics using electrical stimulation: a feasibility study. *J Neuroeng Rehabil.* 2020;17(1):51.

[55]   Wu GJ, Xu F, Sun XM, and Chen JDZ. Transcutaneous neuromodulation at ST36 (Zusanli) is more effective than transcutaneous tibial nerve stimulation in treating constipation. *J Clin Gastroenterol.* 2020;54(6):536–44.

[56]   Atkinson DA, Sayenko DG, D'Amico JM, *et al.* Interlimb conditioning of lumbosacral spinally evoked motor responses after spinal cord injury. *Clin Neurophysiol.* 2020;131(7):1519–32.

[57]   Benavides FD, Jo HJ, Lundell H, Edgerton VR, Gerasimenko Y, and Perez MA. Cortical and subcortical effects of transcutaneous spinal cord stimulation in humans with tetraplegia. *J Neurosci.* 2020;40(13):2633–43.

[58]   Kumru H, Rodriguez-Canon M, Edgerton VR, *et al.* Transcutaneous electrical neuromodulation of the cervical spinal cord depends both on the stimulation intensity and the degree of voluntary activity for training. A pilot study. *J Clin Med.* 2021;10(15):3278.

[59]   Sasaki A, de Freitas RM, Sayenko DG, *et al.* Low-intensity and short-duration continuous cervical transcutaneous spinal cord stimulation intervention does not prime the corticospinal and spinal reflex pathways in able-bodied subjects. *J Clin Med.* 2021;10(16):3633.

[60]   Ward AR. Electrical stimulation using kilohertz-frequency alternating current. *Phys Ther.* 2009;89(2):181–90.

[61]   Dalrymple AN, Hooper CA, Kuriakose MG, Capogrosso M, and Weber DJ. Using a high-frequency carrier does not improve comfort of transcutaneous spinal cord stimulation. *J Neural Eng.* 2023;20(1):016016.

[62]   Tanner JA. Reversible blocking of nerve conduction by alternating-current excitation. *Nature.* 1962;195:712–3.

[63]   Bowman BR, and McNeal DR. Response of single alpha motoneurons to high-frequency pulse trains: firing behavior and conduction block phenomenon. *Stereotact Funct Neurosurg.* 1986;49(3):121–38.

[64]   Massey S, Konig D, Upadhyay P, *et al.* The effects of transcutaneous spinal cord stimulation delivered with and without high-frequency modulation on spinal and corticospinal excitability. *Artif Organs.* 2024;48(3):297–308.

[65]   Linderoth B, and Foreman RD. Conventional and novel spinal stimulation algorithms: hypothetical mechanisms of action and comments on outcomes. *Neuromodulation.* 2017;20(6):525–33.

[66]   Massey S, Konig D, Upadhyay P, *et al.* The effects of transcutaneous spinal cord stimulation delivered with and without high-frequency modulation on spinal and corticospinal excitability. *Artif Organs.* 2024;48(3):297–308.

[67]   Cook AW, and Weinstein SP. Chronic dorsal column stimulation in multiple sclerosis. Preliminary report. *N Y State J Med.* 1973;73(24):2868–72.

[68]   Illis LS, Oygar AE, Sedgwick EM, and Awadalla MA. Dorsal-column stimulation in the rehabilitation of patients with multiple sclerosis. *Lancet.* 1976;1(7974):1383–6.

[69]   Dooley DM, and Sharkey J. Electrostimulation of the nervous system for patients with demyelinating and degenerative diseases of the nervous system and vascular diseases of the extremities. *Appl Neurophysiol.* 1977;40(2–4): 208–17.

[70]   Mansour NM, Pena Pino I, Freeman D, *et al.* Advances in epidural spinal cord stimulation to restore function after spinal cord injury: history and systematic review. *J Neurotrauma.* 2022;39(15–16):1015–29.

[71]   Pino IP, Nightingale TE, Hoover C, *et al.* The safety of epidural spinal cord stimulation to restore function after spinal cord injury: post-surgical complications and incidence of cardiovascular events. *Spinal Cord.* 2022;60 (10):903–10.

[72]   Darrow D, Balser D, Netoff TI, *et al.* Epidural spinal cord stimulation facilitates immediate restoration of dormant motor and autonomic supraspinal pathways after chronic neurologically complete spinal cord injury. *J Neurotrauma.* 2019;36(15):2325–36.

[73]   Rejc E, Angeli C, and Harkema S. Effects of lumbosacral spinal cord epidural stimulation for standing after chronic complete paralysis in humans. *PLoS One.* 2015;10(7):e0133998.

[74]   Grahn PJ, Lavrov IA, Sayenko DG, *et al.* Enabling task-specific volitional motor functions via spinal cord neuromodulation in a human with paraplegia. *Mayo Clin Proc.* 2017;92(4):544–54.

[75]   Gill ML, Grahn PJ, Calvert JS, *et al.* Neuromodulation of lumbosacral spinal networks enables independent stepping after complete paraplegia. *Nat Med.* 2018;24(11):1677–82.

[76]   Wagner FB, Mignardot JB, Le Goff-Mignardot CG, *et al.* Targeted neurotechnology restores walking in humans with spinal cord injury. *Nature.* 2018; 563(7729):65–71.

[77]   Boakye M, Ball T, Dietz N, *et al.* Spinal cord epidural stimulation for motor and autonomic function recovery after chronic spinal cord injury: a case series and technical note. *Surg Neurol Int.* 2023;14:87.

[78]   Barolat G, Myklebust JB, and Wenninger W. Enhancement of voluntary motor function following spinal cord stimulation–case study. *Appl Neurophysiol.* 1986;49(6):307–14.

[79]   Herman R, He J, D'Luzansky S, Willis W, and Dilli S. Spinal cord stimulation facilitates functional walking in a chronic, incomplete spinal cord injured. *Spinal Cord.* 2002;40(2):65–8.

[80]   Huang H, He J, Herman R, and Carhart MR. Modulation effects of epidural spinal cord stimulation on muscle activities during walking. *IEEE Trans Neural Syst Rehabil Eng.* 2006;14(1):14–23.

[81]   Carhart MR, He J, Herman R, D'Luzansky S, and Willis WT. Epidural spinal-cord stimulation facilitates recovery of functional walking following incomplete spinal-cord injury. *IEEE Trans Neural Syst Rehabil Eng.* 2004;12 (1):32–42.

[82]   DiMarco AF, Geertman RT, Tabbaa K, Nemunaitis GA, and Kowalski KE. Effects of lower thoracic spinal cord stimulation on bowel management in

individuals with spinal cord injury. *Arch Phys Med Rehabil.* 2021;102 (6):1155–64.

[83]   Ganley KJ, and Powers CM. Intersegmental dynamics during the swing phase of gait: a comparison of knee kinetics between 7 year-old children and adults. *Gait Posture.* 2006;23(4):499–504.

[84]   Powell MP, Verma N, Sorensen E, *et al.* Epidural stimulation of the cervical spinal cord for post-stroke upper-limb paresis. *Nat Med.* 2023;29(3):689–99.

[85]   Mesbah S, Herrity A, Ugiliweneza B, *et al.* Neuroanatomical mapping of the lumbosacral spinal cord in individuals with chronic spinal cord injury. *Brain Commun.* 2023;5(1):fcac330.

[86]   Hofstoetter US, Perret I, Bayart A, *et al.* Spinal motor mapping by epidural stimulation of lumbosacral posterior roots in humans. *iScience.* 2021;24 (1):101930.

[87]   Angeli CA, and Gerasimenko Y. Combined cervical transcutaneous with lumbosacral epidural stimulation improves voluntary control of stepping movements in spinal cord injured individuals. *Front Bioeng Biotechnol.* 2023;11:1073716.

[88]   Hoglund BK, Zurn CA, Madden LR, *et al.* Mapping spinal cord stimulation-evoked muscle responses in patients with chronic spinal cord injury. *Neuromodulation.* 2023;26(7):1371–80.

[89]   Barolat-Romana G, Myklebust JB, Hemmy DC, Myklebust B, and Wenninger W. Immediate effects of spinal cord stimulation in spinal spasticity. *J Neurosurg.* 1985;62(4):558–62.

[90]   Barolat G, Singh-Sahni K, Staas WE, Jr., Shatin D, Ketcik B, and Allen K. Epidural spinal cord stimulation in the management of spasms in spinal cord injury: a prospective study. *Stereotact Funct Neurosurg.* 1995;64 (3):153–64.

[91]   Jilge B, Minassian K, Rattay F, and Dimitrijevic MR. Frequency-dependent selection of alternative spinal pathways with common periodic sensory input. *Biol Cybern.* 2004;91(6):359–76.

[92]   Sherwood AM, McKay WB, and Dimitrijevic MR. Motor control after spinal cord injury: assessment using surface EMG. *Muscle Nerve.* 1996;19(8):966–79.

[93]   Minassian K, McKay WB, Binder H, and Hofstoetter US. Targeting lumbar spinal neural circuitry by epidural stimulation to restore motor function after spinal cord injury. *Neurotherapeutics.* 2016;13(2):284–94.

[94]   Rath M, Vette AH, Ramasubramaniam S, *et al.* Trunk stability enabled by noninvasive spinal electrical stimulation after spinal cord injury. *J Neurotrauma.* 2018;35(21):2540–53.

[95]   Sharma P, Panta T, Ugiliweneza B, *et al.* Multi-site spinal cord transcutaneous stimulation facilitates upper limb sensory and motor recovery in severe cervical spinal cord injury: a case study. *J Clin Med.* 2023;12(13):4416.

[96]   Ganley K, Willis W, Carhart M, He J, and Herman R. Epidural spinal cord stimulation improves locomotor performance in low ASIA C, wheelchair-dependent, spinal cord-injured individuals: insights from metabolic response. *Top Spinal Cord Inj Rehabil.* 2005;11:50–63.

[97] Pena Pino I, Hoover C, Venkatesh S, *et al.* Long-term spinal cord stimulation after chronic complete spinal cord injury enables volitional movement in the absence of stimulation. *Front Syst Neurosci.* 2020;14:35.

[98] Al'joboori Y, Massey SJ, Knight SL, Donaldson NN, and Duffell LD. The effects of adding transcutaneous spinal cord stimulation (tSCS) to sit-to-stand training in people with spinal cord injury: a pilot study. *J Clin Med.* 2020; 9 (9):2765.

[99] Moritz C, Field-Fote EC, Tefertiller C, *et al.* Non-invasive spinal cord electrical stimulation for arm and hand function in chronic tetraplegia: a safety and efficacy trial. *Nat Med.* 2024;30(5):1276–83.

[100] Rowald A, Komi S, Demesmaeker R, *et al.* Activity-dependent spinal cord neuromodulation rapidly restores trunk and leg motor functions after complete paralysis. *Nat Med.* 2022;28(2):260–71.

[101] Lorach H, Charvet G, Bloch J, and Courtine G. Brain-spine interfaces to reverse paralysis. *Natl Sci Rev.* 2022;9(10):nwac009.

[102] Vodovnik L. Therapeutic effects of functional electrical stimulation of extremities. *Med Biol Eng Comput.* 1981;19(4):470–8.

[103] Taylor PN, Burridge JH, Dunkerley AL, *et al.* Clinical use of the Odstock dropped foot stimulator: its effect on the speed and effort of walking. *Arch Phys Med Rehabil.* 1999;80(12):1577–83.

[104] Donaldson N, Perkins TA, Fitzwater R, Wood DE, and Middleton F. FES cycling may promote recovery of leg function after incomplete spinal cord injury. *Spinal Cord.* 2000;38(11):680–2.

[105] McDonald JW, Becker D, Sadowsky CL, Jane JA, Sr., Conturo TE, and Schultz LM. Late recovery following spinal cord injury. Case report and review of the literature. *J Neurosurg.* 2002;97(Suppl 2):252–65.

[106] Lovely RG, Gregor RJ, Roy RR, and Edgerton VR. Effects of training on the recovery of full-weight-bearing stepping in the adult spinal cat. *Exp Neurol.* 1986;92(2):421–35.

[107] Riddoch G. The reflex functions of the completely divided spinal cord in man, compared with those associated with less severe lesions. *Brain.* 1917; 40(2–3):264–402.

[108] Wernig A, Muller S, Nanassy A, and Cagol E. Laufband therapy based on 'rules of spinal locomotion' is effective in spinal cord injured persons. *Eur J Neurosci.* 1995;7(4):823–9.

[109] Barbeau H, Norman K, Fung J, Visintin M, and Ladouceur M. Does neurorehabilitation play a role in the recovery of walking in neurological populations? *Ann N Y Acad Sci.* 1998;860:377–92.

[110] Wessels M, Lucas C, Eriks I, and de Groot S. Body weight-supported gait training for restoration of walking in people with an incomplete spinal cord injury: a systematic review. *J Rehabil Med.* 2010;42(6):513–9.

[111] Behrman AL, Bowden MG, and Nair PM. Neuroplasticity after spinal cord injury and training: an emerging paradigm shift in rehabilitation and walking recovery. *Phys Ther.* 2006;86(10):1406–25.

[112] Hodgson JA, Roy RR, de Leon R, Dobkin B, and Edgerton VR. Can the mammalian lumbar spinal cord learn a motor task? *Med Sci Sports Exerc.* 1994;26(12):1491–7.

[113] Jones ML, Evans N, Tefertiller C, *et al.* Activity-based therapy for recovery of walking in chronic spinal cord injury: results from a secondary analysis to determine responsiveness to therapy. *Arch Phys Med Rehabil.* 2014; 95(12):2247–52.

[114] Roy RR, Harkema SJ, and Edgerton VR. Basic concepts of activity-based interventions for improved recovery of motor function after spinal cord injury. *Arch Phys Med Rehabil.* 2012;93(9):1487–97.

[115] Morrison SA, Lorenz D, Eskay CP, Forrest GF, and Basso DM. Long-itudinal recovery and reduced costs after 120 sessions of locomotor training for motor incomplete spinal cord injury. *Arch Phys Med Rehabil.* 2018; 99(3):555–62.

[116] Field-Fote EC. Combined use of body weight support, functional electric stimulation, and treadmill training to improve walking ability in individuals with chronic incomplete spinal cord injury. *Arch Phys Med Rehabil.* 2001; 82(6):818–24.

[117] Tefertiller C, and Gerber D. Step ergometer training augmented with functional electrical stimulation in individuals with chronic spinal cord injury: a feasibility study. *Artif Organs.* 2017;41(11):E196–E202.

[118] Sadowsky CL, Hammond ER, Strohl AB, *et al.* Lower extremity functional electrical stimulation cycling promotes physical and functional recovery in chronic spinal cord injury. *J Spinal Cord Med.* 2013;36(6):623–31.

[119] Duffell LD, Paddison S, Alahmary AF, Donaldson N, and Burridge J. The effects of FES cycling combined with virtual reality racing biofeedback on voluntary function after incomplete SCI: a pilot study. *J Neuroeng Rehabil.* 2019;16(1):149.

[120] Yasar E, Yilmaz B, Goktepe S, and Kesikburun S. The effect of func-tional electrical stimulation cycling on late functional improvement in patients with chronic incomplete spinal cord injury. *Spinal Cord.* 2015; 53(12):900.

[121] Megia Garcia A, Serrano-Munoz D, Taylor J, Avendano-Coy J, and Gomez-Soriano J. Transcutaneous spinal cord stimulation and motor reha-bilitation in spinal cord injury: a systematic review. *Neurorehabil Neural Repair.* 2020;34(1):3–12.

[122] Freyvert Y, Yong NA, Morikawa E, *et al.* Engaging cervical spinal circuitry with non-invasive spinal stimulation and buspirone to restore hand function in chronic motor complete patients. *Sci Rep.* 2018;8(1):15546.

[123] Alam M, Ling YT, Rahman MA, *et al.* Restoration of over-ground walking via non-invasive neuromodulation therapy: a single-case study. *J Clin Med.* 2023;12(23):7362.

[124] Lu DC, Edgerton VR, Modaber M, *et al.* Engaging cervical spinal cord networks to reenable volitional control of hand function in tetraplegic patients. *Neurorehabil Neural Repair.* 2016;30(10):951–62.

[125] Belci M, Catley M, Husain M, Frankel HL, and Davey NJ. Magnetic brain stimulation can improve clinical outcome in incomplete spinal cord injured patients. *Spinal Cord.* 2004;42(7):417–9.

[126] Bunday KL, and Perez MA. Motor recovery after spinal cord injury enhanced by strengthening corticospinal synaptic transmission. *Curr Biol.* 2012;22(24):2355–61.

[127] Jo HJ, and Perez MA. Corticospinal-motor neuronal plasticity promotes exercise-mediated recovery in humans with spinal cord injury. *Brain.* 2020; 143(5):1368–82.

[128] Glover IS, and Baker SN. Both corticospinal and reticulospinal tracts control force of contraction. *J Neurosci.* 2022;42(15):3150–64.

[129] Germann M, and Baker SN. Testing a novel wearable device for motor recovery of the elbow extensor triceps brachii in chronic spinal cord injury. *eNeuro.* 2023;10(7).

[130] Alder G, Signal N, Olsen S, and Taylor D. A systematic review of paired associative stimulation (PAS) to modulate lower limb corticomotor excitability: implications for stimulation parameter selection and experimental design. *Front Neurosci.* 2019;13:895.

[131] Nicoll RA. A brief history of long-term potentiation. *Neuron.* 2017; 93(2):281–90.

[132] Vints WAJ, Levin O, Fujiyama H, Verbunt J, and Masiulis N. Exerkines and long-term synaptic potentiation: mechanisms of exercise-induced neuroplasticity. *Front Neuroendocrinol.* 2022;66:100993.

[133] Smallwood N, Spriggs MJ, Thompson CS, *et al.* Influence of physical activity on human sensory long-term potentiation. *J Int Neuropsychol Soc.* 2015;21(10):831–40.

[134] Pons DJ, Vaughan CL, and Jaros GG. Cycling device powered by the electrically stimulated muscles of paraplegics. *Med Biol Eng Comput.* 1989; 27(1):1–7.

[135] Arora T, Desai N, Kirshblum S, and Chen R. Utility of transcranial magnetic stimulation in the assessment of spinal cord injury: current status and future directions. *Front Rehabil Sci.* 2022;3:1005111.

[136] Smith AC, and Knikou M. A review on locomotor training after spinal cord injury: reorganization of spinal neuronal circuits and recovery of motor function. *Neural Plast.* 2016;2016:1216258.

[137] Ladenbauer J, Minassian K, Hofstoetter US, Dimitrijevic MR, and Rattay F. Stimulation of the human lumbar spinal cord with implanted and surface electrodes: a computer simulation study. *IEEE Trans Neural Syst Rehabil Eng.* 2010;18(6):637–45.

[138] Lorach H, Galvez A, Spagnolo V, *et al.* Walking naturally after spinal cord injury using a brain-spine interface. *Nature.* 2023;618(7963):126–33.

[139] Chen S, Andary M, Buschbacher R, *et al.* Electrodiagnostic reference values for upper and lower limb nerve conduction studies in adult populations. *Muscle Nerve.* 2016;54(3):371–7.

[140] Knikou M. The H-reflex as a probe: pathways and pitfalls. *J Neurosci Methods*. 2008;171(1):1–12.
[141] Knikou M, and Mummidisetty CK. Locomotor training improves pre-motoneuronal control after chronic spinal cord injury. *J Neurophysiol*. 2014;111(11):2264–75.
[142] Baker SN, and Perez MA. Reticulospinal contributions to gross hand function after human spinal cord injury. *J Neurosci*. 2017;37(40):9778–84.

Chapter 5

Upper extremity rehabilitation

Naaz Desai[1,2,3], Cesar Marquez-Chin[2,4,5],
Matija Milosevic[6,7,8] and Milos R. Popovic[2,4,5,9]

5.1 Introduction: impact of upper extremity paralysis on quality of life

Activities of daily living largely depend on arm function [1], particularly for personal activities such as feeding, dressing and grooming. Upper extremity (UE) paralysis can be a sequela of various conditions, including stroke, spinal cord injury (SCI), multiple sclerosis, traumatic brain injury and many more. Irrespective of its cause and the degree of impairment, paralysis interferes with the ability to perform activities of daily living, reducing the quality of life (QOL) of individuals living with the condition. UE (i.e., arm, hand and/or fingers) motor impairment following stroke is often persistent and disabling [2], with only half of all stroke survivors regaining some useful UE function six months after stroke [3]. One year after stroke persistent arm motor impairment is associated with anxiety [4], poorer perception of health-related QOL [5] and subjective well-being [6]. Similarly, multiple studies investigating the impact of loss of UE function in individuals with SCI have reported that regaining arm and hand function is the highest priority for individuals with tetraplegia and would improve their QOL [7–9]. These findings did not differ by gender or number of years post SCI suggesting that even partial recovery of UE function may have a significant impact on the independence of many individuals with SCI [7]. UE function is also one of the affected domains in people with Parkinson's disease. Schrag *et al.* reported that patients with PD are 2–10 times more likely to have difficulties with self-care, social functioning, mobility, anxiety

[1]Krembil Research Institute, University Health Network, Canada
[2]CRANIA, University Health Network and University of Toronto, Canada
[3]Department of Physical Therapy, University of Toronto, Canada
[4]The KITE Research Institute, Toronto Rehabilitation Institute – University Health Network, Canada
[5]Institute of Biomedical Engineering, University of Toronto, Canada
[6]The Miami Project to Cure Paralysis, University of Miami, USA
[7]Department of Neurological Surgery, University of Miami, USA
[8]Department of Biomedical Engineering, University of Miami, USA
[9]Rehabilitation Sciences Institute, University of Toronto, Canada

and depression and have significantly lower overall health-related QOL scores [10]. Rehabilitation is one of the core treatment strategies adopted to restore lost upper limb function following many, if not all, of these debilitating conditions.

5.2   Upper extremity movement and its anatomical and neural bases

Individuals living with UE deficits undergo rehabilitation to increase their functional abilities, enhance their QOL and reduce burden of care [7]. Clinicians and researchers recognise that some sensory and motor gains may occur following various neurological conditions. The extent of these gains is modulated by several factors, including the disease condition and its natural progression, extent of damage, timing and extent of intervention – both medical/surgical and rehabilitative – and so on and so forth. There is significant literature around recovery curves defining the spatial and temporal recovery following many common neurological conditions above and beyond the ones discussed in more detail below.

When we think about retraining UE function, we parse UE movements into reaching, grasping and manipulation movements. In a functional context, a working group of UE rehabilitation experts [11] defined the construct of reaching, grasping and manipulation as

> Reaching, grasping and manipulation are the important components of upper extremity function that allow individuals to use the sensorimotor integrity of their arm and hand to develop abilities and perform activities that meet their personal needs, and enable them to explore and participate in their external environment in meaningful ways.

Although this definition was created in relation to individuals with SCI, it can very easily be applied to any other population.

5.2.1   Anatomical bases of movement

Reaching alone refers to the ability to position the arm in space to carry out the desired task, whereas reaching and manipulation involves interaction with an external object. Therefore, reaching essentially involves movements at the scapulothoracic, shoulder and elbow joints, whereas grasping and manipulation primarily involves movements of the wrist and fingers. Physiologically, the muscle recruitment at any one of these joints is dictated by the direction of the movement, i.e., the plane in which the movement occurs. This information is critical as these are muscles that one would ideally target with functional electrical stimulation (FES), especially when the goal of treatment is to retrain lost function. Table 5.1 provides information on muscles involved during some of the UE movements commonly performed during activities of daily living.

5.2.2   Neural bases of movement

Regarding reaching, visual information on the target's location informs the arm direction and movement amplitude. For hand-related actions, to grasp an object based

Table 5.1 Upper extremity movements and muscles involved

Functional movement	Joint involved and anatomical movement	Muscles involved during movement execution
Forward reaching	Shoulder forward flexion and elbow extension	Upper trapezius, lower trapezius, serratus anterior, anterior deltoid and triceps
Sideways reaching	Shoulder abduction and elbow extension	Upper trapezius, lower trapezius, serratus anterior, middle deltoid and triceps
Reaching over opposite shoulder	Shoulder adduction and elbow flexion	Upper trapezius, lower trapezius, serratus anterior, pectoralis major and biceps
Reaching opposite knee	Shoulder adduction and elbow extension	Pectoralis major and triceps opposite knee
Hand to mouth	Shoulder flexion and elbow flexion	Upper trapezius, lower trapezius, serratus anterior, anterior deltoid and biceps
Palmar grasp	Finger flexion and thumb opposition	Flexor digitorum superficialis and flexor digitorum profundus and opponens pollicis
Two finger pinch	Index finger flexion at MCP and 1st PIP and thumb opposition	First dorsal interosseous, 1st lumbrical and opponens pollicis
Lumbrical grasp	MCP flexion at all four fingers and thumb opposition	1–4 lumbricals and opponens pollicis

MCP = metacarpophalangeal joint; PIP = proximal interphalangeal joint

on visual information, object features such as shape, size and orientation must be transformed in the type of grip and wrist orientation appropriate for that object [12]. Several functional magnetic resonance imaging experiments in humans have identified the existence of a well-defined fronto-parietal network for reaching movements [13,14]. In particular, visually guided reaching activates a dorso-dorsal pathway that includes several key areas: the medial parietal sulcus, the superior parietal occipital cortex, the precuneus, the supplementary motor area and the dorsal premotor area [14,15]. When compared with pure reaching, grasping leads to higher activation of a dorso-ventral pathway [16] including anterior sector of the intraparietal sulcus [17,18] and ventral pre motor cortex [15]. Grol *et al.* showed that the dorsal premotor cortex is involved not only in reaching, but also in grasping [19]. Beyond the cortical level, a set of subcortical structures playing a fundamental role during reach to-grasp actions is represented by the basal ganglia. Vaillancourt *et al.* conducted a functional magnetic resonance imaging study and found that performing a task requiring precision grip force control was associated with changes in neural activity of the putamen, caudate nucleus, globus pallidus and the subthalamic nucleus [20]. The cerebellum is another subcortical structure involved in several aspects of motor control of grasping actions. Several studies have reported the activation of deep cerebellar nuclei, including both the dorsal and ventral sectors of the dentate nucleus, which constitutes the main output channel of the cerebellum to the parietal and premotor areas via the thalamus [21]. The function of the dentate nucleus appears to be correlated with voluntary coordination or movement correction, even when sensory feedback is absent [22].

5.3   Neurological conditions affecting upper extremity movements

5.3.1   Stroke

5.3.1.1   Epidemiology and clinical presentation

Stroke or cerebrovascular accident is defined as a sudden non-convulsive focal neurologic deficit. It remains the second-leading cause of death, and the third-leading cause of death and disability combined (as expressed by disability-adjusted life years lost) in the world [23]. The economic impact of stroke is substantial; the estimated global cost of stroke is over US$721 billion (0.66% of the global GDP) [23].

The clinical presentation of individuals who have suffered a stroke can vary widely. The neurologic deficit reflects the location and the size of the infarct or haemorrhage. Hemiplegia stands as the most typical sign of cerebrovascular diseases, whether in the cerebral hemisphere or brainstem, but there are many other manifestations occurring in recognisable combinations. Here, we provide an overview of the UE deficits commonly seen in individuals presenting with hemiplegia. Common UE impairments after stroke include paresis, loss of fractionated movement, abnormal muscle tone and/or changes in somatosensation [24]. Although each of these impairments can occur in isolation, more often, they exist in combinations. Lang *et al.* reported that paresis across the entire limb is the primary contributor to decreased function, and the severity of paresis at three or more weeks post-stroke is the strongest indicator of present and eventual UE function [24].

5.3.1.2   Rehabilitation strategies for upper extremity rehabilitation following stroke

Over the past two decades, there has been a tenfold increase in UE rehabilitation research. Scientists are constantly trying to develop and improve therapies leading to better outcomes. Conventional therapeutic manoeuvres used in the rehabilitation of the arm include Bobath techniques, neuro-developmental techniques, strengthening exercises, repetitive task specific training, constraint-induced movement therapy, bilateral arm training, electrical stimulation, mirror therapy, training for compensatory techniques using the unaffected UE and many more. Alongside these long-standing practices, there are various newer interventions that are available or still being researched like electrical stimulation specifically in the functional electrical stimulation embodiment, brain–computer interface (BCI)-controlled FES therapy, transcranial magnetic stimulation therapy (TMS) [25], transcranial direct current stimulation (tDCS) [26] and others. Whereas a detailed discussion of individual therapies for stroke is outside the scope of this document and can be found in published best practice guidelines [27] we provide a detailed discussion on FES in the later part of this chapter.

5.3.2   Spinal cord injury

5.3.2.1   Epidemiology and clinical presentation

A SCI can be traumatic or non-traumatic in aetiology. Traumatic SCI occurs when there is a sudden, traumatic impact on the spine that compromises the spinal column and/or fractures or dislocates vertebrae, subsequently causing damage or compression to the spinal cord [28]. Non traumatic SCI refers to compression of the spinal cord related to a tumour, infection or degeneration of the spinal column. SCIs are typically divided into two broad functional categories: tetraplegia and paraplegia. Tetraplegia refers to partial or complete paralysis of all four extremities and the trunk, including the respiratory muscles, and results from cervical cord injury. Paraplegia refers to partial or complete paralysis of all or part of the trunk and both lower extremities resulting from lesions of the thoracic or lumbar spinal cord or sacral roots. Injuries above T12 result in an upper motor neuron type paralysis of the lower limbs whereas injury below T12 or that of the cauda equina will result in lower motor neuron type of presentation. Globally, as of 2019, there were 900,000 incident cases, 20.6 million prevalent cases and 6.2 million years lived with disability of total SCI [29]. Since then, the age-standardised prevalence rate has increased, while the incidence rate and years lived with disability have decreased. Tetraplegia results in higher age-standardised years lived with disability compared to paraplegia [29].

Clinically, traumatic and non-traumatic SCI have a presentation of motor and sensory deficits coupled with functional loss. In other words, they manifest with a degree of paralysis or paresis because of an upper motor neuron lesion. Over and above the motor and sensory deficits, there are a myriad of other clinical manifestations, including an initial period of spinal shock, impaired temperature control, respiratory impairment, spasticity, bladder and bowel dysfunction, sexual dysfunction and secondary complications like pressure sores, autonomic dysreflexia, postural hypotension, contractures, deep venous thrombosis, osteoporosis and neuropathic pain. Overall, the extent of the damage determines the severity of SCI, and the severity of the injury is the greatest factor in predicting the recovery profiles [30,31].

5.3.2.2 Rehabilitation strategies for upper extremity rehabilitation following spinal cord injury

Rehabilitation for patients with SCI has evolved over the years and is undergoing a paradigm shift. In the past, physical and occupational therapy focused on teaching compensatory strategies. However, with an increasing understanding of neuroplasticity, goals have shifted towards neuromuscular re-education and recovery of lost function. Worldwide, there are published guidelines related to rehabilitation protocols for individuals with SCI [32]. Literature supports the use of therapies targeting muscle strength and function to improve task performance and QOL in individuals with SCI. Therapy-based interventions consist of exercise therapy, including strength and endurance training exercises, range of motion exercises, stretching exercises and functional goal-oriented exercises. Technology-based interventions in SCI are either used to compensate for lost function, like the use of orthotic devices, or as therapeutic devices. Similar to stroke rehabilitation, there are many therapeutic modalities in a typical SCI rehabilitation toolbox. In this chapter, we discuss FES application in detail and highlight some combinational approaches such as non-invasive brain stimulation, including TMS, tDCS and BCI technologies with FES.

5.3.3   Parkinson's disease

5.3.3.1   Epidemiology and clinical presentation

Parkinson's disease is a chronic and progressive neurodegenerative disorder that primarily affects the motor system [33]. The disease affects millions of people worldwide and is the second most prevalent neurodegenerative condition next to Alzheimer's disease [33]. It is estimated that the number of affected patients may double by 2030. Age-standardised Parkinson's disease incidence estimates for ages 65 and older range from 108 to 212 per 100,000 person-years [34].

Clinically, patients with Parkinson's disease typically present with tremor, bradykinesia, rigidity, postural instability and a variety of nonmotor symptoms such as sleep and mood disorders. In early onset Parkinson's disease, unilateral upper limb motor deficits, including decreased writing velocity and impaired coordination, are common [35]. As the disease progresses, impairments progressively affect upper limb function, manifesting abnormal force generation [36], impaired fine manual dexterity [37] and poor bimanual coordination [35]. The ability to perform activities of daily living becomes compromised, and patients often adopt compensatory strategies or avoid tasks entirely.

5.3.3.2   Rehabilitation strategies for upper extremity rehabilitation in Parkinson's disease

Physical therapy is advocated as an adjuvant therapy, and while the primary focus of most studies has been to improve gait and balance or bradykinesia there is evidence related to the use of task-specific training, virtual reality and sensorimotor training for improving UE function [38,39]. FES in Parkinson's disease is less researched. However, it has been shown to achieve tremor reduction comparable or superior to first-line pharmacotherapies, but technical and safety barriers specifically for lower extremity tremor may limit its clinical translation [40].

Researchers have explored various implementations of closed-loop FES, including varying the stimulation site, alternate feedback signals (e.g., electromyography) and creating new models to drive the stimulation [41,42], and have achieved 67–84% acute tremor reduction.

5.3.4   Other central nervous system disorders

Other central nervous system conditions that may also benefit from FES as a rehabilitation modality include, but are not limited to, traumatic brain injury [43], cerebral palsy [44], multiple sclerosis [45]. There is limited research into the application and potential benefits of FES in these conditions, and most studies have focused on FES for gait improvements. Nonetheless, these remain active areas of research for the application of FES for UE rehabilitation.

5.3.5   Peripheral nerve injuries

5.3.5.1   Epidemiology and clinical presentation

Peripheral nerve injuries commonly occur with concomitant limb injuries or polytrauma but may also occur as an iatrogenic complication of medical or surgical care episodes. Based on the extent of nerve damage, the regeneration mechanisms can be complex and may lead to poor prognosis and slow recovery, often resulting in motor and sensory deficits and, occasionally, lifelong disability. Although there are no worldwide statistics on the incidence and prevalence of peripheral nerve injuries, studies have reported statistics from some countries. For example, the mean incidence of UE peripheral nerve injuries in the United States in 2014 was 16.9 per 100,000 persons [46]; in Sweden, the incidence was 13.9 per 100,000 persons between 1998 and 2006 [47] and peripheral nerve injuries represented 3.3% of cases of severe trauma with UE involvement in a German trauma multi-centre database between 2002 and 2015 [48]. One study reported the incidence of peripheral nerve injuries in South Korea as 7.88 per 100,000 in 2018 [49].

Clinical presentation following a peripheral nerve injury largely depends on the nerves involved and the extent of damage. If the damaged nerve is a mixed nerve, then there may be both motor and sensory symptoms, whereas a sensory nerve will cause sensory symptoms. The Sedden classification system, which is based on the extent of nerve damage, classifies peripheral nerve injuries into (a) Neurapraxia, where there is no disruption of the axon or its sheath and the injury tends to be reversible with recovery occurring within days to weeks; (b) Axonotmesis, wherein there is axonal disruption, but the surrounding connective tissue remains intact and recovery occurs at a rate of 1 mm per day; and (c) Neurotmesis which is the most severe form of nerve injury in which both the axon and surrounding connective tissue are disrupted – recovery typically requires surgical intervention.

5.3.5.2   Rehabilitation strategies for upper extremity peripheral nerve injuries

Recovery from peripheral nerve injuries depends on various factors, including the nerve involved, the degree of injury to the nerve and surrounding tissues and the mechanism of injury [50]. Rehabilitation therapies are critically important following peripheral nerve injuries. Even after surgical management, recovery is typically incomplete and may be associated with significant ongoing physical dysfunction and pain. Rehabilitation plans are designed to reduce physical disability and/or enable compensatory functional adaptation based on individual patient presentation. In the early phase, therapeutic modalities include range of motion exercises and stretching, strategies for control of oedema and pain management. Electrical stimulation plays a critical role in minimising muscle degeneration and maintaining neuronal integrity for future reinnervation. It is a promising technique that has been shown to promote muscle reinnervation by increasing the expression of structural protective proteins and neurotrophic factors [51]. Optimisation of stimulation settings and evaluation of this technique in specific clinical scenarios is still required [51]. Chapter 8 discusses in great depth use of electrical stimulation in cases with severely damaged peripheral nerves resulting in denervation of muscle or muscle groups.

5.4   FES and its evolution over past several decades

In this section, we present a few FES systems that have been important in the development and adoption of FES techniques for restoring UE voluntary movement.

5.4.1   Neuroprostheses for upper extremity function

5.4.1.1   Restoration of grasping

In the early 1970s, Rebersek and Vodovnik developed a three-channel stimulator at the University of Ljubljana in Slovenia [52]. The stimulator could be controlled using sensors for electromyographic (EMG) recording and a sliding resistor for intensity control. One of its significant contributions is the design for selecting an optimal interface to match users' physical abilities.

The Neuromuscular Electrical Stimulation System (NESS) H200 [53] is a neuromuscular electrical stimulation system with three channels that stimulate finger extensors, flexors, and the thumb. It uses a carbon fibre splint for support, and the stimulation intensity and duration can be adjusted using an external control box. An updated FDA-approved version is available as the H200 Wireless Hand Rehabilitation System by Bioness, Inc., California, USA.

The Bionic Glove is a fingerless glove with a wrist position sensor and an electrical stimulator [54]. It is designed for individuals with SCI who have active wrist flexion and extension. The glove triggers stimulation to produce a grasp when the wrist is extended and opens the hand when the wrist is flexed.

The Freehand System is an FDA-approved implanted neuroprosthesis developed at Case Western Reserve University [55]. It assists individuals with SCIs by controlling finger and thumb movements. The system uses external power and control signals and provides sensory feedback through one of the electrodes.

5.4.1.2   Restoration of reaching and grasping

The Belgrade Grasping-Reaching System was a non-invasive three-channel neuroprosthesis that allowed users to generate palmar and lateral grasps as well as opening the hand [56]. Additionally, a fourth channel triggered the triceps brachii muscle according to the angular velocity of the shoulder.

The COMPEX Motion Neuroprosthesis for Reaching and Grasping was a programmable four-channel device designed to assist individuals with SCI in reaching and grasping [57]. The system allowed for customisable stimulation parameters and could be triggered by various sensors to match the user's functional abilities and requirements.

The MyndMove is a commercially available stimulator designed to restore upper limb function after stroke and SCI using FES therapy. It offers over 30 reaching and grasping protocols and has been successful in restoring function in various conditions, including severe hemiplegia [58], SCI, traumatic brain injury, cerebral palsy, cervical myelopathy and brachial plexus injuries.

5.4.2   Therapeutic use of FES for restoring upper extremity movement

Starting in the 1960s, several reports have suggested the potential for using FES to restore unassisted voluntary function [59–62]. Since then, multiple trials have verified the efficacy of FES therapy for movement rehabilitation. Tables 5.2 and 5.3 list some

Table 5.2 Clinical evidence for FES in stroke

Year	Reference	Population	Study design	Intervention	Summary of findings
1998	[63]	Subacute, moderate to severe hemiplegia	Double-blind, placebo-controlled, randomised design	60 min/session 15 sessions NMES	Significantly greater gains in Fugl-Meyer scores for the treatment group.
2001	[64]	Chronic stroke with spastic paresis resulting in upper extremity dysfunction	Exploratory, uncontrolled trial	20–60 min/day 8 weeks Home-based	Statistically significant improvement in muscle score and tone, which decreased after therapy. Patients with higher initial scores benefited the most.
2002	[65]	Chronic stroke with mild to moderate hemiparesis	Three-group randomly controlled, pre-post test Group 1: coupled EMG-triggered NMES and bilateral movement Group 2: EMG-triggered NMES and unilateral movement Group 3: control	90 min 4 days Completed over 2 weeks	Hemiparesis decreased in the wrist and fingers. Results favoured bilateral movement retraining with EMG-triggered NMES
2004	[66]	Chronic hemiplegia resulting from stroke	Before-after trial, with 2-month follow-up	60 min/session 12 consecutive days	Significant improvements in hand function in all six participants, with possible long-term benefits.
2004	[67]	Subacute and chronic hemiplegia with some wrist and finger extension	Cross-over study FEST vs. non-FEST	30 min/session Daily 3 weeks FES-assisted grasping while using objects functionally	Significant increases in the Upper Extremity Functioning and drawing tests were recorded when FEST was applied earlier

(Continues)

Table 5.2 (*Continued*)

Year	Reference	Population	Study design	Intervention	Summary of findings
2005	[68]	Subacute and chronic severe hemiplegia resulting from stroke	RCT (FEST vs. best-practice physical and occupational therapy)	45 min/session 2–5 days/week 12–16 weeks	FEST resulted in greater improvements in arm and hand function.
2007	[69]	Subacute hemiplegia resulting from stroke	Single-blind, randomly controlled comparison of high- and low-intensity treatment	High intensity: 60 min/session 15–20 consecutive days Low intensity: 15 min/day 4 days + 1 hour FEST on 5th day	Significantly greater improvement in high-intensity group on Wolf Motor Function Test.
2007	[70]	Chronic hemiparesis resulting from stroke	Non-concurrent, multiple-baseline, single-subject design	Home-based 15 min 2–3 times/day 8 weeks Task-specific exercises	Statistically significant changes in Action Research Arm Test (6/10 participants), Stroke Rehabilitation Assessment of Movement (5/10), and =>10% improvement in Spasticity (Modified Ashworth Assessment of Spasticity; 4/10)
2008	[71]	Subacute severe hemiplegia resulting from stroke Chronic severe hemiplegia	RCT (FEST vs. best-practice physical and occupational therapy)	45 min 5 days/week 12–16 weeks	FEST improvements were significantly greater than control in object manipulation, palmar grip torque, pinch grip pulling force, Barthel Index, Upper Extremity Fugl-Meyer scores and Upper Extremity Chedoke-McMaster Stages of Motor Recovery. Participants with chronic stroke improved in most categories; change was not statistically significant.

(*Continues*)

Year	Ref	Condition	Protocol	Outcomes
2013	[72]	Chronic severe hemiplegia resulting from stroke	2 × 60 min/day 12 weeks	Ability to initiate and stop movement was regained in several upper extremity muscles. Reduction in arm spasticity, indicated by a reduction in H-reflex in wrist flexors and decreased Modified Ashworth Scale score (from 3 to 2 for the hand and 4 to 3 for the arm) Improved coordination between shoulder and elbow joints (circle drawing)
2014	[73]	Paediatric chronic severe hemiplegia resulting from stroke	Single-subject, pre-post design 60 min/day 3 days/week 16 weeks Task-oriented activities for reaching and grasping Each task was repeated 10+ times for each motion in each session.	All four participants improved their hand function (REL-HFT) (object manipulation, wooden blocks). All four participants improved on all domains of all secondary measures (Quality if Upper Extremity Skills Test, Pediatric Evaluation of Disability Inventory, Assisting Hand Assessment Test), but not statistically significant due to small sample size.

Case study

NMES = neuromuscular electrical stimulation; EMG = electromyography; FEST = functional electrical stimulation therapy; RCT = randomised controlled trial;
REL–HFT = Rehabilitation Engineering Laboratory Hand Function Test.

Table 5.3   Clinical evidence for FES in spinal cord injury

Year	Reference	Population	Study design	Intervention	Summary of findings
2006	[74]	Subacute and chronic, complete and incomplete SCI, C3-C7	RCT FEST vs. best-practice OT	45 min/day 5 days/week 12 weeks	Differences between groups were observed. Non-statistical significance was attributed to the small number of participants. Participants reported a positive perception of FEST and improved ADL and satisfaction.
2011	[57]	Subacute, incomplete SCI, C4-C7	RCT FEST vs. best-practice OT	60 min/day (BPCOT) 60 min/day (FEST or BPCOT) 5 days/week 8 weeks	The FET + COT group showed a significant improvement in the mean FIM self-care subscore compared to the COT group. Additionally, the FET + COT group demonstrated greater improvements in SCIM and TRI-HFT scores.
2013	[75]	Chronic, incomplete SCI, C4-C7	RCT FEST vs. best-practice PT AND OT	60 min/day 3 days/week 13 weeks	FES therapy group showed greater improvement in FIM-SC, SCIM-UE, and TRI-HFT total score for manipulation of objects.

RCT = Randomised controlled trial; OT = Occupational Therapy; ADL = Activities of Daily Living; COT = Conventional Occupational Therapy; FIM = Functional Independence Measure; SCIM: Spinal Cord Injury Independence Measure; FIM-SC = Self-care subscores of the Functional Independence Measure; SCIM-UE = Upper extremity component of the Spinal Cord Independence Measure; TRI-HFT = Toronto Rehabilitation Institute Hand Function Test.

of our contributions to the field and other representative clinical works. Other reviews can be found elsewhere [76–78].

5.5   Hardware considerations/device design

5.5.1   Components of an FES device

The main parts of a neuroprosthesis are an electrical stimulator, electrodes for stimulation, sensors for control and possibly an orthosis to aid movement [79]. Along with the characteristics of the electrical pulses used, these components determine the movement produced by the stimulation. Readers are directed to Chapter 1 of this book, which provides a detailed description of electrical characteristics and their effect on stimulation of nerves and muscles.

5.5.1.1   Electrical stimulator

The electrical stimulator generates electrical discharges to produce muscle contractions. Complex stimulation pulses are delivered through a pair of electrodes, forming a stimulation channel. Multiple stimulation channels in a stimulator can activate unique muscle groups using distinct settings. A programmable stimulator with multiple channels can facilitate various functional movements by allowing the sequence of each channel's activity to be specified [80]. Coordinating the activation and deactivation of multiple channels makes it possible to produce synergistic contractions required during complex functional movements.

5.5.1.2   Stimulation electrodes

The stimulation is delivered through electrodes, which can be invasive and non-invasive. Invasive electrodes can be fully implanted or percutaneous, with only part of the electrodes being implanted, while the rest can be accessed externally through the skin. Implanted electrodes are suitable for permanent applications (i.e., when used permanently as an assistive device) and when high precision is necessary (e.g., when stimulating small and deep muscles). As with other invasive procedures, implanting these electrodes requires specialised personnel and perioperative care. Non-invasive electrodes are placed on the skin over the area to be stimulated. They can be self-adhesive, or another mechanism can be used for securing them to the body (i.e., adhesive tape). Generally speaking, stimulation is less precise than implanted systems, often making it difficult to stimulate small and deep muscles. However, they are ideal for short-term interventions that require frequent repositioning of the electrodes to, for example, change a facilitated movement, as is often required in response to the changes experienced by a person undergoing therapy for voluntary movement rehabilitation.

5.5.1.3   Pulse shape

Pulses can be monophasic or biphasic. Monophasic pulses are characterised by a single component (phase), while biphasic pulses have two components of opposite polarity (i.e., positive and negative). Bipolar pulses can be symmetric, in which both phases have equal amplitude and duration, or asymmetric, in which the second (negative) phase can be of lower amplitude and longer duration. The use of biphasic pulses gives the opportunity to balance the energy (electrical charges) applied to the body, ensuring safe extended use of FES [81,82]. Most modern stimulators use biphasic pulses. In the subsequent sections, we also discuss some technical limitations of using different stimulating pulse shapes (Section 5.8.1).

5.5.1.4   Stimulation sequence

The stimulation sequence defines the order and duration in which channels become active (or inactive). The sequence determines the movements facilitated by FES, whether they are simple or require a complex orchestration of multiple muscles. For example, a three-step sequence could start delivering stimulation to finger and wrist extensor muscles to open a hand, then stimulating finger flexor muscles to close a hand, and finally using the stimulation to open the hand again. In addition to specifying the order of active channels, the stimulation sequence may also include the mechanisms by which the transition between the stimulation states occurs.

Transition could happen automatically after a predetermined period or in response to, for example, pressing a button.

5.6   Clinical applications

There is a need for long-term FES orthotic solutions for UE function, although a growing trend towards more incomplete injury profiles over the last decades has created a need for therapeutic FES applications for SCI and other neurological conditions. The use of FES as a therapeutic intervention for improving and restoring voluntary UE function has been termed FES therapy and is discussed extensively in subsequent sections. In this context, FES serves as a temporary, short-term intervention aiming to assist the neuromuscular system in relearning how to perform functions affected by neurological injury or disorder. The primary aim of this intervention is to facilitate the recovery of voluntary function.

We have used transcutaneous FES to retrain reaching and grasping in individuals with both SCI and stroke [58,71,75,83]. The results obtained in both patient populations indicate functional clinically relevant improvements after 8–14 weeks of therapy (20–48 h of stimulation) [58,71,75,83]. The utilised FES system comprised a four-channel surface stimulation device with software, a portable stimulator integrated with a programmed chip card, self-adhesive stimulation electrodes and various interfaces, including push buttons, sliding potentiometers [59], accelerometers [84], electromyography/biofeedback sensors, joysticks [85], foot switches [80], gait phase detection systems [86] and BCI technology [87].

Various stimulation parameters can be adjusted to optimise effectiveness and ensure a successful FES session [81,88], while safety should be ensured through use of medically certified devices for on-label applications and by following the device manuals by trained professionals. Stimulation parameters settings include:

1. Amplitude/intensity, which refers to the strength of the electrical current delivered to the muscles. Most FES systems can deliver stimulating amplitudes up to 100 mA and should typically not exceed 150 mA in most applications. Typical values used for UE FES sessions in our clinical trials ranged from 8 mA to 30 mA. Please note that these amplitude/intensity levels are applicable to current regulated stimulators, which presently dominate the field. The new and emerging technology provides new generation of stimulation pluses where steady state amplitudes are a fraction of these values resulting in more comfortable stimulation experience [89].
2. Frequency, which defines the rate at which the electrical pulses are delivered per second. Common frequencies for FES therapy range from 20 Hz to 50 Hz. Typical values used for UE FES sessions are around 30–40 Hz.
3. Pulse width/duration, which represents the duration of each electrical pulse that allows for controlling the duration of muscle contractions. Typical FES systems use 100 to 400 μs pulse width stimulation waveforms. Typical values used for UE FES sessions are around 300–400 μs.
4. Ramp time, which defines the stimulation increase or decrease time to maximum or minimum intensity to improve patient comfort by preventing abrupt muscle contractions.

Optimising stimulation parameters requires expertise/experience and finding the ideal settings for each individual can be time-consuming.

The FES technology and clinical studies in stroke and spinal cord injury conducted by our lab are summarised elsewhere [81,90], while the focus of the subsequent sections is to provide practical considerations and general guidelines for implementing FES therapy.

5.6.1 Considerations when selecting patients for FES UE therapy

Limited evidence exists regarding the precise criteria for identifying patients who could benefit from FES therapy. The following factors serve as a general guide and are not meant to be strict criteria. Responses to FES therapy can differ significantly among individuals. Evaluating the appropriateness and efficacy of FES should consistently occur on a case-by-case basis and be conducted by healthcare practitioners experienced in using FES for rehabilitation.

5.6.1.1 Why use FES?

FES can be used to improve muscle strength, prevent muscle atrophy, reduce spasticity or muscle tone, and notably improve voluntary motor function through intensive training. The most significant portion of the literature concerning FES therapy focuses on UE function after stroke, traumatic spinal cord injury and traumatic brain injury [90]. Regarding lower motor neuron lesions such as cervical radiculopathies and other peripheral neuropathies, recent evidence suggests a potential positive impact of FES therapy [91,92], although larger systematic randomised trials are necessary to establish more robust results.

Conducting a comprehensive evaluation involving physical or occupational therapists, rehabilitation specialists and other relevant healthcare professionals helps in determining the potential benefits of FES therapy for an individual. A thorough evaluation to identify the functional movements to be trained is a critical component of designing the therapeutic regimen. Typically, an assessment is performed by a licensed therapist to assess and understand patient's clinical presentation and goals, and its implication on devising a therapy plan. This clinical assessment may include assessments of a patient's motor and sensory deficits, coordination deficits, functional deficits and more. The assessment will be tailored to the neurological condition. It is not the intent nor within the scope of this chapter to provide detailed assessments for individual conditions. However, the authors recommend that beyond the traditional clinical assessment, clinicians should include a standardised outcome assessment tool specific to the disease condition and one that is sensitive to change following FES therapy in a specific neurological condition [93–95]. The premise of recovery of function following FES therapy is based on the patient combining their voluntary effort, which FES superimposes. Hence, patients must be able to follow the therapist's instructions during therapy sessions.

5.6.1.2   When to use FES
Evidence from therapeutic FES clinical trials in acute stroke, traumatic spinal cord injury and traumatic brain injury populations indicates that early application, potentially as soon as the patient is medically stable, tends to produce better results [96]. However, ongoing research into neuroplasticity and its mechanisms indicates that individuals beyond the subacute phase post-injury might still experience improvements, albeit at a slower pace, even into the chronic stages [96] and sometimes up to 30 years post stroke [58].

5.6.2   *Practical considerations for therapists prior to applying FES therapy*
In our prior UE clinical trials, we applied self-adhesive surface stimulation electrodes for FES sessions lasting 45–60 minutes. These sessions were administered 3–5 times per week over 8–16 weeks, for an approximate total of 40 sessions [81,90]. Our findings revealed that patients typically benefit the best from a maximum of one 60-minute session of activity-based therapy with FES per day. If more than one session was delivered to a patient per day they would experience fatigue and tone in the second session of the day. During one 60-minute session, we achieved approximately 10–15 repetitions of a specific movement pattern before the onset of fatigue. It is crucial to note that individual tolerance for this frequency can vary due to factors such as injury severity, duration and the state of the neuromuscular system as well as the stage of the therapy. People in later stages of the therapy are able to generate more movement repetitions in a shorter time period.

FES therapy should be supervised by licensed physiotherapists or occupational therapists. During FES therapy, the initiation of each phase of FES is controlled by the treating therapist through a push button. Active participation during functional task performance is essential and requires focus and active effort during task execution on the part of the patient while being assisted by the therapist triggered FES.

By adhering to the following practical aspects, therapists can enhance the safety, effectiveness and overall outcomes of FES therapy for their patients.

5.6.2.1 Assessment and evaluation

Conduct a comprehensive assessment of the patient's condition, including medical history, current physical status, muscle strength, sensation and functional limitations. This evaluation will help determine the suitability of FES and design an appropriate treatment plan.

5.6.2.2 Informed consent

Ensure the patient understands the FES procedure, its goals, potential benefits and possible risks or discomforts, and allow time for questions and answers to respond to any concerns. Informed consent must be obtained before initiating therapy.

5.6.2.3 Skin integrity

Ensure the skin where electrodes will be placed is intact and free from lesions, wounds or irritation. Assess skin sensitivity to avoid potential skin breakdown due to prolonged stimulation.

5.6.2.4 Electrode placement

Precisely position electrodes based on the target muscles or nerves prior to initiating the therapy. Correct electrode placement is crucial for optimal stimulation and to avoid discomfort or adverse effects. Validated protocols should be followed to minimise potential risks and to ensure safety.

5.6.2.5 Stimulation parameters

Set appropriate stimulation parameters (intensity, frequency, duration) based on the individual's tolerance and response. Gradually increase the intensity to a comfortable level within the patient's tolerance threshold. Manufacturer protocols and guidelines should be followed and not exceeded to minimise potential risks and to ensure safety.

5.6.2.6 Monitoring and adjustment

Continuously monitor the patient's response to FES during sessions. If needed, adjust stimulation settings to optimise comfort and effectiveness while avoiding muscle fatigue or overstimulation.

5.6.2.7 Muscle warm-up

Encourage patients to relax and, if appropriate, perform warm-up exercises before starting FES to prepare muscles for stimulation and minimise discomfort or spasms.

5.6.2.8   Other safety precautions

Ensure that safety measures, such as emergency stop mechanisms, are in place during FES sessions and respond promptly to any adverse reactions or discomfort reported by the patient.

5.6.2.9   Documentation

Maintain records of FES sessions, including stimulation parameters used, patient responses, progress and any observed changes in muscle function or sensation.

5.6.2.10   Therapist training and experience

Therapists should undergo proper training and have expertise and experience using FES devices and techniques. Stay updated with the latest research and guidelines related to FES therapy.

5.6.3   Conducting an upper extremity FES training session

The following general steps should be taken into consideration to conduct a UE FES training session with a surface multi-channel FES system [97]:

- Identify the functions to be trained (reaching or grasping for UE applications).
- Select the order of the tasks to be re-trained: Typically, start with gross motor tasks (proximal muscles) in the early stages of therapy, followed by fine motor tasks (distal muscles) as the therapy progresses.
- For each task, identify the muscles to be stimulated: based on the number of channels available for a stimulator, either simple reaching or grasping tasks such as touching the mouth or palmar grasp, or more complex tasks such as reaching and grasping can be trained at a given time (see Section 5.8.1).
- Identify optimal electrode positioning: For a given function, find the motor point, the electrode position where a maximal contraction is obtained with minimum stimulation current delivered. Finding the motor point using a smaller electrode can be done by trying several positions on the bulk of the muscles. This allows for finding an electrode position with minimal secondary and unintentional stimulation of other muscles and nerves. Once you find the optimal electrode position(s) for a muscle, mark it with a pen or marker and identify the position(s) for other muscles [98].
- Apply self-adhesive electrodes over the motor points of the muscles identified and the ground (return) electrode. It should be noted that the type of FES device used will have implications on where the ground electrode is placed for optimal activation (see Section 5.8.1).
- Identify and document different stimulation thresholds: Identify sensory threshold (the intensity in which the patient first feels the stimulation), motor threshold (when a palpable or a visible contraction is produced), functional threshold (when the desired functional movement is produced) and maximum threshold (beyond which the patient does not tolerate an increase in current amplitude). Define the thresholds with the same current characteristics (pulse width and frequency) as those used during FES therapy because it impacts the stimulation's comfort and efficiency.

- Turn on the stimulator and adjust the current intensities for all muscles to the previously determined levels. Intensity should not exceed the maximum threshold.
- Explain to the patient what to expect both in terms of sensory stimulation and motor response when the FES is turned on. For example: 'First you may feel a tingling or a pins and needles kind of sensation and as the current intensity will increase your hand will close, and then it will open'.
- Trigger the FES protocol several times so the patient understands what to expect with each phase of FES. Once the patient has experienced the protocol, if appropriate, select the functional object to be used during training. If needed, assist the patient in bringing their hand close to/around the object to be manipulated.
- Instruct the patient to make an active attempt to perform the intended movement. For example, for a grasp/release task, ask the patient to close the hand to grasp the object and, after the patient has attempted for about 5–10 s, assist with FES. Once the patient can grasp the object with assistance from FES, and completes the intended ask, instruct the patient to release the object and after about 5–10 s of attempting the task, trigger the FES sequence for hand opening.
- Repeat the protocol 10–15 times. Then, select another protocol and perform the next task for 5–7 min or as appropriate for that task. Execute 3–6 different protocols in a 60-min session, with active stimulation for 30-40 min (depending on the patient's fatigue level and therapist's expertise with the system). The one-hour therapy session includes positioning of the electrodes and all preparations for therapy initiation and therapy completion.
- Rest time should be given when the patient requests it, when the therapist evaluates that the patient is not actively engaging, or when muscle fatigue sets in.
- When the therapy session is completed, turn off the stimulator, remove the electrodes and inspect the skin underneath for any redness or irritation. Redness may occasionally be present from the electrode sticking to the skin, but it should dissipate within 24 hours. Ask the patient to monitor the area and inspect at the next session. If persistent, do not continue the therapy until it is deemed safe to do so.

Use of home-based FES systems has the potential to increase therapy volume especially in cases where compliance to hospital or clinic-based therapies is limited by virtue of lack of sufficient care giver support to bring patients to a rehabilitation centre or where distance and travel time are prohibitive. There is limited systematic research to date to evaluate the effectiveness of home- based FES therapy alone. In our experience, the expertise of the physio/occupational therapist is critical for ongoing evaluations and progression of the FES therapy, however in light of the limitations to inpatient sessions we advocate for comparative trials which look at efficacy of home-based FES programs vs. in clinic sessions. Especially, sessions that are conducted in a telerehabilitation setting where the therapists have the ability to remotely monitor sessions may result in better therapeutic efficacy and higher compliance ratios. To date to the best of our knowledge other than the NESS H200 and MyndMove there are no other sophisticated FES devices with regulatory approvals for home use. Whereas the NESS H200 has advantages there are several limitations of this device published in literature including: (1) the system requires sufficient arm function, especially shoulder and elbow function to stabilise and position the arm in the working space (2) the system lacks sufficient flexibility to vary the position of the electrodes for stimulation of the finger flexors for grasp and the fact that it is a stiff orthosis that fixes the wrist joint angle and prevents full supination of the forearm [99]. As for MyndMove, it can be delivered at home, but it requires a physiotherapist to deliver the treatment. Patient cannot use the stimulator independently (FDA approval – https://finance.yahoo.com/news/myndtec-receives-fda-510-k-120000494.html).

5.6.4   FES UE protocols

Over the years the FES-reaching protocols have expanded to cover various functional reach and grasp patterns some of which we discussed in Table 5.1. Some example FES protocols include:

1. Sideways reaching
2. Sideways reaching with hand opening
3. Forward reaching and retrieving
4. Forward reaching and retrieving with hand opening
5. Reaching over opposite shoulder
6. Reaching over opposite shoulder to forward reaching to sideways reaching
7. Reaching over opposite knee
8. Hand to mouth
9. Palmar Grasp (holding a ball)
10. Lateral Grasp (holding a tray)
11. Tripod grip (thumb, index and middle finger: holding a pen)
12. Two finger opposition (thumb and index finger: holding a peg)
13. Lateral Pinch (thumb and index finger: holding a credit card)
14. Two finger lateral pinch (index and middle finger: smoker's grip)
15. Lumbrical grip (all four fingers with the thumb: holding a closed book).

Figure 5.1 shows the electrode placement and stimulation sequence for producing a grasp and release movement and Figure 5.2 shows the electrode placement and stimulation sequence for producing a forward reaching movement.

5.6.5   Therapy progression

FES therapy progression is dynamic and personalised, focusing on optimising functional outcomes while adapting to the individual's responses and goals throughout the treatment. Training of reaching functions could progressively include distal function or include unilateral or bilateral tasks, as needed. Typically, once the patient shows the ability to use a particular muscle against gravity (Grade

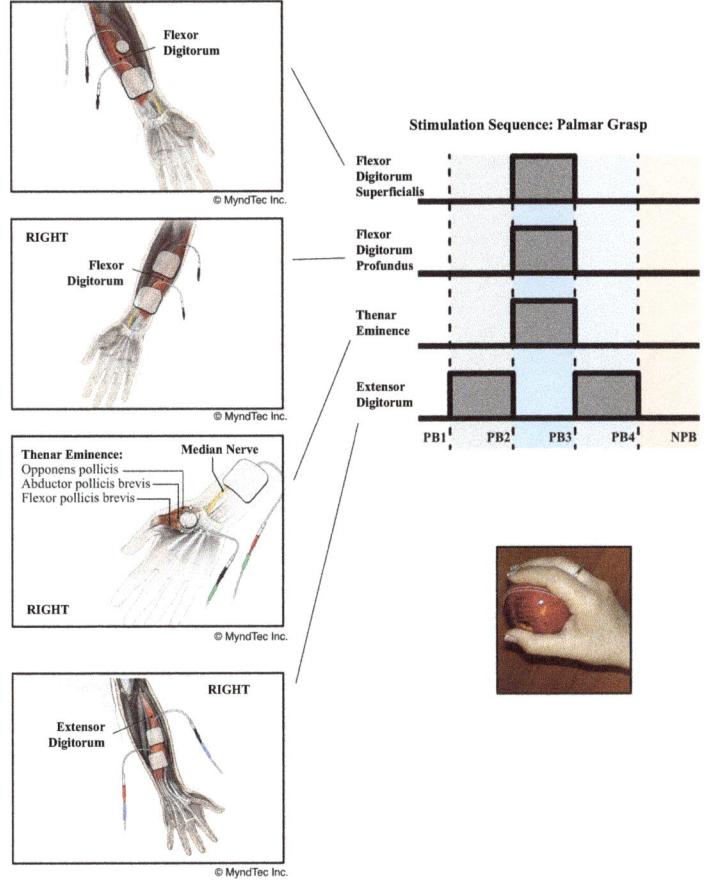

Figure 5.1   *Electrode location and stimulation sequence for facilitating palmar grasp and hand opening. Electrode placement (left portion): The exact positioning of electrodes is determined by exploring different locations on the patient and observing the resulting action. The top two panels illustrate the use of electrodes of different sizes; smaller electrodes can be used to accommodate smaller anatomy. Stimulation sequence (right portion): The diagram shows four stimulation channels, each one displayed on a horizontal axis. The activation of a channel is illustrated by a grey block. An initial activation of a push button (PB1) activates the channel that delivers stimulation to the extensor digitorum, which produces hand opening. The second activation of the button (PB2) interrupts the stimulation to the extensor digitorum and triggers stimulation of the flexor digitorum and thenar eminence muscles, resulting in palmar grasp. The next button activation (PB3) stops the stimulation to the flexor digitorum and thenar eminence and restarts stimulation of the extensor digitorum to produce hand opening again. The final button activation (PB4) ceases stimulation of the extensor digitorum. This stimulation sequence could allow, for example, to practice the manipulation of different objects requiring palmar grasp (a ball is shown as an example). Electrode placement illustrations used with permission from MyndTec Inc.*

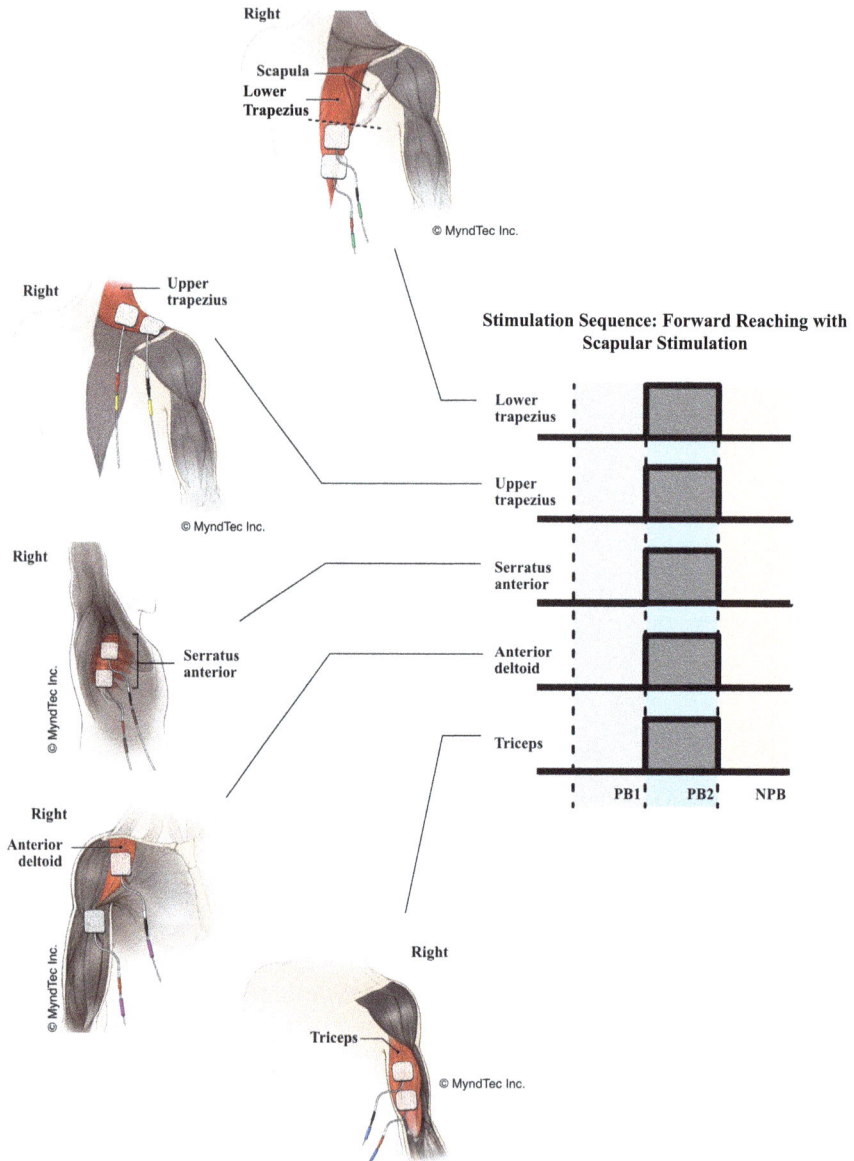

*Figure 5.2   Forward reaching with scapular stimulation. All channels become
active after the first activation of a push button (PB1). The combined
stimulation results in a forward reaching movement of the proximal
upper extremity, i.e., flexion of the shoulder and extension of the
elbow, respectively, along with scapular muscles, which are
synergistic stabilisers of the deltoid. A second push-button activation
(PB2) stops stimulation in all channels. Electrode placement
illustrations used with permission from MyndTec Inc.*

3 on manual muscle test) FES can be withdrawn from that muscle. Therapy progression is an important component of FES therapy and should be carefully carried out and planned by the therapist to challenge the patient's functional capacity. Any gains the patient experiences with FES should be incorporated in day-to-day activities as best as the patient can. This is an important and necessary step to retain newly acquired movement/function. So, for example, a patient training for palmar grasp, should, if possible, attempt to grasp larger objects like a bottle at home during day-to-day life. This training can be graded by initially trying to lift an empty bottle and then filling it up gradually as the function improves.

5.6.6 Post-treatment assessment

While it may not be necessary to conduct a comprehensive post-treatment functional assessment after every FES session, vigilant monitoring of changes is recommended. This monitoring serves not only to guide therapy progression but also to help the therapist determine whether FES therapy is suitable for the specific patient. A detailed assessment, encompassing motor, sensory and functional evaluations, is advised after approximately 20 FES sessions (delivered 3–5 times per week), during which clinically significant functional changes are anticipated. The number of sessions suggested here are based on our research experience with FES for UE training and maybe based on individual patient presentation. As an example, a patient with subacute high level (C4-C5) traumatic incomplete cervical spinal cord injury may present with flaccid paralysis of the upper arm muscles with strength of muscles around the shoulder and elbow no more than grade 1–2 on manual muscle testing and may need to undergo more than 20 sessions of FES therapy before functional gains in reaching might be realised.

Also, it is worth noting that there is a lack of systematic literature around frequency and dosing for FES for UE function retraining. In a clinical setting these parameters would be driven by practical factors beyond disease condition and level of impairment. The therapist doing the intervention might be best suited to identify the optimal timing for assessment and consecutively increasing or decreasing the duration between assessments based on patient impairment at baseline, patient progress and the goals to be achieved. This approach allows for flexibility as deemed necessary for optimal patient care and therapy advancement.

5.6.7 FES as an orthotic functional device

The orthotic applications of FES could aid patients in performing essential functions and require that the FES system be used whenever these functions are needed, serving as a permanent orthotic solution. Implanted FES systems primarily serve as permanent neuroprostheses, while surface FES systems function both as neuroprostheses and therapeutic platforms. Orthotic FES applications could include applications to assist, support and improve neuromuscular function. Primarily, FES application can assist movement by stimulating muscles to produce specific actions. For example, FES applied to UE muscles can assist with generating functional grasping supporting self-feeding or grooming tasks. FES can also assist in improving joint stability by activating muscles around a specific joint to aid in maintaining proper alignment, preventing contractures and to enhance the overall joint function. Orthotic FES applications can assist in improving muscle strength, improving coordination and conditioning of weakened or paralysed muscles. Similarly, FES could help prevent muscle atrophy by providing regular electrical stimulation to inactive muscles, which can slow down muscle wasting and maintain muscle size and strength. A list of some approved UE orthotic devices and their functions is summarised in our previous reports [90].

5.7   Precautions and contraindications to therapeutic FES application

Some important limitations to FES technology include: (1) The muscles intended for FES treatment must be accessible for placing the stimulation electrodes [100]. (2) There should not be a major degree of lower motor neuron injury or nerve-root damage of the stimulated muscle. In a number of patients with spinal cord injury, there may be a variable amount of peripheral nerve damage [101] (motoneurons and nerve-roots) that restricts the application of FES. (3) The patient should not have any contraindications for FES application like metal implants at the site of stimulation, a pacemaker close to the site of stimulation (precaution), an open wound or rash at the site of electrode placement, uncontrolled or poorly managed autonomic dysreflexia and other conditions that may be affected by the stimulation. Although spasticity and tone should be considered as it may limit the ability of individuals to perform some activity-based training tasks, FES could also potentially reduce tone and spasticity.

5.8   Shortcomings of therapeutic FES

5.8.1   Technical limitations

The availability of FES systems and their technical capacity can limit the application of FES therapy. Addressing and understanding these technical limitations is crucial in optimising FES therapy, enhancing its efficacy and managing expectations during rehabilitation programs. Pertaining to FES technology, it is important to consider:

5.8.1.1   Number of channels available

A four-channel stimulation can support simple reaching or grasping tasks such as touching the mouth or palmar grasp. To train more complex reaching and grasping tasks multiple muscles need to be stimulated and this may require sometimes up to eight channels. Two-channel stimulation, which is most frequently available in clinical settings, may also be appropriate for delivering very simple (e.g., hand opening/closing) tasks.

5.8.1.2 Galvanically isolated stimulation channels to enable a 'common ground'

If a stimulator has galvanically isolated stimulation channels, it is possible to apply all electrodes on one aspect of the arm to be 'grounded' using a single return/anode electrode. For example, all muscles on the palmar aspect of the forearm can be grounded using one electrode just proximal to the ventral aspect of the wrist joint and similarly, all electrodes on the dorsal aspect of the forearm can be grounded using one electrode over the dorsal aspect of the wrist. If the stimulator does not have galvanically isolated stimulation channels, one should not use this strategy.

5.8.1.3 External triggering

During FES therapy, therapists use external triggering to manually control or initiate FES stimulation at specific moments to complement therapeutic exercises or movements. External triggering capabilities enable user-initiated triggers, such as a button press from a handheld controller, allowing therapists to control the timing of muscle stimulation as needed during FES therapy. Although not necessary, lack of external triggering may pose a challenge in executing an optimal therapy session.

5.8.1.4 Stimulation pulses

The availability of different stimulating pulses may limit and dictate how the electrodes are positioned. Specifically, the use of asymmetrical pulse waveforms (i.e., with the negative depolarising pulse on the same electrode and the positive balancing pulse at a lower amplitude) requires the 'active' electrode to be positioned on the motor point, and a 'passive' or return/anode electrode under which there is no effective stimulation could be placed distal to the motor point. This setting could enable smaller muscles to be activated more effectively and ensure greater specificity of the muscle activation. Alternatively, the use of symmetrical pulse waveform results in both electrodes acting as 'active' and depolarising continuously, which could be appropriate for larger muscles.

5.8.1.5 Equipment size

Equipment size impacts device portability, which may limit the practicality for its continuous use outside conventional therapy settings.

5.8.2 Accessibility and knowledge translation challenges

Beyond the technical challenges, often in clinical practice there are logistical challenges related to budgeting, therapist training, access to a device, space considerations and others. A recent study that conducted focus groups to understand the challenges in the adoption of electrical stimulation in rehabilitation practice identified that there is a need for a clinician-friendly practice guide to support the use of electrical stimulation as a rehabilitation modality, and that a lack of such a guide resulted in inconsistent use of electrical stimulation across practitioners and settings [102]. Further, the therapists in the focus group identified specific information that they would like included in such a guide, including (1) an overview of electrical stimulation; (2) specific intervention guidelines grouped by sensory, muscle strengthening and functional activity applications; and (3) guidelines for patient indicators, parameter settings, electrode placement, conditions and rationale for parameter settings, troubleshooting suggestions, functional outcomes measures [102]. A few studies have explored the experiences and perceptions of rehabilitation specialists' use of FES in their practice and found that therapists quote various reasons, including support from colleagues, access to theoretical and practical education, access to equipment, objective results and family buy-in as facilitators for FES implementation [103,104], and access and cost of FES treatment, family apprehension, time-consuming treatment, time to learn, equipment failures and lack of time with clients as barriers to adoption of FES in clinical practice [103,104].

We have tried to address some of these barriers in this chapter. Similarly, rehabilitation researchers worldwide are focused on dealing with these issues, with some countries now developing and/or updating their clinical practice guidelines specifically for use of FES [105].

Other challenges besides technical and accessibility issues include identifying responders a priori, preventing fallout of patients and ensuring home practice. Whereas identifying responders might still be somewhat challenging, using clinical judgement in combination with sophisticated diagnostic tools [106,107] and clinical prediction rules [108] might be helpful. In order to prevent fall out, motivating the patient to engage in therapy is important. Patients who are actively engaged and motivated in their rehabilitation process tend to benefit more from FES therapy. Having specific functional goals such as improving self-care may not only improve participation but also maximise functional gains. Another important consideration is building realistic expectations. The patient should understand their condition well and its implications on movement/functional restrictions. It is also essential that the patient clearly understands the scope of FES therapy. Lastly, in order to get patients engaged in home practice, it is imperative that the caregivers are educated on its importance and trained in how they can encourage the patients to use the functional improvements in day-to-day life.

5.8.3   Limitations specific to orthotic FES applications

UE FES orthoses have received some acceptance in the clinical setting and amongst patients; however, there remain limitations to this approach that further need to be tackled to make this technology mainstream, especially in cases where functional recovery might be limited by virtue of the disease process. Some of these challenges are similar to the ones explained under therapeutic FES. Over and above those, the complexity and reliability of the mechanical components and the stimulation systems, the need for assistance with donning and doffing, and limited resources for troubleshooting make their wide-scale acceptance prohibitive. Future directions need to address wearability through improvements in component technologies and user interfaces. In case of FES orthotics combined with exoskeletons, increased integration of the control action between FES and robotic subsystems to reanimate the limb should be pursued [109]. Standardised reporting of system performance and expanded clinical assessments of these systems are also needed. All these advancements are critical to facilitate translation from laboratory to home.

5.9 Combination therapies

During FES therapy, participants are asked to actively attempt each movement before the therapist triggers muscle activations using FES to assist completion of the intended tasks [81,88,90]. The success of the therapist-assisted FES approach emphasises that associative activation of cortical and peripheral circuits can elicit Hebbian plasticity [110], resulting from the simultaneous activation at the pre- and postsynaptic circuits. Subsequently, associative stimulation techniques that combine cortical and peripheral activations were developed by combining non-invasive brain stimulation, voluntary muscle contractions, motor imagery, BCI control of FES and others. Most of these approaches are investigational and remain focused on understanding neuroplasticity mechanisms. Nerve or tendon transfer surgery is the most clinically promising approach that could benefit from FES before and/or after the surgery [111] and has the potential to improve functional outcomes. In a review by Friden *et al.*, the authors have discussed the influence of FES on muscle architecture, power output, cross-sectional area and muscle fibre type adaptation and how that may positively impact preparation for surgery [112,113]. After surgery, FES can be applied to retrain the transferred muscle to perform a new function that may induce neuroplastic changes supporting the motor learning. In nerve transfers, FES testing of potential donor and recipient nerves is a simple and reliable method to determine if upper and/or lower motor neuron lesions are present.

5.9.1 Non-invasive brain stimulation combined with FES

Non-invasive brain stimulation can activate cortical networks. The associative stimulation techniques emphasise the importance of brain state activation for obtaining effects when using FES. Non-invasive brain stimulation techniques are summarised in comprehensive reviews elsewhere (e.g., [114]), and the objective of the following section is to provide a preview of why some of these techniques could be combined with FES in therapeutic applications:

1. Paired associative stimulation combines single-pulse transcranial magnetic stimulation (TMS) with single-pulse electrical stimulation of the periphery and involves multiple cortical and peripheral stimuli delivered at specific inter-stimulus intervals, can induce long-term potentiation and long-term depression plasticity following Hebbian principles of associative activation [115]. The mechanisms underlying paired associative stimulation align with spike timing-dependent plasticity (STDP), where the temporal sequence of pre- and post-synaptic terminal activity determines synaptic efficacy and polarity [116,117].
2. Repetitive TMS (rTMS) intermittent theta burst stimulation over the primary motor cortex can drive corticospinal excitability [118], and it was shown to

facilitate corticospinal excitability if applied before, but not after FES [119]. The cortical facilitation priming protocol delivered by intermittent theta burst stimulation was effective in facilitating corticospinal excitability when followed by FES, but not the inhibitory protocol applied through continuous theta burst stimulation [120]. Notably, frequency synchronisation through intermittent theta burst stimulation and FES was also more effective when intermittent theta burst stimulation and FES were applied simultaneously, while nonsynchronised activation was less effective in modulating corticospinal excitability [121].

3. Transcranial direct current stimulation (tDCS) could also be effective for facilitating the cortical networks and when was applied over the contralateral motor cortex during FES, corticospinal excitability facilitation was larger compared to when tDCS was applied alone [122]. It is important to consider that current evidence around tDCS is not consistent with recommendations of Level A (definite efficacy) for any indication [123].

4. Transcranial focused ultrasound stimulation is a novel non-invasive brain stimulation technique with superior spatial resolution and an ability to target superficial and deep structures of the brain. Studies have shown that transcranial focused ultrasound generated plasticity after sonication lasted from minutes to hours, possibly even days, and likely reflects structural and functional neuroplastic changes [124]. Transcranial focused ultrasound delivered in a theta burst pattern increased M1 excitability, potentiated intracortical facilitation and attenuated intracortical inhibition for up to 60 min in healthy adults. A recent study using theta burst transcranial focused ultrasound immediately followed by FES showed intracortical inhibition secondary to homeostatic metaplastic effects [125]. The effects of this combination for motor rehabilitation in neurological populations needs to be explored.

5.9.2 Voluntary muscle contractions

While electrical stimulation alone offers a means of artificially producing muscle contractions without the cortical drive, its effectiveness may be enhanced through voluntary contractions to a greater extent compared to FES alone [126]. For instance, it was shown that delivery of electrical stimulation at the onset of muscle electromyographic activity during wrist extension was successful in facilitating corticospinal motor evoked potential responses, while electrical stimulation alone was not [126]. Using functional magnetic resonance imaging, the magnitude of cortical activation changes relative to rest was shown to be larger during voluntary contractions of upper-limb muscles compared to FES-induced movements in the motor and sensory cortical areas and combining voluntary and FES-induced contractions produced even larger activations compared to FES alone [127].

5.9.3 Motor imagery and FES

Motor imagery can also offer a means to activate the cortical circuits during FES. It was shown that combined delivery of FES and motor imagery resulted in stronger cortical desynchronisation compared to FES alone or motor imagery applied prior to delivery of FES [128]. When motor imagery was delivered by means of watching and imagining actions shown on a pre-recorded video of grasping, while not producing the movements, it was shown that concurrent electrical stimulation facilitated corticospinal excitability and that either motor imagery or FES alone did not elicit any effects [129]. Preliminary results with chronic stroke patients also suggest that applying electrical stimulation in combination with motor imagery over the course of ten days may improve upper-limb function [130].

5.9.4   Brain–computer interfaces for controlling FES

Motor imagery and motor attempts can give rise to brain activity, which can be detected using non-invasive electroencephalography to trigger FES. Studies focused on investigating the underlying mechanisms of BCI-FES have provided insights into the effectiveness of BCI-FES compared to FES alone by demonstrating that associative activation can be more effective in modulating corticospinal excitability compared to FES [131] and that activation of the motor cortex is essential for the effectiveness [132]. Of combinational therapies, BCI-FES has been most successfully utilised in therapeutic applications. Most clinical BCI-FES applications in rehabilitation are in patients with stroke, with many early investigations comprising case studies and series [133–136]. These early trials offer crucial insights into the feasibility of clinical adoption and hypotheses concerning recovery mechanisms. Recently, several clinical trials showed more significant upper limb functional improvements after BCI-FES compared to FES training alone [137,138]. The most comprehensive clinical trial in stroke patients also compared BCI-FES and FES alone interventions for upper limb rehabilitation and demonstrated the superiority of the BCI-FES approach [139]. Together, the stroke trials present evidence that functional motor improvements are associated with enhanced cortical activations in the affected hemisphere, and all point to BCI-FES being more effective in producing functional and cortical changes than FES delivery alone. Fewer clinical trials have tested the efficacy of using UE BCI-FES in spinal cord injury patients. Most early applications were focused on orthotic/neuroprosthetic applications [140,141]. A more recent study that applied BCI-FES as a rehabilitation intervention after incomplete spinal cord injury demonstrated feasibility in eliciting cortical neuroplastic changes [142] and improving motor function after a therapeutic intervention [134]. A scoping review of recent studies combining BCI and FES technologies for rehabilitation of voluntary movement can be found in [143].

5.10   Take home message

Although FES in both therapeutic and orthotic applications has shown tremendous progress in both technological and clinical application realms over the past 50–60 years, there still remains a considerable amount of work that needs to be done on several fronts from technological improvements to knowledge translation. From a technical standpoint, developing FES devices that allow users to configure the stimulation parameters and externally trigger the device is one of the challenges that remains. Further technological development may also explore a closer integration of different neurotechnologies that play a neuromodulatory role. Future development of multifaceted closed-loop systems that integrate and synchronise efferent and afferent activity at different temporal and spatial scales have an enormous potential to advance rehabilitation for both restorative and orthotic applications.

From a knowledge translation perspective, resources need to be created around continuing education including preliminary training sessions as well as brainstorming sessions for application of FES in complex clinical situations. This material needs to be clinically-oriented, whereas in this chapter we have tried to address the need for a systematic step-wise approach to FES application, more needs to be done to keep therapists and clinicians engaged and up to date with this technology.

On a positive note, FES has now been included as an adjunct therapy in clinical practice guidelines across many countries for the rehabilitation of patients with stroke and spinal cord injury. The FES stimulators that are marketed tend to be more user-friendly, which makes training of therapists and adoption of the technology easier. Clinical research related to FES application is now more robust, with more randomised clinical trials showing a positive effect of FES. Hence, based on current evidence, we advocate the early use of FES in combination with other rehabilitation tools to maximise recovery of function in individuals with neurological disorders. While orthotic applications remain important for restoring UE movements, we also demonstrated the utility of FES therapy lasting 45–60 minutes, administered 3–5 times per week over 8–16 weeks, for an approximate total of 40 sessions for UE neuromotor recovery [81,90].

References

[1]   Sveen U, Bautz-Holter E, Sødring KM, Wyller TB, and Laake K. Association between impairments, self-care ability and social activities 1 year after stroke. *Disability and Rehabilitation*. 1999;21(8):372–7.
[2]   Lai SM, Studenski S, Duncan PW, and Perera S. Persisting consequences of stroke measured by the Stroke Impact Scale. *Stroke*. 2002;33(7):1840–4.
[3]   Kwakkel G, Kollen BJ, van der Grond J, and Prevo AJ. Probability of regaining dexterity in the flaccid upper limb: impact of severity of paresis and time since onset in acute stroke. *Stroke*. 2003;34(9):2181–6.
[4]   Morris JH, van Wijck F, Joice S, and Donaghy M. Predicting health related quality of life 6 months after stroke: the role of anxiety and upper limb dysfunction. *Disability and Rehabilitation*. 2013;35(4):291–9.
[5]   Franceschini M, La Porta F, Agosti M, and Massucci M. Is health-related-quality of life of stroke patients influenced by neurological impairments at one year after stroke? *European Journal of Physical and Rehabilitation Medicine*. 2010;46(3):389–99.

[6]   Wyller TB, Sveen U, Sødring KM, Pettersen AM, and Bautz-Holter E. Subjective well-being one year after stroke. *Clinical Rehabilitation*. 1997;11 (2):139–45.

[7]   Anderson KD. Targeting recovery: priorities of the spinal cord-injured population. *Journal of Neurotrauma*. 2004;21(10):1371–83.

[8]   Lo C, Tran Y, Anderson K, Craig A, and Middleton J. Functional priorities in persons with spinal cord injury: using discrete choice experiments to determine preferences. *Journal of Neurotrauma*. 2016;33(21):1958–68.

[9]   Snoek GJ, IJzerman MJ, Hermens HJ, Maxwell D, and Biering-Sorensen F. Survey of the needs of patients with spinal cord injury: impact and priority for improvement in hand function in tetraplegics. *Spinal Cord*. 2004;42 (9):526–32.

[10] Schrag A, Jahanshahi M, and Quinn N. How does Parkinson's disease affect quality of life? A comparison with quality of life in the general population. *Movement Disorders: Official Journal of the Movement Disorder Society*. 2000;15(6):1112–8.

[11] Kalsi-Ryan S, Kapadia N, Gagnon DH, *et al*. Development of reaching, grasping & manipulation indicators to advance the quality of spinal cord injury rehabilitation: SCI-high project. *The Journal of Spinal Cord Medicine*. 2021;44(Suppl 1):S134–S46.

[12] Errante A, Ziccarelli S, Mingolla G, and Fogassi L. Grasping and manipulation: neural bases and anatomical circuitry in humans. *Neuroscience*. 2021;458:203–12.

[13] Cappadocia DC, Monaco S, Chen Y, Blohm G, and Crawford JD. Temporal evolution of target representation, movement direction planning, and reach execution in occipital-parietal-frontal cortex: an fMRI study. *Cerebral Cortex (New York, NY: 1991)*. 2017;27(11):5242–60.

[14] Gallivan JP, Cavina-Pratesi C, and Culham JC. Is that within reach? fMRI reveals that the human superior parieto-occipital cortex encodes objects reachable by the hand. *The Journal of Neuroscience: The Official Journal of the Society for Neuroscience*. 2009;29(14):4381–91.

[15] Cavina-Pratesi C, Monaco S, Fattori P, *et al*. Functional magnetic resonance imaging reveals the neural substrates of arm transport and grip formation in reach-to-grasp actions in humans. *The Journal of Neuroscience: The Official Journal of the Society for Neuroscience*. 2010;30(31):10306–23.

[16] Rizzolatti G, and Matelli M. Two different streams form the dorsal visual system: anatomy and functions. *Experimental Brain Research*. 2003;153 (2):146–57.

[17] Culham JC, Danckert SL, DeSouza JF, Gati JS, Menon RS, and Goodale MA. Visually guided grasping produces fMRI activation in dorsal but not ventral stream brain areas. *Experimental Brain Research*. 2003;153(2):180–9.

[18] Frey SH, Vinton D, Norlund R, and Grafton ST. Cortical topography of human anterior intraparietal cortex active during visually guided grasping. *Brain Research Cognitive Brain Research*. 2005;23(2–3):397–405.

[19] Grol MJ, Majdandzić J, Stephan KE, *et al*. Parieto-frontal connectivity during visually guided grasping. *The Journal of Neuroscience: The Official Journal of the Society for Neuroscience*. 2007;27(44):11877–87.

[20] Vaillancourt DE, Yu H, Mayka MA, and Corcos DM. Role of the basal ganglia and frontal cortex in selecting and producing internally guided force pulses. *NeuroImage*. 2007;36(3):793–803.

[21] Dimitrova A, de Greiff A, Schoch B, *et al*. Activation of cerebellar nuclei comparing finger, foot and tongue movements as revealed by fMRI. *Brain Research Bulletin*. 2006;71(1–3):233–41.

[22] Weeks AC, Ivanco TL, Leboutillier JC, Racine RJ, and Petit TL. Sequential changes in the synaptic structural profile following long-term potentiation in the rat dentate gyrus: I. The intermediate maintenance phase. *Synapse (New York, NY)*. 1999;31(2):97–107.

[23] Feigin VL, Brainin M, Norrving B, *et al*. World stroke organization (WSO): global stroke fact sheet 2022. *International Journal of Stroke: Official Journal of the International Stroke Society*. 2022;17(1):18–29.

[24] Lang CE, Bland MD, Bailey RR, Schaefer SY, and Birkenmeier RL. Assessment of upper extremity impairment, function, and activity after stroke: foundations for clinical decision making. *Journal of Hand Therapy: Official Journal of the American Society of Hand Therapists*. 2013;26 (2):104–14;quiz 15.

[25] Smith MC, and Stinear CM. Transcranial magnetic stimulation (TMS) in stroke: ready for clinical practice? *Journal of Clinical Neuroscience: Official Journal of the Neurosurgical Society of Australasia*. 2016;31:10–4.

[26] Pollock A, Farmer SE, Brady MC, *et al*. Interventions for improving upper limb function after stroke. *The Cochrane Database of Systematic Reviews*. 2014;2014(11):Cd010820.

[27] Teasell R, Salbach NM, Foley N, *et al*. Canadian stroke best practice recommendations: rehabilitation, recovery, and community participation following stroke. Part one: rehabilitation and recovery following stroke; 6th edition update 2019. *International Journal of Stroke: Official Journal of the International Stroke Society*. 2020;15(7):763–88.

[28] Oyinbo CA. Secondary injury mechanisms in traumatic spinal cord injury: a nugget of this multiply cascade. *Acta Neurobiologiae Experimentalis*. 2011; 71(2):281–99.

[29] Ding W, Hu S, Wang P, *et al*. Spinal cord injury: the global incidence, prevalence, and disability from the global burden of disease study 2019. *Spine*. 2022;47(21):1532–40.

[30] Alizadeh A, Dyck SM, and Karimi-Abdolrezaee S. Traumatic spinal cord injury: an overview of pathophysiology, models and acute injury mechanisms. *Frontiers in Neurology*. 2019;10:282.

[31] Khorasanizadeh M, Yousefifard M, Eskian M, *et al*. Neurological recovery following traumatic spinal cord injury: a systematic review and meta-analysis. *Journal of Neurosurgery Spine*. 2019;30(5):683–99.

[32]   Fehlings MG, Tetreault LA, Aarabi B, *et al.* A clinical practice guideline for the management of patients with acute spinal cord injury: recommendations on the type and timing of rehabilitation. *Global Spine Journal.* 2017;7(Suppl 3):231s–8s.

[33]   Zesiewicz TA. Parkinson disease. *Continuum (Minneapolis, Minn).* 2019;25 (4):896–918.

[34]   Willis AW, Roberts E, Beck JC, *et al.* Incidence of Parkinson disease in North America. *NPJ Parkinson's Disease.* 2022;8(1):170.

[35]   Ponsen MM, Daffertshofer A, van den Heuvel E, Wolters E, Beek PJ, and Berendse HW. Bimanual coordination dysfunction in early, untreated Parkinson's disease. *Parkinsonism & Related Disorders.* 2006;12(4):246–52.

[36]   Ingvarsson PE, Gordon AM, and Forssberg H. Coordination of manipulative forces in Parkinson's disease. *Experimental Neurology.* 1997;145(2 Pt 1):489–501.

[37]   Wagle Shukla A, Ounpraseuth S, Okun MS, Gray V, Schwankhaus J, and Metzer WS. Micrographia and related deficits in Parkinson's disease: a cross-sectional study. *BMJ Open.* 2012;2(3):e000628.

[38]   Popa L, and Taylor P. Functional electrical stimulation may reduce bradykinesia in Parkinson's disease: a feasibility study. *Journal of Rehabilitation and Assistive Technologies Engineering.* 2015;2:2055668315607836.

[39]   Sujith OK. Functional electrical stimulation in neurological disorders. *European Journal of Neurology.* 2008;15(5):437–44.

[40]   Pascual-Valdunciel A, Rajagopal A, Pons JL, and Delp S. Non-invasive electrical stimulation of peripheral nerves for the management of tremor. *Journal of the Neurological Sciences.* 2022;435:120195.

[41]   Dosen S, Muceli S, Dideriksen JL, *et al.* Online tremor suppression using electromyography and low-level electrical stimulation. *IEEE Transactions on Neural Systems and Rehabilitation Engineering: A Publication of the IEEE Engineering in Medicine and Biology Society.* 2015;23(3):385–95.

[42]   Popović Maneski L, Jorgovanović N, Ilić V, *et al.* Electrical stimulation for the suppression of pathological tremor. *Medical & Biological Engineering & Computing.* 2011;49(10):1187–93.

[43]   Milosevic M, Nakanishi T, Sasaki A, *et al.* Cortical re-organization after traumatic brain injury elicited using functional electrical stimulation therapy: a case report. *Frontiers in Neuroscience.* 2021;15:693861.

[44]   Moll I, Marcellis RGJ, Fleuren SM, *et al.* Functional electrical stimulation during walking in children with unilateral spastic cerebral palsy: a randomized cross-over trial. *Developmental Medicine and Child Neurology.* 2024; 66(5):598–609.

[45]   Sampson P, Freeman C, Coote S, *et al.* Using functional electrical stimulation mediated by iterative learning control and robotics to improve arm movement for people with multiple sclerosis. *IEEE Transactions on Neural Systems and Rehabilitation Engineering: A Publication of the IEEE Engineering in Medicine and Biology Society.* 2016;24(2):235–48.

[46] Tapp M, Wenzinger E, Tarabishy S, Ricci J, and Herrera FA. The epidemiology of upper extremity nerve injuries and associated cost in the US emergency departments. *Annals of Plastic Surgery*. 2019;83(6):676–80.

[47] Asplund M, Nilsson M, Jacobsson A, and von Holst H. Incidence of traumatic peripheral nerve injuries and amputations in Sweden between 1998 and 2006. *Neuroepidemiology*. 2009;32(3):217–28.

[48] Huckhagel T, Nüchtern J, Regelsberger J, and Lefering R. Nerve injury in severe trauma with upper extremity involvement: evaluation of 49,382 patients from the TraumaRegister DGU® between 2002 and 2015. *Scandinavian Journal of Trauma, Resuscitation and Emergency Medicine*. 2018;26(1):76.

[49] Kim SJ, Kwon YM, Ahn SM, Lee JH, and Lee CH. Epidemiology of upper extremity peripheral nerve injury in South Korea, 2008 to 2018. *Medicine*. 2022;101(48):e31655.

[50] Barnes SL, Miller TA, and Simon NG. Traumatic peripheral nerve injuries: diagnosis and management. *Current Opinion in Neurology*. 2022;35(6):718–27.

[51] Ni L, Yao Z, Zhao Y, *et al*. Electrical stimulation therapy for peripheral nerve injury. *Frontiers in Neurology*. 2023;14:1081458.

[52] Rebersek S, and Vodovnik L. Proportionally controlled functional electrical stimulation of hand. *Archives of Physical Medicine and Rehabilitation*. 1973; 54(8):378–82.

[53] IJzerman MJ, Stoffers T, Groen FIT, *et al*. The NESS Handmaster orthosis: restoration of hand function in C5 and stroke patients by means of electrical stimulation. *Journal of Rehabilitation Sciences*. 1996;9(3):86–9.

[54] Prochazka A, Gauthier M, Wieler M, and Kenwell Z. The bionic glove: an electrical stimulator garment that provides controlled grasp and hand opening in quadriplegia. *Archives of Physical Medicine and Rehabilitation*. 1997;78 (6):608–14.

[55] Mulcahey MJ, Betz RR, Kozin SH, Smith BT, Hutchinson D, and Lutz C. Implantation of the Freehand system during initial rehabilitation using minimally invasive techniques. *Spinal Cord*. 2004;42(3):146–55.

[56] Popovic MB, Popovic DB, Sinkjaer T, Stefanovic A, and Schwirtlich L. Restitution of reaching and grasping promoted by functional electrical therapy. *Artificial Organs*. 2002;26(3):271–5.

[57] Popovic MR, Kapadia N, Zivanovic V, Furlan JC, Craven BC, and McGillivray C. Functional electrical stimulation therapy of voluntary grasping versus only conventional rehabilitation for patients with subacute incomplete tetraplegia: a randomized clinical trial. *Neurorehabilitation and Neural Repair*. 2011;25(5):433–42.

[58] Hebert DA, Bowen JM, Ho C, Antunes I, O'Reilly DJ, and Bayley M. Examining a new functional electrical stimulation therapy with people with severe upper extremity hemiparesis and chronic stroke: a feasibility study. *British Journal of Occupational Therapy*. 2017;80:651–9.

[59] Mangold S, Keller T, Curt A, and Dietz V. Transcutaneous functional electrical stimulation for grasping in subjects with cervical spinal cord injury. *Spinal Cord*. 2005;43:1–13.

[60] Vodovnik L, Kralj A, Stanic U, Acimovic R, and Gros N. Recent applications of functional electrical stimulation to stroke patients in Ljubljana. *Clinical Orthopaedics and Related Research*. 1978;(131):64–70.

[61] Long C, 2nd. An electrophysiologic splint for the hand. *Archives of Physical Medicine and Rehabilitation*. 1963;44:499–503.

[62] Popović D, Stojanović A, Pjanović A, *et al*. Clinical evaluation of the bionic glove. *Archives of Physical Medicine and Rehabilitation*. 1999;80(3):299–304.

[63] Chae J, Bethoux F, Bohine T, Dobos L, Davis T, and Friedl A. Neuromuscular stimulation for upper extremity motor and functional recovery in acute hemiplegia. *Stroke*. 1998;29(5):975–9.

[64] Hendricks HT, IJzerman MJ, de Kroon JR, in 't Groen FA, and Zilvold G. Functional electrical stimulation by means of the 'Ness Handmaster Orthosis' in chronic stroke patients: an exploratory study. *Clinical Rehabilitation*. 2001;15(2):217–20.

[65] Cauraugh JH, and Kim S. Two coupled motor recovery protocols are better than one: electromyogram-triggered neuromuscular stimulation and bilateral movements. *Stroke*. 2002;33(6):1589–94.

[66] Gritsenko V, and Prochazka A. A functional electric stimulation-assisted exercise therapy system for hemiplegic hand function. *Archives of Physical Medicine and Rehabilitation*. 2004;85(6):881–5.

[67] Popovic DB, Popovic MB, Sinkjaer T, Stefanovic A, and Schwirtlich L. Therapy of paretic arm in hemiplegic subjects augmented with a neural prosthesis: a cross-over study. *Canadian Journal of Physiology and Pharmacology*. 2004;82(8–9):749–56.

[68] Popovic MR, Thrasher TA, Zivanovic V, Takaki J, and Hajek V. Neuroprosthesis for retraining reaching and grasping functions in severe hemiplegic patients. *Neuromodulation*. 2005;8(1):58–72.

[69] Kowalczewski J, Gritsenko V, Ashworth N, Ellaway P, and Prochazka A. Upper-extremity functional electric stimulation-assisted exercises on a workstation in the subacute phase of stroke recovery. *Archives of Physical Medicine and Rehabilitation*. 2007;88(7):833–9.

[70] Sullivan JE, and Hedman LD. Effects of home-based sensory and motor amplitude electrical stimulation on arm dysfunction in chronic stroke. *Clinical Rehabilitation*. 2007;21(2):142–50.

[71] Thrasher TA, Zivanovic V, McIlroy W, and Popovic MR. Rehabilitation of reaching and grasping function in severe hemiplegic patients using functional electrical stimulation therapy. *Neurorehabilitation and Neural Repair*. 2008; 22:706–14.

[72] Kawashima N, Popovic MR, and Zivanovic V. Effect of intensive functional electrical stimulation therapy on upper-limb motor recovery after stroke: case study of a patient with chronic stroke. *Physiotherapy Canada Physiotherapie Canada*. 2013;65(1):20–8.

[73] Kapadia NM, Nagai MK, Zivanovic V, *et al*. Functional electrical stimulation therapy for recovery of reaching and grasping in severe chronic pediatric stroke patients. *Journal of Child Neurology*. 2014;29(4):493–9.

[74]   Popovic MR, Thrasher TA, Adams ME, Takes V, Zivanovic V, and Tonack MI. Functional electrical therapy: retraining grasping in spinal cord injury. *Spinal Cord*. 2006;44(3):143–51.

[75]   Kapadia N, Zivanovic V, and Popovic MR. Restoring voluntary grasping function in individuals with incomplete chronic spinal cord injury: pilot study. *Topics in Spinal Cord Injury Rehabilitation*. 2013;19:279–87.

[76]   Glanz M, Klawansky S, Stason W, Berkey C, and Chalmers TC. Functional electrostimulation in poststroke rehabilitation: a meta-analysis of the randomized controlled trials. *Archives of Physical Medicine and Rehabilitation*. 1996;77(6):549–53.

[77]   Kapadia N, and Popovic M. Functional electrical stimulation therapy for grasping in spinal cord injury: an overview. *Topics in Spinal Cord Injury Rehabilitation*. 2011;17:70–6.

[78]   Popović DB, and Sinkjær T *Control of Movement for the Physically Disabled: Control for Rehabilitation Technology*. Berlin: Springer; 2000.

[79]   Stein RB, Peckham PH, and Popović DP. *Neural Prostheses: Replacing Motor Function after Disease or Disability*. Oxford: Oxford University Press; 1992.

[80]   Popovic MR, Keller T, Pappas IPI, Dietz V, and Morari M. Surface-stimulation technology for grasping and walking neuroprostheses: improving quality of life in stroke/spinal cord injury subjects with rapid prototyping and portable FES systems. *IEEE Engineering in Medicine and Biology Magazine*. 2001;20:82–93.

[81]   Marquez-Chin C, and Popovic MR. Functional electrical stimulation therapy for restoration of motor function after spinal cord injury and stroke: a review. *BioMedical Engineering Online*. 2020;19(1):34.

[82]   Mayr W. Role of electrical parameters in functional electrical stimulation. In: Schick T, (ed.), *Functional Electrical Stimulation in Neurorehabilitation: Synergy Effects of Technology and Therapy*. Cham: Springer International Publishing; 2022. p. 29–41.

[83]   Kapadia NM, Zivanovic V, Furlan JC, Craven BC, McGillivray C, and Popovic MR. Functional electrical stimulation therapy for grasping in traumatic incomplete spinal cord injury: randomized control trial. *Artificial Organs*. 2011;35(3):212–6.

[84]   Widjaja F, Shee CY, Au WL, Poignet P, and Ang WT (eds). Using electromechanical delay for real-time anti-phase tremor attenuation system using functional electrical stimulation. In: *IEEE International Conference on Robotics and Automation*, Shanghai, China, 2011, pp. 3694–99.

[85]   Sayenko D, Robinson M, Milosevic M, Masani K, and Popovic M. Video game-based approach in calf muscle training: lessons from pilot study. *Clinical Research on Foot & Ankle*. 2013;1:110.

[86]   Pappas IPI, Keller T, Mangold S, Popovic MR, Dietz V, and Morari M. A reliable gyroscope-based gait-phase detection sensor embedded in a shoe insole. *IEEE Sensors Journal*. 2004;4:268–74.

[87]   Márquez-Chin C, Popovic MR, Cameron T, Lozano AM, and Chen R. Control of a neuroprosthesis for grasping using off-line classification of electrocorticographic signals: case study. *Spinal Cord*. 2009;47:802–8.

[88]   Milosevic M, Marquez-Chin C, Masani K, *et al*. Why brain-controlled neuroprosthetics matter: mechanisms underlying electrical stimulation of muscles and nerves in rehabilitation. *BioMedical Engineering Online*. 2020; 19:1–30.

[89]   Garcia-Garcia MG, Jovanovic LI, and Popovic MR. Comparing preference related to comfort in torque-matched muscle contractions between two different types of functional electrical stimulation pulses in able-bodied participants. *The Journal of Spinal Cord Medicine*. 2021;44(Suppl 1):S215–S24.

[90]   Popovic MR, Masani K, and Milosevic M. Functional electrical stimulation therapy: mechanisms for recovery of function following spinal cord injury and stroke. In: Reinkensmeyer DJ, Marchal-Crespo L, and Dietz V (eds), *Neurorehabilitation Technology*. Cham: Springer; 2022. pp. 401–27.

[91]   Wang TJ, Sung K, Wilburn M, and Allbright J. Russian stimulation/functional electrical stimulation in the treatment of foot drop resulting from lumbar radiculopathy: a case series. *Innovations in Clinical Neuroscience*. 2019;16(5–6):46–9.

[92]   Popovic MR, Zivanovic V, and Valiante TA. Restoration of upper limb function in an individual with cervical spondylotic myelopathy using functional electrical stimulation therapy: a case study. *Frontiers in Neurology*. 2016;7:81.

[93]   Kapadia N, Jovanovic L, Musselman K, Wang R, Marquez-Chin C, and Popovic MR. Preliminary evaluation of the reliability and validity of the 3D printed Toronto rehabilitation institute-hand function test in individuals with spinal cord injury. *The Journal of Spinal Cord Medicine*. 2021;44(Suppl 1): S225–S33.

[94]   Kapadia N, Myers M, Musselman K, Wang RH, Yurkewich A, and Popovic MR. 3-dimensional printing in rehabilitation: feasibility of printing an upper extremity gross motor function assessment tool. *Biomedical Engineering Online*. 2021;20(1):2.

[95]   Kalsi-Ryan S, Beaton D, Ahn H, *et al*. Responsiveness, sensitivity, and minimally detectable difference of the graded and redefined assessment of strength, sensibility, and prehension, version 1.0. *Journal of Neurotrauma*. 2016;33(3):307–14.

[96]   Khan MA, Fares H, Ghayvat H, *et al*. A systematic review on functional electrical stimulation based rehabilitation systems for upper limb post–stroke recovery. *Frontiers in Neurology*. 2023;14:1272992.

[97]   Kapadia N, Moineau B, and Popovic MR. Functional electrical stimulation therapy for retraining reaching and grasping after spinal cord injury and stroke. *Frontiers in Neuroscience*. 2020;14:718.

[98]   Perotto AO, Delagi EF, Iazzetti JM, and Morrison D *Anatomical Guide for the Electromyographer: The Limbs and the Trunk*. 4th edn. Springfield, IL: Charles C Thomas Publisher Ltd; 2005.

[99]   Snoek GJ, IJzerman MJ, in 't Groen FA, Stoffers TS, and Zilvold G. Use of the NESS handmaster to restore handfunction in tetraplegia: clinical experiences in ten patients. *Spinal Cord.* 2000;38(4):244–9.

[100]   Popovic MR, Curt A, Keller T, and Dietz V. Functional electrical stimulation for grasping and walking: indications and limitations. *Spinal Cord.* 2001;39(8):403–12.

[101]   Doherty JG, Burns AS, O'Ferrall DM, and Ditunno JF. Prevalence of upper motor neuron vs lower motor neuron lesions in complete lower thoracic and lumbar spinal cord injuries. *The Journal of Spinal Cord Medicine.* 2002;25 (4):289–92.

[102]   Dionne TP, and Lenker JA. Expert recommendations for a clinical guide supporting the use of electrical stimulation in occupational therapy treatment of spinal cord injury. *The American Journal of Occupational Therapy: Official Publication of the American Occupational Therapy Association.* 2023;77(4):7704205130.

[103]   Musselman KE, Provad E, Djuric A, Bercovitch D, Yuen I, and Kane KJ. Exploring the experiences and perceptions of pediatric therapists who use functional electrical stimulation in their clinical practice. *Physical & Occupational Therapy in Pediatrics.* 2023;43(6):759–79.

[104]   Tedesco Triccas L, Donovan-Hall M, Dibb B, and Burridge JH. A nation-wide survey exploring the views of current and future use of functional electrical stimulation in spinal cord injury. *Disability and Rehabilitation Assistive Technology.* 2023;18(6):752–62.

[105]   Bulley C, Burridge J, Adonis A, *et al.* Delphi consensus study and clinical practice guideline development for functional electrical stimulation to support upright mobility in people with an upper motor neuron lesion. *Artificial Organs.* 2024;48(3):285–96.

[106]   Bersch I, Krebs J, and Fridén J. A prediction model for various treatment pathways of upper extremity in tetraplegia. *Frontiers in Rehabilitation Sciences.* 2022;3:889577.

[107]   Arora T, Desai N, Kirshblum S, and Chen R. Utility of transcranial magnetic stimulation in the assessment of spinal cord injury: current status and future directions. *Frontiers in Rehabilitation Sciences.* 2022;3:1005111.

[108]   Ariji Y, Hayashi T, Ideta R, *et al.* A prediction model of functional outcome at 6 months using clinical findings of a person with traumatic spinal cord injury at 1 month after injury. *Spinal Cord.* 2020;58(11):1158–65.

[109]   Dunkelberger N, Schearer EM, and O'Malley MK. A review of methods for achieving upper limb movement following spinal cord injury through hybrid muscle stimulation and robotic assistance. *Experimental Neurology.* 2020;328:113274.

[110]   Hebb DO. *The Organization of Behavior.* New York: John Wiley and Sons, Inc.; 1949.

[111]   Fridén J, and Lieber RL. Reach out and grasp the opportunity: reconstructive hand surgery in tetraplegia. *The Journal of Hand Surgery, European Volume.* 2019;44(4):343–53.

[112] Gordon T. Brief electrical stimulation promotes recovery after surgical repair of injured peripheral nerves. *International Journal of Molecular Sciences.* 2024;25(1):665.

[113] Bersch I, and Fridén J. Role of functional electrical stimulation in tetraplegia hand surgery. *Archives of Physical Medicine and Rehabilitation.* 2016;97(Suppl 6):S154–9.

[114] Huang Y-Z, Lu M-K, Antal A, *et al.* Plasticity induced by non-invasive transcranial brain stimulation: a position paper. *Clinical Neurophysiology.* 2017;128:2318–29.

[115] Suppa A, Quartarone A, Siebner H, *et al.* The associative brain at work: evidence from paired associative stimulation studies in humans. *Clinical Neurophysiology.* 2017;128:2140–64.

[116] Bunday KL, and Perez MA. Motor recovery after spinal cord injury enhanced by strengthening corticospinal synaptic transmission. *Current Biology.* 2012;22:2355–61.

[117] Taylor JL, and Martin PG. Voluntary motor output Is altered by spike-timing-dependent changes in the human corticospinal pathway. *The Journal of Neuroscience.* 2009;29:11708–16.

[118] Huang Y-Z, Edwards MJ, Rounis E, and Rothwell JC. Theta burst stimulation on human motor cortex. *Clinical Neurophysiology.* 2007;118:e151.

[119] Yamaguchi T, Fujiwara T, Lin SC, *et al.* Priming with intermittent theta burst transcranial magnetic stimulation promotes spinal plasticity induced by peripheral patterned electrical stimulation. *Frontiers in Neuroscience.* 2018;12:339497.

[120] Cao N, Sasaki A, Yuasa A, Popovic MR, Milosevic M, and Nakazawa K. Short-term facilitation effects elicited by cortical priming through theta burst stimulation and functional electrical stimulation of upper-limb muscles. *Experimental Brain Research.* 2022;240:1565–78.

[121] Cao N, Sasaki A, Yuasa A, Popovic MR, Milosevic M, and Nakazawa K. Effective corticospinal excitability neuromodulation elicited by short-duration concurrent and synchronized associative cortical and neuromuscular stimulations. *Neuroscience Letters.* 2022;790:136910.

[122] Rizzo V, Terranova C, Crupi D, Sant'Angelo A, Girlanda P, and Quartarone A. Increased transcranial direct current stimulation after effects during concurrent peripheral electrical nerve stimulation. *Brain Stimulation.* 2014;7:113–21.

[123] Lefaucheur JP, Antal A, Ayache SS, *et al.* Evidence-based guidelines on the therapeutic use of transcranial direct current stimulation (tDCS). *Clinical Neurophysiology.* 2017;128(1):56–92.

[124] Darmani G, Bergmann TO, Butts Pauly K, *et al.* Non-invasive transcranial ultrasound stimulation for neuromodulation. *Clinical Neurophysiology.* 2022;135:51–73.

[125] Desai N, Grippe T, Arora T, Gunraj C, and Chen R. Effects of low intensity focused ultrasound stimulation combined with functional electrical

stimulation on corticospinal excitability and upper extremity fine motor function. *Submitted to Under review in Brain and Behavior, July* 2024.

[126]   de Kroon J, IJzerman M, Chae J, Lankhorst G, and Zilvold G. Relation between stimulation characteristics and clinical outcome in studies using electrical stimulation to improve motor control of the upper extremity in stroke. *Journal of Rehabilitation Medicine*. 2005;37:65–74.

[127]   Joa KL, Han YH, Mun CW, *et al*. Evaluation of the brain activation induced by functional electrical stimulation and voluntary contraction using functional magnetic resonance imaging. *Journal of NeuroEngineering and Rehabilitation*. 2012;9:1–10.

[128]   Reynolds C, Osuagwu BA, and Vuckovic A. Influence of motor imagination on cortical activation during functional electrical stimulation. *Clinical Neurophysiology*. 2015;126:1360–9.

[129]   Yasui T, Yamaguchi T, Tanabe S, *et al*. Time course of changes in corticospinal excitability induced by motor imagery during action observation combined with peripheral nerve electrical stimulation. *Experimental Brain Research*. 2019;237:637–45.

[130]   Okuyama K, Ogura M, Kawakami M, *et al*. Effect of the combination of motor imagery and electrical stimulation on upper extremity motor function in patients with chronic stroke: preliminary results. *Therapeutic Advances in Neurological Disorders*. 2018;11:1756286418804785.

[131]   Suzuki Y, Jovanovic LI, Fadli RA, *et al*. Evidence that brain-controlled functional electrical stimulation could elicit targeted corticospinal facilitation of hand muscles in healthy young adults. *Neuromodulation: Technology at the Neural Interface*. 2023;26:1612–21.

[132]   Fadli RA, Yamanouchi Y, Jovanovic LI, *et al*. Effectiveness of motor and prefrontal cortical areas for brain-controlled functional electrical stimulation neuromodulation. *Journal of Neural Engineering*. 2023;20:056022.

[133]   Daly JJ, Cheng R, Rogers J, Litinas K, Hrovat K, and Dohring M. Feasibility of a new application of noninvasive brain computer interface (BCI): a case study of training for recovery of volitional motor control after stroke. *Journal of Neurologic Physical Therapy*. 2009;33:203–11.

[134]   Jovanovic LI, Kapadia N, Lo L, Zivanovic V, Popovic MR, and Marquez-Chin C. Restoration of upper limb function after chronic severe hemiplegia: a case report on the feasibility of a brain–computer interface-triggered functional electrical stimulation therapy. *American Journal of Physical Medicine and Rehabilitation*. 2020;99:e35–e40.

[135]   Marquez-Chin C, Marquis A, and Popovic MR. EEG-triggered functional electrical stimulation therapy for restoring upper limb function in chronic stroke with severe hemiplegia. *Case Reports in Neurological Medicine*. 2016;2016:1–11.

[136]   Mukaino M, Ono T, Shindo K, *et al*. Efficacy of brain–computer interface-driven neuromuscular electrical stimulation for chronic paresis after stroke. *Journal of Rehabilitation Medicine*. 2014;46:378–82.

[137] Kim T, Kim S, and Lee B. Effects of action observational training plus brain–computer interface-based functional electrical stimulation on paretic arm motor recovery in patient with stroke: a randomized controlled trial. *Occupational Therapy International*. 2016;23:39–47.

[138] Li M, Liu Y, Wu Y, Liu S, Jia J, and Zhang L. Neurophysiological substrates of stroke patients with motor imagery-based brain–computer interface training. *International Journal of Neuroscience*. 2014;124:403–15.

[139] Biasiucci A, Leeb R, Iturrate I, *et al*. Brain-actuated functional electrical stimulation elicits lasting arm motor recovery after stroke. *Nature Communications*. 2018;9:1–13.

[140] Müller-Putz GR, Scherer R, Pfurtscheller G, and Rupp R. EEG-based neuroprosthesis control: a step towards clinical practice. *Neuroscience Letters*. 2005;382:169–74.

[141] Müller-Putz GR, Scherer R, Pfurtscheller G, and Rupp R. Brain–computer interfaces for control of neuroprostheses: from synchronous to asynchronous mode of operation. *Biomedizinische Technik*. 2006;51:57–63.

[142] Osuagwu BCA, Wallace L, Fraser M, and Vuckovic A. Rehabilitation of hand in subacute tetraplegic patients based on brain computer interface and functional electrical stimulation: a randomised pilot study. *Journal of Neural Engineering*. 2016;13(6):065002.

[143] Jovanovic LI, Jervis Rademeyer H, Pakosh M, Musselman KE, Popovic MR, and Marquez-Chin C. Scoping review on brain–computer interface-controlled electrical stimulation interventions for upper limb rehabilitation in adults: a look at participants, interventions, and technology. *Physiotherapy Canada*. 2023;75:276–90.

Chapter 6

Electrical stimulation in musculoskeletal populations

Louise C Burgess[1], Thomas W Wainwright[2,3,4] and Ian Swain[5]

6.1 Introduction

Musculoskeletal disorders encompass a diverse range of conditions affecting joints, bones, muscles, and connective tissues [1]. They are typically characterised by pain and limitations to function, and include conditions such as arthritis, back and neck pain, fibromyalgia, and osteoporosis. Globally, their impact is disabling and costly, with people with musculoskeletal disorders less likely to work and more likely to suffer from additional health conditions, for example, diabetes, obesity, or a mental health problem [2]. Long-term, many people with musculoskeletal disorders, such as arthritis, may progress to needing surgery when non-surgical treatment options become ineffective at managing pain and function, creating an economic and rehabilitation burden on healthcare systems. To alleviate this burden, research endeavours have sought to identify strategies to improve pain and function in common musculoskeletal conditions. This has included studies investigating non-pharmacological (such as therapeutic exercise, weight management, and education) and pharmacological strategies (such as topical and oral medication and intra-articular injections). While therapeutic exercise, broadly defined as bodily move-ment prescribed to correct impairment, improve musculoskeletal function, or maintain a state of well-being [3] can help to manage symptoms, many individuals with chronic joint pain or poor mobility may be reluctant or unable to perform voluntary exercise. As a result, people with musculoskeletal conditions are more likely to be inactive and overweight than their counterparts.

The relationship between musculoskeletal conditions and inactivity is complex and exists in part due to shared risk factors such as lifestyle factors but also due to

[1]Department of Rehabilitation and Sports Sciences, Bournemouth University, UK
[2]Orthopaedic Department, Lanzhou University, China
[3]Physiotherapy Department, University Hospitals Dorset, NHS Foundation Trust, UK
[4]Rehabilitation Department, Lanzhou University, China
[5]Orthopaedic Research Institute, Bournemouth University, UK

causative risk factors, such as pain and limited range of motion. The relationship between causative factors can be reciprocal, whereby pain and obesity worsen inactivity, and inactivity worsens pain and obesity [4]. Research has shown that people with a long-term musculoskeletal condition are around twice as likely to be physically inactive than those without [5] and this poses a threat to the maintenance of skeletal muscle mass, strength, and power [6]. As a result, muscle atrophy is likely to occur, driven by deregulation of muscle protein synthesis rates to protein breakdown rates [7]. Over time, individual muscle fibres decrease in size due to reduced protein content. Muscle fibres can lose both contractile proteins (actin and myosin) and structural proteins, leading to a decrease in muscle size and strength [8]. Reduced muscle activity also leads to lower neural activation, resulting in muscle fibre deconditioning and reduced force production, and increased muscle fatigue, due to changes in the metabolic properties of the tissue [9]. The combined effects of increased protein degradation, decreased protein synthesis, muscle fibre size and neural activation, and increased fatigue, contribute to gradual loss of muscle mass, identified as muscle atrophy.

Muscle atrophy may also occur because of joint injury, such as ligament rupture, and surgery (ligament reconstruction surgery, joint replacement) and the mechanisms behind this are distinct from atrophy due to disuse [10]. There is a growing body of evidence to suggest that a combination of unique disruptions to morphological, phenotypic, and signalling pathways can alter the ability of an individual to complete voluntary muscle strengthening exercise [10]. Trauma to a joint can cause disruption of neural signalling that is required to activate the muscle due to swelling, pain, and the loss of mechanoreceptors [11]. This disruption of neural activity compromises muscle tissue volume and quality because of the decreased ability to effectively discharge action potentials, thereby limiting complete or sustained voluntary muscle contractions [12]. Due to a lack of activation, muscle fibres may decrease in size and number, leading to muscle atrophy and weakness [13], contributing to the mobility challenges faced in people with musculoskeletal disorders. Regardless of the cause, muscle atrophy due to disuse or joint trauma creates a significant rehabilitation challenge in musculoskeletal populations, particularly in those who require treatment with surgical interventions, yet present with persistent muscle atrophy and weakness [14].

Neuromuscular electrical stimulation (NMES) can be used to induce involuntary muscle contractions by delivering electrical impulses to peripheral nerves and can therefore be used to preserve or restore muscle strength or function following atrophy or loss of neuromuscular control. Where individuals are experiencing pain during joint loading due to musculoskeletal conditions, joint injury, or immobilisation due to surgery, NMES offers unique advantages where voluntary contractions are limited. During voluntary muscle contraction, muscles receive an electrical signal from the central nervous system through motor nerve cells that connect with individual muscle fibre. During electrical muscle stimulation, a stimulation pulse is delivered through the electrode that causes depolarisation of the motor nerve and this elicits a nerve impulse that causes a muscle contraction [14], detailed further in Chapter 2. With repeated and prolonged use, it is possible that NMES can elicit strengthening benefits like those of voluntary exercise, if prescribed with a sufficient stimulation dose. Two mechanisms have been suggested to explain the training effects resulting from NMES. The first proposes that augmentation of muscle strength through NMES may occur as a similar response to that of voluntary exercise, if prescribed with sufficient repetitions of high external load, and a high intensity of muscle contraction [15]. The second mechanism suggests that strength gains are a result of a reversal of voluntary recruitment order with a selective augmentation of type II muscle fibres, which have a higher specific force than type I fibres, which in turn increases the overall strength of the muscle [15]. However, as some studies have shown that nerve recruitment is random in that electrical stimulation is as likely to excite a muscle fibre connection to a type 1 fibre as a type II fibre [16], this theory has been challenged [17]. An alternative theory has been proposed, suggesting that motor unit recruitment during NMES reflects a nonselective, spatially fixed, and temporally synchronous pattern rather than in a reversal of the physiological voluntary recruitment order [18].

Deciding whether to recommend NMES or voluntary exercise will depend on the specific needs and conditions of the individual undergoing rehabilitation. In many cases, a combination of both voluntary exercise and NMES may optimise outcomes. While it is commonly agreed that the strength gains elicited from NMES do not exceed those from voluntary exercise, NMES can offer unique benefits that may for those unable to independently perform strengthening movements, such as while immobilised, due to injury or pain. In addition, NMES allows targeted muscle activation for precise strengthening of specific muscle groups and can often be tolerated in early stage rehabilitation, when intense voluntary strengthening may be uncomfortable. Other advantages include muscle strengthening with minimal joint stress, re-education of voluntary movements where disruption has occurred in the neuromuscular system, and time-efficient strengthening for those unable or unwilling to participate in voluntary strengthening sessions. However, unlike voluntary movements, there is a lack of evidence that demonstrates the ability of electrical stimulation to improve proprioception in musculoskeletal populations, which is an important factor to aid balance, movement control and injury prevention. Other limitations may exist if the individual is unable to tolerate stimulation intensities sufficient to induce muscle strengthening benefits. The evidence for the effectiveness of electrical stimulation in musculoskeletal populations has increased in recent years and is summarised and critiqued in this chapter.

6.2   Survey of previous work

6.2.1   Surgical populations

6.2.1.1   Lower limb joint replacement

Current practice

Prior to and following major orthopaedic surgery, patients are at risk of muscle atrophy due to immobilisation because of pain, long durations of bed rest and reluctance to exercise. Muscle atrophy occurs when protein degradation exceeds protein synthesis, leading to reduced muscle cross-sectional area of myofibers and decreased muscle strength [6]. In addition, alterations in both neural and structural components of the neuromuscular system can determine muscle weakness following orthopaedic surgery, particularly in the early postoperative period [19]. To counteract muscle weakness, patients may receive pre- and postoperative rehabilitation interventions to ameliorate physiological responses to bed rest and immobilisation. Over the last ten years, numerous systematic reviews and meta-analyses have been conducted to compare outcomes for patients receiving preoperative exercise and those receiving standard care, however the level of evidence has remained low [20,21]. It remains that there is some evidence to support the effectiveness of prehabilitation in improving outcomes pre- and postoperatively, yet this is not conclusive, and further evidence is required [14]. Authors of a systematic review in this area called for more innovative interventions that are effective in improving muscle strength and function, and that can be well tolerated by older adults awaiting joint replacement, and highlighted the potential benefits of NMES and blood-flow restriction [22].

Current postoperative rehabilitation strategies in orthopaedic surgery include early mobilisation, whereby patients are encouraged to sit, stand, and walk as soon as possible, and this rehabilitation technique has demonstrated benefits for reducing the risk of postoperative complications and length of stay in hospital [23–25]. Upon discharge, patients are often provided with exercises in the form of patient information leaflets and told to progress independently until their six week follow up. Patient information leaflets commonly contain recommendations on recovery from surgery and exercise prescription in the form of isometric bed exercises and movements to facilitate the return to daily activities (sit-to-stand, stair climbing). Although useful to guide patients through their postoperative recovery, research has found that these patient resources are often designed on a 'one size fits all' basis, and rarely offer advice on exercise frequency, intensity, or progression [26]. In addition, studies have shown that isometric bed exercises are insufficient at producing a level of neuromuscular activation required to induce a muscle strength adaption [27] and have little influence on patient function or quality of life in the six weeks or 12 months following surgery [28,29].

Voluntary resistance training confers substantial benefits to the musculoskeletal system for those with health disorders and in healthy individuals [30] such as muscle hypertrophy, increased strength, enhanced metabolism, and better functional capacity. Despite this, studies that have investigated the potential benefits of pre- and postoperative voluntary resistance training on recovery from orthopaedic surgery conclude that pre- and postoperative exercise have minimal functional effect on the recovery from joint replacement [31,32]. It is possible that individuals with end-stage osteoarthritis may not be able to tolerate the recommended dosage of strength exercise due to pain during joint loading, therefore limiting musculoskeletal benefits pre-surgery. Alternatively, it is possible that exercise dose prescribed in the evidence-base is not sufficient to evoke strengthening effects. Other possible explanations include lack of personalisation or progression in exercise plans, limited adherence to a rehabilitation programme, inadequate exercise type or high levels of heterogeneity across exercise programmes included in systematic reviews. Regardless, NMES offers advantages to preserve or restore skeletal muscle mass and function, before, during, and after a period of disuse due to pain or surgery, where voluntary exercise is difficult or not possible [33]. The following section describes the existing evidence for NMES use in specific lower limb orthopaedic populations.

6.2.1.2　Knee replacement surgery

Despite the supporting evidence, adoption of electrical stimulation devices into clinical orthopaedic practice can be slow, and some questions around optimal protocol and dose remain unanswered. The use of electrical stimulation can be limited by access to stimulators, concerns regarding patient tolerance, restrictions on the time available to mentor therapists on devices and the delay between research evidence and translation into clinical practice. In knee replacement surgery, the evidence for using NMES in rehabilitation protocols has increased over the last ten years [14]. A Cochrane review conducted in 2008 and updated in 2010 found just two studies suitable for inclusion in the evidence synthesis and from this limited evidence were unable to make any conclusions on the application of NMES for purposes of strengthening quadriceps (Figure 6.1) pre- or post knee replacement surgery [34]. In the two included studies, no differences were reported between the NMES and control groups for maximum voluntary isometric torque or endurance, but significantly better quadriceps muscle activation was noted in one study [35,36]. A systematic review published in 2015 was largely in agreement with earlier work, concluding from four studies with a moderate risk of bias, and a low statistical power, that NMES is less effective than conventional rehabilitation yet may offer advantages to improve muscle activation in the early postoperative stage [37]. Similarly, Kittelson conducted a critical review in 2015, and found mixed evidence for the benefits of NMES, with the included studies varying considerably in regard to their methodology and dose of NMES [38].

More recently, a meta-analysis of six studies (496 patients), including some overlap in studies with previous literature reviews, evaluated NMES use in patients

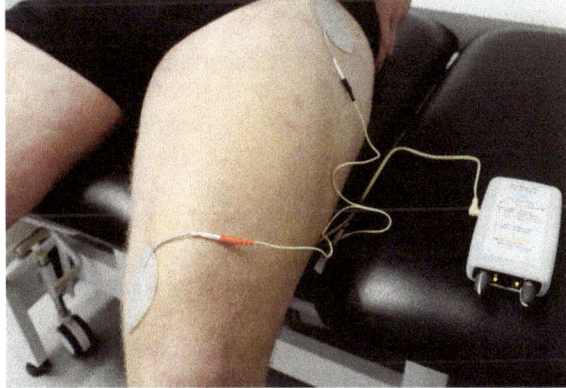

Figure 6.1　NMES applied to the knee extensors (quadriceps)

undergoing knee replacement in comparison to patients in a control group. The characteristics of the NMES interventions varied with a frequency of 40–100 Hz, a pulse duration of between 250 and 400 μs and variety of waveforms. The intensity of the stimulation was consistent across the studies, described as the maximum tolerated by the patient, sufficient to induce a tetanic contraction of the muscle. Both the NMES and the control group followed a standard rehabilitation protocol that involved strength and flexibility exercises, and functional workouts. The meta-analysis reported benefits for the NMES group in timed up and go, stair climb, and walk test scores, particularly in those with severe muscle activation deficits [39]. The benefits of NMES were strongest in the first postoperative weeks/months and gradually diminished, suggesting the NMES may allow better functional recovery in the immediate postoperative phase [39]. A randomised controlled trial of 66 participants (NMES use = 44, no NMES = 22) evaluated 12 weeks of NMES use for quadriceps strengthening for an average of 200 min per week starting at post-operative week one [40]. The NMES group experienced quadriceps strength gains at three-, six- and twelve-weeks following surgery, which were significant when compared with the control group at three ($p = 0.050$) and six weeks ($p = 0.015$). In addition, improvements in timed up and go time were found when compared to the control group at six ($p = 0.018$) and 12 weeks ($p = 0.003$) postoperatively. However, no differences were observed between groups regarding pain, range of motion, length of stay, number of outpatient physiotherapy sessions, mental component score, pain catastrophising score or knee injury, and osteoarthritis outcome score (KOOS) [40]. Overall, the risk of bias in the included studies was considered high, given it was not possible to blind the participants or evaluators to intervention type [39]. In addition, the authors highlighted that the small sample sizes of the included studies, and heterogeneity of NMES settings and protocols should be considered as limitations of the existing evidence. Nonetheless, it was concluded that NMES use, applied in the immediate postoperative period following knee replacement, is safe and may accelerate functional recovery in terms of joint function and physical performance, providing promising results for clinicians [39].

6.2.1.3   Hip replacement surgery

In hip replacement surgery, there is less research available investigating the potential benefits of NMES added to the rehabilitation protocol [14]. A review conducted in 2019 found just two studies investigating the benefits of NMES on quadriceps strength following hip replacement surgery [41]. The studies included differed in terms of NMES intervention, study population, outcome measures and main findings, and therefore it is not possible to offer best-practice recommendations for clinical rehabilitation using NMES in hip replacement. Nonetheless, the available research offers suggestions on how NMES can be used to support recovery from those undergoing hip replacement surgery and can be used by clinicians to guide NMES protocols. One study investigated unilateral NMES of the operated limb (quadriceps) in comparison to standard home-based rehabilitation or unilateral resistance training in older adults [42]. All three groups received standard, home-based rehabilitation which included six bed exercises and knee extension, hip abduction, knee flexion, step training and calf stretching while standing, for one hour a day, for 12 weeks. The NMES intervention involved one hour of stimulation a day for 12 weeks, at a pulse rate of 40 Hz, a pulse width of 250 μs, and an on/off cycle of 10/20 seconds. The resistance training group performed daily knee extension exercises (3 × 10 repetitions) in a seated position with sandbags strapped to the ankle during hospitalisation. As soon as possible, training was performed on adjustable leg press and knee-extension machines. The protocol included a 10-min warm-up on a stationary bike, seated knee extensions and leg presses performed in the supine position. Training intensity was progressively increased in intensity from 20-repetition maximum (RM) ($\sim$ 50% of 1 RM) the first week to 15 RM ($\sim$ 65% of 1 RM) during weeks two to four to 12 RM ($\sim$ 70% of 1 RM) during weeks five to six and finally to 8 RM ($\sim$ 80% of 1 RM) the last six weeks.

Data were analysed between training groups and from pre- to post intervention. Mean $\pm$ standard error length of stay was shorter for the resistance training group (10.0 $\pm$ 2.4 days) than the electrical stimulation group (12 $\pm$ 2.8 days) and the standard rehabilitation group (16.0 $\pm$ 7.2 days) (p <0.05). Resistance training increased maximal gait speed by 30% ($p < 0.001$), stair climbing performance by 28% ($p < 0.005$) and sit-to-stand score by 30% ($p < 0.001$) from the pre-intervention assessment to the post-intervention assessment. Electrical stimulation increased maximal gait speed by 19% ($p < 0.05$), stair climbing performance by 21% (p <0.001), and sit-to-stand score by 21% ($p < 0.001$) from baseline to the post-intervention assessment. No improvements were seen in these measures for the group receiving standard, home-based rehabilitation. A second evaluation of the same participants included measures to evaluate muscle morphology, architecture and function [43]. After 12 weeks, the authors found that resistance training led to improvements in maximal dynamic muscle strength, muscle fibre area, muscle fibre pennation angle, muscle thickness and stair walking power. Interestingly however, there were no improvements in outcome measures for participants receiving electrical stimulation or standard rehabilitation [43].

In one study, low-frequency electrical stimulation of the quadriceps and calf muscles, in addition to conventional physiotherapy, was compared to conventional physiotherapy alone following hip replacement surgery [44]. Stimulation was applied to the quadriceps and calf muscles bilaterally using two portable dual-channel stimulators. Each delivered a 10-Hz biphasic current, with a pulse width of 200 μs and each cycle was on and off alternatively for 20 seconds. As the rehabilitation intervention progressed, the stimulation intensity applied to each muscle was increased to the maximum value tolerated by the patients. The stimulation occurred for a one-hour session, five days a week for five weeks in addition to two hours of physiotherapy. All included participants were evaluated at baseline, and 45 days later. Low-frequency electrical muscle stimulation of the quadriceps and calf muscles in addition to conventional physiotherapy, in elderly subjects (mean age 78 years) was well tolerated and led to a significant increase in muscle strength in the operated limb from baseline to 45 days later. There was a significant increase from baseline in maximal isometric strength of the knee extensors on the operated (77% increase ($p < 0.01$)) and non-operated (15% increase ($p < .005$)) sides in the electrical stimulation group, and in the operated limb of the control group (23% increase ($p < 0.05$)). When compared to the control group, the improvement in strength was significantly greater for the electrical stimulation group for the operated limb ($p < 0.05$) but not for the non-operated limb. An example of an NMES protocol, designed to improve muscle endurance before or after joint replacement surgery, is presented in Case Study 1 (Table 6.1)

Table 6.1   Clinical case study of a patient with knee or hip osteoarthritis awaiting joint replacement and suffering from muscle atrophy and low resistance to fatigue

Case Study 1 – Pre- or post joint replacement

Body chart	Electrodes Patient is awaiting or recently had joint replacement surgery in their hip or knee for the treatment of end-stage osteoarthritis. Pain, swelling and limited mobility. Muscle atrophy due to disuse and disruption to neuromuscular function.
Clinical indication	Increase knee extensor (quadricep) muscle endurance and function
Indication for NMES	Along with standard pre- and post-operative rehabilitation, the patient is provided with NMES to help improve the endurance capacity of the knee extensions following a prolonged period of disuse due to pain, swelling and surgery. NMES is used with a low frequency and long on times with the aim of building muscular endurance to support the completion of daily activities.
Electrode positioning	To stimulate the knee extensors (quadriceps), electrodes should be positioned proximally over the vastus lateralis and more distally over the vastus medialis muscles.
Stimulation settings	To induce changes to muscle endurance.
	20 Hz, 300 µs, 0.5 seconds ramp, 10 seconds on 3 seconds sec off, simultaneous stimulation.
NMES protocol	To improve muscle endurance, a lower frequency stimulation with a longer on period is used to induce a change in muscle fibre properties from fast to slow, with the aim of improving fatigue resistance muscle fibres. A typical training session could be 15–30 min long, progressed from three to five times a week as tolerance to stimulation improves. The patient should be asked to adjust the stimulation current to achieve a visible muscle contraction and then an involuntary isotonic movement as tolerance builds.

Other benefits

The feasibility of NMES to improve recovery after lower limb joint replacement continues to be explored, and studies from this patient population have reported additional benefits of NMES to blood flow, oedema, deep vein thrombosis (DVT) risk, and pain [45–47]. In orthopaedic surgeries such as hip and knee replacement, increased coagulation and inflammatory reactions have been observed most noticeably in the first three postoperative days [48]. NMES offers an alternative DVT prophylaxis option and works by applying an electrical stimulus to motor points on the body, via surface electrodes, to elicit involuntary muscle contractions. Commonly, stimulation is applied to the common peroneal nerve or calf muscle, as contraction of the lower leg muscles increases blood flow from the lower limbs back to the heart, increasing venous return, local blood circulation and help to prevent venous thrombosis [49]. Furthermore, electrical stimulation has demonstrated benefits to pain in this patient population through transcutaneous electrical nerve stimulation (TENS), although the evidence is mixed [47,50,51]. The surgical procedure causes trauma to the surrounding tissues, such as ligaments, tendons and muscles, and nerves, leading to pain during the immediate postoperative phase and during the recovery phase, as the healing process begins. The underlying basis of TENS is based around the pain-gate theory and assumes that the electrical stimulus from the TENS unit can interfere with and block the transmission of pain signals to the brain and thus reduce the perception of pain. While there is some evidence to support TENS use in lower limb joint replacement [47], other studies have reported no benefit of TENS when compared to sham stimulation [51]. Further research is needed in this area, and in electrical stimulation for strength improvement, to confirm the role of NMES as an adjunct to broader rehabilitation programmes in lower limb joint replacement.

6.2.1.4   Shoulder surgery

Following shoulder surgery, rehabilitation protocols vary based on procedure but typically involve protection of the joint, pain and swelling management, followed by gentle range of motion to prevent stiffness and promote circulation. As healing progresses, active range of motion can be introduced and gradual strengthening exercise can begin, typically focusing on the rotator cuff muscles, the deltoid muscles, the scapular stabilisers, the biceps brachii, pectoralis major and latissimus dorsi. At present, there is emerging, but limited evidence, to suggest that NMES could be used to improve muscle strength and restore function in shoulder surgeries. For example, in reverse shoulder replacement, a recent study was the first to investigate NMES of the deltoid following reverse shoulder arthroplasty ($n = 43$) compared to no NMES use ($n = 33$) [52]. NMES was used for one month following surgery, with electrodes placed over the middle and posterior deltoid muscles. Shoulder pain and function, and deltoid thickness were compared between groups at 3, 6, and 12 months postoperatively. At three months, the range of motion and power of external rotation of the NMES group were significantly greater than for those not receiving NMES (ROM, $36° \pm 14°$ vs. $29° \pm 12°$; $p = 0.003$; power, 4.8 kg ± 1.8 kg vs. 3.8 kg ± 1.0 kg; $p < 0.002$) [52]. Furthermore, at six months, the external rotation range of motion of the NMES group was greater than the non-NMES group (41° ± 12° vs. 34° ± 11°; $p = 0.013$). However, there were no significant differences in pain, self-reported function, or deltoid thickness between groups.

In rotator cuff repair, it is generally agreed that pain and function are improved following surgery, however complications are frequent, with some patients suffering from retear and stiffness. In addition, muscle weakness following surgery is often prominent, and limits range of motion. For example, atrophy of the deltoid muscle commonly impedes shoulder flexion, and weakness of the infraspinatus and teres minor muscles may limit external rotation. Research has demonstrated that resisted isometric contraction of the external shoulder rotators produced significantly greater peak force with NMES applied to the infraspinatus muscle belly in 39 patients who had undergone rotator cuff repair surgery around 10 days previously [53]. Peak force production was on average 3.75 kg when tested with NMES and 3.08 kg ($p < 0.001$) when tested without, equating to a 22% higher external rotation force, suggesting benefits of NMES used concomitantly with exercise to enhance rehabilitation and minimise the inhibition of the rotator cuff after surgery [53]. In another study, patients were randomised to receive either NMES or TENS of the deltoid muscle for six weeks following arthroscopic rotator cuff repair and measured pain, range of motion, abduction strength and function [54]. No significant differences were found between the two groups in pain, function, range of motion or abduction strength at 12 weeks. However, atrophy of the anterior, middle, and posterior deltoid was greater in the TENS group than the NMES group (-14%, -2.6%, and 8.2% vs. -2.5%, -0.7%, and -6.8%, $p = 0.016$, $p = 0.677$, and $p = 0.791$, respectively) [54]. These findings suggest NMES may play a role in reducing atrophy after rotator cuff repair, however further research is needed in this patient population to confirm this.

6.2.1.5 Knee ligament reconstruction

As participation in sport and physical activity continues to increase, so does the incidence of sporting injuries. Anterior cruciate ligament (ACL) injuries are frequent in sport, more so in female athletes [55], and in athletes participating in sports such as basketball, football, rugby and skiing due to the specific demands of these activities [56]. Surgeries to repair ruptured ligaments in the knee, such as ACL reconstruction, are common, particularly in those looking to return to competitive sport, however, re-injury rates are high [57]. Factors associated with reducing reinjury risk include a later return to sport and more symmetrical quadriceps strength and loading [57]. As such, it is agreed that progressive rehabilitation is imperative to return an individual to their pre-injury state and sporting activities. However, consensus statements and reviews conclude there is a very low level of certainty for most components of ACL rehabilitation [58], and that there is no current gold standard rehabilitation [59].

Despite a multitude of randomised controlled trials and systematic reviews, there is mainly a low level of evidence for the effectiveness of various rehabilitation interventions, such as open vs. closed kinetic chain exercises (similarly effective), home-based vs. in-person rehabilitation (no difference) and knee bracing for physical function and knee laxity (ineffective), when examined in a best-evidence synthesis of systematic reviews [60]. In this review, the highest level of evidence for ACL rehabilitation was observed for NMES to improve quadriceps strength. In four systematic reviews (16 RCTs) [61–64] evaluating the effects of NMES following ACL reconstruction, there was a mix of evidence from low to moderate certainty regarding the benefits of including NMES as an adjunctive therapy method. While the evidence is limited by heterogenous protocols of stimulation, and no consistent dose-dependent effect, the evidence suggests a large effect size for the addition of NMES (2 to 6 sessions per weeks) to standard rehabilitation to improve quadriceps strength in the first 12 postoperative weeks.

Research in this area has demonstrated the effectiveness of NMES applied to the quadriceps either on the muscle belly or femoral nerve of the rectus femoris or vastus intermedius, or motor point or muscle belly of the vastus medialis, while knees are flexed at around 65 degrees [65]. Frequencies of between 30 and 50 Hz and pulse durations of 250–400 μs have been investigated, and often, patients are asked to increase the current intensity to the maximum they can tolerate and to ensure they can produce an involuntary muscle contraction. Work–rest cycles have also been investigated and can be modified dependent on the strength of the existing muscle (lower duty cycle to reduce muscle fatigue) or aim of the intervention (higher duty cycle with lower frequency to improve muscle endurance). Three sessions a week, initiated within one week post-operatively have been recommended, with 12–15 contractions per session included [65]. In one study, NMES of the quadriceps muscles superimposed on repeated sit-to-stand exercises lead to a higher muscle strength of the knee extensors, lower perception of pain and higher symmetry in lower extremity loading compared to sit-to-stand only or no treatment [66]. Often voluntary exercise in the early phases following surgery is not sufficient to produce strengthening benefits, as it's not possible to overload an operated joint with high external loads, and due to ongoing atherogenic muscle inhibition, where neuromuscular signalling is impaired [66]. The addition of NMES in this study facilitated an effective eccentric contraction in the early postoperative phase, that may not have been possible without it. In addition, NMES allows selective muscle strengthening with the activation of both slow and fast-twitch motor units by using two stimulation frequencies, leading to an increase in muscle activation and strength. Similar research efforts conclude that NMES application can lead to greater strength and self-reported and physical function, however further research is required to determine an optimal NMES dose and protocol [60].

6.2.2   Non-surgical populations

6.2.2.1   Osteoarthritis

In non-surgical knee osteoarthritis patients, a meta-analysis of nine studies (409 participants) concluded that inconsistent evidence exists regarding the impact of NMES on measures of pain, function, and quadriceps femoris muscle strength [67]. However, the results from a pooled analysis found that NMES improved self-reported pain and function, but not objective measures of function [67]. It should be noted, however, that questions have been raised regarding the methodology of this work, suggesting there was a lack of consistency amongst the NMES interventions included in the analysis [68]. The first review to provide standardised clinical treatment parameters for NMES to improve strength and pain in patients with knee osteoarthritis was published in 2020 [69]. Nine RCTs were included, the authors recommended a frequency of at least 50 Hz and no more than 75 Hz with a pulse duration between 200 and 400 μs and a treatment duration of 20 min in order to achieve a successful treatment to improve muscle strength [69]. This review was the first of its kind, providing an important stepping-stone for future clinical work and research endeavours involving NMES [14]. An example NMES protocol for an individual presenting with end-stage knee osteoarthritis is demonstrated in Case Study 2 (Table 6.2).

In individuals with hip osteoarthritis, not waiting joint replacement surgery, very little research has investigated the potential benefits of NMES for improving muscle strength [14]. A study investigating the acceptability and feasibility of NMES for stimulating the hip abductors and knee extensors in people with hip osteoarthritis found that stimulation of the hip abductors (Figure 6.2) was less acceptable and effective at producing an involuntary muscle contraction when compared to stimulation of the knee extensors [70]. Therefore, although the hip abductor muscles are frequently atrophied in individuals with hip osteoarthritis, interventions of NMES may have more success if designed to strengthen the knee extensors. Several factors are critical in determining if a stimulating current is sufficient to cause neural excitation, including impedance, which is the sum of resistive, capacitive, and inductive tissue components that resist the current [71]. It is possible that the higher percentage of fatty infiltration in the gluteal muscles, when compared to the quadriceps, may affect the contractile force produced by NMES [72]. Due to the high resistively of subcutaneous fat tissue, high stimulus currents are required where there is higher skeletal muscle fat infiltration, which can lead to pain or discomfort. Furthermore, it has previously been observed that persons with hip osteoarthritis may present with a decrease in contractile tissues of the gluteal muscles because of the disease [73,74], therefore limiting the potential effects of NMES [14].

6.2.2.2   Rheumatoid arthritis

Rheumatoid arthritis is a chronic, multifactorial autoimmune disease, primarily affecting the joints. Due to its complex presentation, management requires a multidisciplinary approach with the aim of reducing inflammation, controlling symptoms, preventing joint damage, and enhancing quality of life for those with the condition. Management strategies may include a combination of pharmacological treatment (such as disease modifying anti-rheumatic and anti-inflammatory drugs), physical rehabilitation, diet advice, and lifestyle modifications. Physical rehabilitation programmes aim to improve joint function, flexibility and muscle strength and may involve low impact exercise, gentle stretching, and education on balancing activity with rest to manage fatigue and joint pain. People with rheumatoid arthritis

Table 6.2   Clinical case study of a patient with knee osteoarthritis, presenting with pain and loss of knee extensor muscle function

Case Study 2 – End-stage knee osteoarthritis	
Body chart	2-year history of increasing knee pain and stiffness. Osteoarthritis confirmed on x-ray. Patient presents for advice on how to self-manage and improve symptoms. Significant pain and loss of knee extensor muscle function
Clinical indication	Increase knee extensor (quadricep) muscle strength
Indication for NMES	Along with education and advice on self-management of osteoarthritis, the patient is provided with NMES to help strengthen knee extensors. In this case, the pain and restricted movement caused by the osteoarthritis, can prevent patients from completing standard resistance training exercises. Therefore, NMES is used to stimulate muscle contractions to strengthen the muscle.
Electrode positioning	To stimulate the knee extensors (quadriceps), electrodes should be positioned proximally over the vastus lateralis and more distally over the vastus medialis muscles.
Stimulation settings	To induce changes to muscle strength and power. 40–60 Hz, 300 µs, 200 ms ramps, 3.5 seconds on 4.5 seconds off, simultaneous stimulation
NMES protocol	To improve strength of a muscle a higher frequency of NMES with short bursts of high intensity stimulation to induce muscle fatigue is required. A typical training programme is 3 × 10 min sessions, with at least five min of rest in between. Initially use twice per week, building up to three or four times per week. The patient should be asked to adjust the level to the highest level they can tolerate. The idea behind this mode is to cause muscle fatigue to increase strength.

commonly experience substantial deceases in skeletal muscle mass, which is characterised by the depletion of protein storage and the accumulation of fat within the muscle [75]. Electrical muscle stimulation has been suggested as a rehabilitation intervention to enhance muscle performance. In a Cochrane review, published in 2002 [76], just one study met the inclusion criteria [77], and compared the effects of two electrostimulation protocols on hand function in people with rheumatoid

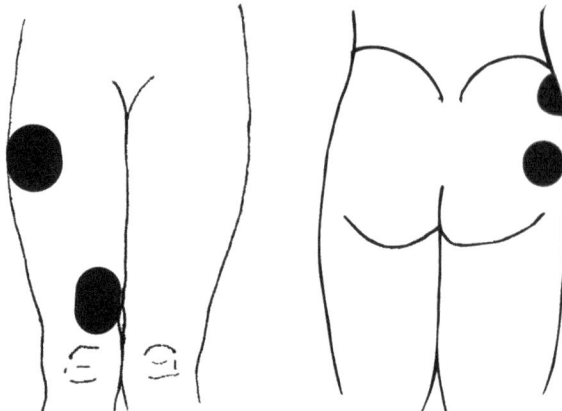

Figure 6.2   Electrode positioning on the knee extensors and hip abductors during participant acceptability testing

arthritis. The results showed that the stimulation had a significant benefit when compared to a control group in terms of muscle strength and fatigue resistance of the first dorsal interosseous [77].

Since then, research in NMES and rheumatoid arthritis has developed, with a pilot randomised controlled trial reporting similar benefits of NMES ($n = 13$) when compared to high-intensity voluntary resistance training ($n = 28$) to improve quadriceps muscle structure and function in people with rheumatoid arthritis, delivered in 36 sessions over 16 weeks [78]. The NMES protocol involved 15 electrically elicited contractions, delivered at a pulse rate of 75 pps, a pulse duration of 450 microseconds and an on/off time of 12 seconds and 48 seconds, to minimise muscle fatigue. Voluntary resistance training included leg extension and leg press on respective machines. The within-group changes from baseline to post-intervention were significant for all measures of muscle structure and function, and muscle strength increased on average 14 Nm in both intervention groups. The between group differences were not statistically significant, suggesting equal benefits of strength improvement through NMES and voluntary resistance training for individuals with rheumatoid arthritis [78]. A secondary analysis of this data investigated the dose–response relationship between NMES use and muscle function [75]. This study reported the minimum NMES intensity required to elicit significant gains in muscle function was around 15% of the individual's maximal voluntary isometric contraction (MVIC), and therefore recommended training intensities of between 15% and 50% of the MVIC [75].

6.2.2.3   Osteoporosis

Electrical stimulation can also have an effect on osteoporosis as the result of muscle contractions which will cause loading of the bones and hence an increase in bone density [79]. This is discussed in more detail in Chapter 14 where its effect can be particularly beneficial to people after spinal cord injuries.

6.2.2.4   Ankle sprain

Ankle sprains are amongst the most common musculoskeletal injuries that occur in both athletic and non-athletic populations. They are recurrent in nature in that once an initial ankle sprain has occurred, there is a subsequent higher risk for ankle sprain due to ligament damage, persistent muscle weakness and altered proprioception. Rehabilitation protocols following ankle sprain are varied, with heterogenous and limited research evidence preventing agreement on optimal rehabilitation techniques [80]. Studies have suggested that NMES can be used to restore the stability and function of the ankle joint following sprain through swelling reduction, targeted muscle activation, and strengthening and enhanced proprioception. However, a systematic review conducted in 2015 concluded that the use of electrical stimulation could not be recommended as a method to improve function, reduce oedema, or decrease pain after synthesising four randomised controlled trials in lateral ankle sprain [81]. Studies of ankle kinematics have reported that electrical stimulation of the peroneal muscles can resist a simulated ankle sprain motion [82,83]. One study found that a pair of small electrodes, fitted the width of the muscle belly of the peroneal muscles, attached to the upper quarter along the lateral shank resulted in a significantly lower peak inversion angular velocity, demonstrating a 21% drop from 273.3 to 215.8 degrees/s in a sudden simulated ankle supination spraining motion [82]. The intervention tested in this study could therefore be used to design an ankle sprain injury prevention programme using electrical stimulation.

In studies comparing NMES to other interventions as a method of reducing oedema following ankle sprain, the research evidence is mixed. In one study of participants recruited from an A & E department where they presented with ankle sprain and were referred to physiotherapy, NMES plus standard care ($n = 11$) was compared to standard care alone ($n = 11$) (education, manual therapy, and personalised exercise prescription) [84]. The NMES device delivered electrical stimulation at a frequency rate of 1 Hz and promoted venous return by stimulating the peroneal nerve, thereby increasing blood flow and subsequently reducing stasis and oedema. While the study found no between group differences in figure of 8 circumference measurements, function, or pain, the NMES had significantly reduced oedema as measured by the volumetric displacement method (mean change for standard care group: $-0.8 \pm 3.8\%$, NMES group: $8.1 \pm 11.2\%$) [84]. In contrast, in another study comparing NMES ($n = 15$) and Kinesio taping ($n = 16$) for the reduction of swelling following ankle sprain, neither intervention demonstrated effectiveness in improving ankle volume [85]. However, NMES was applied for just 30 min per day in this study, and in the previously described study, participants were required to wear the device for a minimum of eight and maximum of 16 hours per day and combined with standard rehabilitation care. An example application of NMES applied to improve recovery from an ankle sprain in demonstrated in Case Study 3 (Table 6.3).

6.2.2.5   Shoulder instability

Electrical muscle stimulation can also be considered as part of a rehabilitation programme for people with shoulder instability and there is emerging research to

Table 6.3   Clinical case study of a patient with an ankle sprain, presenting with pain, swelling, and loss of movement

Case Study 3 – Acute ankle sprain	
Body chart	Electrodes Inversion injury 2 days ago, presenting with pain, swelling, and loss of movement.
Clinical indication	Decrease oedema.
Indication for NMES	Patient treated with a multi-modal approach of education, exercise, and advice on how to reduce oedema utilising NMES. In this case, the electrodes are positioned to increase venous return in the lower leg by facilitating the normal 'calf muscle pump'.
Electrode positioning	Blood flow is increased by stimulating the common peroneal nerve to cause dorsiflexion of the foot by stimulation of the common peroneal nerve to activate the calf muscle pump. The active electrode is placed over the nerve where it is most superficial, just distal to the head of fibula, and the second electrode is placed distally on the muscle belly.
Stimulation settings	There are two options for stimulation. 1. 300 μs 1 Hz continuous. This provides a single impulse to cause a twitch of the foot. 2. 300 μs pulse width, 40 Hz, 0.5 seconds burst then rest for 2 seconds and then repeat, continuous. This causes a more sustained contraction and movement of the foot to replicate squeezing the calf.
NMES protocol	NMES is ideally used in the first seven days after injury, as part of a multi-modal approach to help accelerate the reduction of oedema. The patient should be advised to use NMES daily whilst resting for one or two sessions a day of 20–30 min each.

support its use. Treatment for shoulder instability varies, with surgical and non-surgical approaches considered. Non-surgical treatment typically aims to strengthen the muscles surrounding the shoulder joint, and NMES used in conjunction with these exercises can enhance muscle activation and strength. NMES can be used to target specific muscle groups that may be weakened or inhibited due to instability (e.g., Figure 6.3). As a result, muscle strength and neuromuscular

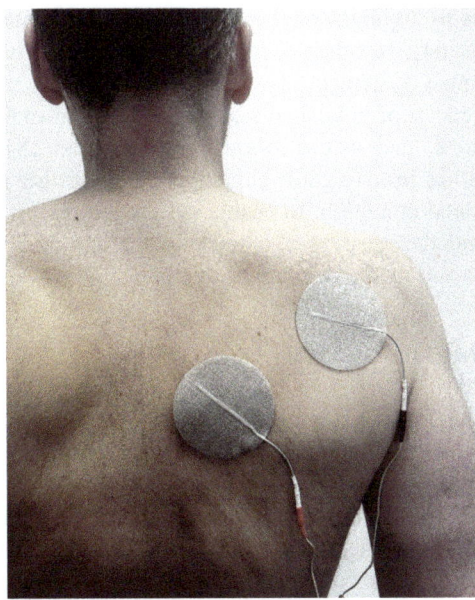

Figure 6.3   Example electrode placement for stimulation of the external rotators and scapula retractors

control may be improved, which can contribute to better shoulder stability. In a recent randomised controlled trial ($n = 49$), patients with functional posterior shoulder instability were allocated to either six weeks of physical therapy with simultaneous motion triggered NMES, or physical therapy alone [86]. The NMES was delivered at a frequency of 35 Hz to the external rotators and scapula retractors of the affected shoulder, at varying intensity based on the angle of the arm and the maximum intensity level set by the patient. The stimulation protocol was designed to produce a contraction while avoiding excessive muscle fatigue with the potential to induce supraspinal neural adaptions. The group that received NMES demonstrated a significantly better main outcome measurement in terms of the Western Ontario Shoulder Instability Index at three months ($64 \pm 16\%$ vs. $51 \pm 24\%$) [86]. The frequency of the instability episodes showed a significant improvement in the group that received NMES whereas the physical therapy group alone observed no significant difference [86].

6.2.3   Conclusion

The research evidence in NMES in musculoskeletal populations continues to increase, with emerging evidence demonstrating its benefit when compared to, or combined with, standard rehabilitation protocols. In total knee replacement, research has progressed sufficiently to inform clinical guideline publications, providing guidance on NMES settings and protocol [69]. Likewise, in ACL reconstruction, sufficient research has been published to enable systematic syntheses of

NMES interventions to highlight the benefit of devices and support translation to clinical practice [62–64]. In other conditions or surgeries, the lack of available evidence has prevented systematic evaluation, such as hip replacement surgery [41], or the heterogeneity of published studies has resulted in a lack of agreement regarding effectiveness, such as use following ankle sprain [81]. While the available research can still be used to guide clinical practice, further research is required in each musculoskeletal condition, to enable systematic review and meta-analyses to be conducted, and the creation of clinical guidelines, to further enhance the strength of evidence and recommendations available.

6.3   Current state of the art

The choice of equipment depends upon the type of service being delivered, the conditions of the patients being treated and what is available. Whatever the aim, it is essential the equipment chosen is fit for purpose and will deliver the stimulation parameters required. More details on equipment and the essential requirements and regulations that need to be met are provided in Chapter 3. However, it is worth noting here that there is some confusion when using neuromuscular stimulators in that not all are built to the medical device standard. Many of the lower-priced devices are sold as sports technology devices and as such do not meet the requirements of standards such as the Medical Device Regulations, CE marking or FDA approval. While such standards often give users peace of mind, the process of obtaining them is expensive, which is in part why there is a wide range of stimulator prices. In addition, one of the requirements of a medical device is that there is scientific evidence to support its effectiveness for the result it aims to achieve. Where a device claims benefits without evidence to support its effectiveness, it must be questioned whether it is likely to be effective.

Most people with a musculoskeletal condition seen in an electrical stimulation clinic will be referred for muscle strengthening and as such will require a simple exercise stimulator, working on a cyclical programme to provide the necessary amount of muscular activation. These exercises will normally be undertaken while at rest for the prescribed amount of time. However, on some occasions it might be necessary or preferred to link the muscle contractions to a specific action, such as at a given point in the gait cycle or a functional arm activity. In such cases the stimulation needs to be timed to these actions and triggered by a sensor on the user. This is like the use of FES in people with neurological disabilities where devices such as dropped foot stimulators are used to initiate foot dorsiflexion in people with stroke or multiple sclerosis (see Chapter 7). Therefore, the same equipment used for these people with neurological problems can be used in a musculoskeletal clinic. One possible application could be in people with repeated ankle sprains where the foot evertors can be stimulated, timed to the gait cycle, or in people with poor knee control where the vastus medialis can be stimulated at heel strike to provide greater knee stability [87,88]. A range of stimulators are available for such use and can be found on the Electrotherapy website (www.electrotherapy.org/manufacturers).

6.3.1 Practical clinical considerations

6.3.1.1 Clinical service considerations

As discussed in Chapters 1 and 2, muscles respond to the demands they are placed under, hence the difference in both the training regimes and the appearance of the musculature between marathon runners and sprinters. The same is true if the method of muscle activation is through electrical stimulation rather than voluntary contraction. Therefore, at the outset of treatment it is necessary to determine what electrical stimulation is being used to achieve, the final clinical outcome and how the success of the treatment goal is being measured. As the muscle response to stimulation is dose dependent, unless the treatment plan is based on sound physiological principles and the user follows that treatment plan, the chances of success are considerably reduced.

6.3.1.2 Example parameters

An example of a treatment goal may be to improve strength and function in an individual following joint replacement surgery or tendon or ligament reconstruction, to improve disuse atrophy and muscle weakness (Case Study 2, Table 6.2). Most commonly this will be for the lower limb conditions; however, the training principles are equally applicable to upper limb problems and the evidence for NMES use in these populations is increasing. For this treatment aim, a waveform of 40 pps is often a good place to start, with a pulse width of up to 300 μs and of sufficient amplitude to cause the required muscle contraction. This is a suitable stimulation regime for general muscle conditioning. If the aim is to increase strength, then higher frequencies are required of greater intensity and used less often, with stimulation until fatigue sets in (see Case Study 2). If the aim is to increase endurance, then stimulating at a lower frequency will minimise fatigue, and with a long on time, a short rest time and a longer stimulation period, muscle endurance can be improved (see Case Study 1).

It is difficult to be specific on choice of pulse width as some stimulators use pulse width adjustment of stimulation intensity, whereas others use adjustment of the stimulation current. However, the important thing is to get a visible, involuntary muscle contraction. The current/pulse width necessary to achieve this will depend upon; the size of the muscle, the condition of the skin, and how much subcutaneous fat there is. The electrodes will usually be positioned over the motor points of the muscle concerned and the distance between the electrodes will determine the depth the stimulating current penetrates. The size of electrodes needs to be chosen depending upon the muscle being treated with electrodes 70 mm or more in circumference often being used for large leg muscles such as quadriceps or gluteals, 50 mm diameter for lower leg and upper arm and as small as 25 mm for the forearm and hand. For a guide to muscle motor points, see Perotto AO 2011, *Anatomic guide for the electrode myographer (Fifth Edition)*, Pub Charles C Thomas Springfield, USA.

An on:off cycle of 1:1 with the stimulation envelopes ramping up and down, typically over one second, should be used with an on and off time of between 3 and 10 seconds, depending on the treatment aim (see Case Studies 1 and 2). Some

Table 6.4   Treatment targets and example outcome measures

Treatment target	Example outcome measures
Maximal muscle strength	Dynamometry – maximal voluntary contraction
Muscle endurance	Dynamometry – fatigue test
Function	Walking speed, sit-to-stand test, stair negotiation test, the timed up and go test, six-min walk test
Muscle size	Ultrasound assessment – cross-sectional area or muscle depth measurement. Limb circumference
Range of motion	Goniometry – active or passive range of motion
Swelling	Volumetric displacement test, limb circumference
Pain	Numeric rating scale, visual analogue scale
Symptoms	Patient reported outcome measures (PROMs) specific to condition

stimulators have such waveforms built in, whereas other will need to be manually configured. Example treatment programmes, including frequency and duration of individual sessions and overall stimulation protocol are provided in the following session. In general, duration will vary based on the condition being treated, the treatment goals and outcome measures of interest. Some example outcome measures to determine effectiveness of NMES protocols are provided in Table 6.4.

Selection of NMES programme will depend on their baseline levels and general physical status. One such treatment programme, running over a 12-week period is given below (Table 6.5). If a person is young, healthy and motivated with general good muscle tone, apart from the cause of the weakness, then a six-week protocol could be used with the time intervals scaled accordingly. If an average person is going through joint replacement surgery, a 12-week protocol may be more beneficial, for example, starting before surgery to improve muscle condition and then continuing for around six weeks after. For people with long-term chronic problems, or if any neurological involvement, then it's likely a longer protocol (e.g., six months) may be beneficial, with time intervals between each appointment scaled accordingly. The decision which protocol to use will be greatly determined by the attitude of the subject, their motivation and likelihood of compliance and their tolerance of NMES.

6.3.1.3   Adherence

While new technologies have the potential to revolutionise how we manage musculoskeletal health conditions, and recovery from orthopaedic surgery, successful implementation of new devices can only be achieved once widespread adoption has occurred. Successful implementation may be influenced by the strength of research evidence available, recommendations from clinical guidelines, and availability of training resources. It is likely that if a new technology appears difficult to implement, clinicians will be hesitant to adopt it into their practice. To consider whether patients are less likely to adhere to an intervention of electrical stimulation than

Table 6.5 Example treatment protocol – 12 weeks

Visit		
1	Initial assessment	Determine if muscles can be stimulated and that the patient is happy with the sensation. In most cases in MSK this and the set up can be undertaken at the same appointment. Teach the patient to use the equipment and ensure that they understand it and have realistic expectations.
2	Set up	Determine outcome measures and make baseline measurements. Start with low levels of stimulation to get a muscle contraction but not generating any great force. You should, however, be able to see a muscle contraction. Look at how the muscle contracts with stimulation, i.e. any fasciculation. Days 1–3 – 2 periods of 15 min per day Days 4–8 – 2 periods of 20 min Days 8–14 – 2 periods of 25 min Weeks 3 and 4 – 2 periods of 30 min
3	2 weeks	Phone to check no problems. If there are, arrange to see the patient. Adjust treatment regime accordingly.
4	4 weeks	Review treatment. Look at how the muscle contracts with stimulation, i.e. any fasciculation. Watch over a period of 10 min to see if any fatigue. Complete MRC scale with and without NMES. Ask about compliance. Repeat outcome measures. Increase stimulation intensity, possibly with knee over a roll and get heel to clear the bed when semi-recumbent. This will vary from patient to patient but just try and increase the strength of contraction/load. Can use weights on the ankle as well if the person is making good progress. Can ask them to work with the stimulation.
5	8 weeks	Review treatment. Look at how the muscle contracts with stimulation, i.e. any fasciculation. Watch over a period of 10 min to see if any fatigue. Do MRC scale with and without NMES. Ask about compliance. Repeat outcome measures. If patient is making good progress, and you want to specifically increase muscle strength, you could now use a higher stimulation frequency and intensity and you could ask the patient to work with the stimulator and look to apply an additional load. Examples of this could be in the case of a knee to have the leg over the edge of the bed initially with just the weight of the leg but then adding ankle weights. Alternately, if endurance is your main goal, reduce the frequency and increase the stimulation time. Adjust your outcome measures appropriately so that you can see if it is working. Will be very patient dependent.
6	12 weeks	Review treatment. Look at how the muscle contracts with stimulation, i.e. any fasciculation. Watch over a period of 10 min to see if any fatigue. Do MRC scale with and without NMES. Ask about compliance. Repeat outcome measures. Make this the last session and stop stimulation, asking patient to monitor function and strength over the next month.

(Continues)

Table 6.5   (Continued)

Visit
This should be the last day but it depends if patient is back to normal values. If not then continue treatment for a further 4 weeks, modified accordingly.
At the end of the 12 weeks, as the patient to have 2 weeks 'No Stimulation'. Get patient to replace NMES with normal exercise to maintain gains. Do not take stimulator from patient at this stage. (if rapid deterioration ask patient to phone in and restart with stimulation at once). Then reassess.
Phone patient after two weeks to check progress and ask for stimulator back if all going well.

traditional rehabilitation techniques (such as voluntary exercise or education), a systematic review recently examined adherence levels in studies of NMES in participants with hip or knee arthritis [89]. The review found that over the duration of the study period, mean adherence in the participants receiving NMES was $85\pm12\%$ (range: 55–99%), and $84\pm9\%$ (range: 78–97%) in the participants receiving exercise or education interventions [89]. Study retention rate in the NMES participants was $83\pm13\%$ (range: 55–100%) and $81\pm15\%$ in the participants receiving standard care, laser-therapy, sham stimulation, education, or voluntary exercise (range: 46–100%) [89]. There were no differences between the NMES and comparison/control groups in terms of adherence ($p = 0.97$) or retention rate ($p = 0.64$). In addition, no significant differences were observed in adherence rates between participants receiving supervised and unsupervised NMES, or between surgical or non-surgical participants [89].

Participants who were non-compliant reported that they did not like the device or did not want to be inconvenienced whilst recovering from surgery [89]. Other reasons for non-adherence and attrition related to the device included discomfort, dizziness, and pain. In some cases, it was not possible to determine the reason for study attrition, and therefore it is possible patients dropped out due to medical necessity, or other commitments, rather than the success or failure of the prescribed intervention [89]. While this review focused on patient-related factors, one study included in the synthesis discussed that therapists may be reluctant to push patients to tolerate uncomfortable doses of stimulation, which may limit the potential benefits of the treatment [90]. The following potential contributors to protocol adherence were discovered and can be used to inform clinical practice [89]:

1. Comprehensive NMES training to teach the participant how to use the stimulator and where to place the electrodes.
2. A familiarisation period with a clinician/researcher.
3. Supervision during stimulation.
4. Home-based monitoring/training.
5. Monitoring pain levels during stimulation.

6. A simple and clear NMES training protocol.
7. Built in adherence trackers.
8. Participant diaries/logbooks.
9. Written instructions to use devices in the home environment.
10. Setting stimulation intensity based on patient tolerance.
11. Marking electrode placement.

6.3.2 *Precautions and contraindications*

Looking at the instruction manual of any electrical stimulation device will show a long list under the heading of 'Precautions and Contraindication'. These may include the following that apply to the use of electrical stimulation in MSK:

- Not to use on people with a pacemaker, implanted defibrillator or any other implanted medical device.
- Do not stimulate over open or infected wounds.
- Do not stimulate in proximity to any cancerous tissue.
- Do not use in the presence of electronic monitoring equipment such as cardiac monitors.
- Risk of strangulation from the wires.
- Do not apply electrodes across the chest where the current path will flow through the heart
- Do not use in pregnancy.
- Do not use in people with poorly controlled epilepsy.
- Take extra care when applying stimulation to people with high level spinal cord injuries who suffer from autonomic dysreflexia. This is covered in more depth in Chapter 4.
- Do not use whilst sleeping.
- Do not use in the presence of high frequency surgical equipment.
- Do not use over implanted orthopaedic fixation systems.

The most commonly met contraindications in using electrical stimulation in musculoskeletal clinical practice are those patients with pacemakers, a history of epilepsy, and those with implanted or external metalwork Therefore, when taking a medical history and planning treatment, it is a good idea to have specific questions relating to these so that they are not overlooked. In addition to these specific patient groups, which will be considered in more detail below, it is essential to always consider the current path when applying electrical stimulation and to ensure that a direct line between the electrodes does not pass through any structures to be avoided such as the heart, cancerous tissue.

6.3.2.1 Pacemakers and implanted defibrillators

Because many of the people with a musculoskeletal condition seen in an electrical stimulation clinic are elderly, such as after joint replacement, they will often have multiple pathologies and hence it is possible that they have been fitted with an implanted device. As a result, there have been several studies and literature reviews in order to determine whether or not problems are likely to occur [65,91]. Again, it is the current path that is important and such studies have shown that with modern pacemakers and defibrillators stimulation of the limbs when both electrodes are on the same limb are safe and hence FES for drop foot could be considered safe in patients with pacemakers and ICDs. Several short-term safety studies investigating lower limb NMES have concluded that it is safe to use in patients with pacemakers and ICDs; and there have been no case studies describing electromagnetic inter-ference (EMI) when stimulation is applied to the lower limb. However, a definitive safety study that is intervention specific, long term, and high powered is needed to identify the exact risk. Until such a study is realised, electrical stimulation should be used with caution in patients with pacemakers and ICDs. We recommend that pacemaker and ICD function should be monitored for EMI when stimulation is first used. This is more important when stimulation is applied to the upper limb or torso as the risk of interference increases the closer the electrodes are to the active device.

6.3.2.2   Uncontrolled epilepsy

Taking a detailed medical history is always essential to identify if a given patient has a history of epilepsy, and if so, if that epilepsy is well controlled. It is recom-mended that people need to have been seizure free for at least three months. Given this provision, a history of epilepsy itself is not a contraindication, but such people will require more reassurance. Asking them if they have any advanced warning of a coming seizure might also be useful and they should be carefully observed at the initial assessment when first trying stimulation.

6.3.2.3   Implanted metal work

An important consideration, particularly in musculoskeletal applications of elec-trical stimulation, is that the person might have implanted metalwork. Any implanted metalwork for fracture fixation will alter the current path if that metal-work is between the two electrodes, as the metal will be more conductive than the surrounding tissues. In such cases, careful consideration and alterative electrode placements will be necessary. For people with joint replacements however, elec-trical stimulation will usually be used on the limb segment either above or below the implant and hence the prosthesis will not be in the current path. Areas of external mechanical fixation should be avoided.

6.4   Suggestions for future work

Further clinical research is required in NMES for strength improvement in specific musculoskeletal populations. At present, there is emerging research to justify and support its use across a range of patient populations, however, the small volume of research and heterogeneous nature of published studies currently prevents clinical practice guidelines being created for most conditions. For example, evidence for NMES use has progressed sufficiently to inform clinical guideline publications and enable systematic analyses in procedures such as total knee replacement and ACL reconstruction, yet in procedures such as hip replacement, should replacement, rotator cuff repair and conditions such as ankle sprain, shoulder instability, rheumatoid arthritis and osteoporosis, further research is required. To produce reliable results, randomised controlled trials could be considered to compare electrical stimulation to current standard care in a longitudinal study design, to evaluate if the effects of NMES can be maintained long-term. In addition, in hospitals where there is currently no standard therapy practice intervention, NMES could be implemented and monitored longitudinally to assess its impact on recovery targets. Furthermore, including measures of physical activity participation may help to understand whether strength gains in the early postoperative period translate to greater participation in physical activity in the months and years following surgery. Finally, evaluations of the cost effectiveness of NMES therapy compared to current practice may also be useful to inform implementation and adoption into healthcare settings.

References

[1]   National Academies of Sciences Engineering and Medicine, Health and Medicine Division, Board on Health Care Services, Treatment CoIDMCLtlw. *Musculoskeletal Disorders. Selected Health Conditions and Likelihood of Improvement with Treatment.* Washington (DC): National Academies Press; 2020.

[2]   Swain S, Sarmanova A, Coupland C, Doherty M, and Zhang W. Comorbidities in osteoarthritis: a systematic review and meta-analysis of observational studies. *Arthritis Care Res (Hoboken)*. 2020;72(7):991–1000.

[3]   Kottke F, Stillwell G, and Lehmann J. *Krusen's Handbook of Physical Medicine and Rehabilitation*. 3rd edn. Philadelphia: WB Saunders Co; 1982.

[4]   Versus Arthritis. *The State of Musculoskeletal Health 2023*. 2023. https://www.versusarthritis.org/media/duybjusg/versus-arthritis-state-msk-musculoskeletal-health-2023pdf.pdf

[5]   NHS Digital. *Health Survey for England 2018: Longstanding Conditions.* Health and Social Care Information Centre; 2018.

[6]   Fanzani A, Conraads VM, Penna F, and Martinet W. Molecular and cellular mechanisms of skeletal muscle atrophy: an update. *J Cachexia Sarcopenia Muscle*. 2012;3(3):163–79.

[7]   Kandarian SC, and Stevenson EJ. Molecular events in skeletal muscle during disuse atrophy. *Exerc Sport Sci Rev*. 2002;30(3):111–6.

[8]   Boonyarom O, and Inui K. Atrophy and hypertrophy of skeletal muscles: structural and functional aspects. *Acta Physiol (Oxf)*. 2006;188(2):77–89.

[9]   Bogdanis GC. Effects of physical activity and inactivity on muscle fatigue. *Front Physiol*. 2012;3:142.

[10] Lepley LK, Davi SM, Burland JP, and Lepley AS. Muscle atrophy after ACL injury: implications for clinical practice. *Sports Health*. 2020;12(6):579–86.

[11]   Rice DA, and McNair PJ. Quadriceps arthrogenic muscle inhibition: neural mechanisms and treatment perspectives. *Semin Arthritis Rheum.* 2010;40 (3):250–66.

[12]   Lepley LK, Stoneback L, Macpherson PCD, and Butterfield TA. Eccentric exercise as a potent prescription for muscle weakness after joint injury. *Exerc Sport Sci Rev.* 2023;51(3):109–16.

[13]   Sartori R, Romanello V, and Sandri M. Mechanisms of muscle atrophy and hypertrophy: implications in health and disease. *Nat Commun.* 2021;12 (1):330.

[14]   Burgess LC. *Neuromuscular Electrical Stimulation to Improve Muscle Weakness in Hip Osteoarthritis: A Feasibility Study.* Poole: Bournemouth University; 2023.

[15]   Lake DA. Neuromuscular electrical stimulation. An overview and its application in the treatment of sports injuries. *Sports Med.* 1992;13(5):320–36.

[16]   Jubeau M, Gondin J, Martin A, Sartorio A, and Maffiuletti NA. Random motor unit activation by electrostimulation. *Int J Sports Med.* 2007;28 (11):901–4.

[17]   Bickel CS, Gregory CM, and Dean JC. Motor unit recruitment during neuromuscular electrical stimulation: a critical appraisal. *Eur J Appl Physiol.* 2011;111(10):2399–407.

[18]   Gregory CM, and Bickel CS. Recruitment patterns in human skeletal muscle during electrical stimulation. *Phys Ther.* 2005;85(4):358–64.

[19]   Paravlic AH, Kovac S, Pisot R, and Marusic U. Neurostructural correlates of strength decrease following total knee arthroplasty: a systematic review of the literature with meta-analysis. *Bosn J Basic Med Sci.* 2020;20(1):1–12.

[20]   Wallis JA, and Taylor NF. Pre-operative interventions (non-surgical and non-pharmacological) for patients with hip or knee osteoarthritis awaiting joint replacement surgery–a systematic review and meta-analysis. *Osteoarthritis Cartilage.* 2011;19(12):1381–95.

[21]   Wang L, Lee M, Zhang Z, Moodie J, Cheng D, and Martin J. Does preoperative rehabilitation for patients planning to undergo joint replacement surgery improve outcomes? A systematic review and meta-analysis of randomised controlled trials. *BMJ Open.* 2016;6(2):e009857.

[22]   Almeida GJ, Khoja SS, and Zelle BA. Effect of prehabilitation in older adults undergoing total joint replacement: an overview of systematic reviews. *Curr Geriatr Rep.* 2020;9(4):280–7.

[23]   Pearse EO, Caldwell BF, Lockwood RJ, and Hollard J. Early mobilisation after conventional knee replacement may reduce the risk of postoperative venous thromboembolism. *J Bone Joint Surg Br.* 2007;89(3):316–22.

[24]   Jorgensen CC, Jacobsen MK, Soeballe K, *et al.* Thromboprophylaxis only during hospitalisation in fast-track hip and knee arthroplasty, a prospective cohort study. *BMJ Open.* 2013;3(12):e003965.

[25]   Mak JC, Fransen M, Jennings M, *et al.* Evidence-based review for patients undergoing elective hip and knee replacement. *ANZ J Surg.* 2014;84 (1–2):17–24.

[26]   Wainwright TW, and Burgess LC. To what extent do current total hip and knee replacement patient information resources adhere to enhanced recovery after surgery principles? *Physiotherapy*. 2018;104(3):327–37.

[27]   Gavin J, Immins T, Burgess LC, and Wainwright, TW. Functional sit-to-stands evoke greater neuromuscular activation than orthopaedic bed exercises in healthy older adults. *Isokinet Exerc Sci*. 2018;26(2):10.

[28]   Smith TO, Mann CJ, Clark A, and Donell ST. Bed exercises following total hip replacement: 1 year follow-up of a single-blinded randomised controlled trial. *Hip Int*. 2009;19(3):268–73.

[29]   Smith T, Mann C, Clark C, and Donnell S. Bed exercises following total hip replacement: a randomised controlled trial. *Physiotherapy* 2008;94(4):286–91.

[30]   Maestroni L, Read P, Bishop C, *et al.* The benefits of strength training on musculoskeletal system health: practical applications for interdisciplinary care. *Sports Med*. 2020;50(8):1431–50.

[31]   Saueressig T, Owen PJ, Zebisch J, Herbst M, and Belavy DL. Evaluation of exercise interventions and outcomes after hip arthroplasty: a systematic review and meta-analysis. *JAMA Netw Open*. 2021;4(2):e210254.

[32]   Su W, Zhou Y, Qiu H, and Wu H. The effects of preoperative rehabilitation on pain and functional outcome after total knee arthroplasty: a meta-analysis of randomized controlled trials. *J Orthop Surg Res*. 2022;17(1):175.

[33]   Jones S, Man WD, Gao W, Higginson IJ, Wilcock A, and Maddocks M. Neuromuscular electrical stimulation for muscle weakness in adults with advanced disease. *Cochrane Database Syst Rev*. 2016;10(10):CD009419.

[34]   Monaghan B, Caulfield B, and O'Mathuna DP. Surface neuromuscular electrical stimulation for quadriceps strengthening pre and post total knee replacement. *Cochrane Database Syst Rev*. 2010;20(1):CD007177.

[35]   Oldham J, Howe T, Petterson T, Smith G, and Tallis R. Electrotherapeutic rehabilitation of the quadriceps in elderly osteoarthritis patients. A double blind assessment of patterned neuromuscular electrical stimulation. *Clin Rehabil*. 1995;9:10–20.

[36]   Stevens JE, Mizner RL, and Snyder-Mackler L. Quadriceps strength and volitional activation before and after total knee arthroplasty for osteoarthritis. *J Orthop Res*. 2003;21(5):775–9.

[37]   Volpato HB, Szego P, Lenza M, Milan SL, Talerman C, and Ferretti M. Femoral quadriceps neuromuscular electrical stimulation after total knee arthroplasty: a systematic review. *Einstein (Sao Paulo)*. 2016;14(1):77–98.

[38]   Kittelson AJ, Stackhouse SK, and Stevens-Lapsley JE. Neuromuscular electrical stimulation after total joint arthroplasty: a critical review of recent controlled studies. *Eur J Phys Rehabil Med*. 2013;49(6):909–20.

[39]   Bistolfi A, Zanovello J, Ferracini R, *et al.* Evaluation of the effectiveness of neuromuscular electrical stimulation after total knee arthroplasty: a meta-analysis. *Am J Phys Med Rehabil*. 2018;97(2):123–30.

[40]   Klika AK, Yakubek G, Piuzzi N, Calabrese G, Barsoum WK, and Higuera CA. Neuromuscular electrical stimulation use after total knee arthroplasty improves early return to function: a randomized trial. *J Knee Surg*. 2022;35(1):104–11.

[41]   Burgess LC, Swain ID, Taylor P, and Wainwright TW. Strengthening quadriceps muscles with neuromuscular electrical stimulation following total hip replacement: a review. *Curr Phys Med Rehabil Rep*. 2019;7(3):275–83.

[42]   Suetta C, Magnusson SP, Rosted A, *et al*. Resistance training in the early postoperative phase reduces hospitalization and leads to muscle hypertrophy in elderly hip surgery patients–a controlled, randomized study. *J Am Geriatr Soc*. 2004;52(12):2016–22.

[43]   Suetta C, Andersen JL, Dalgas U, *et al*. Resistance training induces qualitative changes in muscle morphology, muscle architecture, and muscle function in elderly postoperative patients. *J Appl Physiol (1985)*. 2008;105 (1):180–6.

[44]   Gremeaux V, Renault J, Pardon L, Deley G, Lepers R, and Casillas JM. Low-frequency electric muscle stimulation combined with physical therapy after total hip arthroplasty for hip osteoarthritis in elderly patients: a randomized controlled trial. *Arch Phys Med Rehabil*. 2008;89(12):2265–73.

[45]   Wainwright TW, Burgess LC, and Middleton RG. A single-centre feasibility randomised controlled trial comparing the incidence of asymptomatic and symptomatic deep vein thrombosis between a neuromuscular electro-stimulation device and thromboembolism deterrent stockings in post-operative patients recovering from elective total hip replacement surgery. *Surg Technol Int*. 2020;28(36):289–98.

[46]   Zhao Y, Zhang B, Wang Y, and Liu J. Effectiveness of neuromuscular electrical stimulation for enhanced recovery after total hip replacement surgery: a randomized controlled trial. *Geriatr Orthop Surg Rehabil*. 2022; 13:21514593221129528.

[47]   Demircioglu DT, Paker N, Erbil E, Bugdayci D, and Emre TY. The effect of neuromuscular electrical stimulation on functional status and quality of life after knee arthroplasty: a randomized controlled study. *J Phys Ther Sci*. 2015;27(8):2501–6.

[48]   Hughes SF, Hendricks BD, Edwards DR, Bastawrous SS, and Middleton JF. Lower limb orthopaedic surgery results in changes to coagulation and non-specific inflammatory biomarkers, including selective clinical outcome measures. *Eur J Med Res*. 2013;18(1):40.

[49]   Tucker A, Maass A, Bain D, *et al*. Augmentation of venous, arterial and microvascular blood supply in the leg by isometric neuromuscular stimulation via the peroneal nerve. *Int J Angiol*. 2010;19(1):e31–7.

[50]   Kadi MR, Hepguler S, Atamaz FC, *et al*. Is interferential current effective in the management of pain, range of motion, and edema following total knee arthroplasty surgery? A randomized double-blind controlled trial. *Clin Rehabil*. 2019;33(6):1027–34.

[51]   Beckwee D, Bautmans I, Lefeber N, Lievens P, Scheerlinck T, and Vaes P. Effect of transcutaneous electric nerve stimulation on pain after total knee arthroplasty: a blind randomized controlled trial. *J Knee Surg*. 2018;31 (2):189–96.

[52]   Lee JH, Chun YM, Kim DS, Lee DH, and Shin SJ. Effects of neuromuscular electrical muscle stimulation on the deltoid for shoulder function restoration after reverse total shoulder arthroplasty in the early recovery period: a prospective randomized study. *Arch Orthop Trauma Surg.* 2023;143(6):3037–46.

[53]   Reinold MM, Macrina LC, Wilk KE, Dugas JR, Cain EL, and Andrews JR. The effect of neuromuscular electrical stimulation of the infraspinatus on shoulder external rotation force production after rotator cuff repair surgery. *Am J Sports Med.* 2008;36(12):2317–21.

[54]   Lee GJ, Cho H, Ahn BH, and Jeong HS. Effects of electrical muscle stimulation for preventing deltoid muscle atrophy after rotator cuff repair: preliminary results of a prospective, randomized, single-blind trial. *Clin Shoulder Elb.* 2019;22(4):195–202.

[55]   Ireland ML. The female ACL: why is it more prone to injury? *Orthop Clin North Am.* 2002;33(4):637–51.

[56]   Moses B, Orchard J, and Orchard J. Systematic review: annual incidence of ACL injury and surgery in various populations. *Res Sports Med.* 2012;20 (3–4):157–79.

[57]   Grindem H, Snyder-Mackler L, Moksnes H, Engebretsen L, and Risberg MA. Simple decision rules can reduce reinjury risk by 84% after ACL reconstruction: the Delaware-Oslo ACL cohort study. *Br J Sports Med.* 2016;50(13):804–8.

[58]   Kotsifaki R, Korakakis V, King E, *et al.* Aspetar clinical practice guideline on rehabilitation after anterior cruciate ligament reconstruction. *Br J Sports Med.* 2023;57(9):500–14.

[59]   Badawy CR, Jan K, Beck EC, *et al.* Contemporary principles for post-operative rehabilitation and return to sport for athletes undergoing anterior cruciate ligament reconstruction. *Arthrosc Sports Med Rehabil.* 2022;4(1): e103–e13.

[60]   Culvenor AG, Girdwood MA, Juhl CB, *et al.* Rehabilitation after anterior cruciate ligament and meniscal injuries: a best-evidence synthesis of systematic reviews for the OPTIKNEE consensus. *Br J Sports Med.* 2022;56 (24):1445–53.

[61]   Arna Risberg M, Lewek M, and Snyder-Mackler L. A systematic review of evidence for anterior cruciate ligament rehabilitation: how much and what type? *Phys Ther Sport.* 2004;5(3):125–45.

[62]   Kim KM, Croy T, Hertel J, and Saliba S. Effects of neuromuscular electrical stimulation after anterior cruciate ligament reconstruction on quadriceps strength, function, and patient-oriented outcomes: a systematic review. *J Orthop Sports Phys Ther.* 2010;40(7):383–91.

[63]   Gatewood CT, Tran AA, and Dragoo JL. The efficacy of post-operative devices following knee arthroscopic surgery: a systematic review. *Knee Surg Sports Traumatol Arthrosc.* 2017;25(2):501–16.

[64]   Hauger AV, Reiman MP, Bjordal JM, Sheets C, Ledbetter L, and Goode AP. Neuromuscular electrical stimulation is effective in strengthening the quadriceps muscle after anterior cruciate ligament surgery. *Knee Surg Sports Traumatol Arthrosc.* 2018;26(2):399–410.

[65]   Nussbaum EL, Houghton P, Anthony J, Rennie S, Shay BL, and Hoens AM. Neuromuscular electrical stimulation for treatment of muscle impairment: critical review and recommendations for clinical practice. *Physiother Can.* 2017;69(5):1–76.

[66]   Labanca L, Rocchi JE, Laudani L, *et al.* Neuromuscular electrical stimulation superimposed on movement early after ACL surgery. *Med Sci Sports Exerc.* 2018;50(3):407–16.

[67]   Giggins O, Fullen B, and Coughlan G. Neuromuscular electrical stimulation in the treatment of knee osteoarthritis: a systematic review and meta-analysis. *Clin Rehabil.* 2012;26(10):867–81.

[68]   Li H, Zeng C, and Lei GH. Comment on 'Neuromuscular electrical stimulation in the treatment of knee osteoarthritis: a systematic review and meta-analysis'. *Clin Rehabil.* 2014;28(11):1145–6.

[69]   Novak S, Guerron G, Zou Z, Cheung G, and Berteau JP. New guidelines for electrical stimulation parameters in adult patients with knee osteoarthritis based on a systematic review of the current literature. *Am J Phys Med Rehabil.* 2020;99(8):682–8.

[70]   Burgess LC, Taylor P, Wainwright TW, and Swain ID. Lab-based feasibility and acceptability of neuromuscular electrical stimulation in hip osteoarthritis rehabilitation. *J Rehabil Assist Technol Eng.* 2021;8:2055668320980613.

[71]   Benton LA, Baker LL, Bowman BR, and Waters RL. *Functional Electrical Stimulation–A Practical Clinical Guide.* 2nd edn. California: USA Rancho Los Amigos; 1981.

[72]   Doheny EP, Caulfield BM, Minogue CM, and Lowery MM. Effect of subcutaneous fat thickness and surface electrode configuration during neuromuscular electrical stimulation. *Med Eng Phys.* 2010;32(5):468–74.

[73]   Zacharias A, Pizzari T, English DJ, Kapakoulakis T, and Green RA. Hip abductor muscle volume in hip osteoarthritis and matched controls. *Osteoarthritis Cartilage.* 2016;24(10):1727–35.

[74]   Rasch A, Bystrom AH, Dalen N, and Berg HE. Reduced muscle radiological density, cross-sectional area, and strength of major hip and knee muscles in 22 patients with hip osteoarthritis. *Acta Orthop.* 2007;78(4):505–10.

[75]   Almeida GJ, Khoja SS, and Piva SR. Dose-response relationship between neuromuscular electrical stimulation and muscle function in people with rheumatoid arthritis. *Phys Ther.* 2019;99(9):1167–76.

[76]   Brosseau LU, Pelland LU, Casimiro LY, Robinson VI, Tugwell PE, and Wells GE. Electrical stimulation for the treatment of rheumatoid arthritis. *Cochrane Database Syst Rev.* 2002;2002(2):CD003687.

[77]   Oldham JA, and Stanley JK. Rehabilitation of atrophied muscle in the rheumatoid arthritic hand: a comparison of two methods of electrical stimulation. *J Hand Surg Br.* 1989;14(3):294–7.

[78]   Piva SR, Khoja SS, Toledo FGS, *et al.* Neuromuscular electrical stimulation compared to volitional exercise for improving muscle function in rheumatoid arthritis: a randomized pilot study. *Arthritis Care Res (Hoboken).* 2019;71 (3):352–61.

[79] Zhang W, Luo Y, Xu J, *et al.* The possible role of electrical stimulation in osteoporosis: a narrative review. *Medicina (Kaunas).* 2023;59(1):121.

[80] Gaddi D, Mosca A, Piatti M, *et al.* Acute ankle sprain management: an umbrella review of systematic reviews. *Front Med (Lausanne).* 2022; 9:868474.

[81] Feger MA, Goetschius J, Love H, Saliba SA, and Hertel J. Electrical stimulation as a treatment intervention to improve function, edema or pain following acute lateral ankle sprains: a systematic review. *Phys Ther Sport.* 2015;16(4):361–9.

[82] Fong DT, Wang D, Chu VW, and Chan KM. Myoelectric stimulation on peroneal muscles with electrodes of the muscle belly size attached to the upper shank gives the best effect in resisting simulated ankle sprain motion. *J Biomech.* 2013;46(6):1088–91.

[83] Fong DT, Chu VW, and Chan KM. Myoelectric stimulation on peroneal muscles resists simulated ankle sprain motion. *J Biomech.* 2012;45(11):2055–7.

[84] Wainwright TW, Burgess LC, and Middleton RG. Does neuromuscular electrical stimulation improve recovery following acute ankle sprain? A pilot randomised controlled trial. *Clin Med Insights Arthritis Musculoskelet Disord.* 2019;12:1179544119849024.

[85] Mazloum V, Akbari H, and Gholampour A. The comparison of the effects of neuromuscular electrical stimulation and Kinesio Taping on ankle swelling in athletes with lateral ankle sprain. *J Exp Orthop.* 2023;10(1):63.

[86] Moroder P, Karpinski K, Akgun D, *et al.* Neuromuscular electrical stimulation-enhanced physical therapist intervention for functional posterior shoulder instability (type B1): a multicenter randomized controlled trial. *Phys Ther.* 2023.

[87] Bamber ZA, Sun W, Menon RS, Wheeler PC, Swain ID, and Fong DTP. Functional electrical stimulation of peroneal muscles on balance in healthy females. *Cyborg Bionic Syst.* 2021;2021:9801097.

[88] Bamber ZA, Wheeler PC, Swain ID, and Fong DT. Effect of 8-week treadmill running with peroneal muscle functional electrical stimulation on laterally deviated centre of plantar pressure position and star excursion balance test performance. *J Rehabil Assist Technol Eng.* 2021;8:20556683211021526.

[89] Burgess LC, Taylor P, Wainwright TW, Bahadori S, and Swain ID. Adherence to neuromuscular electrical stimulation interventions for muscle impairment in hip and knee osteoarthritis: a systematic review. *Clin Med Insights Arthritis Musculoskelet Disord.* 2021;14:11795441211028746.

[90] Stevens-Lapsley JE, Balter JE, Wolfe P, Eckhoff DG, and Kohrt WM. Early neuromuscular electrical stimulation to improve quadriceps muscle strength after total knee arthroplasty: a randomized controlled trial. *Phys Ther.* 2012; 92(2):210–26.

[91] Badger J, Taylor P, and Swain I. The safety of electrical stimulation in patients with pacemakers and implantable cardioverter defibrillators: a systematic review. *J Rehabil Assist Technol Eng.* 2017;4:2055668317745498.

Chapter 7

FES to support walking in people with UMN lesions

Marietta L. van der Linden[1], Christine Singleton[2] and Paul Taylor[3]

7.1 Introduction

The most common use of functional electrical stimulation (FES) in clinical practice is to aid walking in people with an upper motor neuron (UMN) lesion [1]. UMN lesions are due to strokes, multiple sclerosis, cerebral palsy, head injuries or Parkinson's disease. These conditions are commonly associated with muscle weakness, impaired motor coordination, spasticity and altered balance mechanisms leading to a range of gait impairments. The peripheral nerves in people with UMN lesions are mostly intact. FES is a treatment option to manage gait impairments in this population through the stimulation of a muscle or muscles at an appropriate time in the gait cycle. FES can improve walking through the stimulation of just one muscle, using a single channel FES device or two or more muscles using a multi-channel FES device. Which device is most suitable for a patient depends on the nature and complexity of their walking difficulties. FES can support walking by stimulation applied to muscles in one or both legs.

Over the last two decades, FES to support walking has developed from a research tool to become a widely used clinical treatment in over 70 countries with a range of commercial devices. There are no published data on the number of FES trained therapists nor on the number of patients treated worldwide. However, it is estimated that by the end of 2023 over 3000 clinicians had attended FES training courses and over 27,000 people with UMN lesions had been provided with FES devices in the UK alone [2].

In this chapter, the term 'functional electrical stimulation' is used to mean electrical stimulation of a muscle or muscles to achieve a functional movement or purpose such as walking. Where the primary objective is muscle conditioning

[1]Centre for Health, Activity and Rehabilitation Research, Queen Margaret University, Edinburgh, UK
[2]Functional Electrical Stimulation (FES) Clinic, Birmingham Community Healthcare NHS Foundation Trust, UK
[3]Bournemouth University and Odstock Medical Limited, The National Clinical FES Centre, Salisbury District Hospital, UK

rather than directly assisting a functional movement, we use the term Neuromuscular Electrical Stimulation (NMES). Although the focus of this chapter is on FES, it is worth mentioning that NMES aimed at muscle conditioning may be considered prior to providing FES for some patients. Especially in people with spinal cord injuries (SCIs), muscles may be severely atrophied through disuse due to loss of normal muscle activation which means that FES may not elicit the required response [3]. In children with CP, using NMES prior fitting FES at short period of time may also serve to get them used to the sensation of FES [4].

7.2   Normal gait

It is useful to have a basic knowledge of what non-pathological ('normal') gait looks like to understand the gait abnormalities experienced by people living with an UMN lesion. Normal gait is often illustrated through the description of the joint kinematics i.e., how the ankle, knee and hip joint angles change over the duration of one gait cycle. A gait cycle is often defined as starting at the time of initial contact, which is in normal gait the instant the heel strikes the ground and ends with the same foot striking the ground again (Figure 7.1). A gait cycle consists of a

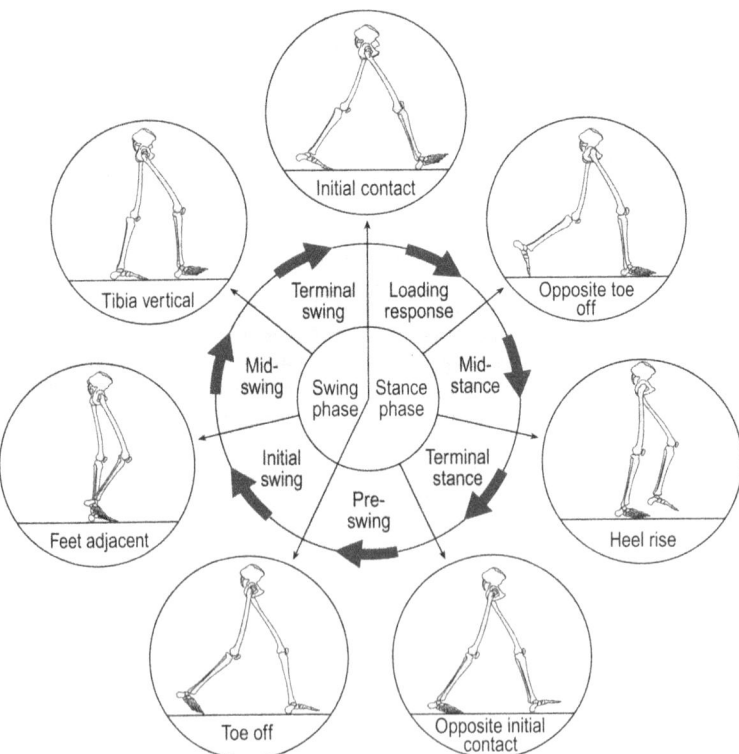

Figure 7.1   Position of the leg during the phases in a normal gait cycle (adapted from [7])

stance phase (the foot is in contact with the floor) and a swing phase (the foot is lifted in the air).

Currently, FES is mostly used to correct dropped foot (or foot drop or drop foot). Taking this gait impairment as an example, Figure 7.2 shows the ankle joint kinematics in normal gait and in a person with dropped foot for whom dorsiflexion of the ankle during the swing phase is absent.

Most of the lower limb movements during gait, such as dorsiflexion of the ankle and flexion of the knee and hip occur in the sagittal plane. The sagittal plane is the vertical plane in the direction of walking which divides the body into a left section and a right section. However, movements out with the sagittal plane are equally important for an efficient gait pattern and are also relevant for the application of FES to aid walking. For example, FES to the muscles around the ankle can influence the amount of ankle inversion and eversion while FES.

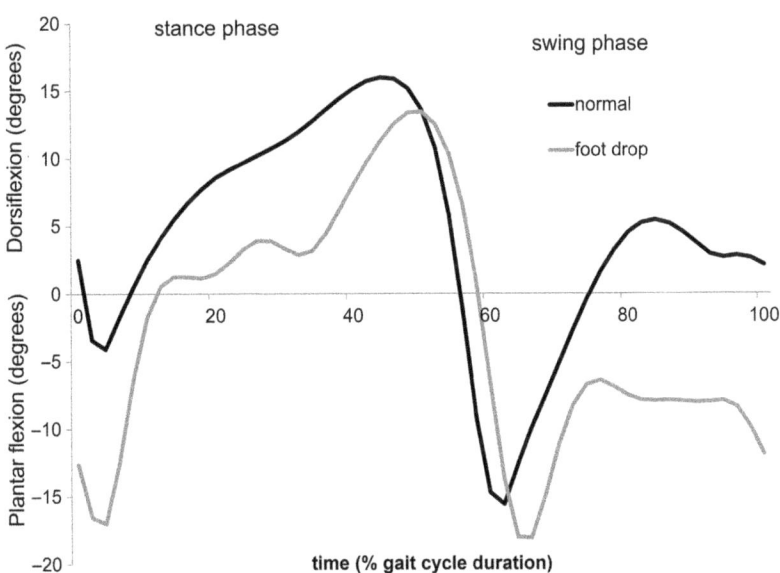

Figure 7.2 Example ankle angle kinematics of a person experiencing dropped foot (grey line) and a healthy control (black line). The horizontal x axis represents the duration of one gait cycle, i.e. from initial contact of the foot with the ground (0%) through the stance phase until approximately 60% when the toe leaves the ground to 100% when the same foot strikes the ground again. The ankle of the person with dropped foot is plantar flexed at initial contact (time 0%) and stays plantar flexed during the swing phase. In contrast, the ankle of the healthy control is dorsiflexed during the latter part of the swing phase to ensure ground clearance and prepare for initial contact by the heel.

Electromyography (EMG) is another type of gait analysis and provides information about when in the gait cycle a certain muscle is activated. Figure 7.3 shows the activation pattern of the major muscle groups in the lower limbs during normal gait. The muscle actions that are most relevant to the scope of this chapter are (1) the activation of the tibialis anterior muscle which is active throughout the swing phase, (2) the action of the gluteus maximus, hamstrings and quadriceps muscles which act to stabilise the hip and knee during the stance phase and (3) the action of the hamstrings during the swing phase of gait which regulates the amount of knee flexion in swing.

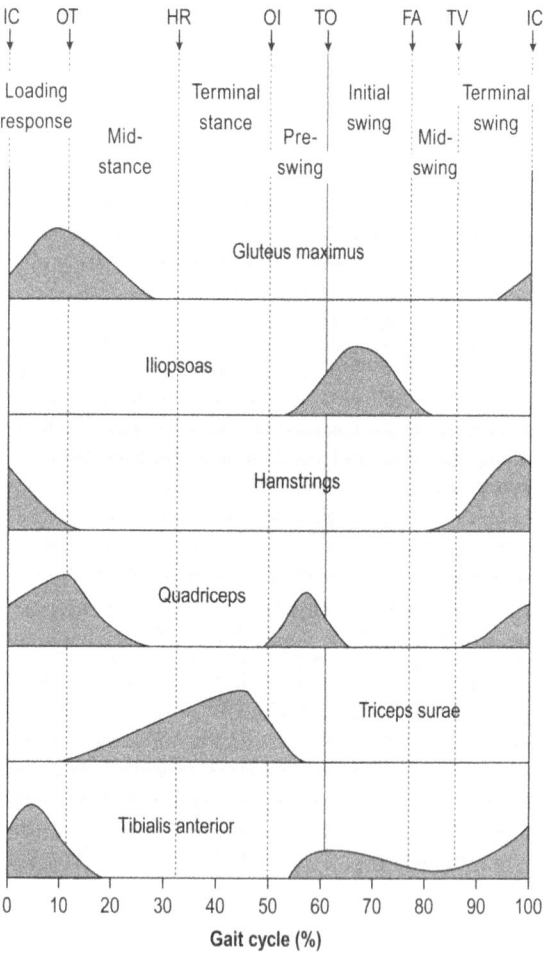

Figure 7.3   Typical activity of the major muscle groups during the gait cycle (adapted from [7]). The timings of the gait cycle events are typical and not derived from a single individual. IC: Initial contact, OT: Opposite foot contact, HR: heel rise, OI: opposite initial contact, FA: feet adjacent, TV: Tibia Vertical.

A detailed description of the joint kinematics and muscle action during gait is out with the scope of this chapter but can be found in textbooks on gait and gat analysis [5–7].

7.3 Common motor symptoms in people living with an UMN lesion

Spasticity is common in most health conditions due to a UMN lesion and is often described as a resistance to stretch of a muscle during a certain movement. The resistance to stretch increases if the speed of the joint movement increases. If spasticity is present in their leg muscles, it can interfere with the person's gait. Some authors distinguish tone from spasticity where tone is resistance to stretch which is not dependent on the speed of the stretching movement [8].

Weakness or impaired muscle strength in people with an UMN lesion can be a direct result of the UMN lesion manifesting as impaired voluntary motor control, reduced neural drive or increased spasticity in the antagonist. However, loss of muscle strength can also be due to muscle atrophy secondary to other UMN lesion motor symptoms [3].

Reduced selective motor control is a common negative sign in people with an UMN lesion and can be defined as the inability or reduced ability to isolate the activation of a muscles or muscle group in a selected pattern [8]. A loss of selective voluntary motor control can also manifest as an associated movement pattern in the joints of the same limb (e.g., flexing the hip when aiming to flex only the knee) and mirror movements (e.g., dorsiflexing the left ankle when aiming to dorsiflex only the right ankle) [9].

Co-contractions of antagonists, i.e. muscles with opposing actions such as flexing or extending the knee in normal gait aid stability and efficient force production. However, in people with an UMN lesion co-contraction can also be a sign of reduced motor control and result in a stiff, inefficient gait pattern which may lead to an increased energy cost of walking [10].

7.4 Walking impairments in populations with a UMN lesion and how FES can help

This section includes a short description of the most common health conditions involving an UMN lesion, gait impairments commonly observed in people with these health conditions and how these gait impairments can be managed by FES. For a detailed description of these FES applications in terms of stimulation characteristics and electrode placement, see sections on stimulation for Drop Foot below.

7.4.1 Stroke

A stroke is caused by a clot or bleed within the brain or spine (spinal stroke). Reduced blood flow causes death of brain cells which can lead to altered motor and sensory function as a consequence. It is estimated that every year 15 million people worldwide experience a new stroke [11] and that of those, 80% experience reduced motor control and more than 30% are not able to walk unsupervised in the community [12]. Most of the natural recovery is thought to occur within the first six months after stroke, although this may depend on the individual and the difficult of the activity assessed [13].

Dropped foot in stroke survivors is a common gait deficit affecting approximately 20% of the stroke survivors [12]. In people experiencing dropped foot, there is a reduced or absence of ankle dorsiflexion during the swing phase of gait. The dropped foot can lead to trips and falls due to insufficient foot clearance (the foot is too close to the ground). Dropped foot can result from muscle weakness, spasticity in the calf muscle, impaired motor control or a combination of these. FES applied to the dorsiflexors during the swing phase can be used to manage dropped foot and is the most common application of FES to support walking in stroke survivors.

Some strokes survivors may benefit from stimulation of another muscle group in addition to the dorsiflexors. For example, hamstring stimulation to increase knee flexion during the swing phase of gait will further increase foot clearance. Other multichannel applications to aid walking in stroke survivors include stimulating the gluteal muscles during the stance phase to improve gait stability and the stimulation of the calf muscles to improve propulsion during the latter part of the stance phase.

7.4.2 Cerebral palsy

Cerebral palsy (CP) is an umbrella term used to describe children with a group of disorders associated with injury to the brain, before, during or shortly after birth. In Europe, the incidence of CP ranges from 1.5 to 3 per 1000 live births [14]. Around two-thirds of the children are ambulant but may need walking aids. People with CP demonstrate one or more motor types which are spasticity, ataxia, and dyskinesia. Spasticity is the most common and is seen in about 70%–85% of the population. Those with the spasticity motor type may also have reduced selective voluntary motor control with the ankle joint most often affected [9].

Dropped foot which often manifests as toe gait in people with diplegic CP is mostly due to spasticity of the calf muscles and resulting calf muscle contractures (fixed shortening of the muscle). However, weakness of the dorsiflexors and impaired selective motor control also often contribute to the severity of this gait impairment.

People with CP have FES mostly applied to the dorsiflexors to reduce dropped foot or toe gait. In some people stimulation of the calf muscles as well as the dorsiflexors can be added to improve propulsion with stimulation of the calf muscles.

7.4.3 Multiple sclerosis

Multiple sclerosis (MS) is a neurological disorder which is characterised by demyelination within the central nervous system affecting an estimated 2.8 million people worldwide [15]. Progressive destruction of the myelin layer, which insulates the axons within the CNS, impairs neural transmission along the spinal cord. This often results in motor deficits such as weakness, spasticity and reduced motor control which affect both the lower and upper extremities. About 80% of people with MS report gait and balance impairments as their main complaint [16].

The most common gait deficits in people with MS (pwMS) are dropped foot due to dorsiflexor weakness, weak hip flexors, impaired motor control or increased calf muscle spasticity or combination of all. Dropped foot in people living with MS can be permanent or induced by exercise. In the latter case, dropped foot is induced by a certain prolonged activity such as walking or running [17]. However, after a period of rest, which can be as short as a couple of minutes, the person's normal gait may be restored. This type of dropped foot is also sometimes referred to as 'fatigable foot drop'.

FES in pwMS also can activate a reflex mechanism called the flexor withdrawal reflex which can assist with hip and knee flexion even when electrodes to apply the stimulation are positioned for dorsiflexion activation. This mechanism is described later in the chapter (Section 7.5.4).

Some PwMS may benefit from stimulation to other muscle groups in addition to the dorsiflexors such as the hamstrings to increase knee flexion in swing and the gluteal muscles during the stance phase to improve hip extension during the stance phase [e.g. 18]. Stimulation to the calf muscles during the stance phase to improve propulsion can also be effective.

7.4.4   Parkinson's disease

Parkinson's disease or Parkinson's is a common neurodegenerative disorder leading to motor and non-motor symptoms, affecting around 6 million people worldwide [19]. The gait of people with Parkinson's (pwP) is characterised by bradykinesia (slowness of movement), rigidity and tremor. People living with Parkinson's also experience reduced balance increasing the risk of falling. Bradykinesia is associated with reduced muscle activation, which is often greater in more distal muscle groups, leading to reduced movement of the ankle [20]. Difficulty with gait initiation ('freezing') is common in those with more advanced disease and is characterized by an inability to generate or sustain an effective walking pattern. Freezing of gait is frequently experienced during functional activities such as turning or gait initiation [21].

FES to support walking is currently less common in pwP compared to stroke survivors and people with MS. However, studies have shown that FES to the dorsiflexors during steady state may benefit walking in pwP [22,23]. Further, bilateral stimulation of the tibialis anterior muscle may also improve the capacity for pwP to generate anticipatory postural adjustments without losing balance during gait initiation [24].

7.4.5   Traumatic brain injury

Traumatic brain injury (TBI) is one of the leading causes of motor and cognitive deficits in adults. In 2016, there were 27 million new cases of TBI worldwide with age-standardised incidence rates of 369 per 100,000 population [25]. In those with severe TBI long-term reduced motor and postural control are common often resulting in slower gait speed and balance impairments. Secondary deficits include muscle weakness and reduced range of joint of movement. FES can aid walking in people with TBI by targeting muscle groups affected by weakness such as the dorsiflexors during the swing phase of gait [26,27]. In people with TBI cognitive impairment may be a limiting factor for using FES and therefore needs careful consideration before providing FES.

7.4.6   Spinal cord injury

The incidence of SCI shows a large variation between countries ranging from 8 to 49 cases per million [24]. SCI can be described as complete or incomplete with those with incomplete injuries still having some sensation and motor function below the site of their injury. A survey of people with SCI in the UK showed that 15% of the respondents were marginal walkers and may therefore benefit from interventions aimed at gait restoration such as FES [28]. Muscle weakness of the dorsiflexors, knee extensors (quadriceps) and hip extensors (gluteal muscles) can result in gait impairments such as inadequate knee and hip extension during stance and reduced ankle dorsiflexion during swing [29]. FES is suitable for those with an incomplete SCI above T12 and is mostly applied to the dorsiflexors during swing with or without stimulation to the quadriceps to improve knee extension or to the gluteal muscles to improve hip extension during the stance phase.

In some cases, if weakness of the gluteus medius muscles (hip abductors) is more pronounced on one side, resulting in a so-called Trendelenburg gait. In people with Trendelenburg gait, hip abductor weakness of the standing leg results in the pelvis to drop down to the site of the non-weight bearing leg (see Figure 7.4). This

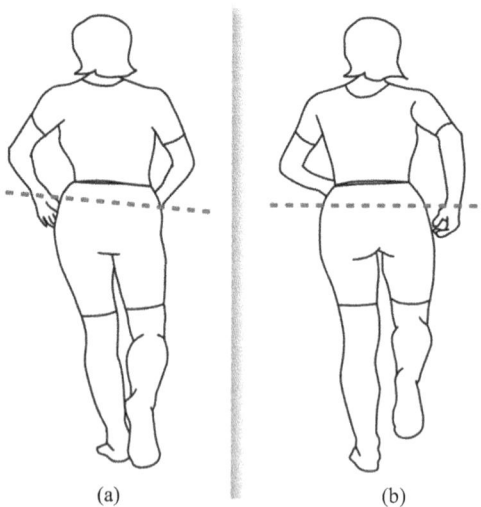

(a)   (b)

Figure 7.4   Trendelenburg gait (a) and normal gait with adequate pelvic stability (b). Adapted from physio-pedia (https://www.physio-pedia.com/index. php?curid=11189).

can lead to reduced foot clearance during the swing phase. FES to the gluteus medius muscle of the more affected leg during the stance phase will aid hip abduction and thus reduce Trendelenburg gait symptoms.

7.4.7　Familial spastic paraplegia

Familial spastic paraplegia (FSP) or hereditary spastic paraplegia refers to a group of inherited disorders. Prevalence is estimated at 3–10 cases per 100,000 in the European population [30].

People with FSP experience weakness, spasticity and loss of selective control resulting in gait impairments that increase in severity over time. FES to the dorsiflexors during the swing phase alone or combined with stimulation of the knee or hip extensors during the stance phase can support walking in people with FSP. People with FSP who demonstrate a Trendelenburg gait pattern may benefit from FES applied to the gluteus medius muscle during the stance phase as described above.

7.5　Neurophysiology relevant to FES

When applying electrical stimulation to a person with an UMN lesion it is important to have some understanding of the neurological process that may be interacted with and what effect the stimulation may have on their gait [31].

7.5.1　The stretch reflex

The stretch reflex is a natural reflex in the control of normal movement of the body and can be described as a closed loop mechanism to maintain the position of a limb. Muscle spindles are natural 'sensors' that are present within muscles, respond to the lengthening of a muscle. When stretched, nerve impulses are generated in the type Ia and type II afferent nerves (see also Chapter 2), which relay the signals to cell body of alpha motor neurones that supply the same muscle that is being stretched. This excites the motor neurone, which then cause the muscle to contract, opposing the stretch. Hence the position of the limb is maintained and may effectively be more rigid. The stretch reflex is velocity dependant meaning the faster the stretch, the more activity is produced in the Ia nerves and hence the greater the contraction in the muscle. This has a consequence for FES application, particularly with the many neurological patients who have spasticity, as it means that rapid contractions produced by FES may induce stretch reflexes, resulting in reduced range of motion. Ramping (increasing stimulation over time) can reduce the reflex response and in general, the slowest ramp that is sufficient to keep with the required movement is often the most effective. It is also likely to increase the comfort of the stimulation as sudden increases of stimulation intensity are perceived as more uncomfortable.

7.5.2　Reciprocal inhibition

The Ia afferent nerves involved in the stretch reflex also act upon the muscles on the opposite side of the joint (antagonistic muscles) via the inhibitory interneuron.

This interneuron is analogous to a NOT gate (i.e. an inverter) in electronics and turns the excitatory stimulus in the Ia afferent nerve into and inhibitory stimulus. This has the effect of reducing the activity in alpha motor neurons connected to the antagonist muscle and this again assists in maintaining the position of the limb or allows the muscle to be stretched. If that antagonist muscle has spasticity, it may, in effect reduce the spastic activity and hence allow a greater range of motion of the joint.

7.5.3   Ia facilitation

The cell body of an alpha motor neurone has many synaptic connections acting upon it including descending nerves from the central nervous system relaying voluntary signals to cause movement. The alpha motor neurone will be caused to produce a nerve impulse when the total activity from all the synapses reaches a certain threshold. If the Ia afferent nerves add to this activity, it means descending voluntary signals are more likely to elicit a response from the alpha motor neurone. The Ia afferent nerve fibres are large diameter fibres and hence can be excited by electrical stimulation at a lower threshold than other nerves. Stimulating over target muscles while attempting voluntary movements may help facilitate these movements, without causing directly stimulated movement.

7.5.4   The withdrawal reflex

The withdrawal reflex is a naturally occurring protective reflex. A noxious stimulus in the lower limb, for example from standing on a pin, will cause the limb to pull away from the stimulus, i.e. lifting the foot off the pin. The position of the stimuli will determine the reflex movement produced and this reflex can be a useful pattern of movement in FES applications. Stimulation of the common peroneal nerve will produce dorsiflexion and eversion through the direct stimulation of motor nerves and can also, through the withdrawal reflex, produce knee and hip flexion with external rotation of the hip. These movements can assist in taking a step, for example, improving knee flexion in stroke, hip flexion in MS or addressing a scissor (internal hip rotation) gait in CP or FSP. While the reflex can be obtained from stimulating the common peroneal nerve at the head of fibular, a greater response is often obtained by stimulating the nerve as it passes through the popliteal fossa, running next to the biceps femoris tendon. Reflexes can be subject to temporal summation, and this means higher stimulation frequencies (50–60 Hz) may be more effective at producing a reflex. However, clinical experience shows that higher frequencies can increase knee stiffness in some cases and a frequency of 30 Hz may be more effective. In some cases, the flexor withdrawal reflexes are subject to habituation, meaning the effect wares off after several minutes. However, it can also be maintained to a lesser degree, and this may be sufficient for useful function for those with limited mobility. The flexor withdrawal reflex has also been used in gait training in stroke, elicited by stimulating the sole of the foot [32].

7.5.5   Hebbian learning and Rushton's theory

Donald Hebb [33] postulated a mechanism underlying learning in the nervous system. The mechanism involved the formation of new synaptic connections between adjacent neurones in response to activity occurring in them at the same time; a principle often referred to as 'nerves that fire together, wire together'. When using FES or NMES systems, the nervous system is subjected to extra sensory input, directly from stimulated nerves and indirectly from sensory systems involved in the movement produced by FES. If this is associated with the user's voluntary efforts to produce the action, the prerequisites for Hebbian learning are set up. More connections will form between nerves active in desired volitional movement and this means that movements may in time become easier.

Rushton [34] suggested an extension to Hebb's theory. When an alpha motor neurone is stimulated, a nerve impulse is generated which travels down to the muscle. At the same time a so-called antidromic impulse is also generated, which travels up the nerve towards the cell body and this impulse can re-excite the cell, causing a new impulse to be generated, referred to as a 'f' wave. If this extra excitation of the alpha motor neurone occurs at the same time as voluntary generated activity, new synaptic connections may be produced. Consequently, the next time voluntary activity occurs, the alpha motor neurone may be a little easier to recruit. It, therefore, follows that encouraging voluntary effort synchronised to FES is a likely strategy to improve the training effect from an FES device.

7.6   FES to correct dropped foot

7.6.1   Introduction

As described above, the most common use of FES is the correction of dropped foot. People with dropped foot have a loss or insufficient dorsiflexion of the ankle during the swing phase of gait resulting in increased plantarflexion of the ankle joint, i.e., the foot drops. This will also impact the foot contact pattern with individuals striking the floor with their whole foot or even toes first instead of with the heel as in normal gait. Dropped foot during the swing phase resulting in reduced clearance can lead to trips and falls with individuals sometimes compensating through ipsilateral hip hiking, flexion of the trunk to the contra-lateral side, circumduction (Figure 7.5) or increased plantar flexion (vaulting) of the contralateral weight bearing leg. Such compensations can increase the energy expenditure of gait and may increase the risk of injury to the muscles and joints involved in gait compensations. Trips and falls or fear of falling will often result in a lower walking confidence which is often associated with reduced participation in activities of daily living, ultimately negatively affecting physical and mental health [35].

The use of FES to treat dropped foot was first described in a paper published in 1961 by Liberson *et al* [36] who described a system that electrically stimulated the common peroneal nerve of a man with hemiparesis, and thus enabling his ankle to dorsiflex and his foot to be lifted off the ground during gait. Liberson *et al.* referred

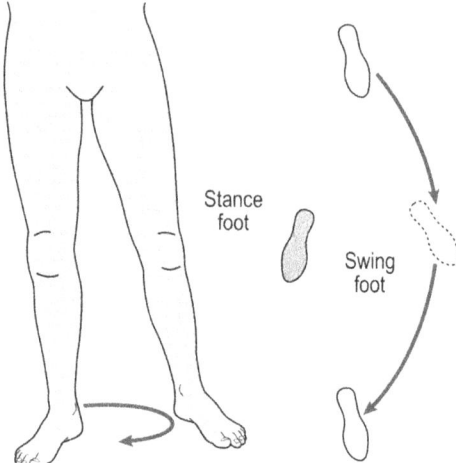

Figure 7.5   Circumduction, adapted from [7]

to this treatment option as Functional Electrotherapy. The term FES was introduced
6 years later by Moe and Post [37] in their patent entitled 'Electrical stimulation of
muscle deprived of nervous control with a view of providing muscular contraction
and producing a functionally useful moment'.

7.6.2   Electrode placement

The common peroneal nerve is a branch of the sciatic nerve and supplies the
muscles that produce ankle dorsiflexion and eversion along with toe extension. The
nerve passes down the back of the thigh becoming superficial at the popliteal fossa,
running parallel to the biceps femoris tendon (Figure 7.6). It then passes just distal
to the head of the fibular bone, where it bifurcates into two main branches. The
superficial branch supplies the peroneus brevis and peroneus longus, which primarily evert the ankle while the deep branch supplies the tibialis anterior muscle
which is the main ankle dorsiflexor, and the muscles that extend the toes; extensor
digitorum brevis and longus and extensor halluces. The deep branch also innervates
the peroneus tertius, which is an ankle evertor. The superficial nature of the common peroneal nerve makes it a good target for FES as it can be stimulated with
relative ease. By choosing the position of the electrodes the relative proportion of
each nerve branch that is stimulated can be determined and hence the FES intervention can be tailored to the requirements of each FES user.

The tibialis anterior muscle, while primarily producing dorsiflexion, also has a
tendency to cause inversion at the ankle. Hence, to produce a straight lift of the foot
it is necessary to compensate by adding eversion from the peroneus brevis and
longus. Indeed, a small amount of eversion with dorsiflexion at first contact is
advisable as the foot is better positioned for weight acceptance through the stance
phase. However, the peroneus and longus and brevis can also produce an element

of plantarflexion, so some care is required to balance these movements. Further variables can also be present. For example, people with calf spasticity will generally have increased inversion of the ankle through both the swing and stance phases of gait. They may also experience excessive toe flexion leading to clawing of the toes while walking. Long-term spasticity and atrophy will often lead to shortening of soft tissues resulting in reduced range of ankle movements. Finally, there is variation in anatomy, with the precise path of nerves or tendons differing between individuals. Hence it is the case that a single FES strategy will not be suitable for all, and a flexible approach is required. Figure 7.6(a)–(g) shows examples of electrode position that can be used for dropped foot correction for a variety of presentations.

7.6.3 Sensors used for timing of muscle stimulation

Although there is a range of commercially available stimulators, they all work on the principle that the dorsiflexors need to be activated during the swing phase of gait and sometimes during the first part of the stance phase to prevent foot slap. A mechanism is therefore required to trigger stimulation, either by detecting heel rise on the stimulated side or heel strike on the contralateral side. The moist commonly used mechanisms are a foot switch, which is inserted into the shoe, or a tilt sensor attached to the shank. Most modern footswitches are wireless.

Tilt sensors, usually based on an Inertial Movement Unit (IMU), have the advantage that they can be included in the stimulation device making a single self-contained unit mounted on the stimulated leg [Figure 7.7]. IMUs can give information on movement in three dimensions and allow gait event detection algorithms to detect changes in both direction and speed. However, the shank angle is a proxy to transition between gait phases and hence may be less than precise in its timing compared to foot switches. The use of IMU devices also requires the person with a dropped foot to have a sufficient knee flexion and this is not always the case with people with UMN lesions. Tilt sensors can only be used on the stimulated side and also create a bulky device mounted on the leg, which may limit the choice of clothing.

Foot switches, in contrast, require hardware within the shoe and hence are not suitable for barefoot walking. However, they can provide more precise information about the transition from the stance to the swing phase. Footswitches, which are often based on force sensitive resistors that detect the pressure under the foot, use adaptive algorithms to detect the switching point appropriate to each individual and hence can be used with a light child or heavy adult. The adaptive algorithm makes the footswitch very sensitive to the gait phase transition, but this may also produce false positives in response to shifting weight between feet. Clinical experience indicates that this is less disruptive to the device user than failure to detect a genuine phase transition. Foot switches can be placed in different positions on the sole of either leg, changing the time point of switching on and allowing adaptation to the loading pattern of the individual. Foot switches can be placed either in the shoe of the stimulated leg to detect heel rise or the contralateral leg to detect heel strike.

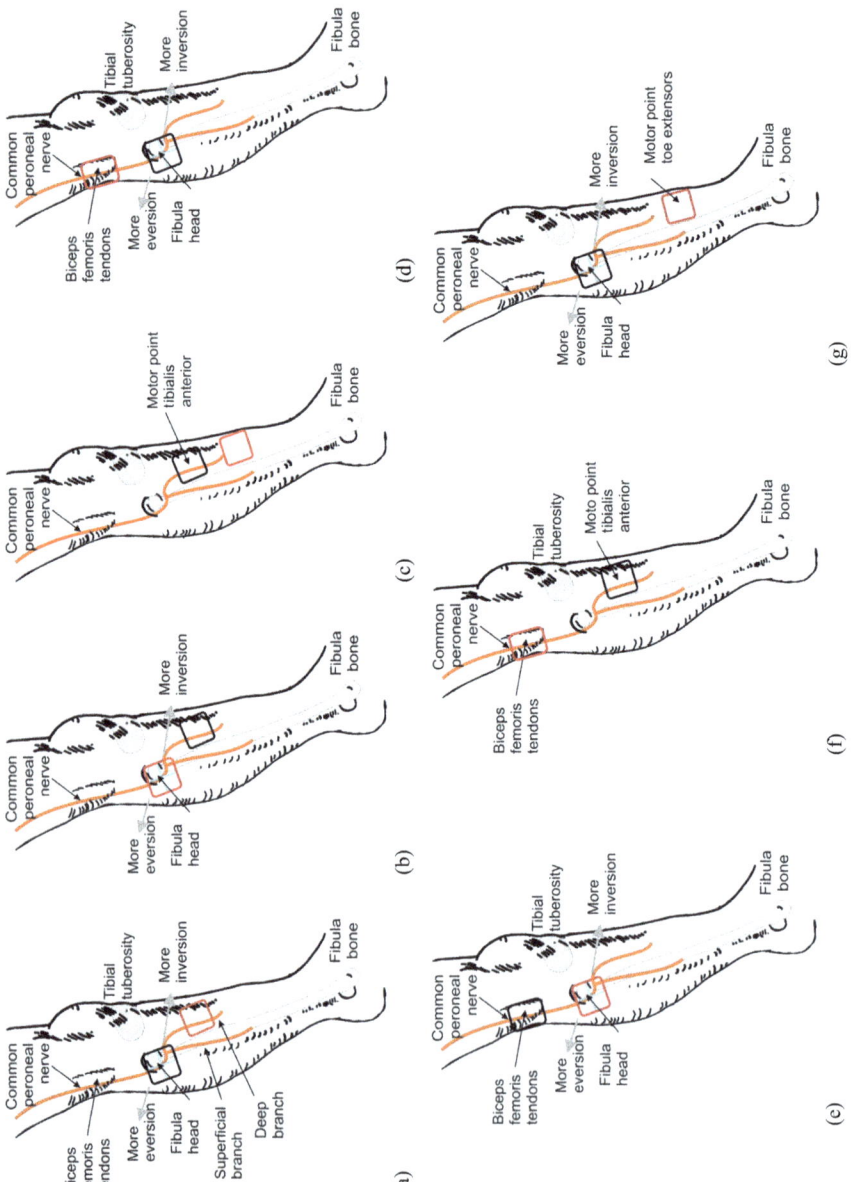

(a) (b) (c) (d) (e) (f) (g)

Figure 7.6 (a)–(g) Electrode placement options to manage dropped foot. (a): The anatomy of the common peroneal nerve with standard electrode positions (50 × 50 mm). The top forward quadrant of the active (cathode or negative, shown with black outline) electrode is placed over the head of the fibula. The common peroneal nerve will therefore pass diagonally under electrode. To verify the correct location, press the four quadrants of the electrode. The one over the head of fibular will feel hard and the other three will be softer. The indifferent electrode (anode or positive, shown with red outline) is placed over the motor point of the tibialis anterior muscle, at its fattest part. Ensure the electrode does not cross over the tibia bone as this tends to sting. If too much inversion occurs, move the active electrode backwards and upwards a little. If there is too much eversion, move the electrode forwards a little. Note the tibial tuberosity, this may be mistaken for the head of the fibula but is always higher and further forward than the fibula head, which is the bony prominence lowest down and furthest back.

(b) and (c): If the standard electrode position (a) produces dorsiflexion with too much eversion, reversing the polarity so that the active electrode (black) is over the motor point of the tibialis anterior and the indifferent electrode (red) is over the common peroneal nerve (b) will generally reduce the eversion. This is because the active electrode has a stronger stimulation effect than the indifferent electrode. If reversing the polarity produces dorsiflexion with inversion, use a symmetrical biphasic waveform. In this type of waveform, the polarity of each pulse is reversed every other pulse. This causes both electrodes to have the same stimulation effect. If this position still produces too much eversion, use position (c), tibialis anterior motor point stimulation. This position is commonly used in paediatrics.

(d), (e) and (f): If greater dorsiflexion is needed or more eversion, knee flexion, hip flexion and external rotation is required the common peroneal nerve can be stimulated in the popliteal fossa. Ensure the electrode is on the lateral side, wrapped around the biceps femoris tendon. If the electrode is too medial, the tibial nerve may be recruited causing plantarflexion from the calf muscles. For a moderate effect place the active electrode over the head of the fibula and the indifferent in the popliteal fossa (d). For the strongest effect, revers the polarity (e). Because the common peroneal nerve is superficial in the popliteal fossa, often a lower stimulation intensity is required to produce an effect and sometimes has a greater sensory effect. However, it is a useful position to use if the lower leg is oedematous or of excessive size. (f): Sometimes stimulation in the popliteal fossa will cause excessive eversion. This can be moderated by moving the head of fibula electrode to the tibialis anterior motor point (f).

(g): If a patient experiences toe clawing due to spasticity, moving the indifferent electrode to over the motor point of the toe extensors can reduce the clawing. The toe extensors may be paired with the head of fibula, popliteal fossa positions or tibialis anterior motor point stimulation.

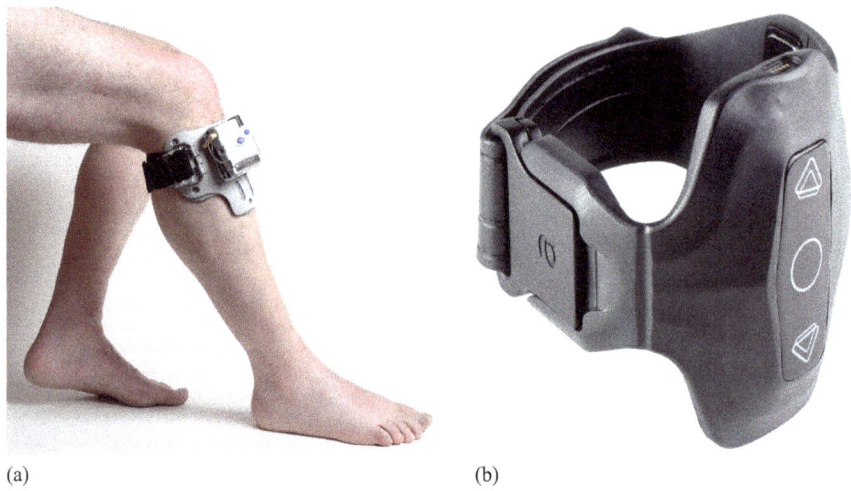

(a)   (b)

Figure 7.7   (a) Example of FES user wearing a cuff that houses the electrodes.
(b) Example of a system using a cuff to house the electrodes (© Allard
Support for Better Life! ALFESS FES).

The latter arrangement is preferred if the user is unable to achieve a consistent heel strike. Parameter settings on the FES device, must of course be adjusted according to footswitch position. In conclusion, tilt sensors may provide greater convenience for some users, but foot switches may be more reliable and usable in a wider range of gait styles and disabilities.

7.6.4   Timing of stimulation

Figure 7.8 shows a typical stimulation envelope and its relationship to normal EMG activity in the tibialis anterior. The envelope shows the rise and fall of intensity of the stimulation over time. The intensity can be modulated using the current level or modulated using the pulse duration. Both parameters will control the strength of the muscle contraction produced. In this example, the envelope is controlled using a foot switch under the heel. The envelope begins with a rising ramp, initiated by the heel leaving the ground. A rising ramp is used to prevent a sudden dorsiflexion of the ankle, reducing the risk of inducing a stretch reflex. It also allows time, if the FES user is able, for push off using the calf muscles. A ramp is also perceived as more comfortable by many FES users and hence may improve compliance. The stimulation is then maintained at the level of the top of the ramp throughout the swing phase until heel strike occurs. This initiates the end of the stimulation envelope but as can be seen from the EMG in normal gait (Figure 7.8 in yellow), the tibialis anterior is still active at this point and so an 'extension' is added to mimic the EMG pattern. In the extension period the tibialis anterior contracts eccentrically, lowering the foot to the ground and also helps prevent the knee extensors from pulling the knee into hyperextension. If this 'extension' period is

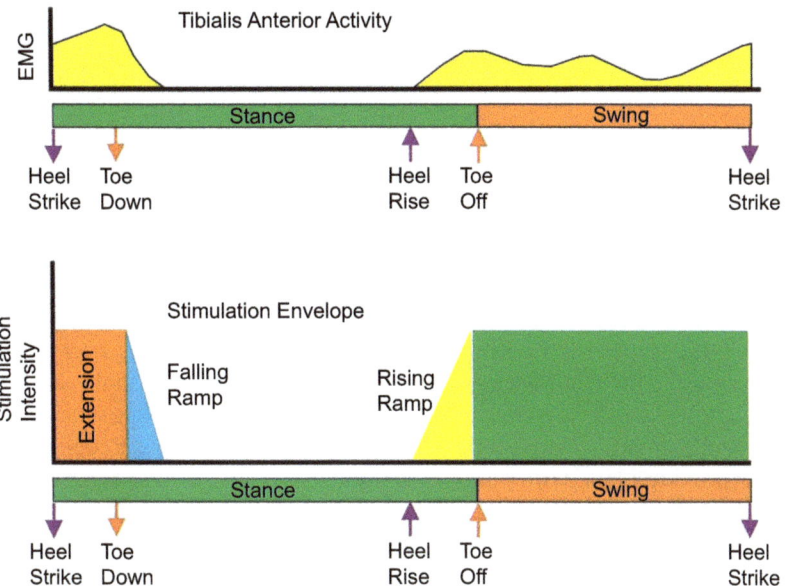

Figure 7.8   Stimulation envelope used for dropped foot correction shown relative to normal EMG

not included the foot will fall rapidly, causing 'foot slap'. The extension also provides some ankle stability through the initial loading response by maintaining some eversion of the ankle (the superficial branch of the common peroneal nerve innervates the peronei group), maintaining weight bearing along the midline of the foot. FES users with a high degree of calf tone spasticity may benefit from a slightly longer extension time for this reason. The envelope ends with a short falling ramp.

While the main period of the envelope is timed by the action of the foot switch, the ramps and extensions have fixed times and do not adapt to changes in walking speed. Extension times must be short enough to allow some rest in the stance phase to avoid muscle fatigue. Hence these timing parameters need to be adjusted to suit the walking speed and style of the device user. Fast walkers will require shorter ramps and extension times. Slower walkers with more calf spasticity or perhaps a tendency to hyperextend the knee will require longer ramps. Note that normal EMG activity is not constant while the tibialis anterior is active and in fact, reaches its peak when lowering the foot to the ground after initial contact. Some FES systems have modulated the stimulation envelope to reproduce this effect [38]. However, it must be remembered that the EMG shown here is from young adults without neurological deficit and most FES users will have compromised control of movement and altered biomechanics due to spasticity and soft tissue shortening. This may mean the normal pattern of tibialis anterior activity will not be optimal to support their new pattern of gait. Another example of using FES to compensate for abnormal biomechanics is the peronei muscles. In normal gait they are not normally active through the swing phase of gait, but because of abnormal positioning of the foot are required to be active in dropped foot correction to align the foot for weight acceptance at heel strike.

7.6.5   *Characteristics of currently available stimulators*

In some FES systems wires connect the foot switch and the electrodes with the stimulator itself, while other designs only use a wire connecting electrodes with the stimulator or do not use any external wires at all. Using separate electrodes require the user and / or their carers to manually place electrodes, a process that requires some care and knowledge of anatomical landmarks. The electrodes are maintained in position using their self-adhesive properties but can become caught on clothing and become detached and are therefore sometimes covered using Tubigrip or a cloth cuff. Long-term clinical follow-up surveys demonstrate electrode-based systems can provide a long-term practical solution [39]. Cuff based systems mount the electrodes within the cuff, and so only require the cuff to be placed in the correct position instead of attaching two electrodes. Two basic cuff-based designs are currently available, those that require the clinician to determine the position of the electrodes within the cuff and those that have a fixed array of electrodes. The latter requires a system to select the stimulation to each electrode. Cuffs can increase the convenience for the user and speed of donning and doffing the device. However, care is still needed to place the cuff in the correct place, and this is best aided by educating the user about the relevant anatomical markers and the position of the common peroneal nerve and the head of the fibula. Some systems, use an electrode array, movement sensors and a closed loop control auto-tuning process to compensate for the imprecision of cuff placement. However, cuffs can be bulky on the leg, limiting the choice of clothing. By keeping the cuffs as small as can be practically achieved the choice of electrode positions is restricted and therefore cuffs may not provide a solution for all patients as described in Section 7.6.2. Finally, because the electrodes are 'hidden' in the cuff, greater diligence is required from the user to monitor the condition of the electrodes and to replace them when they require renewing.

7.7   **Implanted FES systems**

It has often been suggested that using implanted FES would remove or reduce the practical difficulties of using FES for dropped foot such as those associated with correctly placing the electrodes before use. Implanted electrodes are inherently more stable and hence will provide a more consistent response than skin surface electrodes, which need to be placed and removed again each time the device is used. External electrode users may also experience skin irritation from using electrodes with the reported incidence varying between 4.5% [40] and 50% [41]. As the time of writing, two implanted systems had been brought to the market for a limited time, the STIMuSTEP and the Actigait. Both devices used external foot switches to control the stimulation envelope, and both stimulated the common peroneal nerve. The STIMuSTEP used two electrode pairs placed within the epineurium of the superficial and deep branches of the common peroneal nerve. By choosing the relative intensity of stimulation to the two branches, the proportion of dorsiflexion and eversion could be determined. The stimulation was supplied by a passive implanted receiver mounted behind the fibular bone. This was powered by an external mounted transmitter unit worn on a strap directly over the implant. The Actigait used four electrodes mounted on a cuff placed around the common peroneal nerve just proximal to the popliteal fossa. The intensity of stimulation to each electrode could be chosen, again allowing the amount of dorsiflexion and eversion to be controlled. The passive implanted controller was mounted subcutaneously in the thigh and powered using an external transmitter coil.

Both implanted devices were shown to work effectively. In a case series of 71 at Salisbury District Hospital, the STIMuStep was received by 30 people with multiple sclerosis (3 bilateral), 29 stroke survivors, 4 people with SCI, 3 people with TBI, 3 people with Transverse Myelitis and 1 each of cerebral palsy (bilateral), hereditary spastic paraplegia (bilateral) and Neurofibromatosis. The device was successfully used in 69 cases, with 1 device removed due to an infection and the other being unsuccessful due to unusual neural anatomy. However, neither the STIMiSTEP nor the ActiGait proved to be commercially viable due to the high costs associated with the device and the need for surgery to fit it.

7.8   Other applications of FES to support walking

Although the FES to the dorsiflexors during the swing phase on its own is the most common use of FES to support walking, there are several other examples of applications of FES described in the literature.

Tenniglo *et al.* [42] explored the use of FES to increase knee flexion in 16 stroke survivors with stiff knee gait. In this study, a two-channel stimulator system with foot switches was used to stimulate the hamstrings between during late stance, i.e. between heel off and toe-off. Self-adhesive skin surface electrodes were placed on the mediolateral aspect of the hamstrings. The indifferent electrode was placed approximately 5 cm above the knee crease and the active electrode was placed approximately 10 cm above the indifferent electrode. Results showed an immediate positive effect with FES increasing knee flexion during swing compared to no FES. After using FES for a period of 5 weeks participants had increased knee flexion in swing without the assistance of FES compared to that at baseline. Those with lower neurological impairments showed the biggest improvements.

Kim *et al.* [43] examined the feasibility of applying FES to the dorsiflexors in swing and the gluteus medius during stance in 36 stroke survivors. The electrode for the stimulating the tibialis anterior was placed parallel to and just lateral to the medial shaft of the tibia, at approximately one quarter to one-third of the distance between the knee and ankle for the tibialis anterior. The gluteus medius electrode was placed over a line connecting the highest point of the iliac crest and greater trochanter of the femur. Gait events were detected by foot switches attached to the heel of the affected limb. The stimulation to the tibialis anterior was triggered when the heel lifted from the floor while stimulation of the gluteus medius was started when the heel made contact with the floor. Results indicated improved double support time and gait symmetry compared to the non-FES condition.

Kunkel *et al.* [44] explored the feasibility of applying FES to the quadriceps and gluteal muscles during balance exercises in stroke survivors in the acute phase. Although FES was not triggered during gait, the authors provided detailed information on the stimulator set-up including electrode placements that may be of interest to readers. FES was applied using a two-channel stimulator through surface electrodes with two electrodes each positioned over the gluteus maximus and over quadriceps femoris muscles (Figure 7.9). Stimulation was delivered at 40 Hz using an asymmetrical biphasic waveform and an average pulse width of 162 μs. In addition, a rising and falling edge ramp time was set to help avoid stretch reflexes.

Another multi-channel option is to stimulate the plantarflexors to aid forward propulsion in addition to the dorsiflexors to manage dropped foot during swing. In a study by Kesar [45], the plantarflexors of 13 stroke survivors were stimulated during terminal stance in. Adding the plantarflexor stimulation resulted in

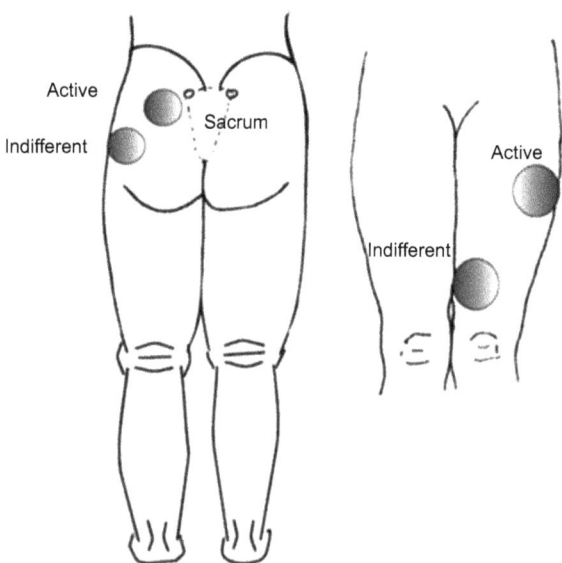

Figure 7.9   Electrode placement for FES to the quadriceps and gluteal muscles described in [43]

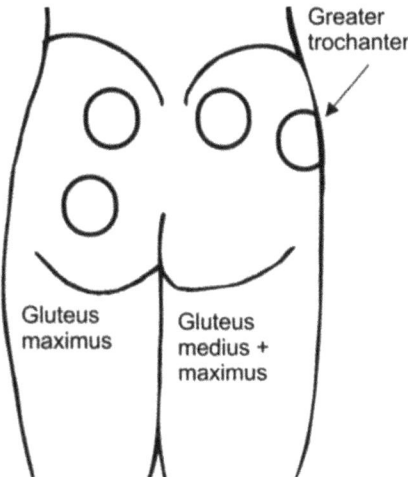

Figure 7.10 *Electrode positions for hip extension; gluteus maximus alone (left leg). Electrode position for hip extension with hip abduction; gluteus maximus with gluteus medius (right leg). In each case, the active electrode (negative electrode) is placed over the muscle or nerve from which the strongest response is required. Adapted from [18].*

increased knee flexion, plantarflexion at toe-off and a higher propulsion force compared to stimulating the dorsiflexors only. However, stimulating both muscles also reduced the dorsiflexion during swing. The results of this study suggest that if both plantarflexors and dorsiflexors are stimulated, the clinician will need to carefully adjust the stimulation parameters and timings to optimize the patient's gait pattern.

Finally, Taylor *et al.* [18] reported on stimulating the gluteal muscles to aid hip extension and abduction during stance in addition to the stimulation of the dorsiflexors during the swing phase in a study involving people with MS. The stimulation and electrode placement of the tibialis anterior was as described in Section 7.6. The electrodes for the gluteal muscles were placed over the motor points of gluteus maximus and gluteus medius or, if abduction was not required, over the gluteus maximus alone (Figure 7.10). Stimulation was provided using multi-channel device using a foot switch placed under the heel to provide the timing of the stimulation. The gluteal stimulation began at heel strike and continued until heel off on the ipsilateral side.

7.9 Definitions of orthotic and training effects of FES

FES to support walking can result in an orthotic effect, or a training (or therapeutic) effect (Table 7.1). The term 'orthotic effect' is used in this chapter to describe an immediate difference in walking outcome such as speed when FES is

Table 7.1   Comparisons used to measure changes in performance due to FES.

Effect	No stimulation		Stimulation	
	Day 1	Follow-up	Day 1	Follow-up
Initial orthotic	✓		✓	
Continuing orthotic		✓		✓
Total orthotic	✓			✓
Training (therapeutic)	✓		✓	

turned on compared to walking without FES, at the same assessment point. This means that the orthotic effect of FES can be evaluated in a single assessment. Some authors differentiate between the initial orthotic effect which is the orthotic effect at baseline and the ongoing orthotic effect which is measured after the patient has used FES for certain period. A training effect, also referred to as a therapeutic or carryover effect, is the improvement in walking related outcome, measured without stimulation, following a period of sustained use of the stimulator. The total orthotic (or combined) effect of FES is the change in the walking related outcome with stimulation following a period of use compared to without stimulation prior to the intervention. It is crucial to be aware of the differences between these potential effects of FES when appraising the evidence for FES improving gait. Unfortunately, a recent umbrella review highlighted that some authors fail to clearly define which type of effect of FES was assessed [46]. Further, the terms 'treatment' and 'training' effect are sometimes also used to describe the change in walking outcomes between baseline without stimulation and follow-up with stimulation, which in this chapter is defined as the total orthotic effect.

7.10   Potential mechanisms responsible for a training effect

As described above, a training effect of FES is defined in this chapter as an improvement in a certain outcome, often walking speed, without the assistance of FES due to a sustained use of the FES device over a period of time.

Neurophysiological mechanisms, through which FES can induce a training effect such as Hebbian learning, and mechanisms described by Rushton's theory are described above. Another potential mechanism explained in this section is the reduction of spasticity through reciprocal inhibition. FES to the dorsiflexors may decrease spasticity in its antagonist, the gastrocnemius, and over time this may increase range of motion of the ankle [47]. Similarly, Burridge *et al.* [48] observed the reduced quadriceps spasticity in patients who had used a peroneal stimulator for a 12-week period. They suggested that simulating the peroneal nerve, which can activate the hamstrings by eliciting the flexor withdrawal reflex, could over time reduce quadriceps tone through the process of reciprocal inhibition and neuroplasticity.

Increased muscle size (hypertrophy) due to repeated stimulation and increased blood flow to the muscle is another mechanism, which could explain the training effect of FES on outcomes such as walking speed. To explore this mechanism, Everaert *et al.* [49] collected a range of measures in a group of people with non-progressive disease (mainly stroke) and progressive disease (mainly MS) before and after 3–12 months use of FES to treat dropped foot. These included the (1) motor evoked potential (MEP) in the Tibialis Anterior resulting from transcranial magnetic stimulation over the motor cortex, (2) the maximum motor wave (Mmax) by stimulating the common peroneal nerve and (3) the maximum voluntary contraction (MVC) which is a measure of muscle strength, and walking speed. Both the MVC and MEP were increased significantly after prolonged FES use in both groups but the increases in Mmax due to hypertrophy were small and did not correlate with the changes in MEP. The authors concluded that repeated use of the stimulator may strengthen the activation of motor cortical areas. This increased activation of the motor cortical areas may have contributed to improved walking function with hypertrophy of the muscle being less important. Damiano *et al.* [50] found increased cross-sectional area and thickness of the tibialis anterior in children with CP who had used FES to this muscle for three months. However, although these muscle size gains were preserved at follow-up, ankle motion returned to baseline values, again suggesting a weak association between an increase in muscle size and improved gait function.

Finally, it is also possible that improved walking outcomes after using FES for a sustained period can also be secondary to FES use. For example, if FES allows the user to walk faster, further and more often this increased walking activity in itself can result in a further overall improvement in walking function.

7.11 Common outcome measures when evaluating the efficacy of FES to support walking

7.11.1 Measuring walking speed

The most common outcome measure to assess the impact of FES to support walking is walking speed which is mostly measured over a relatively short distance such as 10 m (10 m walking test, 10 mWT) or 25 feet (Timed 25-foot walk, T25FW) [51]. Short walking speed tests, especially if the average of two tests is taken for analysis, have been shown to be reliable and valid in a range of UMN populations [51–53].

For some patients, the distance of 10 m or 25 feet may be shorter than the distance walked when using FES during a typical day. For this reason, it can be useful to conduct walking tests over a longer duration such as those the 2-min walk test (2 MWT) or the 6-min walking test (6 MWT). The outcome measure in these longer duration walking tests is usually the distance walked without stopping to rest and thus is often regarded as a measure of walking endurance. However, for those with a higher walking ability this test will be simply a measure of walking speed over a longer distance. Protocols used for conducting tests such as the 6 MWT vary especially regarding the characteristics of the pathway. Because walking in a straight line for 6 min is not feasible in clinical facilities, endurance walking tests are often conducted with the individual walking up and down a corridor which therefore involves 180 degree turns. Performing 180 degree turns however requires motor skills other than walking in a straight line and this why some protocols use a figure of 8, ellipse or a rectangle instead. Whatever pathway is used, the space and time required for these walking endurance tests may make them often less suitable for routine clinical practice. Another issue with tests of a longer duration such as the 6 MWT is pacing. Many FES users may not be accustomed to walking for 6 min without stopping or as fast as possible and will start the tests either too fast or too slow. To increase reliability some researchers recommend incorporating a practice trial prior to the real trial or taking the average of two trials [54]. However, this may not always be feasible in clinical or even research settings again because of time constraints or the risk of fatiguing the participant or patient.

To attach a clinical meaning to improvements in walking speed for example as a result of using FES, studies have referred to work by Perry *et al.* [55] and Perera *et al.* [56]. Perry *et al.* [55] proposed six functional categories based on walking speed of 147 people with a history of stroke; (1) physiological walker (average walking speed of 0.1 m/s), (2) most limited household walkers (0.23 m/s), (3) least limited house hold walker (0.27m/s), (4) most limited community walker (0.40 m/s), (5) least limited community walker (0.58 m/s) and (6) community walker (>0.8 m/s). Using these functional walking categories, clinicians or researchers can attach a functional meaning to an improvement observed due to the use of FES, for example a change in walking from 0.7 m/s to 0.81 m/s would indicate a change from a least limited community walker to a community walker. An example of using these functional walking categories in the context of FES to correct dropped foot can be found in the publication by Street *et al.* [57].

Another approach to clinical relevance of changes in walking speed was adopted by Perera *et al.* [56]. In this study, changes in walking speed were compared with changes in the self-reported mobility related quality of life in a group of elderly people with a variety of health conditions. Based on these data, they proposed a minimum meaningful change in walking speed of 0.05 m/s and a change of 0.1 m/s as a substantial meaningful change. Although useful, clinicians and researchers should consider that an increase of 0.05 m/s in a household walker may have a different clinical impact as compared to the same change in walking speed in a community walker.

7.11.1.1   Walking speed in FES research

Possibly partly because of the relative ease to measure walking speed even in clinical settings, most studies have focused on the effect of FES on walking speed over 10 m or 25 feet. A recent umbrella review (review of reviews) focusing on FES to support walking in people with a UMN lesion showed that all included 24 systematic reviews (a total of 183 different studies) synthesized the evidence on the effect of walking speed [46]. The authors of this umbrella review concluded that there is evidence that FES can improve walking speed compared to a control condition not involving an intervention. However, they also concluded but that this effect depends on factors such as the health condition, whether FES was applied to a single muscle or multiple muscles, and whether FES was applied together with another intervention. For example, regarding the type of UMN lesion, training effects of FES are commonly reported in studies including stroke survivors but not for those living with MS [46], likely due to the progressive nature of MS. Even if FES does not improve walking speed over time, using FES may allow PwMS to retain their functional walking category and mobility for longer [57].

7.11.2   Other gait-related outcomes

A review on the outcome measures used in studies assessing the effect of FES used to support walking in people with MS concluded that after walking speed, the physiological cost index (PCI) and gait analysis were the most common measures [51].

The PCI is a measure of energy efficiency of walking. It is calculated by dividing the change in heart rate from during rest to that during walking by the walking speed in min/meter [58] and is thus measured in beats per meter. The use of the PCI assumes a linear relationship between heart rate and oxygen consumption [59]. It is recommended that the PCI is measured over a period of walking which lasts at least 4 min. The PCI is a relatively easy to record in a clinical environment as it only requires a heart rate monitor and a stopwatch unlike the specialised equipment required to directly measure energy or oxygen consumption. Several studies have demonstrated that FES to correct dropped foot can reduce the PCI when compared to no FES [60,61].

Gait analysis can be performed in numerous ways, ranging from direct visual observation of a patient walking in the clinic, visual observation scales [18,62], electrogoniometry [63] or computerised three-dimensional gait analysis [64]. The latter can be camera-based or using inertial sensors attached to the lower limbs and pelvis. Gait analysis allows a direct assessment of whether FES assists in the desired joint movement. For example, studies using a camera-based three-dimensional gait analysis system demonstrated that FES to correct dropped foot increased both peak dorsiflexion and knee flexion in swing in PwMS [64] compared to without FES and increased ankle range of motion during gait in children with CP compared to an AFO [65].

Gait analysis may also assist in the decision-making process on whether FES is the most appropriate treatment option and if so, which muscle group (s) should be targeted. Insufficient foot clearance is often an indication of dropped foot but may also be the result of decreased knee flexion during swing. This function may be difficult to assess through real-time visual observation but will be evident from the graphs obtained from instrumented gait analysis. Unfortunately, camera-based 3D gait analysis systems are often not feasible in clinics, but slow-motion play back of a video of patient's walking can also provide insight in a person's gait kinematics. Further, the application of easy to use, low-cost movement sensors attached to the legs for use in everyday clinic may become more common in the future.

FES may also have a positive impact on walking-related outcomes such as spasticity [47,48], passive joint range of motion [47] and daily step count [66]. The latter is measured by some FES devices when the device is switched on but can also be recorded through a range of commercially available devices and smartphones apps.

7.11.3   Self (patient) reported outcome measures, surveys and qualitative data

Given the relative ease of administrating questionnaires, it is surprising that until recently, self-reported outcomes have not been applied more often to assess the impact of FES. Street and Singleton [57] used Visual Analogue Scale (VAS) scores to record individuals' perception on such as joint pain, walking effort, trips, confidence, and quality of life over a five-year period of FES use. Other examples of patient reported outcomes include falls diaries and validated questionnaires such as Canadian Occupational Performance measure [67], disease specific outcomes such as the MS-12 walking [68], Stroke Impact Scale [69], or universal health-related quality of life questionnaires [38]. A systematic review of eight studies involving people with MS [70] showed that seven studies reported statistically significant positive effects of using FES on a range of self-reported measures such as Canadian Occupational Performance Measure [67] and the Psychosocial Impact of Assistive Devices Scale [71].

7.11.4   User perspectives

Patient surveys and qualitative methodologies such as interviews or focus groups can also provide important information on the acceptability and feasibility of using FES to aid walking. For example, if a user perceives that the barriers of using FES outweigh the benefits or if no benefits are perceived, the device will not be used which is a clear waste of resources. A survey of FES users showed that commonly perceived benefits of FES include a reduction in trips and falls, increased confidence, less effort (both physical and mental), being able to walk further and improved independence [26]. However, some users reported difficulties with the positioning of electrodes and the time needed to set up the equipment. Insight into the FES user experiences also been provided by users' perspectives recorded in interviews or focus groups [72,73]. Themes emerging from these studies included, 'funding or FES', 'finding out about FES', 'less conscious control of walking with FES', 'getting into routine of using FES', 'Weighing pros and cons' and 'professional help'.

7.12   AFO vs. FES to correct dropped foot

An Ankle Foot Orthosis is often prescribed to people with an UMN lesion to correct dropped foot and thus has a similar orthotic function as FES applied to the dorsiflexors in swing. Synthesis of the current evidence suggests that there is no difference between an AFOs and FES regarding their orthotic or total orthotic effect on walking speed [51,74]. However, considering outcomes other than walking speed, statistically significant differences in favour of the FES group have been reported in obstacle performance [75] and user preference [69]. Similarly, in a qualitative study with nine stroke survivors who had used both FES and AFOs, all but one preferred FES. Participants cited a more normal walking pattern, increased movement of the ankle, less bulky under clothing and greater independence as the main reasons for preferring FES [72]. However, some preferred an AFO for occasional use, for example when travelling or walking near water. This highlights the importance of finding out what matters to each individual and under what circumstances.

7.13   Evidence-based clinical guidelines

Since Liberson *et al.* [36] first reported the results of using FES for a person with hemiplegia in 1961, evidence has accumulated regarding the efficacy of FES to treat dropped foot in neurological populations. The UK National Institute for Health and Clinical Excellence (NICE) guideline 'Functional electrical stimulation for drop foot of central neurological origin' [1] was based on the evidence published up to 2008. It supported the use of FES for dropped foot for people with a UMN lesion. However, it was also suggested that further investigations addressing the efficacy of FES on patient-reported outcomes, such as quality of life and activities of daily living would usefully complement the existing evidence which was mostly based on walking speed. At the time of writing, there has been no further update of these NICE guidelines although in 2016 a MedTech Evaluation Briefing of ODFS devices for use by pwMS and those with history of stroke was published [76].

Recently, Johnston *et al.* [77] published a Clinical Practice Guideline focussing on the use of FES and AFO post stroke based on a comprehensive review of the literature. They concluded that strong evidence exists that both AFO and FES can each increase gait speed, mobility and dynamic balance and both have the potential to increase quality of life, walking endurance, activation of the anterior tibial muscle and improve gait kinematics.

In 2023, another clinical practice guideline for FES to support walking in all UMN lesions, not just stroke, was published [78,79]. This guideline was not only based on evidence synthesis [51] but also on the views of a people with lived experience and those of therapists prescribing and fitting FES. This approach led to guidance statements on who should be given the opportunity to attend a FES service, how this service should be designed and provided, and what follow-up and monitoring should be included. The reader may also wish to refer to chapter 16 on Practical Considerations in running a clinical service for this information.

7.14   Future directions

The previous sections have demonstrated that FES to support walking is effective in increasing walking speed and other outcomes of walking performance and can improve user's confidence and independence. The fact that there are an increasing number of commercial providers of FES systems used to correct dropped foot also shows that the application of FES is a feasible and well accepted by users.

Some difficulties with everyday application remain for certain users. These include issues with the identification of the optimal placement of the electrodes to achieve the desired ankle movement. To overcome this issue, researchers have developed devices using electrode arrays instead of single electrode [80]. The electrical array determines the optimal location of the 'virtual electrode' during an automated set-up protocol which is conducted prior to the use of the device. Currently, only a few of these systems have been applied outside the laboratory setting. Kenney *et al.* [81] reported that the ShefStim could be used outside of the laboratory environment It required no clinical input with the automated set-up protocol taking an average of 9 min to complete. It was also shown that gait improvements using the ShefStim were comparable to those of a conventional FES system [83]. The Fesia walk system also has undergone user testing in a rehabilitation setting with both therapists and users reporting high satisfaction with the device. However, some therapists mentioned that they sometimes did not fully agree with the parameters suggested by the automatic calibration and preferred to adjust the settings manually [26].

The use of closed loop FES systems instead of the open loop systems that is used by most current devices is another focus of FES developmental research. Closed loop systems are systems in which the information about the gait pattern is recorded by sensors. This information is then fed back to the control system which will adapt certain stimulation characteristics accordingly. For example, systems have been developed which use the EMG signal from the tibialis anterior to adjust the stimulation output to minimise muscle fatigue [82]. Others have used Inertial Measurement Unit (IMU) sensors attached to the lower limbs to adjust the stimulation output to optimise the desired joint motion [83]. However, although closed-loop FES devices have been shown to improve gait kinematics and walking speed compared to unassisted gait [84], to date none of these devices have demonstrated clinically relevant benefits compared to those observed in current commercially available devices.

It is evident that there is a need for further research aimed at the development of new FES technologies that can further optimise gait kinematics, minimise muscle fatigue and are easy to use. Such technologies will need to consider the muscle and gait characteristics of the individual as well as demands of their environment. Artificial intelligence may assist with further developments for FES utilisation, but it is crucial that this research is conducted in collaboration with therapists, FES users and their carers. If not, newly developed designs will remain a proof of concept, with issues such as complexity and bulkiness of the design prohibiting its use in everyday life outside the laboratory and the clinic.

7.15   Case studies

7.15.1   Case study 1

This case involves a male diagnosed with Ankylosing Spondylitis aged 27 and primary progressive MS in 2009 aged 43. In 2022, his Expanded Disability Status

Scale (EDSS) [85] score was 6, meaning that he required a walking aid to walk 100m. He lives with his wife and daughter in a house with stairs but with a downstairs bathroom and bedroom if needed.

At the time of his initial FES appointment in 2015, he was able to walk unaided between 100 and 500 m but experienced dropped foot. He tried a foot-up orthotic device to manage his dropped foot but found this uncomfortable. Observational gait analysis showed active dorsiflexion but inconsistent heel strike at initial contact. He also showed increased knee and hip flexion during swing to ensure adequate foot clearance. He required frequent rests because of his dropped foot worsening during walking and general fatigue. He was fitted with a dropped foot stimulator (ODFS PACE) which he used 6–7 times a week when going out the house. Without the assistance of FES his walking speed was 1.05 m/s which increased to 1.2 m/s with FES (Figure 7.11). He was set-up for yearly reviews.

In 2017, his FES device was converted to a ODFS PACE XL which uses a wireless connection to the footswitch thus eliminating the wire from the footswitch to the device. He continued to use his device 6–7 times a week.

His mobility deteriorated due to MS progression and by 2018 he used a walking stick outside but continued to use FES to manage his dropped foot. In 2019 his walking speed without FES has decreased to 0.9 m/s but he still experiences an orthotic effect of FES which increases his walking speed to 1.07 m/s (Figure 7.11). He is also set up with NMES ('exercise mode') to his shoulder muscles to reduce his shoulder pain. In 2020, he stopped using upper limb NMES but continued to use FES to manage his dropped foot. In 2021, NMES ('exercise mode') to strengthen his hamstrings and quadriceps is added to his device used to correct dropped foot. In 2022, he still used the same FES and lower limb NMES treatment, but his walking speed has now further reduced (Figure 7.11). However, he is still in employment and mobile (Figure 7.12) and continues to notice the benefits of FES in terms of tripping (less often), increased confidence in his

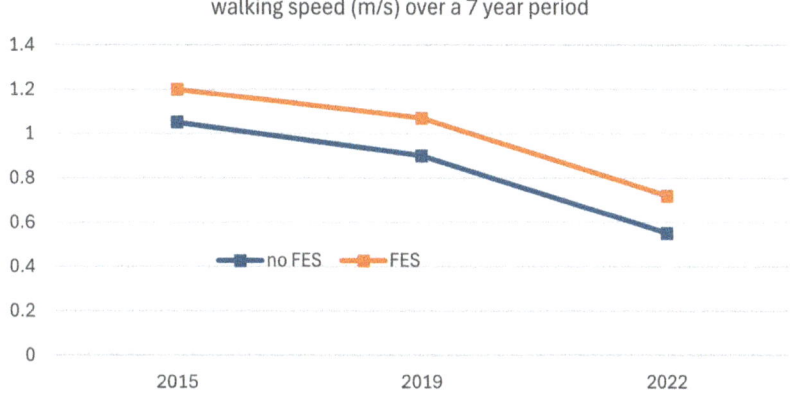

Figure 7.11   Case study 1, individual's walking speed over time

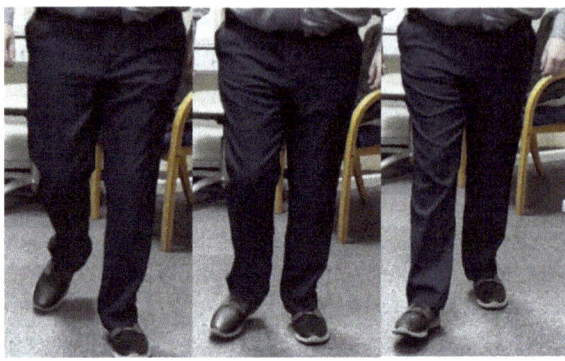

Figure 7.12   Case study 1, individual's foot strike pattern in 2020 with FES, note initial contact by the heel

walking and reduced effort of walking. In his own words: 'FES has been a game changer'.

7.15.2   Case study 2

In 2000 a 35-year-old, employed, right-handed male presented with left hemiplegia from a right basal ganglion lesion at the age of 6 and fluctuating dystonia of his left upper and lower limbs. Dystonia increased with pain, noise, touch, stress, crowds and conversational situations. He fatigued easily and became breathless over a limited walking distance. He circumducted his left leg to ensure clearance and his left foot was inverted (Figure 7.5) with plantarflexion of his ankle and clawing of his toes. He walked slowly with reduced balance, multiple trips and falls and lacked confidence in walking outdoors.

In 2000, he was assessed for FES and provided with a dropped foot stimulator which he wore every day. In 2009, he converted to a new ODFS PACE. In 2010, he stopped using FES as he was able to control his foot during both stance and swing without the assistance of FES. However, in 2011, he restarted FES as ankle inversion had returned due to an increased in his dystonia triggered by social circumstances.

It is evident that individual has benefited from FES intervention as he is walking quicker (Figure 7.13a), more efficient (Figure 7.13b) and with more confidence, less effort pain and safer, leading to a higher quality of life (Figure 7.13c).

7.15.2.1   FES parameters:

ODFS – 50 mA, 0 s rising ramp, 0 s extension, 1 s falling ramp, 5 s time out, asymmetrical waveform, heel rise, adaptive timing, 40 Hz frequency and 60 µs pulse width. In 2004, he changed from 5 cm round Pals Plus electrodes to the 5 cm square Hypoallergenic due to skin irritation. In 2007, an adverse reaction to a BTX injection required a change to heel strike trigger as the alignment in the stance phase of walking

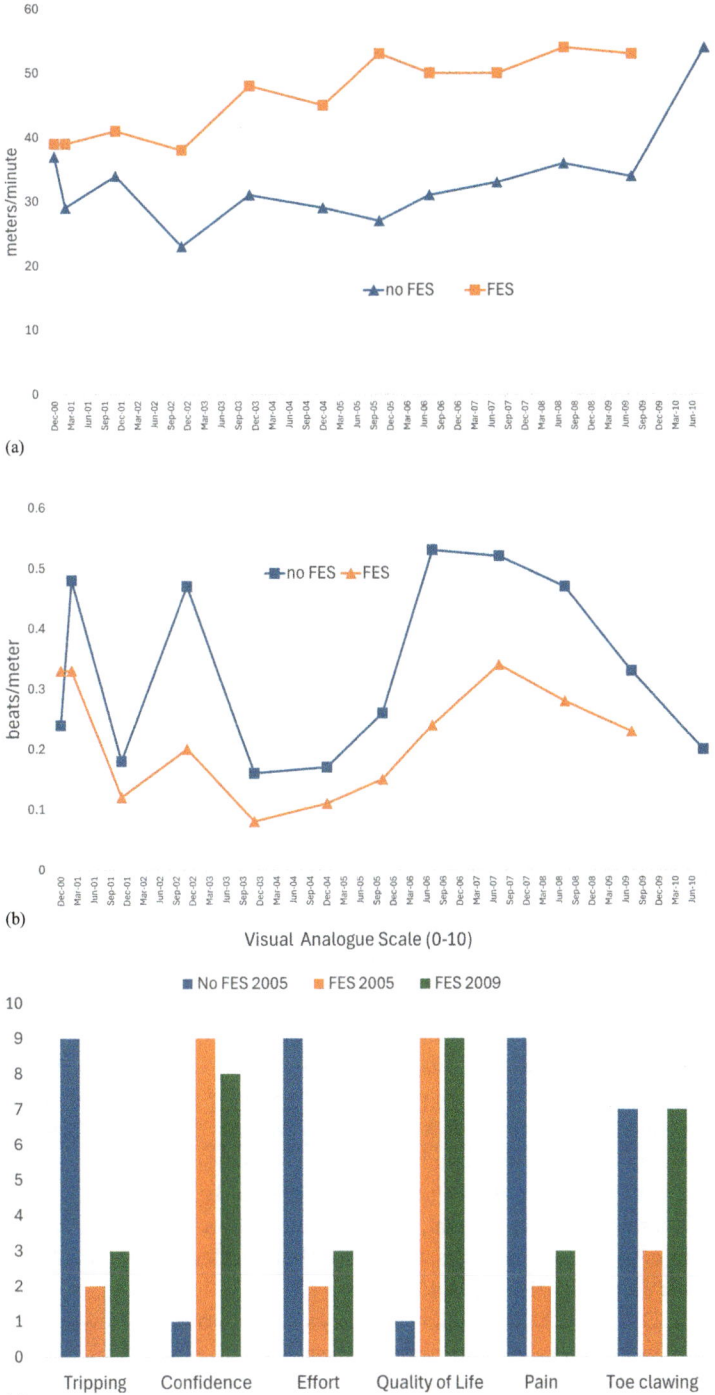

Figure 7.13 Case study 2, (a) walking speed from 2000 to 2010, (b) physiological cost index from 2000 to 2010, (c) self-reported visual analogue scores for no FES (2005), with FES (2005) and with FES (2009) for tripping, confidence, effort, pain, quality of life and toe clawing. Higher ratings for confidence and quality of life and lower ratings for tripping, effort, pain and toe-clawing indicate an improvement.

was difficult to control. In 2009, the ODFS PACE parameters were 74 mA, 0 s rising ramp, 0.4 s extension, 0.6 s falling ramp, 5 s time out, symmetrical waveform, adaptive timing, 40 Hz frequency, 30 µs pulse width and heel strike trigger.

References

[1]   NICE guidelines. http://www.nice.org.uk/Guidance/IPG278 Functional electrical stimulation for drop foot of central neurological origin N1733 1P ISBN 84629-846-6 Jan 09 [accessed 17 April 2024].

[2]   Odstock Medical. *Odstock Medical Research/Evidence* [Internet]. UK. Available from: https://odstockmedical.com/patients/patients-research-evidence/ [accessed 17 April 2024].

[3]   Granat MH, Ferguson ACB, Andrews BJ, and Delargy M. The role of functional electrical-stimulation in the rehabilitation of patients with incomplete spinal-cord injury – observed benefits during gait studies. *Paraplegia*. 1993;31(4):207–15.

[4]   van der Linden ML, Hazlewood ME, Hillman SJ, and Robb JE. Functional electrical stimulation to the dorsiflexors and quadriceps in children with cerebral palsy. *Pediatr Phys Ther*. 2008;20(1):23–9.

[5]   Perry J, and Burnfield JM. *Gait Analysis: Normal and Pathological Function*. Boca Raton, FL: CRC Press; 2nd edn. 2010. pp. 1–551.

[6]   Chockalingam N. *Technologies and Techniques in Gait Analysis: Past, Present and Future*. Stevenage: The Institution of Engineering and Technology; 2022.

[7]   Levine D, Richards J, and Whittle M. *Whittle's Gait Analysis*. 5th edn. Edinburgh: Churchill Livingstone Elsevier; 2012.

[8]   Fleuren JF, Voerman GE, Erren-Wolters CV, *et al.* Stop using the Ashworth Scale for the assessment of spasticity. *J Neurol Neurosurg Psychiatry*. 2010; 81(1):46–52.

[9]   Fowler EG, Staudt LA, and Greenberg MB. Lower-extremity selective voluntary motor control in patients with spastic cerebral palsy: increased distal motor impairment. *Dev Med Child Neurol*. 2010;52(3):264–9.

[10] Unnithan VB, Clifford C, and Bar-Or O. Evaluation by exercise testing of the child with cerebral palsy. *Sports Med*. 1998;26(4):239–51.

[11] Collaborators GBDS. Global, regional, and national burden of stroke, 1990–2016: a systematic analysis for the Global Burden of Disease Study 2016. *Lancet Neurol*. 2019;18(5):439–58.

[12] Wade DT, Wood VA, Heller A, Maggs J, and Langton Hewer R. Walking after stroke: measurement and recovery over the first 3 months. *Scand J Rehabil Med*. 1987;19:25–30.

[13] Skilbeck CE, Wade DT, Langton Hewer R, and Wood VA. Recovery after stroke. *J Neurol Neurosurg Psychiatry* 1983;46:5–8.

[14] Cans C. Surveillance of cerebral palsy in Europe: a collaboration of cerebral palsy surveys and registers. *Dev Med Child Neurol*. 2000;42(12):816–24.

[15] Collaborators GBDMS. Global, regional, and national burden of multiple sclerosis 1990-2016: a systematic analysis for the Global Burden of Disease Study 2016. *Lancet Neurol*. 2019;18(3):269–85.

[16] Larocca NG. Impact of walking impairment in multiple sclerosis: perspectives of patients and care partners. *Patient*. 2011;4:189–201.

[17] Andreopoulou G, Mercer TH, Enriquez JG, *et al*. Exercise-induced changes in gait kinematics in multiple sclerosis with minimal neurological disability. *Mult Scler Relat Disord*. 2021;47:102630.

[18] Taylor P, Barrett C, Mann G, Wareham W, and Swain I. A feasibility study to investigate the effect of functional electrical stimulation and physiotherapy exercise on the quality of gait of people with multiple sclerosis. *Neuromodulation*. 2014;17(1):75–84.

[19] Collaborators GBDMS. Global, regional, and national burden of neurological disorders, 1990–2016: a systematic analysis for the Global Burden of Disease Study 2016. *Lancet Neurol*. 2019;18(5):459–480.

[20] Cioni M, Richards CL, Malouin F, Bedard PJ, and Lemieux R. Characteristics of the electromyographic patterns of lower limb muscles during gait in patients with Parkinson's disease when OFF and ON L-Dopa treatment. *Ital J Neurol Sci*. 1997;18(4):195–208.

[21] Tan DM, McGinley JL, Danoudis ME, Iansek R, and Morris ME. Freezing of gait and activity limitations in people with Parkinson's disease. *Arch Phys Med Rehabil*. 2011;92:1159–65.

[22] Mann GE, Finn SM, and Taylor PN. A pilot study to investigate the feasibility of electrical stimulation to assist gait in Parkinson's disease. *Neuromodulation*. 2008;11(2):143–9.

[23] Taylor PN, Sampson T, Beare B, *et al*. The effectiveness of peroneal nerve functional electrical simulation for the reduction of bradykinesia in Parkinson's disease: a feasibility study for a randomised control trial. *Clin Rehabil*. 2021;35:546–557.

[24] Delafontaine A, Fourcade P, Zemouri A, Diakhate DG, Saiydoun G, and Yiou E. In patients with Parkinson's disease in an OFF-medication state, does bilateral electrostimulation of tibialis anterior improve anticipatory postural adjustments during gait initiation? *Front Hum Neurosci*. 2021;15:692651.

[25] Injury GBDTB, Spinal Cord Injury C. Global, regional, and national burden of traumatic brain injury and spinal cord injury, 1990–2016: a systematic analysis for the Global Burden of Disease Study 2016. *Lancet Neurol*. 2019; 18(1):56–87.

[26] Imatz-Ojanguren E, Sanchez-Marquez G, Asiain-Aristu JR, *et al*. A foot drop compensation device based on surface multi-field functional electrical stimulation-usability study in a clinical environment. *J Rehabil Assist Technol Eng*. 2019;6:2055668319862141.

[27] Taylor PN, Burridge JH, Dunkerley AL, *et al*. Clinical use of the Odstock dropped foot stimulator: its effect on the speed and effort of walking. *Arch Phys Med Rehabil*. 1999;80(12):1577–83.

[28]   Maxwell DJ, Granat MH, Baardman G, and Hermens HJ. Demand for and use of functional electrical stimulation systems and conventional orthoses in the spinal lesioned community of the UK. *Artif Organs*. 1999;23(5):410–2.

[29]   van der Salm A, Nene AV, Maxwell DJ, Veltink PH, Hermens HJ, and IJzerman MJ. Gait impairments in a group of patients with incomplete spinal cord injury and their relevance regarding therapeutic approaches using functional electrical stimulation. *Artif Organs*. 2005;29(1):8–14.

[30]   Salinas S, Proukakis C, Crosby A, and Warner TT. Hereditary spastic paraplegia: clinical features and pathogenetic mechanisms. *Lancet Neurol*. 2008; 7(12):1127–38.

[31]   Rothwell J *Control of Human Movement*. Beckenham: Croom Helm; 1994.

[32]   Spaich EG, Svaneborg N, Jorgensen HR, and Andersen OK. Rehabilitation of the hemiparetic gait by nociceptive withdrawal reflex-based functional electrical therapy: a randomized, single-blinded study. *J Neuroeng Rehabil*. 2014;11:81.

[33]   Hebb DO. *The Organization of Behavior: A Neuropsychological Theory*. New York: Wiley; 1949.

[34]   Rushton DN. Functional electrical stimulation and rehabilitation—an hypothesis. *Med Eng Phys*. 2003;25(1):75–8.

[35]   Coote S, Hogan N, and Franklin S. Falls in people with multiple sclerosis who use a walking aid: prevalence, factors, and effect of strength and balance interventions. *Arch Phys Med Rehabil*. 2013;94(4):616–21.

[36]   Liberson WT. Functional electrotherapy. *Trans Am Soc Artif Intern Organs*. 1962;8:373–7.

[37]   Moe JH, and Post HW. Functional electrical stimulation for ambulation in hemiplegia. *J Lancet*. 1962;82:285–8.

[38]   Hart DJ, Taylor PN, Chappell PH, and Wood DE. A microcontroller system for investigating the catch effect: functional electrical stimulation of the common peroneal nerve. *Med Eng Phys*. 2006;28(5):438–48.

[39]   Taylor P, Humphreys L, and Swain I. The long-term cost-effectiveness of the use of functional electrical stimulation for the correction of dropped foot due to upper motor neuron lesion. *J Rehabil Med*. 2013;45(2):154–60.

[40]   Street T, Taylor P, and Swain I. Effectiveness of functional electrical stimulation on walking speed, functional walking category, and clinically meaningful changes for people with multiple sclerosis. *Arch Phys Med Rehabil*. 2015;96(4):667–72.

[41]   Kluding PM, Dunning K, O'Dell MW, *et al.* Foot drop stimulation versus ankle foot orthosis after stroke: 30-week outcomes. *Stroke*. 2013;44 (6):1660–9.

[42]   Tenniglo MJB, Buurke JH, Prinsen EC, Kottink AIR, Nene AV, and Rietman JS. Influence of functional electrical stimulation of the hamstrings on knee kinematics in stroke survivors walking with stiff knee gait. *J Rehabil Med*. 2018;50(8):719–724.

[43] Kim JH, Chung Y, Kim Y, and Hwang S. Functional electrical stimulation applied to gluteus medius and tibialis anterior corresponding gait cycle for stroke. *Gait Posture*. 2012;36(1):65–7.

[44] Kunkel D, Pickering RM, Burnett M, *et al.* Functional electrical stimulation with exercises for standing balance and weight transfer in acute stroke patients: a feasibility randomized controlled trial. *Neuromodulation*. 2013;16 (2):168–77.

[45] Kesar TM, Perumal R, Reisman DS, *et al.* Functional electrical stimulation of ankle plantarflexor and dorsiflexor muscles: effects on poststroke gait. *Stroke*. 2009;40(12):3821–7.

[46] Andreopoulou G, Busselli G, Street T, *et al.* Is functional electrical stimulation effective in improving walking in adults with lower limb impairment due to an upper motor neuron lesion? An umbrella review. *Artif Organs*. 2024;48(3):210–31.

[47] Pool D, Valentine J, Bear N, Donnelly CJ, Elliott C, and Stannage K. The orthotic and therapeutic effects following daily community applied functional electrical stimulation in children with unilateral spastic cerebral palsy: a randomised controlled trial. *BMC Pediatr*. 2015;15:154.

[48] Burridge J, Taylor P, Hagan SA, Wood DE, and Swain ID. The effect on the spasticity of the quadriceps muscles of stimulation of the common peroneal nerve of chronic hemiplegic subjects during walking. *Physiotherapy*. 1997; 83(2):82–89.

[49] Everaert DG, Thompson AK, Chong SL, and Stein RB. Does functional electrical stimulation for foot drop strengthen corticospinal connections? *Neurorehabil Neural Repair*. 2010;24(2):168–77.

[50] Damiano DL, Prosser LA, Curatalo LA, and Alter KE. Muscle plasticity and ankle control after repetitive use of a functional electrical stimulation device for foot drop in cerebral palsy. *Neurorehabil Neural Repair*. 2013;27(3):200–7.

[51] Andreopoulou G, Mercer TH, and van der Linden ML. Walking measures to evaluate assistive technology for foot drop in multiple sclerosis: a systematic review of psychometric properties. *Gait Posture*. 2018;61:55–66.

[52] Busk H, Holm P, Skou ST, Seitner S, Siemsen T, and Wienecke T. Interrater reliability and agreement of 6 Minute Walk Test and 10 Meter Walk Test at comfortable walk speed in patients with acute stroke. *Physiother Theory Pract*. 2023;39(5):1024–32.

[53] Andersen CW, and Kristensen MT. Performance stability and interrater reliability of culturally adapted 10-meter walking test for Danes with neurological disorders. *J Stroke Cerebrovasc Dis*. 2019;28(9):2459–2467.

[54] Chandra D, Wise RA, Kulkarni HS, *et al.* Optimizing the 6-min walk test as a measure of exercise capacity in COPD. *Chest*. 2012;142:1545–1552.

[55] Perry J, Garrett M, Gronley JK, and Mulroy SJ. Classification of walking handicap in the stroke population. *Stroke*. 1995;26(6):982–9.

[56] Perera S, Mody SH, Woodman RC, and Studenski SA. Meaningful change and responsiveness in common physical performance measures in older adults: meaningful change and performance. *J Am Geriatr Soc* 2006;54:743–9.

[57] Street T, and Singleton C. Five-year follow-up of a longitudinal cohort study of the effectiveness of functional electrical stimulation for people with multiple sclerosis. *Int J MS Care*. 2018;20(5):224–230.

[58] Butler P, Engelbrecht M, Major RE, Tait JH, Stallard J, and Patrick JH. Physiological cost index of walking for normal children and its use as an indicator of physical handicap. *Dev Med Child Neurol*. 1984;26(5):607–12.

[59] Fredrickson E, Ruff RL, and Daly JJ. Physiological cost index as a proxy measure for the oxygen cost of gait in stroke patients. *Neurorehabil Neural Repair*. 2007;21(5):429–34.

[60] Burridge JH, Taylor PN, Hagan SA, Wood DE, and Swain ID. The effects of common peroneal stimulation on the effort and speed of walking: a randomized controlled trial with chronic hemiplegic patients. *Clin Rehabil*. 1997; 11(3):201–10.

[61] Dunning K, O'Dell MW, Kluding P, and McBride K. Peroneal stimulation for foot drop after stroke: a systematic review. *Am J Phys Med Rehabil*. 2015; 94(8):649–64.

[62] Hillman SJ, Hazlewood ME, Schwartz MH, van der Linden ML, and Robb JE. Correlation of the Edinburgh Gait Score with the Gillette Gait Index, the Gillette Functional Assessment Questionnaire, and dimensionless speed. *J Pediatr Orthop*. 2007;27(1):7–11.

[63] van der Linden ML, Andreopoulou G, Scopes J, Hooper JE, and Mercer TH. Ankle kinematics and temporal gait characteristics over the duration of a 6-minute walk test in people with multiple sclerosis who experience foot drop. *Rehabil Res Pract*. 2018;2018:1260852.

[64] van der Linden ML, Scott SM, Hooper JE, Cowan P, and Mercer TH. Gait kinematics of people with multiple sclerosis and the acute application of functional electrical stimulation. *Gait Posture*. 2014;39(4):1092–6.

[65] Moll I, Marcellis RGJ, Fleuren SM, *et al.* Functional electrical stimulation during walking in children with unilateral spastic cerebral palsy: a randomized cross-over trial. *Dev Med Child Neurol*. 2024;66(5):598–609.

[66] Swigchem R, Vloothuis J, den Boer J, Weerdesteyn V, and Geurts AC. Is transcutaneous peroneal stimulation beneficial to patients with chronic stroke using an ankle-foot orthosis? A within-subjects study of patients' satisfaction, walking speed and physical activity level. *J Rehabil Med*. 2010;42 (2):117–21.

[67] Dedding C, Cardol M, Eyssen IC, Dekker J, and Beelen A. Validity of the Canadian occupational performance measure: a client-centred outcome measurement. *Clin Rehabil*. 2004;18(6):660–7.

[68] Hobart JC, Riazi A, Lamping DL, Fitzpatrick R, and Thompson AJ. Measuring the impact of MS on walking ability: the 12-item MS walking scale (MSWS-12). *Neurology*. 2003;60(1):31–6.

[69] Bethoux F, Rogers HL, Nolan KJ, *et al.* The effects of peroneal nerve functional electrical stimulation versus ankle-foot orthosis in patients with chronic stroke: a randomized controlled trial. *Neurorehabil Neural Repair*. 2014;28(7):688–97.

[70] Smith AD, Prokopiusova T, Jones R, *et al.* Functional electrical stimulation for foot drop in people with multiple sclerosis: the relevance and importance of addressing quality of movement. *Mult Scler.* 2020;24:1352458520923958.

[71] Barrett C, and Taylor P. The effects of the Odstock drop foot stimulator on perceived quality of life for people with stroke and multiple sclerosis. *Neuromodulation.* 2010;13(1):58–64.

[72] Bulley C, Mercer TH, Hooper JE, Cowan P, Scott S, and van der Linden ML. Experiences of functional electrical stimulation (FES) and ankle foot orthoses (AFOs) for foot-drop in people with multiple sclerosis. *Disabil Rehabil Assist Technol.* 2015;10(6):458–67.

[73] Miller Renfrew L, Flowers P, Lord AC, *et al.* An exploration of the experiences and utility of functional electrical stimulation for foot drop in people with multiple sclerosis. *Disabil Rehabil.* 2020;42(4):510–518.

[74] Mendes LA, Lima IN, Souza T, *et al.* Motor neuroprosthesis for promoting recovery of function after stroke. *Cochrane Database Syst Rev.* 2020;1(1): CD012991.

[75] van Swigchem R, van Duijnhoven HJ, den Boer J, Geurts AC, and Weerdesteyn V. Effect of peroneal electrical stimulation versus an ankle-foot orthosis on obstacle avoidance ability in people with stroke-related foot drop. *Phys Ther.* 2012;92(3):398–406.

[76] ODFS Pace and Pace XL functional electrical stimulation devices for treating drop foot Medtech innovation briefing [MIB56], Published: 11 March 2016, https://www.nice.org.uk/advice/mib56/chapter/about-this-briefing (accessed 17 April 2024).

[77] Johnston TE, Keller S, Denzer-Weiler C, and Brown L. A clinical practice guideline for the use of ankle-foot orthoses and functional electrical stimulation post-stroke. *J Neurol Phys Ther.* 2021;45(2):112–96.

[78] Bulley C, Burridge J, Adonis A, *et al.* Delphi consensus study and clinical practice guideline development for functional electrical stimulation to support upright mobility in people with an upper motor neuron lesion. *Artif Organs.* 2024;48(3):285–296.

[79] ACPIN Clinical Guideline Working Group. *Evidence Based Clinical Guidelines for the Use of Functional Electric Stimulation to Improve Mobility in Adults with Lower Limb Impairment Due to an Upper Motor Neuron Lesion.* Association of Chartered Physiotherapists in Neurology; 2022. Accessed 9 November 2024. https://www.acpin.net/pdfs/2210%20QMU%20Report.pdf.

[80] Malešević J, Dedijer Dujović S, Savić AM, *et al.* A decision support system for electrode shaping in multi-pad FES foot drop correction. *J Neuroeng Rehabil.* 2017;14(1):66.

[81] Kenney LP, Heller BW, Barker AT, *et al.* A review of the design and clinical evaluation of the ShefStim array-based functional electrical stimulation system. *Med Eng Phys.* 2016;38(11):1159–65.

[82] Chen M, Wu B, Lou X, *et al.* A self-adaptive foot-drop corrector using functional electrical stimulation (FES) modulated by tibialis anterior electromyography (EMG) dataset. *Med Eng Phys.* 2013;35(2):195–204.

[83]   Sijobert B, Azevedo C, Pontier J, Graf S, and Fattal C. A sensor-based multichannel FES system to control knee joint and reduce stance phase asymmetry in post-stroke gait. *Sensors (Basel)*. 2021;21(6):2134.

[84]   Hosiasson M, Rigotti-Thompson M, Appelgren-Gonzalez JP, *et al.* Bio-mechanical gait effects of a single intervention with wearable closed loop control FES system in chronic stroke patients. A proof-of-concept pilot study. *IEEE Int Conf Rehabil Robot*. 2023;2023:1–6.

[85]   Kurtzke JF. Rating neurologic impairment in multiple sclerosis: an expanded disability status scale (EDSS). *Neurology*. 1983;33(11):1444–52.

Chapter 8

Functional electrical stimulation of denervated muscles

Winfried Mayr[1] and Ines Bersch-Porada[2]

8.1 Introduction

Functional electrical stimulation (FES) of denervated muscles has become an effective additional tool for movement rehabilitation that complements nerve and neuromuscular stimulation for cases with peripheral denervation. Eliciting travelling action potentials (APs) in muscle fibres directly can be attained via biphasic long-duration pulses (15–500 ms per phase in contrast to 0.1–1 ms for nerves). The associated higher charge transfer per stimulation pulse across the electrode/skin interface requires some attention in electrode selection and handling. Provided, electrode size, type and contact pressure distribution are appropriate, the method is safe and most useful in maintaining muscle volume, function and metabolic activity over periods of temporary denervation, preserving muscles as functional reinnervation targets. It is also beneficial in cases where there is a chronic loss of nerve supply, where maintenance of healthy soft tissue properties and thus prevention of pressure ulcers is a main priority.

8.2 Terminology

Electrical stimulation (ES) is an umbrella term for methods to induce or alter electrical activity in excitable biological structures by artificial electrical fields. Various types of ES; FES, NMES and TENS are described in Chapter 1 of this book and in most cases are concerned with causing an effect via stimulation of the nerves. For direct stimulation of denervated muscles, which is characterised by inducing travelling APs in the electrosensitive envelop membrane of muscle fibres, we have no specific consensus on a term, and we often face misunderstandings and mixing up of neuromuscular and direct muscle stimulation. This fundamental confusion may be a main reason for the actual underrepresentation of this novel and

[1]Center for Medical Physics and Biomedical Engineering, and Department of Physical Medicine, Rehabilitation and Occupational Medicine, Medical University of Vienna, Austria
[2]International FES Centre®, Swiss Paraplegic Centre, Switzerland

potent therapeutic option in clinical practice. In this chapter, we decided to use 'ES of denervated muscles' for 'Stimulation of denervated muscles', when we explain the phenomena and cite from historic sources. In addition, we use the term 'long-duration stimuli', which characterises the main technical difference to established short-pulse methods and technology for nerve stimulation, which cannot elicit responses in denervated muscles.

8.3   Historic developments

FES stands for a wide range of diagnostic and therapeutic applications, mainly in movement rehabilitation and support or restoration of impaired organ functions. Most approaches rely on eliciting APs in sensory and motor neurons. In the special case of muscle denervation, the original chemical trigger of APs – neurotransmitter reaching the postsynaptic membrane – is lost, but with appropriate electrical stimuli electrical APs can be triggered directly on the muscle fibre surface, which then travel in both fibre directions, induce Ca^2-release and finally protein contraction in the fibre.

Although the principle of direct ES of muscle with long-pulse duration stimuli has been known for decades, it is far less established in clinical application than nerve and neuromuscular stimulation, though efficacy and safety have been proven, and patients, suffering from severe post-denervation complications, could gain substantial benefit beyond any other therapeutic option.

Historically, the book by Ernest Guttmann, *The Denervated Muscle*, first published in 1962, summarises all relevant topic related knowledge of that time. It provides a detailed insight into the physiological and clinical aspects and has remained a valuable key source of specific information ever since. Guttmann concludes on 'evaluation of present-day achievements in clinical therapy and diagnosis of denervated muscles', that those are 'far from satisfactory'. He further rates 'electrotherapy' (at that time long-duration monophasic exponential, i.e. ramp-shaped, stimuli) as 'the most successful method for retarding muscle atrophy' [1].

In the subsequent years various attempts were undertaken to progress beyond just 'retarding atrophy', to accomplishing the restoration and long-term maintenance of muscle tissue and function. Although the number of specialised groups has always remained relatively low, the topic 'ES of denervated muscles' has been constantly presented in the literature and in rehabilitation-related scientific symposia. Over the years we have seen a constant awareness and interest within the 'stimulation community', but due to the lack of major breakthroughs research has been limited and it has not been widely accepted into clinical practice.

8.4   Representative examples for early research reports on ES of denervated muscles

In the triennial 'Advances in External Control of Human Extremities' (ECHE) meeting series in Dubrovnik, FES of denervated muscles starts appearing in the 1981 edition, where Vodovnik presented a joint study by the Faculty of Electrical

Engineering of the University of Ljubljana and the Rehabilitation Institute Ljubljana, aiming in 'development of a clinically useful functional electrical stimulation (FES) method for patients with lower motor neuron lesions'. The group focused on investigating options for improved stimulation waveforms for inducing functionally useable muscle contractions in a more effective, and at the same time, safe manner. As there was a strong concern of causing skin damage by excessive charge transfer and the associated electrochemical tissue reactions, emphasis was placed on ensuring charge balance and minimising that charge, whilst ensuring a functionally sufficient stimuli. Being aware of altered skin properties after denervation, Vodovnik and colleagues added an additional control group of neurologically intact subjects to their study group with peripheral denervation. The same long duration pulse patterns being applied to both groups [2].

In the same 1981 ECHE workshop, Merletti and colleagues presented a study on quantitative assessment of muscle atrophy after complete denervation. The cross section of agonistic and antagonistic muscle groups had been monitored via computerised tomography (CT) scans over a period of 14 months. Part of the study participants applied daily 'electrotherapy' for 10 minutes with monophasic exponential stimuli of 250 ms or 500 ms duration with some, but insufficient effect in retarding the course of atrophy [3].

The 'Vienna International Workshops on Functional Electrical Stimulation' series began in 1983 and were also triennial. In that first workshop proceedings we find a contribution by David, Eichhorn and Schubert reporting on a single case study of a child with complete flaccid paraplegia [4]. They described a regime of the daily application of biphasic long-duration stimuli and a training modality with a home cycle trainer, pedalled by stimulation evoked muscle contractions with some assistance provided by an electric motor. This caused an increase in muscle bulk and an increase in perfusion. They also report on the development of a suitable stimulator, which was at that time point, 1983, under industrial production. They hypothesised that the mechanism of direct muscle activation was 'spatial excitation by current fields resulting in direct release of calcium', but experimental verification of this assumption was not disclosed. In the same event, Carraro *et al.* reported on Isomyosin changes after denervation in the rat model [5] and Zrunek *et al.* on direct electrical stimulation of the posterior cricoarytenoid muscle, the main glottis opener, in sheep after unilateral recurrent nerve neurectomy [6].

A final early example that was important, was a lecture by Terje Lomo, 1985, where he presented a large study on the denervated extensor digitorum longus muscles in rats. Over periods of up to nine months (very long in a rat lifetime of roughly three years) and at different time points after denervation, muscles were assessed in their electrophysiological and biomechanical reaction to biochemical and electrical stimuli. Among many fundamental observations with high practical relevance, were the time course of resting membrane voltages and extra-junctional Acetylcholine (Ach) sensitivity, the contraction force decline after denervation and how those muscles fibres can be reconditioned from different states of atrophy and degeneration. The results support the understanding of clinical observations regarding reinnervation processes, reconditioning of membrane properties in muscle fibres and rebuilding of muscle volume and contractions force by electrical stimulation after denervation [7]. Even though degeneration processes and electrical parameters for rat muscle activation show differences in comparison to human muscle, this fundamental proof-of-principle in the rat model gave substantial evidence that long pulse ES can be an effective tool for restoration and maintenance of functional muscle tissue after denervation. This encouraged further research in humans and led to the development of safe and efficient clinical methods and the associated technical equipment.

8.5   The European initiative RISE

RISE was a Research and Development project in the 5th Framework Program of the European Union, initiated by Winfried Mayr, a Biomedical Engineer and Helmut Kern, a Clinical Specialist, following many years of collaborative research in Vienna, Austria. The acronym 'RISE' standing for the full project name 'Use of electrical stimulation to restore standing in paraplegics with long-term denervated degenerated muscles'. The multi-disciplinary consortium included 20 partner groups, distributed across Europe, and was a collaboration of biomedical engineering, experimental and clinical physiology, clinical rehabilitation, neurology and dermatology [8]. Essential aspects of the ES of denervated muscles were studied in small and large animal experiments, and in a two-year clinical trial with participants with sensory and motor complete flaccid paraplegia, included people who were vastly different times post-injury. Part of RISE project was also the development of stimulation and assessment instrumentation. The project was completed in 2005 after four years with a whole series of emerging publications, definition of clinical guidelines and market transfer of an appropriate stimulator. Nevertheless, it took quite some more years, and it remains an ongoing challenge, to increase awareness and acceptance of this treatment modality in specific clinical environments. Secondary to the study on flaccid paraplegia, there are many more applications for ES of denervated muscles. Examples are the maintenance of facial or laryngeal [9] muscles after peripheral nerve injury, or arm and hand muscles after brachial plexus denervation, or cervical spinal cord injury (SCI) where there is often damage to the peripheral nerves resulting in lower motor neuron (LMN) injury as well as damage to the spinal cord itself [10]. Regular stimulation of target muscles, applied before and then after reconstructive nerve transfers [11], can help maintain those muscle for reinnervation for months, or even years, while nerve are regrowing. These applications have just started to get clinical recognised as support tools for advancing the outcome of reconstructive nerve surgery.

8.6   Physiological developments after denervation

Loss of the nerve supply first results in disuse atrophy in the inactive muscle, meaning that the size of muscle fibres shrinks with time, interfascicular connective tissue grows, electrical excitability gets lower and evocable contraction force and endurance decline. However, in this early period, the first-year post-denervation, the physiological structure of muscle cells remains intact and training by electrical stimulation is an efficient option for maintaining muscle volume and function. In the second year after denervation – the transition point is variable between individuals – we observe a change from atrophy to degeneration of muscle cells and their contractile structures and organelles. Muscle tissue starts to be lost and substituted by fat, collagen and connective tissue. At the same time fibre membrane excitability and conduction speed for APs, perfusion and metabolic activity decline progressively.

In summary, early application of appropriate ES can maintain, or, if necessary, rebuild near normal electrophysiological, metabolic and biomechanical properties. Longer inactivity over periods of more than two years can lead to long delays and limitations in the rebuilding of muscle tissue and restoration of function. However, tissue health can still be positively affected even years after denervation [12].

8.7   Clinical consequences and opportunities

Permanent peripheral motor denervation is generally associated with severe trophic changes in the affected anatomical region. The loss of muscle and the reduction in perfusion and metabolic activity lead to serious secondary complications, above all vulnerability to the development of pressure sores and decubitus ulcers. This can be a continuing source of complications, especially in flaccid paralysis of the lower extremities, and greatly affects quality of life, social functioning and working ability. It can also require frequent protracting medical treatment, often with long periods of admission to hospital. Other peripheral nerve lesions such as plexus brachialis trauma, facial nerve paralysis or recurrent nerve damage [13] can be associated with long waiting time till nerve regeneration, even after the best possible surgical reconstruction. The rapid degradation of muscles when they lose their nerve supply and the alteration of tissue composition and metabolic conditions in affected distal anatomical regions, lead to a loss of appropriate reinnervation targets. This endangers the success of nerve reconstruction surgery and can require secondary operations, such as muscle transfers, with associated limitations in functional restoration.

Electrical stimulation, applied early after injury and regularly in the chronic phase, is at the present time the only efficient option for maintaining muscles and their metabolic condition over longer time periods [14]. This is evident from multiple clinical and experimental studies. However, there is a need for better dissemination of these results to convince sceptical clinicians, and encourage the provision of clinical standards and certified technical equipment.

8.8   Key parameter – pulse width

The fundamental difference in triggering APs in neurons and muscles with electrical stimuli lies in the required pulse duration: Whereas the usual range for activating nerve fibres varies between 50 µs and 1 ms per phase of a biphasic pulse, eliciting APs in muscle fibres can require phase durations of more than 500 ms after longstanding denervation (years after denervation) and 15 ms, shortly after denervation or in well trained muscles [15]. Please note, that after denervation, distal parts of the nerve remain functional for periods of hours or even days, until structural degeneration and exhaustion of neurotransmitters become manifest, and requires the transition from 'neuromuscular stimulation' to 'direct muscle stimulation' to begin.

The pulse duration range for neuronal activation, e.g. NMES or FES, is determined at the lower end by the required increase in amplitude, as the reduction of pulse duration elevates threshold levels for neuron recruitment and is therefore limited by the maximum output of the stimulator. At the upper end an increase beyond approximately 1 ms pulse length is associated with an increase in delivered pulse charge, but not with a further increase in fibre recruitment. For muscle fibres, the minimal pulse duration required is considerably higher due to the lower excitability of the muscle fibre membrane, even a short time after denervation and in well trained muscles. This excitability can get far worse over a period of inactivity after injury and increases the required minimum pulse duration, from 15ms per phase, to several 100ms. On the other hand, the excitability can be conditioned by ES and the necessary pulse length for triggering an AP can be reduced to lower values within rather weeks than months, especially if the elapsed time of inactivity after denervation has not been excessively long.

8.9   Current technical options, limitations

The spectrum of available stimulators for denervated muscle is still limited. There are several stimulators on the market, that allow, by single twitch activation, effective conditioning of small denervated muscles. By doing so their excitability can be improved so that they respond to biphasic pulses of between 15 and 20 ms, which is a precondition to merging single twitch reactions to fused contractions via pulse trains with frequencies of 20 Hz or more. As the repetition period of a 20 Hz patterns is 50 ms, and a 25 Hz waveform repeats pulses every 40 ms, it is apparent that the inter-pulse intervals will also be very short. Unfortunately, most available stimulators do not allow inter pulse intervals shorter than once or twice the pulse duration. However, fused tetanic contractions are essential for building muscle volume and to improve force, endurance and function. Slightly lower frequencies, e.g. 30 ms per phase and hence approximately 15 Hz, can be used as a compromise for muscle building, although contraction patterns appear unfused and overlaid with some ripple. Most available stimulators do not only have limitations in not being able to stimulate at sufficiently high frequency for fused contractions, but also are not able to stimulate at sufficient amplitude for bigger muscles. For most stimulators the maximum amplitude is 80 mA or less, whereas sufficient activation of the denervated quadriceps femoris or gluteus maximus muscles can require up to 300 mA, especially after a long periods of post-injury inactivity. Currently, the

Table 8.1   Essential features of a stimulator for denervated muscles [15]

Pulse form	Symmetric biphasic rectangular Symmetric biphasic ramp	
Amplitude range (per phase)	0 – ≥250 mA for large muscles (quadriceps, gluteus) gradually less for smaller muscles	
Pulse width (per phase)	15 ms – ≥250 ms	
Frequency range	1 Hz – ≥20 Hz	
Pulse train	Tonic: continuous twitching Phasic: 1–5 s ON / 1–10 s OFF	

'Stimulate RISE', offered by the Viennese company Schuhfried, is the only certi-fied medical product capable of delivering stimuli with the necessary reduced inter-pulse pauses, down to 10 ms, and the necessary stimulation intensity for activating larger muscles. The device is powerful enough to even allow standing-up and stepping exercises in persons with flaccid paraplegia, once an appropriate training status has been reached (Table 8.1).

Biphasic rectangular pulses are the first choice and most effective in eliciting muscle contractions, monophasic pulses cannot be recommended, asymmetry of charge transfer involves a substantially higher risk of skin irritations and muscle fibres recruitment is much less efficient. For cases, where intact sensory nerves or motor nerves are unintentionally co-activated, ramp-shaped pulse forms can shift activation thresholds of neurons to be higher than those of muscle fibres. This is due to accommodation effects in the nerve fibre membrane, with a higher basic excitability and faster diffusion currents, counteracting discharge by external electrical fields. This can help in avoiding or reducing unpleasant sensations, or unwanted neuromuscular activation of adjacent muscles. In the case of partial denervation of a muscle, which is often seen in practice, accommodating pulses can be useful for recruiting the denervated fibre population with some selectivity. This is useful for estimating the degree of denervation in single muscles, as well as enabling conditioning of the denervated muscle fibres.

8.10   Safe and effective clinical application

A major concern has been the supposed inhibition of the reinnervation processes by ES with long-duration pulses. However, although the literature contains contra-dictory findings, the evidence of a neutral or even a positive influence of this therapeutic approach is growing. Examples are the experimental work of Tessa Gordon and colleagues, in Toronto, Canada [11], where an experimental study in rats and a clinical study on surgical nerve repair in carpal tunnel syndrome, demonstrated significantly accelerated axonal growth with ES, compared to unsti-mulated controls. Also, a recent retrospective clinical study on facial paralysis in

Jena, Germany [13], showed that stimulation with long-duration pulses led to significantly reduced synkinesis following reinnervation, and that the extent and speed of reinnervation remained unaltered.

Another critical issue is safety in the application of stimulation with a high electrical charge transfer through electrodes on the skin surface. This demands using the correct equipment, as well as clear rules on the application of stimulation to ensure that electrolytic skin lesions are avoided. However, long-term studies do prove feasible and safe in practice, if the stimuli are charge balanced, below identified charge injection limits and that the application of electrodes follows clearly defined procedures.

Careful electrode management – Particular diligence is necessary for the handling and placement of electrodes. As excessive local current density can result in skin injury, care must be taken to assure full surface skin contact with evenly distributed contact pressure and hence, uniform current density. As efficient therapy requires daily home-based application, users need specific training in safe application of electrodes, including cleaning procedures for conductive polymer electrodes, which can easily get damaged, with hidden risk of inhomogeneities in current distribution. Longstanding denervation, particularly when large muscles are involved, require the use of polymer electrodes in wet foam pockets or with gel as the contact medium. In smaller well-conditioned muscles with good skin quality, and for lower charge transfer levels, the widely available hydrogel electrodes are applicable (Figure 8.1).

No implants near or between electrodes – The risks associated with metal implants is a generally often underestimated risk in the clinical practise of ES. In such cases there is a current flow through the metal components, which is an ohmic conductor, embedded in soft tissue, an electrolytic conductor. Therefore, an anode as well as a cathode is formed on the implant surface, and the resulting charge transfer across the metal-tissue interface can lead to both metal corrosion and tissue damage. This risk is also present to some extent with the short pulses required for nerve stimulation but is substantially worse with the dramatically higher charge transfer when stimulating denervated muscle. Therefore, it is of utmost importance to avoid current flow through metal implants, and necessitates verifying beforehand if implants are present and, if they are, ensuring that those implants remain outside the main induced electrical field, i.e. not underneath or between electrodes.

Regular status assessments – An important aspect and essential precondition for successful application is a thorough initial assessment to plan treatment and the regular repetition of tests and outcome measures to monitor response and to modify stimulation parameters accordingly. Part of this initial assessment can be best undertaken by a series of single stimuli with growing intensity. Short pulses, 1 ms and below allow identification of intact motor units and whether there is any partial or total denervation. In the case where a substantial part of a muscle is denervated, the next test should be to identify the minimal pulse duration that can elicit strong twitches. If only part of a muscle is denervated, this part can be more

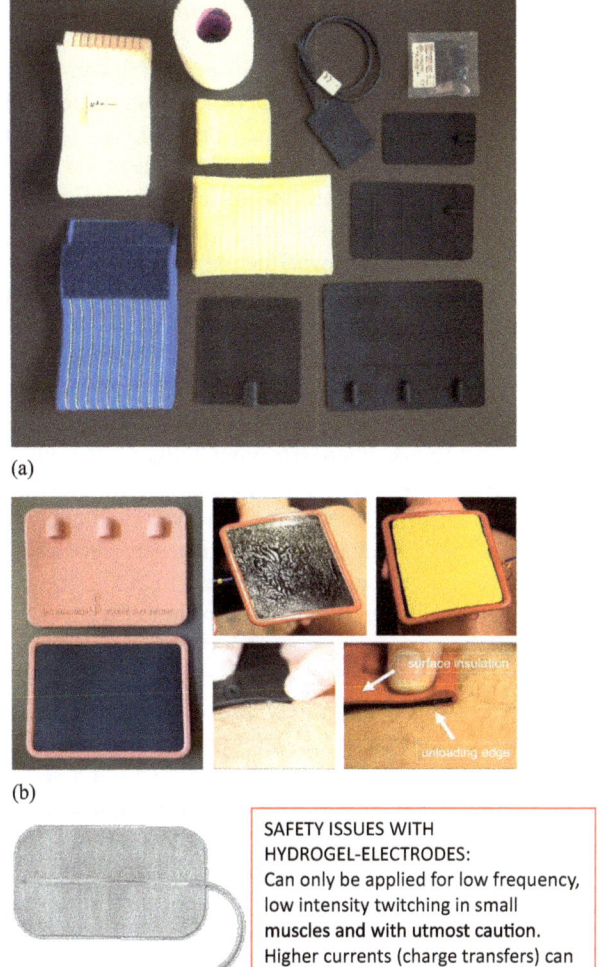

(a)

(b)

SAFETY ISSUES WITH
HYDROGEL-ELECTRODES:
Can only be applied for low frequency,
low intensity twitching in small
muscles and with utmost caution.
Higher currents (charge transfers) can
cause in tissue damage (skin burns)!

(c)

Figure 8.1 *(a) Classical conductive-polymer electrodes and contact and fixation means can be used safely for ES of denervated muscles, if electrode surface conduction is not reduced by ageing and cleaning, contact media (water or gel) and contact pressure are well distributed over the entire contact surface – avoiding current hotspots. (b) Safety electrodes avoid excessive current density at the electrode edges and insulate the electrode's far side against unsafe touch. (c) Though hydrogel electrodes are first choice for most applications for nerve stimulation they bear a considerable risk of skin damage in ES of denervated muscles due to the much higher necessary charge transfer. Those can only be recommended for small muscles, low intensity (few mA) and low frequency twitch activation (>2 per second) in the conditioning phase for improving muscle excitability. Even then meticulous monitoring of eventual adverse skin reactions remains important. Due to many variable application factors exact limit values cannot be defined.*

clearly identified by applying ramp-shaped pulses, which can mask out the neuromuscular reactions of intact motor units. Depending on the identified minimum effective pulse length, now either single twitch conditioning or training with fused contractions can be started. The optimisation of the electrode configuration can also be determined at this time. If there are issues with unpleasant sensory perception or unintended coactivation of adjacent nerve supplied muscles, the use of ramp-shaped pulses can be tried to elevate neural thresholds.

Avoid excessive muscle fatigue – An important additional test for adjustment of training session length is basic fatigue monitoring. If a clear reduction of the twitch or contraction strength is seen, then this should be a reason to end the session, at least in the early treatment phase, and fatigue should not be compensated for by more intensive stimulation. This is to avoid metabolic overload and potential muscle damage. In clinical practise we often see rapid fatigue after just few minutes, especially then starting therapy after a long period of inactivity, but that muscle endurance improves under regular stimulation. However, if stimulation is continued beyond onset of fatigue and administered to exhausted muscles we experience stagnation or further decline in muscle performance. Studies on the daily dose of ES training of freshly denervated rabbit tibialis anterior muscles suggest daily sessions with a maximum of 30 minutes. Longer sessions did not cause adverse effects, but also did not enhance the size and performance of muscles [16].

Monitor skin reactions – Another safety aspect is related to skin reactions. Reddening of skin itself is not critical if it disappears within few hours, usually 30–60 minutes (but 2–4 hours can still be safe) after end of a session, and if the redness appears evenly distributed under the electrode contact area. In this case reddening can usually be attributed to temporary increase in skin perfusion and gets less pronounced over time, basically as a positive component of the therapeutic impact. Of course, any lasting deterioration of the skin condition, in particular small red spots, require greater attention and the identification of the underlying cause. In particular, skin lesions are critical incidences and usually result from inappropriate electrode handling or damaged electrodes, which can, and of course must, be immediately rectified.

Informed patients – As daily homebased application is key for persisting health benefit patients need to be clearly instructed, trained in electrode management, and regularly reminded to be aware of potential risks. If they notice any signs of skin irritation or unusual changes in muscle strength or fatigue, they need to be advised to stop application and contact the supervising clinician.

8.11   Summary on actual clinical opportunities

ES of denervated muscles is an early treatment option cases of temporary and/or chronic peripheral denervation. The best effects are accomplished when the muscle is intact or only slightly atrophied and not yet undergone significant degeneration. However, substantial therapeutic benefits can be expected even after long-term disuse and in the presence of distinct degenerative changes. In cases where there is the possibility of the recovery of the nerve supply, the method can preserve the muscles in near-normal state, ready for future reinnervation. If denervation is permanent, ES of denervated muscle is the only option to maintain muscle tissue and metabolic functioning. This is important for the prevention of pressure injuries and various degenerative developments, generally associated with long-term disuse. In addition, preserving the muscles in a healthy state keeps the applicability of eventual future novel therapy options open, whereas chronic degeneration will lead to irreversible conditions with little or no chance of improvement.

In an early state, long-duration pulses with just 15–20 ms per pulse phase are applicable. Together with short inter-pulse pauses these allow eliciting strong fused contractions with stimulus bursts of 20 Hz or more.

If disuse after denervation has progressed to substantial atrophy and muscle degeneration, later, usually in the second-year post injury, restoration gets more complex, less efficient and requires much longer treatment periods. Depending on the disuse period much longer pulses up to 500 ms per phase are necessary to evoke muscle twitches, as electrical excitability of fibre membranes progressively declines over time. In these conditions fused contractions are not achievable due to the length of the stimulation pulses, and though single twitch conditioning is not enough to build muscle mass and force, it can restore excitability and gradually lowers the necessary minimum pulse duration for eliciting strong twitches. A final goal of this conditioning is to come to applicable pulse phase duration below 20 ms and then rebuilding muscle size and performance towards functional levels.

In conclusion, ES of denervated muscles is an effective therapeutic modality and meanwhile mature enough for clearly beneficial and safe application. It can maintain functional and metabolic properties after temporary or permanent peripheral denervation. In a short or medium time perspective it can improve the course and extent of nerve regeneration as well as the outcome of surgical nerve repair. For chronic denervation it is the only option for maintaining tissue and to avoid or substantially reduce the secondary long-term complications associated with muscular disuse atrophy and degeneration after loss of nerve supply. Given the current state of development, this therapeutic modality deserves more attention in clinical application and offers high potential for serving as a powerful treatment option for a wide range of applications in the restoration of movement and in counteracting tissue degeneration after chronic denervation.

8.12 Insights into the practice of ES of denervated muscles in an established clinical programme, case presentations and critical observations

The clinical application of long pulse stimulation is recommended to be an integral part of the treatment of neurological disorders involving LMN lesions. It is of no importance whether the damage is located in the nucleus of the anterior horn or in the periphery beginning from ventral roots of the spinal nerves reaching to the peripheral nerves and to the motor end plates of a muscle. The stimulation

06.10.2021   13.04.2022   18.01.2023

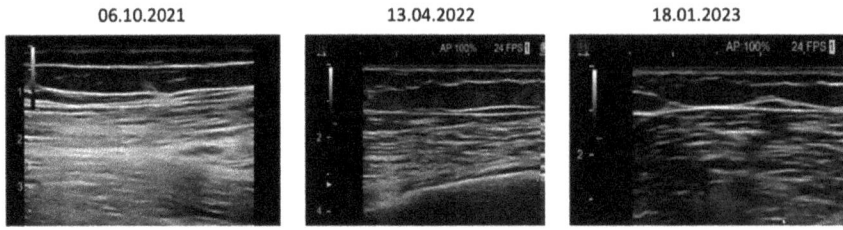

Figure 8.2   Ultrasound image of a quadriceps muscle in the course of long-pulse
stimulation over the stimulation period 14.5 months (five times a week,
33 minutes). At the beginning of the stimulation period (6 October
2021), a pronounced white connective tissue alteration is visible, in
the course of 14.5 months, the muscle fibres become more apparent,
and muscle thickness increases (18 January 2023).

parameters and the stimulation volume, time per week, are the same and depend
primarily on the duration of the LMN lesion.

Common clinical conditions in which LMN damage can be expected include
all plexus injuries, either brachial or lumbar, peripheral nerve injuries such as cuts
or fractures of the extremities, herniated discs in the lumbar region, low level
paraplegia T11/12 or lower, tetraplegia where anterior horn damage occurs at the
level of the injury, most Guillain–Barre syndromes, all types of polyneuropathy,
rarely in patients with multiple sclerosis where plaques affect the anterior horn of
the spinal cord, and the disease pattern of CMT (Charcot-Marie-Tooth).

A careful medical history and diagnosis must be obtained and, if necessary,
neurographic assessments can be used to determine the current morphological
condition of the affected denervated or already degenerated muscles. Clinically, an
ultrasound examination has proven to be simple and informative (Figure 8.2). The
structure, the pennation angle and the thickness of the muscle can be measured and
used subsequently as a parameter of change [10].

8.12.1   Motorpoint integrity testing

In the upper extremity, standardised motor point (MP) integrity testing (Figure 8.3)
can be conducted to verify a LMN lesion. The excitability of the key muscles for
grasping and releasing, as well as those for upper arm functions, are tested by means
of stimulation with short impulses. The classification is based on the Research
Medical Council Scale for manual muscle testing [17,18] If a full range of movement
against gravity is achieved using ES testing with short pulses of 250–300 μs phase
duration, it can be assumed that the LMN is intact. If there is no stimulus response,
there is damage to the LNM and if the muscle contracts, but less noticeably than with
the full range of motion, partial damage to the LMN is presumed [19,20].

Based on the test results, in which the extent of the damage to the LMN was
determined, the stimulation protocol can now be specified. Each stimulation session consists of a warm-up program of three minutes, which comprises single

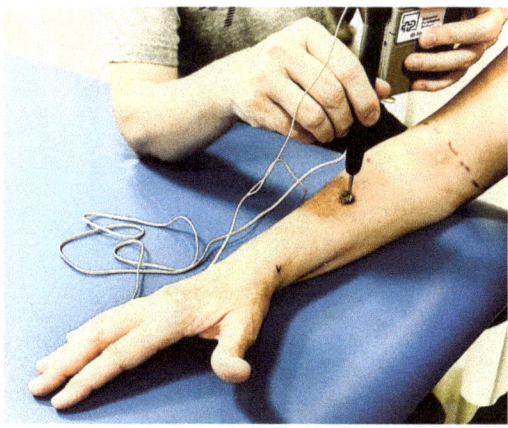

Figure 8.3   MP integrity testing of the dorsal aspect of the forearm, example: extensor pollicis longus

twitches causing temporal or spatial summation, followed by a 30-minute training interval with administering trains of biphasic stimuli, if possible, with 20 Hz stimulation frequency or more to elicit tetanic contractions.

The application protocols and stimulation parameters are determined according to the time after injury (Table 8.2). A differentiation is made between the acute/subacute and chronic phases post-injury [21].

It is advantageous to start stimulation in the acute/subacute phase, i.e. within the first two years after injury. During this period in most cases, denervation atrophy is not yet too pronounced and the alteration of the muscle into fatty and connective tissue is only partial.

The stimulation intensity (mA) depends on the muscle size as well as on the type of muscle, individual physiology and the purpose of stimulation. A muscle contains a composition of slow-twitch (Type I), fast-twitch oxidative-glycolytic (Type IIa) and fast-twitch glycolytic (Type IIb or Type IIX) fibres, depending on its biomechanical function.

In general, larger muscles require higher stimulation amplitudes than smaller muscles for achieving sufficient activation. Larger muscles typically have more motor units and a greater number of muscle fibres, the latter directly after denervation, to be recruited for full contraction. It should be noted that the recruitment characteristic is not strictly linear.

In addition, if sensory perception is maintained, the full intensity of rectangular stimuli, as required for efficient muscle growth, cannot always be applied, at least initially. Adjusting the pulse shape from rectangular to trapezoidal or triangular can be an option, as such pulses have less steep leading edges, which can elevate the threshold of nerve fibres above that of muscle fibres and thus reduce sensory perception of strong stimuli. This accommodation effect is addressed in more detail in the technical section above.

Table 8.2 Recommendations on training protocols for acute, subacute and chronic phase post injury

	LMN integrity	Target	Performance	Parameter
Acute/sub-acute lesion 0–2 years post injury	Denervated	Preservation and/or alteration of the morphology of the muscles to be stimulated Promoting reinnervation Motor learning	Stimulation under gravity-free, limited or anti-gravity movements Task-oriented exercise, defined and assessed after COPM or GAS*	**warm up:** biphasic symmetric ramp-shaped pulses, 1.42 Hz pulse duration 200 ms (100 ms per phase) inter-pulse pauses 500 ms (0.5 s) 11 s bursts (ON) / 11 s breaks (OFF) **training phase 1:** biphasic symmetric rectangular pulses, 12 Hz 65 ms pulse durations (32.5 ms per phase) and 20 ms inter-pulse pauses, 2 s bursts (ON) / 2 s breaks (OFF) **training phase 2:** biphasic symmetric rectangular pulses, 20 Hz 40 ms pulse durations (20 ms per phase) 10 ms inter-pulse pauses 2 s bursts (ON) / 2 s breaks (OFF)
Chronic lesion more than 2 years post injury	Denervated	Restoration of contractile muscle tissue Building up muscle mass Restoring the elasticity of the muscle	Stimulation under gravity-free, limited or anti-gravity movements/ positions	**warm up:** biphasic symmetric ramp-shaped pulses, 0.86 Hz pulse duration 150 ms (75 ms per phase) inter-pulse pauses 1000 ms (1 s) 11 s bursts (ON) / 11 s breaks (OFF) **training phase 1:** biphasic symmetric rectangular pulses, 2 Hz 100 ms pulse durations (50 ms per phase) and 400 ms inter-pulse pauses, 4 s bursts (ON) / 4 s breaks (OFF) **training phase 2:** biphasic symmetric rectangular pulses, 20 Hz 40 ms pulse durations (20 ms per phase) 10 ms inter-pulse pauses 2 s bursts (ON) / 2 s breaks (OFF)
Acute/sub-acute lesion 0–2 years post injury	Partial denervation	Promoting reinnervation Motor learning	Stimulation combined with task-oriented exercise, defined and assessed after COPM or GAS*	Combination of long-pulse direct muscle stimulation and stimulation via nerve NMES
Chronic lesion more than 2 years post injury	Partial denervation	Restoration of contractile muscle tissue Restoring the elasticity of the muscle for optimisation of the tension direction of innervated neighbouring muscles	Stimulation combined with task-oriented exercise defined and assessed by COPM or GAS*	**warm up:** biphasic symmetric ramp-shaped pulses, 0.86 Hz pulse duration 150 ms (75 ms per phase) inter-pulse pauses 1000 ms (1 s) 11 s bursts (ON) / 11 s breaks (OFF) **training phase 1:** biphasic symmetric rectangular pulses, 2 Hz 100 ms pulse durations (50 ms per phase) and 400 ms inter-pulse pauses, 4 s bursts (ON) / 4 s breaks (OFF) **training phase 2:** biphasic symmetric rectangular pulses, 20Hz 40ms pulse durations (20 ms per phase) 10ms inter-pulse pauses 2 s bursts (ON) / 2 s breaks (OFF)

*Canadian Occupational Performance Measure (COPM), Goal Attainment Scaling (GAS)

8.12.2 Partial denervation

In particular, partially denervated muscles constitute a challenge in treatment with ES. To date, there is no diagnostic method that can precisely quantify the proportion of denervation in a muscle. If a partially denervated muscle is stimulated with short pulses (< 1 ms per phase), only intact motoneurons get activated, whereas denervated parts will continue to degenerate over time and convert into connective and fatty tissue. Therefore, the remaining innervated, and often also voluntarily controllable muscle fibres, are unable to replace and compensate for the function of the entire muscle function, with full strength, muscle bulk and coordination.

As described above, nerve fibres are excitable from approximately 50 μs (0.05 ms) to 1 ms pulse phase duration (higher duration can also elicit APs, but with constant threshold, unaltered recruitment and unnecessarily large charge transfer), whereas muscle fibres require longer pulse durations above 15 ms per phase, after longer inactivity the applicable minimum grows progressively to several 100 ms. This might have a negative impact on still functional, i.e. voluntarily innervated muscles in the near neighbourhood. Through their denervation atrophy, and their associated shortening and transition to passive fibrous tissue, they can influence the direction of the tension produced in those parts of the muscle that still have nerve supply or are even still under voluntary control. This will ultimately limit the overall functionality. Consequently, partially denervated muscles should be stimulated with long pulses in order to maintain their contractility and thus elasticity, even if no voluntary activity is to be expected. The incidence of partial denervation is quite common in practice. In 128 new consultations in 2022 at the International FES Centre® at the Swiss Paraplegic Centre, 13% comprised partially denervated fibres and 49% of completely denervated fibres [21]. Hence, muscles with partial denervation require an adapted stimulation regime. Ideally the target muscle or muscle group is stimulated with long pulses and short pulses once a day in separate sessions. Therefore, the denervated, as well as the innervated part of the muscle, remain regularly activated. Between the two different stimulation modalities a temporal interval of at least two hours is recommended to avoid accumulation of metabolic fatigue. If this time exposure for stimulation is not feasible for organisational reasons, at least an alternating schedule should be considered. Hereby the target muscles are treated 1 day for 30 minutes with long-pulse stimulation and the next day for 20 minutes with NMES. The last and not necessarily the best option is, if time or the availability of stimulation devices does not allow otherwise, is the exclusive stimulation of the denervated parts of the muscle with long pulse stimulation. In this way, the parts that have a UMN lesion (i.e. still with a nerve supply to the muscle fibres) are also stimulated, but less efficiently, due to the lower stimulation frequency, while the denervated parts are activated. If only NMES was used the denervated parts of the muscle would never be stimulated.

In the case where a person has residual voluntary control over a part of a denervated muscle, standard rehabilitation methods (voluntary exercises) should be combined with stimulation (Figure 8.4).

Another example shows the use of long-pulse stimulation on the upper extremity with partial denervation. Extensive use of the muscles surrounding the shoulder joint leads to changes in the muscles forming the rotator cuff in people with long-standing paraplegia. Such changes can cause pain and instability. One treatment method is the combination of NMES and long-pulse stimulation with tailored movement and strengthening exercises. For best effect, this can be carried out independently by the patient at home at a correspondingly high training intensity (Figure 8.5).

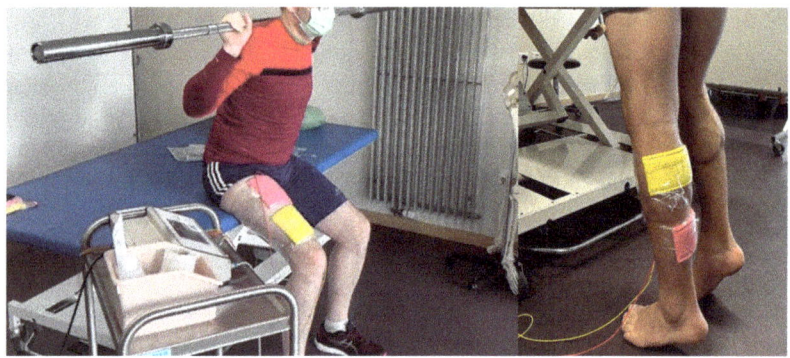

Figure 8.4   *Left side: Stimulation of a partial denervation of the quadriceps muscle in combination with voluntary activation of residual intact motor units. Long-pulse stimulation is combined with actively performed squats, including lifting weights on the shoulders. Right side: Stimulation of partial denervation of the triceps surae muscle, in synchrony with voluntary motor activation of residual controllable motor units. Stepping exercises under the load of body weight are executed with stimulation support.*

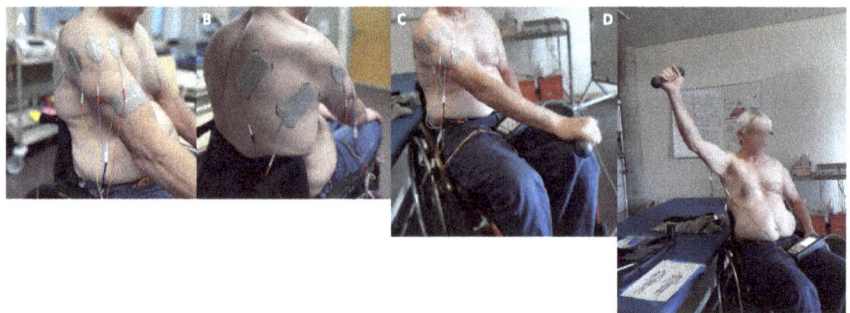

Figure 8.5   *A: Three active electrodes located on different parts of the deltoid muscle to activate each part with the optimal intensity, common reference electrode positioned on the deltoid tuberosity. B: Combined stimulation of teres major and minor as well as infraspinatus and trapezius mainly pars transversus and pars ascendens. C and D: Movement synchronised exercise in terms of proprioceptive neuromuscular facilitation (PNF) with 1 kg load.*

8.12.3   Complete denervation

An example of the impact of denervation regarding functionality is the intrinsic musculature of the hand in patients with tetraplegia. People with incomplete tetraplegia exhibit a very heterogeneous innervation pattern in the hand. It might occur that the flexor digitorum profundus (FDP) and superficialis (FDS), innervated from the spinal segments C7-TH1, presents voluntary motor activity combined with damage to the upper motor neuron (UMN) and consecutive muscle weakness. In contrast, the interossei dorsales and the lumbricales, both innervated from spinal segments C8-TH1 show damage to the LMN. This can either be combined with poor or no voluntary motor function. Hence, the fist closure is hindered by fatty and connective tissue comprising part of the intrinsic musculature (interossei dorsalis and lumbricals) clinically presenting an intrinsic tightness (Figure 8.6).

By administering long-pulse stimulation to the intrinsic hand musculature, the degeneration of the muscle structure can be avoided and even a weak fist closure enabled (Figure 8.7).

How neighbouring degenerated intrinsic muscles can hinder the functionality of fine motor skills is highlighted in the example of a patient with a chronic 17 years LMN lesion. Figure 8.8A emphasises the denervation atrophy of the first dorsal interosseus muscle. The deep finger flexors are voluntarily innervated with reduced force (MRC 3). The same is apparent in the opponens muscle. However, the degenerated interosseous muscles prevent coordinated finger flexion and the first degenerated interosseous inhibits thumb opposition. Therefore, small objects cannot be manipulated. The patient's target was to manipulate objects like credit cards, coins, parking tickets, etc.

Twelve weeks daily stimulation, with a warm up of 3 minutes with 200 ms pulse (100 ms phase) duration, 500 ms inter-pulse pause and 11 s bursts, followed

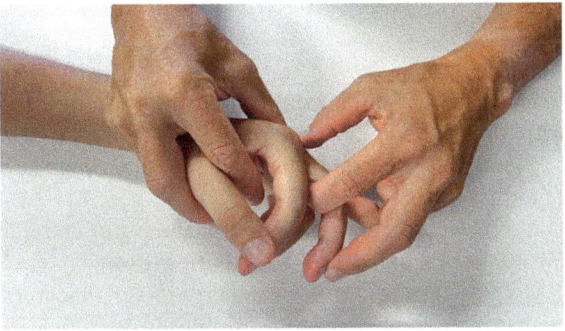

Figure 8.6   *The metacarpophalangeal (MCP) joint is manually locked in extension with simultaneous passive flexion in the proximal interphalangeal (PIP) and distal interphalangeal (DIP) joint. If flexion in the PIP and DIP cannot be executed or can only be executed with difficulty, intrinsic tightness is apparent (Bunnell-test).*

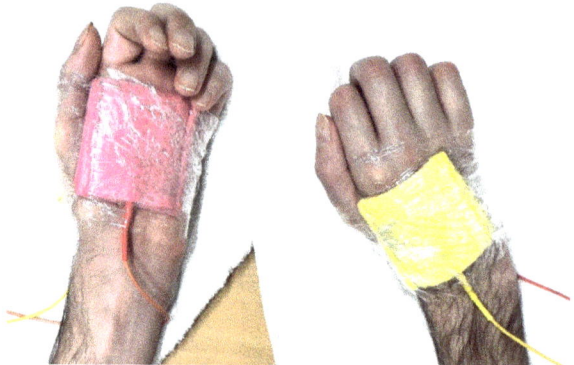

Figure 8.7   Long-pulse stimulation of the intrinsic hand musculature

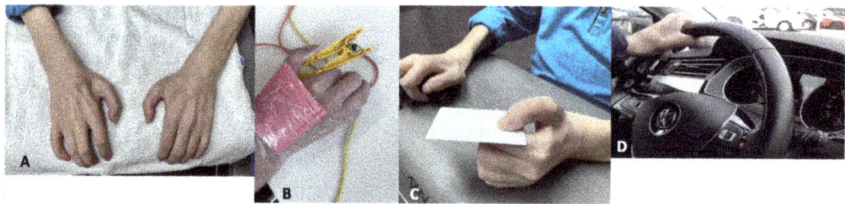

*Figure 8.8   A: visualisation of the denervation atrophy in particular of the first
interosseus dorsalis of the right hand, B: treatment stimulation
combined with clothes peg to increase strength during stimulation,
C: result after 12 weeks of daily stimulation, a credit card/parking
ticket can be manipulated in daily activities, D: after 17 years showing
their ability to grasp a steering wheel for the first time, without using a
'wheel knob'.*

by a treatment stimulation of 30 minutes with 35 ms pulse duration, 10 ms pause
and 2 s bursts with an amplitude range between 20 and 40 mA yielded the desired
function (Figure 8.8C) and even more they enabled grasping the steering wheel
without assistive aids (Figure 8.8D).

A similar clinical picture can be depicted in the lower extremities. In people
with low level incomplete paraplegia L4-S2, the triceps surae might be paralysed as
the result of a LMN lesion, whereas the voluntary motor function in the long toe
flexors could be partially preserved. An example of this could be a functional
independent walker with the impairment of an insufficient terminal stance phase.
This could be due to the elastic recoil of the calf muscles being missing and consequently the facilitation of forward progression into the swing phase. To compensate for the recoil of the calf, the long toe flexors are activated during the
terminal stance phase. This results in the formation of claw toes, which can lead to

pressure injuries on the distal metacarpal joints when wearing shoes. In order to reduce the hyperactivity of the long toe flexors, long pulse stimulation of the calf muscles is recommended. The stimulation alters the elasticity of the denervated muscle so that the lost of recoil can be partially restored. Consequently, the contraction of the toes can be corrected and hence the incidence of pressure injuries to the toes reduced.

Lesions of the brachial plexus caused by traffic accidents are a frequently encountered clinical picture. Timely post-traumatic stimulation of the most important muscles is crucial for the elementary function of the arm and hand. Early stimulation can minimise chronic pain, pronounced subluxation and the denervation atrophy associated with severely altered trophic conditions in the arm. The time required for applying such complex stimulation setups is significant, but the result can be worthwhile (Figure 8.9).

In the case of a brachial plexus lesion, the expected outcome should be clearly defined. Often a primary restoration of function is not to be expected. The ability to maintain contractility of key muscles such as the deltoid, biceps, triceps, finger and wrist extensors and flexors is often the desired effect.

In this clinical example, a completely right-side plexus avulsion after a traffic accident was diagnosed. Surgical repair and the best possible nerve suture were performed. The prognosis was described as moderate to poor by the referring physician. The patient pursued the wish for stimulation promptly after surgery on his own initiative. Three months after the accident and the surgical reconstruction, the patient started long-pulse stimulation, the parameters of which are shown in Table 8.2. Currently, the patient stimulates the deltoid, triceps and biceps, as well as the wrist and finger extensors and the finger flexors once a day, seven days a week. The total time expenditure is 2.5–3 hours per day, including donning and

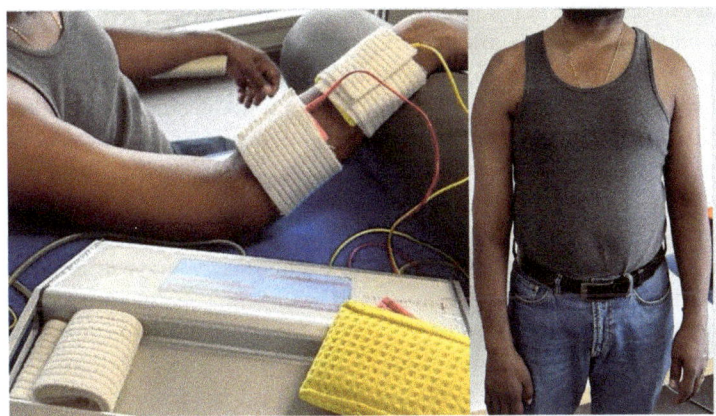

Figure 8.9　Long-pulse stimulation of the finger and wrist extensors (left-side picture), full body image of the current situation at 1.5 years post injury with daily stimulation of the delta, biceps, triceps as well as the finger and wrist extensors and finger flexors (right-side picture).

doffing time. After 11 months, the first voluntary muscle contractions occurred in the biceps muscle. The trophic condition of the arm is good, the neuropathic pain moderate (numeric rating scale 5/10) and the denervation atrophy of the entire arm musculature minimal, as shown in Figure 8.9.

8.12.4   Prevention aspects

A second complication of SCI is pressure injuries, particularly on the buttocks. The worldwide incidence, as reported in a meta-analysis by Chen and colleagues [21], was 0.23 (23% of persons with spinal cord injured). The additional subgroup analysis carried out for European countries revealed an average value of 0.25. This means that between one person in five and one person in four, living with SCI will suffer a pressure induced injury at some point. Data collected in 2020–2022 from the Swiss Paraplegic Centre, Nottwil, show that 674 cases of pressure injuries were hospitalised, 192 of them with a low, flaccid paraplegia. People with paraplegia who have a LMN lesion show clinically severe denervation atrophy and have minimal to no cushioning from the soft tissue on the buttocks and posterior leg muscles along the hamstrings. Consequently, there is insufficient long-term protection while sitting in a wheelchair. Combined with the provision of the appropriate wheelchair, including a protective cushion, stimulation of the gluteal muscles and the hamstrings is an effective treatment method to prevent pressure injuries. In addition to building up muscle mass [23,24], it also increases the epidermal thickness [25] and improves skin trophic and blood circulation. The electrode position is crucial, as at best the glutes and hamstrings should be included in the stimulation (Figure 8.10). Since there is no muscle over the precarious zone of the ischial tuberosity, it is important to allow the adjacent muscles, in this case the gluteal muscles and the hamstrings, to hypertrophy. An increase in the volume of the muscle mass leads to an increase in the seating area and thus to a minimisation of pressure peaks, which occur particularly in exposed areas such as mentioned at the ischial tuberosities.

Depending on the time after the injury, it can take 1–1.5 years in a chronic stage after SCI before there is a measurable increase in muscle volume. The sooner the stimulation is initiated, the faster a pronounced effect can be expected. This is due to the fact that denervation and degeneration take place in four chronologically consecutive steps, which can take up to two years. First fibrillations appear after a few days, followed by a loss of tension during electrical evoked tetanic contraction. After several months, a severe disorganisation of the contractile structure in the muscle occurs and finally ends after years in a transformation of muscle fibres into fat and collagen. To summarise the process first there is an immediate loss of voluntary function and rapid loss of muscle mass, followed by an increasing atrophy and loss of sarcomeric organisation, leading to muscle fibre degeneration and finally replacement of muscle by fibrous connective tissue and fat [12,26]. However, it is still worthwhile to stimulate in the chronic phase, as contractile muscle tissue can be restored even after ten years or more [24] and volume can be increased through long-term stimulation (Figure 8.11).

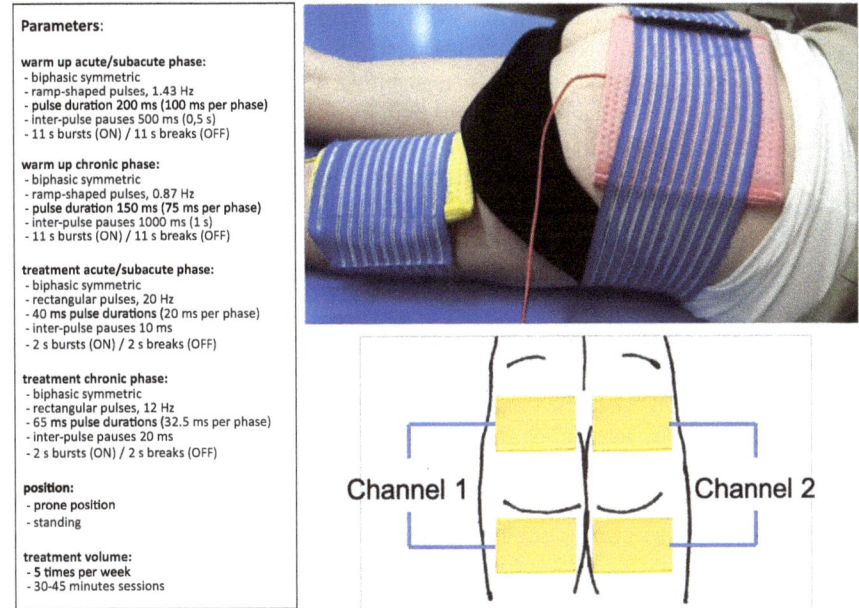

Parameters:

warm up acute/subacute phase:
- biphasic symmetric
- ramp-shaped pulses, 1.43 Hz
- pulse duration 200 ms (100 ms per phase)
- inter-pulse pauses 500 ms (0,5 s)
- 11 s bursts (ON) / 11 s breaks (OFF)

warm up chronic phase:
- biphasic symmetric
- ramp-shaped pulses, 0.87 Hz
- pulse duration 150 ms (75 ms per phase)
- inter-pulse pauses 1000 ms (1 s)
- 11 s bursts (ON) / 11 s breaks (OFF)

treatment acute/subacute phase:
- biphasic symmetric
- rectangular pulses, 20 Hz
- 40 ms pulse durations (20 ms per phase)
- inter-pulse pauses 10 ms
- 2 s bursts (ON) / 2 s breaks (OFF)

treatment chronic phase:
- biphasic symmetric
- rectangular pulses, 12 Hz
- 65 ms pulse durations (32.5 ms per phase)
- inter-pulse pauses 20 ms
- 2 s bursts (ON) / 2 s breaks (OFF)

position:
- prone position
- standing

treatment volume:
- 5 times per week
- 30-45 minutes sessions

Channel 1   Channel 2

Figure 8.10   *Graphical illustration of the recommended electrode positioning to cover the gluteal muscles and hamstrings (right side). Parameters according to time after lower motoneuron damage (Pulse duration is the entire length of a symmetric biphasic pulse, phase duration amounts half).*

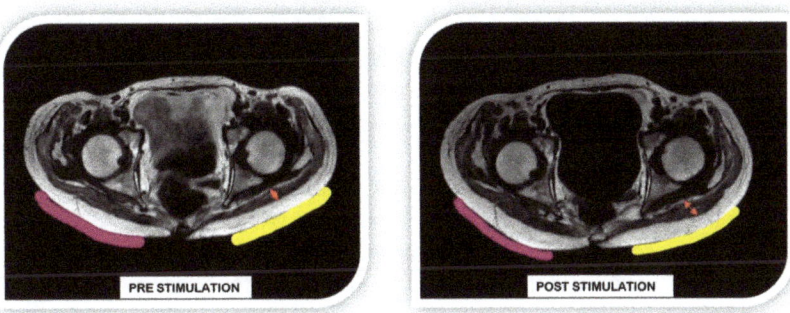

Figure 8.11   *Magnetic resonance imagination pre- and post-stimulation in a L1 paraplegic person two years after injury. The increase in muscle mass after three month of 33 minutes daily stimulation (red arrow on the right picture)*

Consequently, particularly in the patient population with flaccid paralysis, stimulation should be started preventively during the first two years after injury.

Another aspect is the healthcare costs resulting from long hospitalisation times and plastic surgery when treating pressure injuries. If comparing the hospitalisation expenses for ten weeks at the Swiss Paraplegic Centre, Nottwil, Switzerland, with the corresponding costs for an appropriate stimulator, it is only a tenth of the costs to be calculated for hospitalisation.

8.12.5   Conclusions on the clinical significance of ES of denervated muscles

The use of long-pulse stimulation is a promising treatment method that has been shown to change the muscle morphology of denervated muscles. It is relevant where the LMN is damaged, whether in the nucleus of the anterior horn or in the periphery. The treatment can be used initially to improve the structure and function of partially denervated muscles, but also to aid the functionality of the neighbouring synergists. The preservation of contractile muscle tissue favours the possibility of reinnervation in the context of neurological recovery and maintains the trophic level in the tissue. Used in the prevention of pressure injuries, muscle volume can be built up to prevent the development of injuries, thus improving the quality of life of those affected and reducing healthcare costs due to long immobilisation times and hospitalisation.

References

[1]   Guttmann E. *The Denervated Muscle*. Prague: Publishing House of the Czechoslovak Academy of Sciences; 1962.

[2]   Vodovnik L, Valencic V, Stefancic M, Jelnikar T, Gracanin F, and Rebersek S. Functional electrical stimulation of denervated muscles (1981) In: Popovic D (ed.), *Advances in External Control of Human Extremities, Proceedings I-X, Dubrovnik 1962-1990*. Published by the Center for Sensory-Motor Interaction (SMI), Aalborg University, Denmark; 2002 (Originally published by the Yugoslav Committee for ETRAN Kneza Milosa 9, Belgrade, Yugoslavia).

[3]   Merletti R, Burzio M, Granero L, and Rolfo MF. Quantitative evaluation of human muscle atrophy following total denervation (1981) In: Popovic D (ed.) *Advances in External Control of Human Extremities, Proceedings I-X, Dubrovnik 1962–1990*. Published by the Center for Sensory-Motor Interaction (SMI), Aalborg University, Denmark; 2002 (Originally published by the Yugoslav Committee for ETRAN Kneza Milosa 9, Belgrade, Yugoslavia).

[4]   David E, Eichhorn K, and Schubert W. Maintenance, training and functional use of denervated muscles. In: Thoma H (ed.) *Article 2.7, Proceedings of the 1st Vienna International Workshop on Functional electrical Stimulation*; 1983.

[5]   Carraro U, and Catani C. Isomyosin changes in direct electrical stimulation of denervated rat EDL. In: Thoma H (ed.) *Article 1.8, Proceedings of the 1ˢᵗ Vienna International Workshop on Functional electrical Stimulation*; 1983.

[6]   Zrunek M, Streinzer W, Burian K, and Thoma H. The direct electrical stimulation of the p. cricoarytenoid muscle. In: Thoma H (ed.) *Article 12.2, Proceedings of the 1ˢᵗ Vienna International Workshop on Functional electrical Stimulation*; 1983.

[7]   Lømo T, Westgaard RH, Hennig R, and Gundersen K. The response of denervated muscle to long-term electrical stimulation. *Eur J Transl Myol.* 2014;24(1):3300

[8]   Mayr W, Hofer C, Kern H, *et al.* The European R&D project RISE - use of electrical stimulation to restore standing in paraplegics with long-term denervated degenerated muscles (DDM). In: Dössel O, and Schlegel WC (eds) *World Congress on Medical Physics and Biomedical Engineering.* Berlin: Springer; 2009. pp. 540–542.

[9]   Kurz A, Volk GF, Arnold D, Schneider-Stickler B, Mayr W, and Guntinas-Lichius O. Selective electrical surface stimulation to support functional recovery in the early phase after unilateral acute facial nerve or vocal fold paralysis. *Front Neurol.* 2022;13:869900. doi: 10.3389/fneur.2022.869900.

[10]   Bersch I, and Fridén J. Electrical stimulation alters muscle morphological properties in denervated upper limb muscles. *EBioMedicine.* 2021; 74:103737. doi: 10.1016/j.ebiom.2021.103737.

[11]   Gordon T, Brushart TM, and Chan KM. Augmenting nerve regeneration with electrical stimulation. *Neurol Res.* 2008;30(10):1012–22. doi: 10.1179/174313208X362488.

[12]   Carraro U, Rossini K, Mayr W, and Kern H. Muscle fiber regeneration in human permanent lower motoneuron denervation: relevance to safety and effectiveness of FES-training, which induces muscle recovery in SCI subjects. *Artif Organs.* 2005;29(3):187–91. doi: 10.1111/j.1525-1594.2005.29032.x.

[13]   Puls WC, Jarvis JC, Ruck A, Lehmann T, Guntinas-Lichius O, and Volk GF. Surface electrical stimulation for facial paralysis is not harmful. *Muscle Nerve.* 2020;61(3):347–353. doi: 10.1002/mus.26784. Epub 2020 Jan 9.

[14]   Kern H, Carraro U, Adami N, *et al.* One year of home-based daily FES in complete lower motor neuron paraplegia: recovery of tetanic contractility drives the structural improvements of denervated muscle. *Neurol Res.* 2010; 32(1):5–12. doi: 10.1179/174313209X385644.

[15]   Mayr W, Hofer C, Bijak M, *et al.* Existing and Prospective technological solutions. *Basic Appl Myol.* 2002;12(6):287–290,

[16]   Ashley Z, Sutherland H, Lanmüller H, *et al.* Atrophy, but not necrosis, in rabbit skeletal muscle denervated for periods up to one year. *Am J Physiol Cell Physiol.* 2007;292(1):C440–51. doi: 10.1152/ajpcell.00085.2006.

[17]   Research Medical Council *Scale for Manual Muscle Testing.* https://www.ukri.org/publications/aids-to-the-examination-of-the-peripheral-nervous-system.

[18]   Bersch I, Koch-Borner S, and Fridén J. Electrical stimulation – a mapping system for hand dysfunction in tetraplegia. *Spinal Cord*. 2018;56(5):516–522. doi:10.1038/s41393-017-0042-2.

[19]   Bersch I, Koch-Borner S, and Fridén J. Motor point topography of fundamental grip actuators in tetraplegia – implications in nerve transfer surgery. *J Neurotrauma*. 2019;37(3):441–447. doi:10.1089/neu.2019.6444.

[20]   Schick T. ed. (eds.) *Functional Electrical Stimulation in Neurorehabilitation. Vol XXVI*. Cham: Springer; 2022. doi:10.1007/97.

[21]   Bersch I, and Mayr W. Electrical stimulation in lower motoneuron lesions, from scientific evidence to clinical practice: a successful transition. *Eur J Transl Myol. Published online*. 2023;33(2):11230. doi:10.4081/ejtm.2023.11230.

[22]   Chen HL, Cai JY, Du L, *et al.* Incidence of pressure injury in individuals with spinal cord injury: a systematic review and meta-analysis. *J Wound Ostomy Cont Nurs*. 2020;47(3):215–223. doi:10.1097/won.0000000000000633.

[23]   Boncompagni S, Kern H, Rossini K, *et al.* Structural differentiation of skeletal muscle fibers in the absence of innervation in humans. *Proceedings of the National Academy of Sciences of the United States of America*. 2007;104 (49):19339–19344. doi:10.1073/pnas.0709061104.

[24]   Alberty M, Mayr W, and Bersch I. Electrical stimulation for preventing skin injuries in denervated gluteal muscles—promising perspectives from a case series and narrative review. *Diagnostics*. 2023;13(2):219. doi:10.3390/diagnostics13020219.

[25]   Albertin G, Hofer C, Zampieri S, *et al.* In complete SCI patients, long-term functional electrical stimulation of permanent denervated muscles increases epidermis thickness. *Neurol Res*. 2018;40(4):277–282. doi:10.1080/01616412.2018.1436877.

[26]   Kern H, Hofer C, Loefler S, *et al.* Atrophy, ultra-structural disorders, severe atrophy and degeneration of denervated human muscle in SCI and aging. Implications for their recovery by functional electrical stimulation, updated 2017. *Neurol Res*. 2017;39(7):660–666. doi:10.1080/01616412.2017.1314906.

Chapter 9

Prevention and treatment of pressure ulcers using electrical stimulation

Thomas W.J. Janssen[1,2], M. Kristi Henzel[3,4], Boas Wijker[1], Ines Bersch[5] and Kath M. Bogie[3,6]

9.1 The problem and definitions

Skin-related secondary disabilities, especially pressure ulcers (PUs) also referred to as pressure injuries or bed sores, are a common and serious problem for wheelchair users such as individuals with a spinal cord injury (SCI) resulting in great discomfort and significant medical care costs. It is therefore of utmost importance to prevent PUs to maintain quality of life (QoL) in persons with SCI and reduce costs. Although preventive measures such as custom-made wheelchairs, pressure relieving mattresses, wheelchair cushions and performing pressure-relief movements can disperse pressure and forces and reduce PU development, not all PUs are prevented and these wounds still occur frequently.

Electrical stimulation (ES)-induced activation of the paralysed muscles has shown promise in this respect because regular use can improve sitting pressure distribution and positively influence intrinsic factors impacting tissue health, such as muscle size, blood circulation and local oxygenation. ES-induced muscle contractions can thus provide an active pathway to reduce risk factors for developing PU particularly for persons at long-term risk due to impaired mobility, such as individuals with SCI.

When prevention has failed, a chronic wound can occur. A chronic wound is characterised by its failure to follow the normal wound healing process, which typically proceeds through four well-defined and overlapping stages: haemostasis, inflammation, proliferation (closure) and maturation (remodelling). Chronic

[1]Faculty of Behavioural and Movement Sciences, Amsterdam Movement Sciences, Vrije Universiteit Amsterdam, The Netherlands
[2]Reade, Center for Rehabilitation and Rheumatology, The Netherlands
[3]Louis Stokes Cleveland VA Medical Center, Cleveland, OH, USA
[4]Department of Physical Medicine and Rehabilitation, Case Western Reserve University/MetroHealth Medical Center, USA
[5]Schweizer Paraplegiker Zentrum, Nottwill, Switzerland
[6]Department of Orthopaedics and Biomedical Engineering, Case Western Reserve University, USA

wounds are defined as wounds that fail to heal within 4–6 weeks and can be stalled in the inflammatory stage. The most prevalent type of chronic wound is a PU which occurs over the bony prominences, particularly the ischial tuberosities (ITs), sacrum and greater trochanters. Other types, such as diabetic, arterial and venous ulcers are also increasingly common. The failure of these wounds to heal causes pain and distress to the affected individual, thereby reducing their QoL, but chronic wounds are also known to cause fatality [1–4].

Chronic wounds affect patient populations across the board and their incidence continues to increase paralleling the trend of an ageing and increasingly morbid global population. The global prevalence of chronic wounds is estimated at 1.51–2.21 per 1,000 individuals, and the incidence is expected to rise with ageing populations worldwide. Between 2014 and 2019, the number of Medicare beneficiaries in the US with a wound increased from 8.2 million to 10.5 million [5]. The annual increase from 2023 to 2030 is predicted to be 4.1% in the US and 5% globally [6]. The economic impact of managing chronic wounds is substantial, potentially accounting for up to 5% of all healthcare costs. For example, the US spends the most on chronic wounds globally with a total cost for 2022 estimated to be in the region of $145B [7].

The development of even one severe PU has a devastating impact on QoL, often necessitating prolonged hospitalisation or extended care, and may lead to additional complications, and even mortality. Clinical management typically requires frequent outpatient clinic visits and can often lead to hospital admission for inpatient care, adversely affecting QoL over extended periods. Chronic wounds pose a major clinical challenge in the long-term care of many persons with reduced mobility, who often have multiple comorbidities. The growing challenge impacts not only the affected individuals but also their caregivers, the medical team and the healthcare systems. Despite the advancements in medical care and treatments strategies, wound management is one of the few areas where little progress has been made in reducing incidence rates, underscoring the critical need for innovative approaches to prevention.

When prevention fails, PU treatment pathways are dependent on the PU stage. The most widely used PU staging metric in the US has been developed by the NPIAP and is outlined in the combine international clinical practice guidelines (CPG) [7]. The metric comprises four Stages with varying extents of tissue breakdown together with unstageable or suspected deep tissue injuries (DTI), which occur when tissue damage depth cannot be determined. For a person with a Stage 1 PU the skin is intact with sustained erythema and sustained off-loading can be sufficient to reverse the tissue damage. A Stage 2 PU occurs when there is partial-thickness skin loss with exposed dermis, while Stage 3 involve full thickness loss of the dermal layers. Stage 4 PU extend deeper into the soft tissue, i.e. muscle, tendon and sinew, and can involve the underlying bone. These chronic wounds are nearly always colonised by bacteria [8] and require specialised treatment to both treat the infection and promote tissue regeneration. The high rate of bacterial colonisation in these wounds may be due to the prolonged inflammation they undergo, although a direct causative relationship has yet to be established [9,10]. Once a wound infection sets in, it often persists for months, and is responsible for the majority of deaths associated with chronic wounds [11]. The challenge is compounded by the fact that antibiotics are ineffective against biofilm formation [12], a common feature of chronic wound infections, and are showing reduced efficacy in the face of multidrug-resistant bacteria [13]. CPG include recommendations for offloading, patient-centred observation and monitoring, cleansing and dressings to promote hydration and epithelialisation. ES is recommended as a Level I adjunctive therapy, especially for wounds not responding to standard forms of treatment.

9.2 PU prevention

9.2.1 Overview of the available research in the field
9.2.1.1 Introduction
Research into PU prevention by using ES has mainly focused on risk factors such as disuse muscle atrophy, high sustained sitting interface pressure, poor circulation and reduced oxygenation. Various protocols and methods have been used, ranging from implanted to percutaneous ES systems. These have included carbon electrodes and specialised garments with built-in electrodes. Interventions to produce paralysed muscle activation include use of ES during sitting and daily activities to ES-induced exercise using bicycle and rowing ergometers. Preclinical studies have also been reported which seek to refine the stimulation patterns and efficacy of these interventions as described in the next sections.

9.2.1.2 Acute effects
9.2.1.2.1 Sitting pressure distribution
Sustained high sitting pressure under bony prominences such as the ITs is considered a primary risk factor for the development of PUs. ES methods have therefore been employed to reduce these peak pressures. A few studies have shown that using surface ES to activate the paralysed (or paretic) muscles while sitting in a person's own wheelchair can change sitting pressure distribution, relocate pressure more to the legs and subsequently reduce the pressure and shear forces under the ITs during the muscle contraction. Gluteal stimulation alone may lead to offloading of the sacrum allowing a sacral PU to heal once the gluteal muscles, which were almost completely atrophied and concave, bulked back up. A way of reducing the sitting pressure was studied by Ferguson *et al.* [14], who stimulated the quadriceps muscles of wheelchair users with SCI while sitting in their wheelchair with the lower legs restrained to the wheelchair. Activation of these muscles resulted in a knee extension moment and subsequently an improvement in sitting pressure distribution. Although effective to temporarily improve pressure distribution, a drawback of this method is that the legs are inconveniently restrained and, more importantly, the muscles most at risk for PUs, hence the gluteal muscles, are not activated and therefore missing the positive effects of conditioning through repeated use of activation.

Bogie *et al.* developed an implanted gluteal stimulation system that was used daily by recipients for up to 18 hours/day at home [15]. Pressure distributions at the seating support interface tended to change such that although the total pressure acting at the interface did not change, ischial region pressures showed a significant decrease. A case study of one long-term user found that regular daily use had a positive impact on multiple indirect indicators of tissue health, including increases in gluteal muscle bulk and blood flow together with reduced regional interface pressures [16]. Other studies have shown that non-invasive activation of the gluteal muscles can also result in similar sitting pressure distribution results [17]. While it is undesirable to sit on electrodes for a prolonged time, Smit *et al.* [18] placed one electrode at the top of the gluteal muscles and the other one halfway of the hamstring muscles, activating the gluteal and hamstring muscles simultaneously. Intermittently activating the gluteal and hamstring muscles for 30 minutes using ES acutely lowered interface pressures under the tuberosities. They showed that this placement resulted in an even larger pressure reducing effect, probably because of the additional hip extension moments delivered by the activated hamstring muscles, without an increased risk of a PU. In these latter studies, participants wore a Lycra garment equipped with built-in electrodes and utilised a portable stimulator. The gluteal motor point is generally not loaded during sitting, unless the person is a sacral sitter. Flat electrodes in a garment could be relatively safe; only the wire could cause a line of increased pressure.

9.2.1.2.2   Circulation and oxygenation

In inactive paralysed muscles, circulation to the muscles and tissue oxygenation is markedly lower than normal, leading to structural changes in the circulatory system below the lesion and a concomitant higher risk of PU development. However, when these paralysed muscles are activated by ES, circulation to these muscles and surrounding areas may be increased and oxygenation augmented. In a randomised controlled trial (RCT) comparing able-bodied (AB) individuals and individuals with an SCI, Olive *et al.* showed that regardless of muscle size, contraction rate or type of stimulation, circulation significantly increased in activated muscles with ES in both groups [19]. According to the RCT of Kim *et al.* which looked at six people with an SCI and performed four weeks of sub-motor-threshold stimulation for PU prevention superficial to the ITs, a short-term increase in regional blood flow was observed. However, no long-term changes were noted, prompting a recommendation for a contractile muscle response to induce changes [20]. Barton *et al.* illustrated in a study with nine SCI participants that 12 weeks of low-intensity neuromuscular electrical stimulation (NMES) with tetanic contractions of the gluteal and hamstring muscles progressively increased blood flow, as well as skin perfusion [21]. In a review by Smit *et al.* it was shown that even though the methods and outcome measures varied among studies and the findings were somewhat inconsistent, the ES-induced muscle activation induced positive changes in blood flow and/or oxygenation [22].

9.2.1.3   Long-term effects

9.2.1.3.1   Muscle size

Many studies have shown that the atrophy of paralysed muscles can be (partially) reversed by incorporating ES-induced muscle contractions, especially using resistance exercise techniques. Several studies have even shown that quadriceps muscle bulk can return to normal levels in individuals with an SCI after ES training consisting of quadriceps muscle (knee extension) exercise. Not only resistance training, but also ES leg cycling exercise (ES-LCE) training in individuals with an SCI has been shown to result in muscle hypertrophy as indicated by increased thigh circumference and quadriceps muscle area. Most studies have focused on the thigh area, while data on the probably more relevant gluteal muscles is limited. Using an implanted gluteal ES system, Bogie *et al.* showed in one individual with an SCI that over a period of 40 months, muscle hypertrophy occurred, also resulting in a 38% reduction in average interface pressure [16]. Baldi *et al.* showed that ES-LCE exercise training during the first six months after the SCI could prevent the occurrence of gluteal muscle atrophy seen in a non-exercising control group, with this control group losing an average of 27% of gluteal lean mass, while the ES-LCE group showed an increase of 5–10% [23].

In summary, ES-induced muscle activation, along with resistance and cycling exercises, appears to significantly counteract or slow the progression of disuse atrophy in paralysed muscles. Furthermore, initiating ES-LCE promptly after traumatic SCI may even aid in preventing muscle atrophy altogether.

9.2.1.3.2 Muscle fibre type and fatigue

Several studies have explored long-term effects of ES on muscle histology and fatigue [24–26]. It was generally seen that enhanced neuromuscular activity boosts the oxidative capacity of the muscles being exercised. Additionally, electrically stimulating paralysed muscles leads to an increased proportion of Type 1 fibres, an increased capillarity and an effectively reduced susceptibility to fatigue. Despite extended periods of inactivity, it is promising to observe that characteristics such as resistance to fatigue, contraction speed and oxidative capacity quickly return to normal levels when exposed to daily ES treatments. The time after SCI does not seem to markedly influence adversely the retraining of muscle properties.

9.2.1.4 Pressure ulcer incidence

Although studies have shown that ES can significantly enhance factors that reduce the risk of PUs by improving muscle mass and local circulation in skin and muscle of paralysed individuals, there is a notable absence of RCTs that confirm the impact of ES on reducing the incidence or recurrence of PUs. Additionally, the cost-effectiveness of ES in individuals with SCI remains unclear. Despite ES not being a novel intervention, its widespread implementation in clinical SCI rehabilitation practices is limited, primarily due to a lack of comprehensive, well-controlled studies. Currently, a relatively large RCT is being undertaken to evaluate if regular ES of the gluteal and hamstring muscles at home can reduce the incidence of PUs of individuals with an SCI [27].

It is important to consider what conclusions can be drawn for clinical practice. The question arises: given the limited scientific evidence, should ES be used in clinical practice, and if so, to what extent? Our perspective is that it should not be disregarded. When combining the findings of efficacy studies and intervention studies, there is a strong indication that ES has a direct positive impact on tissue health, making it likely to be effective in reducing the incidence of PUs.

It is important to note that ES should be viewed as a complementary measure, rather than as a substitute for standard preventive strategies. Key preventive methods including optimal positioning in a wheelchair, using pressure-reducing cushions, and regularly performing pressure-relief movements remain crucial in PU prevention. Skin assessments should also be an integral part of the individual's rehabilitation program and included in their daily comprehensive assessment.

9.2.2   Practical clinical considerations

For prevention and therapeutic intervention, practical recommendations will be provided concerning stimulators, stimulation parameters, muscle selection, electrodes, placement and therapy protocols. Special attention will be given to ES of paralysed, paretic and denervated muscles. As previously discussed in this chapter, the physiological changes resulting from disuse atrophy are reversible through activation of inactive muscles with ES. In patients with paralysed, paretic and denervated muscles, this intervention becomes lifelong, and the choice of intervention (e.g. FES or NMES) and the selected parameters can significantly influence morphological and physiological changes. The wash-out period for the intervention effects of ES-induced activation is estimated to be around six months (not proven), but it also depends on the therapy received, the duration of treatment and the timing of intervention initiation.

9.2.2.1   Electrical stimulators

There are various stimulators available, primarily differing in the type of stimulation, the number of channels, whether they are pre-programmed or programmable, and the magnitude of the stimulation or current. Most commercially available stimulators typically come with either two or four channels, allowing for the stimulation of two or more muscle groups or the placement of multiple electrodes on the same muscles, depending on the electrode size. High-end stimulators often come with more preprogrammed programs, and the differences are typically found in battery quality, stimulation time and/or other changeable parameters.

As mentioned above, most stimulators are pre-programmed. Unfortunately, not all companies are transparent about the parameters used with each program, making it difficult to discern exactly what the output is. For rehabilitation and research purposes, it is recommended to use a stimulator where the parameters are clearly defined or one that allows you to create your own stimulation program.

The intensity of stimulation and current are also crucial factors to consider when purchasing a stimulator. Most commercially available devices may suffice for an individual with innervated muscles; a higher dosage is generally unnecessary to trigger twitches, tetanic contractions or movement. For individuals with denervated muscles, such as those with lower motor neuron disease or severe nerve injuries, currently the only option is the Euro-RISE [28], albeit at a significantly higher cost of a few thousand dollars.

9.2.2.2 Different types of electrodes or garments

Self-adhesive electrodes, measuring 5 × 9 cm, are the easiest for patients to apply independently. They can be effortlessly attached and removed, providing convenience for users. The adhesive is typically durable for about 40 uses, though this may vary depending on the brand and person. To extend the shelf life, rinse off any remaining skin and hair residue with lukewarm water. Additionally, it is recommended to store the electrodes in a sealed bag in the refrigerator to prolong the lifespan of the water-based gel. Caution should be taken to avoid allergic reactions to silicone, a common material used in self-adhesive electrodes. It is important to note that the sticky part can degrade quickly and may not adhere firmly enough, leading to curling up or movement during transfers or other activities. Additionally, there is a risk of pulling out hairs or leaving small clumps of sticky residue when removing the electrodes, particularly if the patient has a significant amount of hair. In such cases, trimming the hairs may be necessary to ensure proper adhesion and a comfortable experience.

Carbon/rubber electrodes require somewhat more effort for self-application by patients, but they offer reusability, and a longer lifespan compared to self-adhesive electrodes. These electrodes utilise gel for conductivity and may need additional measures to secure them, such as using tape or bandages. However, it is important to note that tape removal can potentially cause skin problems, particularly due to hair removal or sensitivity issues. A cautionary point arises when using these electrodes with devices like RISE, as there is a risk of burns with incorrect use. Carbon electrodes may be used in conjunction with a moisture sponge for enhanced effectiveness. However, it is important to mention that these electrodes are not suitable for lying on, as their firm composition is less comfortable in such positions, especially when lying on the belly, presenting a notable disadvantage.

Garments should only be put on with assistance or by individuals with sufficient trunk stability and arm-hand function. There are disposable or washable versions available, each with its own features. These include garments with a self-adhesive electrode that needs to be placed inside, those with built-in electrodes requiring wetting before usage, or custom electrodes with a polymer pouch placed on top. Garments are versatile and can be worn during various activities such as sleeping, sitting and engaging in daily tasks. However, it is crucial to wear them only when the stimulation is active; otherwise, they become an additional (potentially wet) layer over the skin, increasing the risk of PUs. For garments designed to prevent PUs, special attention should be given to their construction. They should be free of cables, connection pieces, seams and stitching in bony areas to minimise extra pressure. Discomfort may occur if the electrode does not make proper contact with the skin when sitting or lying down. In summary, while these garments offer valuable functionalities, proper usage is essential to mitigate the risk of PUs and ensure user comfort.

Implanted intramuscular electrodes offer the advantage of eliminating the need for external wiring or an external stimulator. Every component, including the batteries, is implanted, requiring replacement every few years [29].

Sacral anterior root stimulation (SARS) with the Brindley stimulator can induce tetanic contractions in the gluteal muscles, potentially enhancing blood flow and skin perfusion more effectively than traditional FES. It is noteworthy that the device's settings must be manually configured to prevent inadvertent bowel or bladder activation [30]. There is anecdotal evidence suggesting an increase in local circulation when the Brindley stimulator is exclusively used for bladder and bowel emptying. However, this observation is currently based on expert opinions, and further studies are warranted to validate these findings.

9.2.2.3   Stimulation parameters

There are currently no known optimal parameters for ES to prevent PUs for both surface and implanted ES. Nevertheless, extensive research has been conducted over the years to mitigate potential risk factors associated with PU development. Investigations have focused on reversing muscle atrophy (due to disuse and/or denervation), enhancing local and microcirculation and reducing pressure distribution. Crucially, when employing ES, achieving a tetanic contraction is essential to activate the muscle pump effectively and increase blood flow. Responses to ES can vary among individuals and may also change over time. It is advisable to consider basic training principles when incorporating daily ES stimulation, gradually increasing the dosage until reaching a plateau, at which point maintenance is the primary goal.

Before delving into stimulation parameters, it is crucial to distinguish between innervated and denervated muscles. In cases of innervated muscles, or an upper motor neuron lesion, a shorter pulse duration is applicable as the nerve fibres remain responsive to the impulses. In contrast, with denervated muscles or individuals affected by a lower-motor neuron condition, the peripheral nerves are no longer functional. Therefore, the only viable option is to stimulate the muscles directly through muscle stimulation which requires much longer stimulation pulse duration.

Table 9.1 provides an overview of the parameters utilised, including frequency, pulse duration, amplitude, duty cycle, schedule, position and the type of electrodes. This compilation encapsulates the current literature on the application of ES for PU prevention or addressing potential risk factors on developing PUs. For innervated nerves, it is recommended to use bi-phasic waveforms with symmetrical pulses. This choice is favoured as it tends to be more comfortable and causes fewer adverse reactions to the skin and possible pain responses.

In the case of denervated muscles, biphasic rectangular, triangular or trapezoid waveforms are suitable. A warming-up phase is advisable, and in the European RISE project, various phases were employed to reactivate denervated muscles. Due to the significantly decreased muscle morphological changes associated with conditions such as cauda equina lesions, lower-motor neuron issues or completely denervated muscles, it is not typical to observe twitches or muscle contractions when using ES for the first time when the condition is chronic. The initial changes occur on a cellular level, and it may take between three and six months before noticeable twitches or contractions can be observed. When working with

Table 9.1 Stimulation parameters for innervated and denervated muscles

	Innervated nerves/ upper motor neuron [28,31–33]	Denervated nerves/lower motor neuron [28,30,34]	Intensive care unit (ICU) [35]	Prolonged daily stimulation [36]	Overnight stimulation [36]
Frequency	20–60 Hz	0.25–60 Hz	100 Hz	17.5 Hz	35 Hz
Pulse duration	64–600 µs	20–200 ms	500 µs	300 µs	
Current amplitude	10–200 mA	10–250 mA	Greatest possible to evoke maximum contractions (sedated people)	Visible muscle contraction	Best muscle activation with steps of 10 mA till tetanic contraction visible without creating discomfort or excessive or painful muscle contractions
Duty cycle	1:6 or 1:5 ramp time 1 to 5 s	4s:2 s, 5:2 s, 3:3 s	5:25 s 1 s ramp up 1s ramp down	10 s contraction every 10 minutes for 12 hours a day	8 hours with 1 hour stimulation cycles each cycle 30 minutes with 36 contractions 10:40 followed by 30 minutes rest
Schedule	2–5 times a week 30–60 min per session	5 times a week 30–45 minutes	25 min once a day 6 times a week	4 days per week	All night stimulation for 2 weeks
Position	4 on the gluteus 8; 4 on the gluteus and 4 on the hamstring. 4: 2 on the gluteus and 2 on the hamstring. Laying in prone, side, back, sitting, standing in form or on a tilt table.	2 on the gluteus, prone or side position standing in a frame or tilt table	laying on side, placement, 2 on each gluteal.		laying
Type of electrode	self-adhesive electrodes	special, large rubber electrodes salt free gel or wet sponges	self-adhesive electrodes 9 × 5 cm	surface electrode inside of a garment	garment wilt built in electrode and hydrophilic granules pockets.

denervated muscles, referring to the European RISE project can serve as a valuable example [28]. More details on the stimulation on denervated muscles can be found in Chapter 8.

9.2.2.4   Muscle selection

For optimal activation of the high-risk area, it is recommended to target the gluteal and hamstring muscles. This choice stems from the fact that the anatomical origin of the hamstrings is the IT, and the gluteus maximus only partially overlaps with this high-risk zone. The goal is to enhance muscle quality, muscle thickness and (micro)circulation in this region to better withstand and/or avoid prolonged pressure peaks around the IT.

Electrode placement can be achieved with either eight (4 left, 4 right side) electrodes—placing one on top of the gluteus maximus and one at the bottom for the gluteal muscles, and one just below the buttocks and one about 5–10 cm above the knee joint for the hamstrings (Figure 9.1). Another option is a more proximal/ distal approach using four electrodes, with one electrode on top of the gluteus and one on the distal part of the hamstrings, resulting in a co-contraction of the gluteal and hamstring muscles. The advantage of this approach is that individuals do not sit on the electrodes and with proper hand function can often self-place the electrodes in their wheelchair, while lying down, or even in a long-sit position. Alternatively, a single muscle group at a time, either the gluteus or the hamstring, can be targeted. For stimulating denervated muscles using the RISE stimulator, larger electrodes can be placed on the gluteal and hamstring muscles (Figure 9.2).

9.2.2.5   Activation cycles

The effectiveness of a stimulation program depends significantly on the intensity and duration of the stimulation. It is crucial to design a protocol that optimally activates muscles and surrounding structures without inducing muscle fatigue. According to Olive *et al.* [37], to prevent fatigue, a recovery time at least three times the activation time is recommended for low-intensity stimulation, resulting in a 1:6 (activation:rest) protocol [14,38]. In continuous protocols, fatigue may set in within 30–45 minutes. To address this, the duty cycle multiplied for example to a 6:18 protocol could be used [27]. For an extended approach, overnight stimulation has been explored, lasting eight hours a night with cycles consisting of 30 minutes and 10 seconds on, followed by 40 seconds off, with a 30 minute rest. This regimen was successfully applied seven days a week for two weeks [36]. Hence, the duty cycle could vary depending on the total duration of the stimulation and intensity.

Studies conducted in the intensive care unit (ICU) on individuals without an SCI have demonstrated prolonged stimulation durations, with some lasting up to 12 hours, delivering a 10 s contraction every 10 minutes for the prevention of DTI, even up to four days a week [39]. Another study in the ICU utilised 25-minute stimulation periods with a 5:25 duty cycle, including a 1 s ramp-up and a 1 s ramp-down, and found it to be effective in preventing PUs. However, there is a need for well-designed clinical studies to investigate the optimal stimulation location and parameters, specifically in individuals with SCI [29].

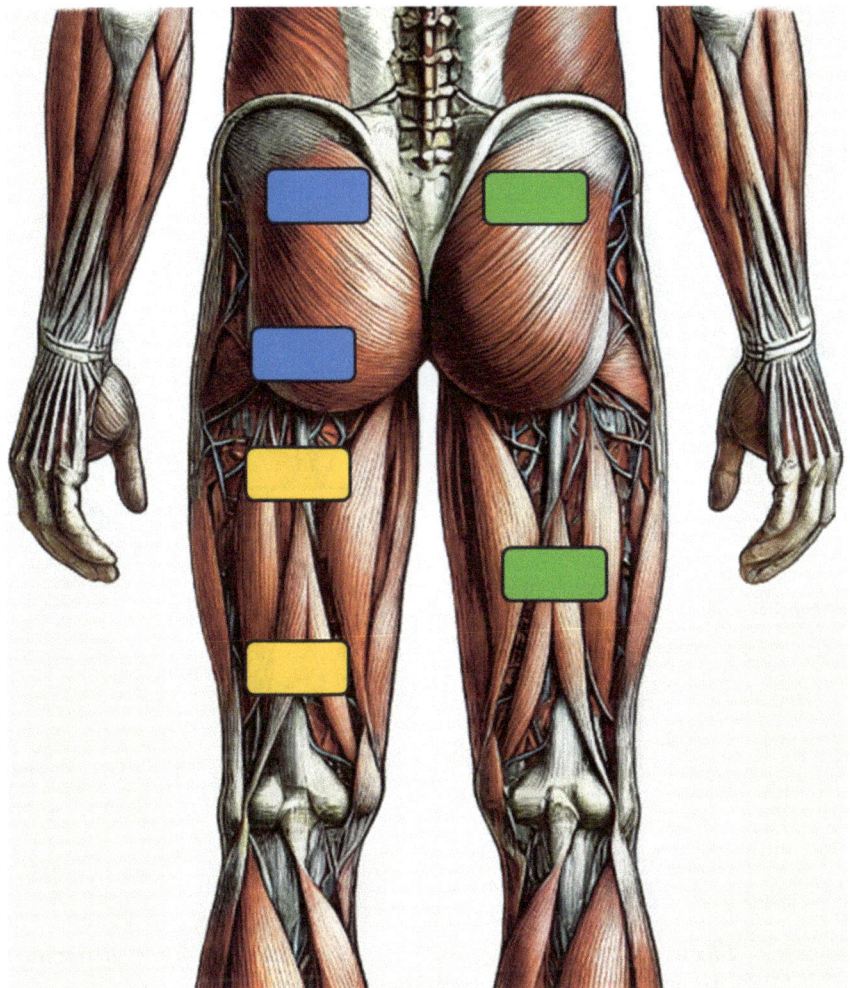

Figure 9.1   Electrode placement for activating gluteal and hamstring muscles. On the left leg, gluteal and hamstring muscles are activated separately by four electrodes, while on the right leg, these muscles are activated simultaneously using two electrodes.

For denervated muscles, the activation-rest cycles differ significantly in parameters, as there may be no occurrence of twitches or muscle contractions at all for persons with chronic SCI, i.e. more than two years post-injury. Due to the infrequency of this scenario, especially for prolonged periods, these parameters are primarily based on the study of Kern *et al.* from the Euro-RISE project and on expert opinion [20]. Bersch *et al.* (2022) suggested a warmup and treatment settings each with their own parameters, with the warm-up mode having a pulse duration of

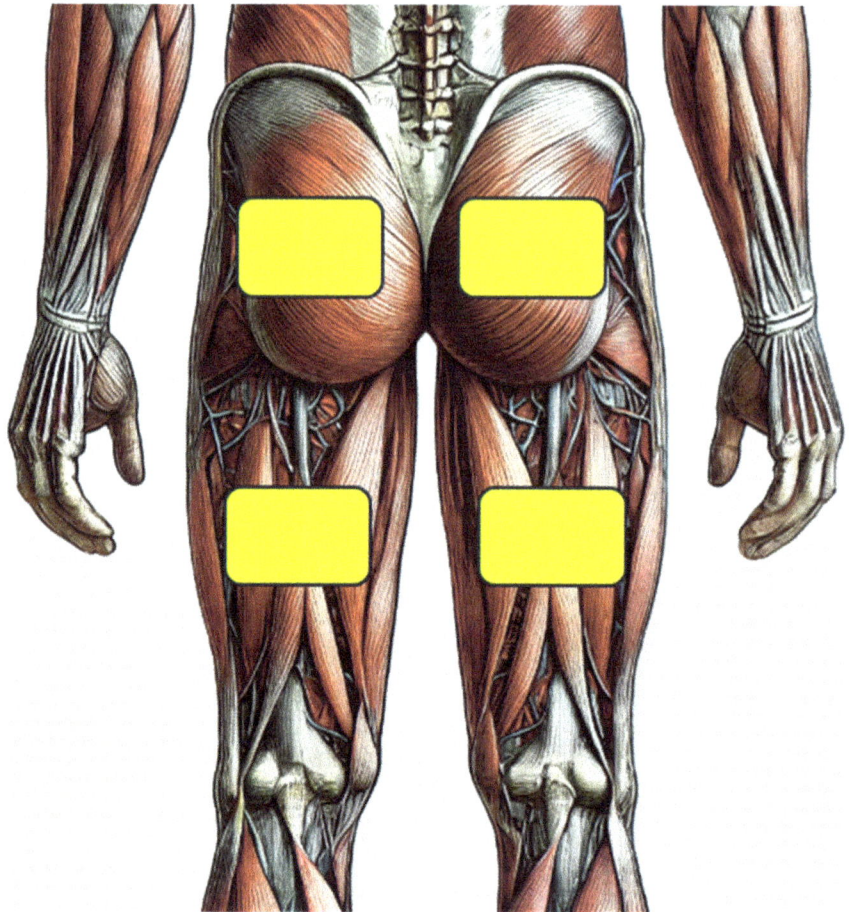

Figure 9.2   Electrode placement for simultaneous activation of denervated gluteal and hamstring muscles using the RISE stimulator.

150/200 ms, pause 1000/500 ms, with a 11 s burst and the treatment mode having the pulse duration adjusted to 100/35 ms, with a pause of 400/10 ms with a burst of 2/2 s [40].

9.2.2.6   Exercise modes

The two main ES-induced exercise modes for individuals with SCI are resistance and endurance exercise. Studies show that resistance training principles, effective for muscle strengthening in able-bodied individuals, can be applied to ES-induced exercise for paralysed muscles [41–43]. These principles include isometric and dynamic contractions, progressive overload, multiple sets with low repetitions at high resistance and a frequency of 2–5 times per week. Most research has focused on the paralysed quadriceps due to their responsiveness to ES and ease of

Figure 9.3   Electrical stimulation-induced hybrid leg cycling exercise by an individual with an SCI

implementation. However, ES resistance exercise can likely be adapted for other paralysed or weakened muscles.

For endurance exercise, a leg cycle ergometer (LCE) was developed in 1982 [44] pedalled through ES-induced contractions of the paralysed lower-limb muscles. The operation of a commercially available ES-LCE by an individual with SCI involves computer-controlled ES to induce contractions in the quadriceps, hamstring and gluteal muscles. A more recent exercise mode is hybrid cycling, using a tricycle with the propulsion coming from both arm and leg power (Figure 9.3). Arm power is voluntary while leg muscles are activated with ES. A similar form of ES-induced exercise is hybrid rowing, with voluntary arm exercise combined with ES-induced leg exercise.

9.2.3   Assessing efficacy

Assessing efficacy of any preventative ES therapy is preferred. Measuring techniques to evaluate changes in muscle size, circulation, sitting pressure distribution, skin condition and wound status will be discussed.

9.2.3.1   Muscle size assessment

The gold standard for assessing skeletal muscle is magnetic resonance imaging (MRI), but alternatives like dual-energy X-ray absorptiometry (DXA) or computed tomography (CT) are used due to cost. These methods provide regional and lean mass estimates with lower radiation. Ultrasound, a more affordable option,

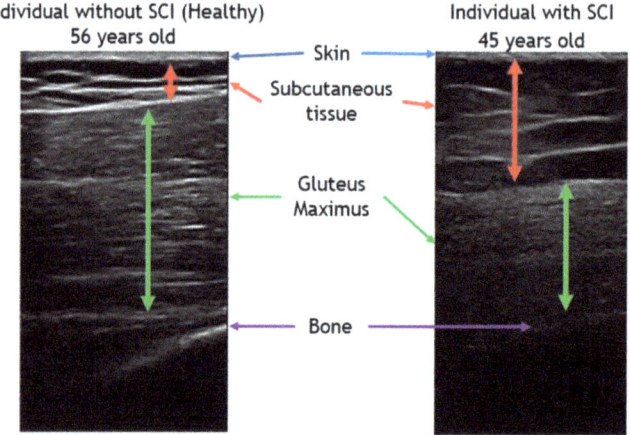

Figure 9.4   Ultrasound measurements in B-mode of the gluteus maximus

measures muscle thickness, cross-sectional area, or performs 3D ultrasound imaging (3DUS). However, its use has challenges in individuals with neuromuscular diseases, including disuse atrophy and image quality issues. Developing adjusted and carefully interpreted scanning protocols, using standardised methods and considering visible landmarks are crucial for consistent measurements in people with SCI. Employing a transducer with at least a frequency of $\geq$ 4–12 MHz, horizontal scanning in B-mode for muscle quality, and longitudinal scanning for cross-sectional area is recommended. Maintaining scanner consistency, creating short video clips and collecting three images at the thickest part enhance reliability. Offline analysis with tools like Matlab, ImageJ or microdicom viewer and interpretation by independent observers who have proper anatomical knowledge are important for accurate assessment. Algorithms to automate these analyses are needed to make this type of analysis feasible in clinical practice. In Figure 9.4, the effect of disuse atrophy is visible by comparing an individual with a SCI and an able-bodied/healthy individual.

9.2.3.2   Circulation

Blood flow calculation involves essential parameters such as artery/vessel diameter and velocity. Duplex Doppler ultrasound, a widely used method, requires specific skills distinct from muscle thickness measurements. Optimisation recommendations include a 10–15-minute pre-assessment period, a supine posture and avoidance of caffeine, exercise (including ES) and tobacco smoking for specified durations [45].

For individuals with SCI, a standardised protocol is crucial due to reduced vessel diameter and velocity compared to individuals without SCI. During scanning, focus on arteries no deeper than 4.5 cm, using an insonation angle of less than 60 degrees. Collect at least a one-minute clip for offline analysis, employing continuous edge detection, wall tracking software and automated mathematical algorithms for diameter and velocity calculations. To assess changes in local circulation in hamstring and gluteal areas, concentrate on the deep femoral artery just after the femoral bifurcation, approximately 1.5–2 cm [21,27]. Standardised placement is vital for consistent follow-up assessments. Note: Individuals with spasms may experience spastic reactions during measurement due to the quadriceps' muscle-tendon junction.

Skin blood flow can be measured using various techniques, each with unique principles and applications [46]. Laser Doppler flowmetry (LDF) and Laser Speckle Contrast Imaging (LSCI) utilise laser light to detect red blood cell movement, while photoplethysmography (PPG) measures blood volume changes using light absorption. Thermal Imaging captures infrared radiation to infer blood flow based on temperature, and transcutaneous oxygen tension measurement (TcPO2) assesses oxygen levels linked to blood flow. Capillary microscopy provides direct visualisation of capillaries, Doppler Ultrasound uses sound waves to measure blood velocity, and near-infrared spectroscopy (NIRS) evaluates tissue oxygenation through light absorption. These methods vary in resolution, depth of penetration and invasiveness, catering to different clinical and research needs. For measuring skin blood flow around the area of interest, i.e. the gluteal area, no standard method has been developed.

9.2.3.3 Sitting pressure distribution

The measurement of sitting pressure distribution utilises a calibrated sitting distribution mat, ideally calibrated at ≥ 250 mmHg. Individuals with SCI often experience high local pressure around the ITs, exceeding 200 mmHg, even with adapted wheelchair cushions. The choice of measurement focus – changes over time or in real-life scenarios – dictates the approach. For tracking changes over time, it is recommended to use separately standardised chairs with specialised anti-PU cushions, adjustable components and a chair per individual. If measuring changes in daily life, individuals should adopt their comfortable sitting position for at least five minutes, considering dynamic factors like posture, weight changes, cushion adjustments and correct footplate height.

Before measurements, allow an adaptation time, particularly with air cushions taking 5–10 minutes to conform fully. Regular calibration of the pressure distribution mat, preferably before each measurement or at least every 1.5 months, minimises errors. A five-minute data collection duration is recommended to neutralise movement during rest. Data analysis involves using a custom-made MATLAB/ Python script to examine key parameters like overall mean interface pressure, maximum pressure, peak pressure, pressure spread and pressure gradient. (IT pressure and pressure gradient are crucial indicators of shear force. Calculating IT interface pressure involves a 3×3 cell selection for both left and right IT. The pressure gradient is determined by subtracting the average of the 16 surrounding cells from the IT interface pressure [17,18].

9.2.4 Precautions and contraindications

Before embarking on ES therapy for PU prevention, understanding both precautions and contraindications is essential to ensure safe and effective treatment. Precautions outline potential risks and discomforts that may arise during the intervention, ranging from mild skin reactions to orthostatic hypotension and autonomic dysreflexia. While these risks are typically small and temporary, they necessitate attention and proactive measures for patient safety. Conversely, contraindications delineate conditions or scenarios where ES should be avoided or approached with caution. Recent research challenges longstanding beliefs regarding certain contraindications, such as pregnancy and cancer, shedding light on potential benefits and safety considerations.

9.2.4.1　Precautions

The known risks/adverse effects of discomfort during the intervention are:

- Red, raised or itchy skin [22,47,48].
- Muscle pain [47,49].
- Feeling of discomfort [47,49].
- Orthostatic hypotension may lead to fatigue, altered mentation, dizziness, light-headedness, blurred vision or syncope [50].
- Autonomic dysreflexia, characterised by slow heart rate, high blood pressure, headache, flushing and sweating, may occur [47].
- An increase in nerve pain is a rare occurrence.
- Pain induced by spasms may occur.

Additionally, in some cases, there may be pain or spasms, minimally dependent on the amount of current. These risks are small and temporary, subsiding once the stimulation has stopped. It is also recommended to empty the bladder before the first use of ES to reduce the risk of autonomic dysreflexia occurring.

9.2.4.2　Contraindications

Below the most contemporary contraindications are summarised and discussed, based on literature and experts. For pregnant women, it is advised not to apply ES on the abdomen or lumbar spine [51]. However, studies have explored the lumbar and sacral spine for managing back pain during pregnancy and labour, yielding positive results [52–54].

Regarding cancer, extensive research over the last two decades has investigated the use of ES for various types of cancer, cancer symptoms and treatment-related issues. It seems beneficial in pain reduction, nausea and vomiting [55]. There are even studies suggesting it can result in an enhanced chemotherapy [56,57]. A review from 2018 suggests that the use of ES may enhance QoL during cancer rehabilitation [58]. They even made a narrative review about the most effective ES parameters [59]. Due to the relatively low grade of evidence and the absence of large RCT's confirming the effectiveness and safety of ES on cancer patients, it is recommended to consult a specialist before initiating ES, especially if cancer is active in the treatment area and/or a patient is still receiving chemotherapy.

When an intact joint is involved, the use of ES to shorten muscles may result in leg shortening, especially in the absence of a functional hip joint. This change in leg length can pose challenges in seating, particularly for individuals reliant on adapted wheelchairs or cushions.

Regarding metal implants, in the past, it was considered a contraindication due to the use of monophasic stimulation, which could lead to a negative charge buildup in the metal or body foreign material. However, with the more commonly used alternating biphasic current nowadays, it is not possible to leave a negative charge, as it utilises both positive and negative charges [60].

When encountering new implanted devices, extensive consultation with the company device specialist and the involved surgical specialist is recommended to rule out any interference between the two devices. A good example of implanted devices are the intrathecal baclofen pump and the Brindley bladder stimulator. In both cases there is no known interference and ES can be used if the field of current does not directly involve the implanted device.

Regarding implantable cardioverter defibrillators (ICDs) or pacemakers, Badger *et al.* conducted a systematic review on the safety of using ES in 2016. They suggested that it is safe to use ES for treating drop foot; however, larger sample sizes are necessary to definitively exclude any risks when ES is employed [61]. It is important to note that this concern is primarily relevant when stimulating around the area of the ICD or pacemaker. A study by Egger *et al.* demonstrated a potential interference between cardiac device therapy and bilateral stimulation (both legs) but not with unilateral stimulation. Therefore, it is recommended to use unilateral stimulation only to avoid potential complications [62].

9.3   Pressure ulcer treatment

9.3.1   Overview of the available research in the field
9.3.1.1   Overview

ES was first proposed to have an impact on wound healing as early as 1843 [63]. The presence of a wound disrupts the body's bioelectric field locally and creates a 'current of injury' [64]. It has long been proposed from in vitro studies that exogenous application of ES can provide a field that guides cell migration and thus promotes healing [65,66]. Specifically, electroceutical therapy, i.e. therapeutic use of ES, for wounds activates the natural bio electrical response to injury [67] by promoting angiogenesis and inducing the migration of endothelial cells to the wound site [68]. Additionally, bacterial growth is inhibited [69], together with disruption of wound biofilms [70]. These well-known effects can synergise to promote healing of chronic wounds using appropriate electroceutical therapy.

It is important to note that in the treatment of chronic wounds, ES promotes activity of the endogenous electrical system essential for tissue development and normal wound repair and, unlike ES for PU prevention, should not produce a muscle contraction.

The 2014 CPG jointly issued by multiple international PU advisory panels, strongly recommended ES for treatment of severe Stage III/IV PUs [71]. In their 2019 guidelines, electroceutical therapy remains the only adjunctive treatment recommended for chronic wound therapy, especially for wounds not responding to standard forms of treatment [72]. Most recently, the Wound Healing Society 2023 updated guidelines for the treatment of PUs also advised that ES used in conjunction with conventional therapy may be useful in the treatment of chronic Stage II–Stage IV PUs [73]. The highest-level recommendation (Level 1) was maintained, with the stated principle being that high voltage monophasic pulsed current (HVMPC) appears to be effective in the treatment of chronic PUs by reducing the wound area, enhancing the likelihood of achieving complete wound healing and decreasing the probability of worsening healing. This recommendation was based on the systematic review and meta-analysis by Girgis and Duarte which examined eleven studies comprising nine RCTs and two case series studies [74]. It was concluded that HVMPC intervention improved the healing of PUs, i.e. reduced wound surface area, with a net effect of 5.4% increased healing per week, representing an increase of 78% over standard wound care alone. HVMPC was considered to increase the probability of complete healing and almost eliminate the probability of worsening of healing.

Despite the wealth of historical knowledge and the acknowledgement of ES as a Level I adjunctive therapy, the utilisation of various stimulation variables has resulted in variable outcomes. Over a decade ago it was considered that ES for ischemic wound therapy is worthy of careful consideration but still not fully understood [75]. While there is a general consensus that repeated and regular delivery of ES to the wound bed is required for effective treatment of PUs, there is limited agreement on which ES protocols reliably enhance wound healing efficiency [76]. A 2020 Cochrane Review revealed moderate certainty evidence suggesting that ES probably increases the proportion of PUs healed and the rate of healing. However, there remains very low certainty evidence regarding the effect on time to complete healing compared with no ES [77]. Furthermore, uncertainty persists regarding whether ES decreases the surface area of PUs. The review recommended further research, particularly large-scale trials, to ascertain the effect of ES on all key outcomes.

A wide variety of ES types, most typically direct current (DC) or pulsed DC have been used. The selection of waveforms and other stimulation parameters has varied both between and within studies [48,78]. A 2015 meta-analysis by Lala *et al.* [79] found nearly 600 papers published on ES wound therapy for persons with SCI alone. However, less than 20 met the criteria for sufficient rigour to be considered in the meta-analysis. It was noted that these studies all used different stimulation variables and different treatment paradigms, highlighting the lack of standardisation in ES wound therapy research.

Khouri *et al.* conducted an effect size meta-analysis to assess the overall efficacy of ES treatment of varying wound types to determine whether efficacy differs depending on the wound characteristics [80]. Their analysis encompassed 29 RCTs. They concluded that ES was more effective in treating PUs compared to venous and diabetic ulcers. Furthermore, efficacy appeared to decrease with larger wound size and increased chronicity. The overall efficacy of ES on wound healing was considered to provide a low quality of evidence. However, a moderate to large effect size was also found. Additionally, the review noted that reversing the polarity of the active electrode did not significantly alter healing outcomes or demonstrate improved healing outcomes. The authors emphasised the necessity for further studies to determine the optimal stimulus parameters of HVPC. They underscored the presence of a non-negligible placebo component in ES and advocated the adoption of a common primary endpoint, sham stimulation and high-quality designed trials in future research endeavours. There is ongoing discussion on whether it would be more clinically meaningful to use healing measurements such as rate of healing or percentage wound closure rather than a single endpoint such as complete healing [81]. The FDA in collaboration with the Association for the Advancement of Wound Care and the Wound Healing Society carried out a review of patients and caregivers. It was found that patients consider the five most useful endpoints for measuring clinical trial success are time to heal, wound size, infection, recurrence and pain. The inability to perform activities of daily living and pain were reported to be major factors that impacted their daily lives.

Chen *et al.* conducted a systematic review and meta-analysis which identified 17 RCT studies. This review indicated that ES significantly reduced PU surface area compared to standard wound care alone or pulsed sham ES [82]. Moreover, nine studies showed that PUs treated with ES were more likely to be completely healed compared to the control group. Adverse effects were rare. However, the authors concluded that while ES could help in PU healing and increase the healing rate, the certainty of evidence for all outcomes was moderate at best due to variations in the stimulation characteristics of ES used, and wide variation in patient demographics.

These meta-analyses collectively concluded that optimal clinical treatment protocols for ES therapy have yet to be elucidated. Optimal stimulation patterns and treatment protocols, including duration and frequency of ES delivery have yet to be determined, highlighting knowledge gaps in evidence-based treatment guidelines for PU electroceutical therapy. Further research is needed, employing standardised and clinically meaningful outcome measures. While benefits have been observed for a wide range of parameters, refinement is necessary to guide the selection of optimal ES paradigms based on underlying physiological effects.

9.3.1.2   Wound bioburden and electroceutical therapy for PUs

There are indications that ES for PUs may have a dual effect by not only promoting healing but also reducing both incident wound colonisation and infection [83], as well as disrupting wound biofilm. Chronic wounds are often considered to be 'stuck' in the inflammatory stage of wound healing [84,85]. Prolonged inflammation has been implicated in the high rate of bacterial colonisation for these wounds, although a causative relationship has not yet been fully defined [86]. Moreover, antibiotics are demonstrating reduced efficacy in the face of multi-resistant bacteria [87] and do not effectively address biofilm formation or control [88]. Therefore, there is a pressing need for alternative and adjunctive approaches to infection control in the management of chronic wounds like PUs.

Standard clinical microbiology analyses typically focus on assessing monospecies planktonic bacterial load to determine infection status. Traditionally, it has been considered unnecessary to treat wounds in the absence of visible infection. However, there is an increasing awareness that bacteria in the wound bed form polymicrobial colonies and biofilms, particularly in chronic wounds [89–91]. Costerton *et al.* originally proposed that 65% of chronic infections are due to biofilms, which may severely inhibit healing [92]. Similarly, Wolcott *et al.* suggested that these biofilms contribute to chronicity, perpetual inflammation and increased antimicrobial tolerance associated with non-healing wounds [93,94]. Furthermore, the presence of biofilm has been shown to delay re-epithelialisation in mouse wounds [94]. Despite these insights, the complexity of the chronic wound microbiome often goes undetected by standard testing methods [95,96]. A meta-analysis revealed that, in general, the chronic PU microbiome is poorly characterised [97]. Little is currently known about the impact of altering the microbiome on the wound healing process [98,99]. However, ongoing research efforts are aimed at addressing this deficit in understanding [100–103].

It has long been known that ES can have *in vitro* bactericidal effects on many strains relevant for wound infection [104,105]. ES increases local metabolic activity and tissue oxygenation, thereby inhibiting initial biofilm formation as well [106,107]. These effects may be due to electrolysis products [108] or to increases in bacterial membrane permeability [109]. Furthermore, sustained ES application is bactericidal when applied to infected but unwounded skin [110], while also enhancing blood flow and capillary density in compromised wounds [111]. The efficacy of ES appears to vary with stimulation profile, with the primary ES factor being current density, implying that the bactericidal effect is electrochemically mediated.

Thus, in addition to promoting wound healing, ES may also address the challenges of increasing antibiotic resistance and managing the chronic wound bioburden. Appropriate and repeated delivery of ES to the wound bed may both *limit incident* planktonic and polymicrobial colonisation and *inhibit* biofilm formation. As noted by Korzendorfer and Hettrick, there is a need for clinically relevant studies to examine wound bioburden in addition to healing progression when developing adjunctive therapeutic technologies [112].

9.3.1.3   Preclinical studies

Appropriate animal models are essential for preclinical studies of novel wound therapies for human health. The numerous origins and mechanisms of chronic wounds, means that having one *in vivo* model will not fit all studies of therapeutic interventions for human health. Rodents, rabbits and pigs are most commonly used, with porcine models being optimal because pig wounds have greater similarity to human wounds in responses to a variety of treatments, including wound dressings and antibiotics.

9.3.1.4   Clinical studies

Current technology for treating chronic wounds requires patients to visit the clinic daily for brief therapy sessions, during which the wound dressing is changed each time [113]. CPGs recommend maintaining a physiologically appropriate moist wound environment [71,72] to minimise pain and enhance healing. In vitro and clinical studies concur in the finding that maintaining the wound moist is crucial for optimal wound healing [114–116]. However, exposing the wound bed can disrupt this moist environment, increasing the risk of drying out. Frequent dressing changes can also cool the wound bed, which may slow or even inhibit the healing process [117]. Maintaining a moist wound environment is vital for the effective electrotherapeutic current delivery to the wound bed. However, covering the wound makes it difficult to monitor its healing status.

An examination of treatment patterns summarised in recent meta-analyses showed that ES parameters are often incompletely reported in clinical studies. Stimulation parameters and treatment paradigms reported for pulsed DC stimulation therapy are highly variable [118]. Optimal treatment parameters for effective ES for PU treatment must be guided by consideration of the underlying physiological effects.

9.3.2   How to apply stimulation for wound healing

Systems using intermittent controlled electroceutical therapy for PU treatment often use disposable surface electrodes placed at the start of each treatment session and removed post-treatment. This introduces the potential for non-uniform treatment delivery due to errors in repeatability of electrode placement and increases infection risk due to repeated wound exposure.

The Exciflex system combines a low-cost disposable bandage substrate with a reusable programmable stimulation and communication module [119–121]. The dry electrodes are supported on an integrated substrate that is placed over the wound so that the electrodes are located over the intact skin around the wound (Figure 9.5) [122]. A 90-V compliance voltage is used because the typical skin preparation for electrode application cannot be performed on the compromised

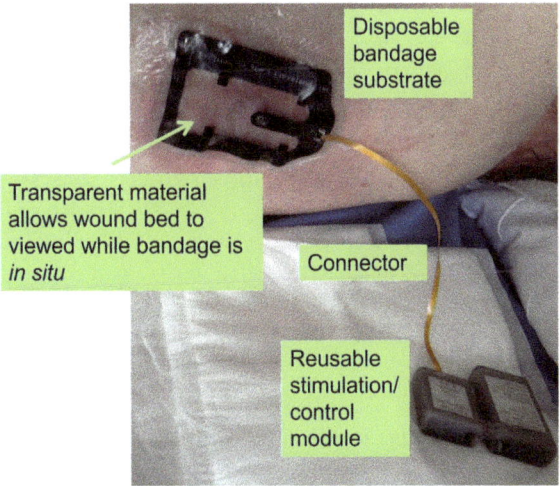

Figure 9.5   Wearable electroceutical device in use

tissues surrounding a wound. Each substrate can deliver reliable electrotherapy with concurrent monitoring of wound status for up to seven days [123]. The substrate is changed when necessary. The power/control module is reusable for the same wound. The electrode layout can be adjusted to fit the wound geometry as the wound heals to maintain optimum ES delivery (Figure 9.6). The wound shown in

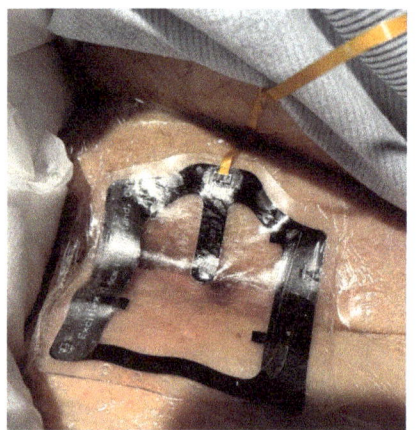
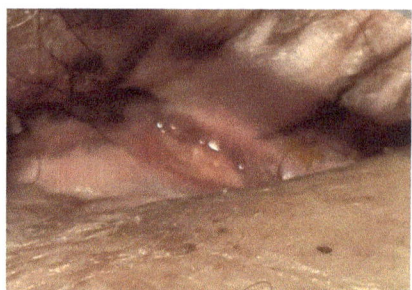

Week 1: 6 cm bandage

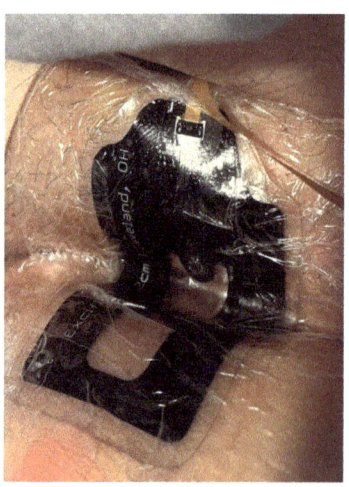
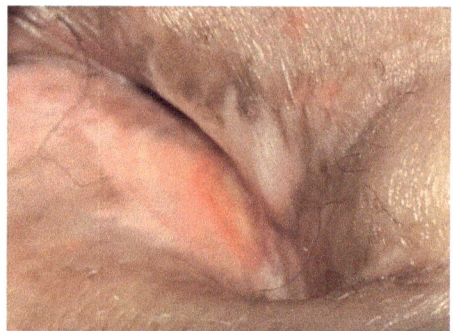

Week 4: 2 cm bandage

Figure 9.6   Daily wound status monitoring showed progressive healing (up to 97%) with no infection from week 1 (a) to week 4 (b). Week 1: 6 cm bandage; wound size at start of treatment: 0.8 × 1.8 × 0.1 cm (l × w × d) paper tape measurement). Week 4: 2 cm bandage; wound size: 0.7 × 0.2 × 0.1 cm

Figure 9.6 had been present and non-healing for over 15 months while full closure was achieved with seven weeks of Exciflex treatment.

Wound size can be measured with different approaches depending on the detail needed. Digital wound images can be obtained using a 35mm camera at a standardised distance and orientation to the wound. However, these digital techniques are restricted to 2D measurements of surface area. Stereophotogrammetry (SPG) is an image analysis technique to determine the 3D coordinates and geometric properties of objects. The application of stereophotogrammetric techniques to non-invasively measure wound dimensions allows wound depth and volume to be determined. Infra-red imaging can also be used to obtain wound region temperatures, including the periwound area as a baseline reference.

9.3.3   Current state of the art

9.3.3.1   Availability/coverage – US

In the US, the apparent disconnect between the strongly positive recommendations of widely used CPG and approval for use in PU treatment is ongoing. Medicare still only provides coverage for use of ES only when all other types of therapy have failed [124]. This determination was made over 20 years ago and has not been revisited since 2004 despite a wealth of research and development in the field. Similarly, as of April 2023, Blue Cross/Blue Shield Association does not provide any coverage because they still consider ES for wound therapy to be investigational and unreliable due to insufficient evidence to determine the best type of device and most effective form of ES for treatment of chronic wounds [125]. It was further noted that there appeared to be no standard type, waveform or frequency of ES used clinically.

9.3.3.2   Availability/coverage –EU and UK

In some countries in the EU, ES is considered an adjunct therapy in rehabilitation care, as outlined in guidelines. Conversely, in other regions, it's seen as a last resort after ineffective treatments, prompting consideration of ES. Interestingly, in certain countries, ES isn't used in clinical practice at all. These disparities reflect differing healthcare philosophies and resource allocations. In most EU countries, ES is not covered by insurance. In the Netherlands, it is primarily because treating PU is deemed an insurance cost rather than prevention of PUs. Additionally, the lack of substantial evidence supporting its efficacy of the prevention of PUs contributes to this decision. However, some insurance providers do cover ES costs for the treatment of chronic wounds. Despite insurance challenges, most stimulators are CE marked for EU use, with many available in the UK.

9.3.3.3   Future concepts in electroceutical therapy for PUs

Unfortunately, the incidence of PUs is continuing to increase across many patient populations, highlighting the urgent need for new approaches to prevention and treatment. Mehmood *et al.* have pointed out the necessity for wound dressings equipped with smart sensing capabilities that enable remote, real-time monitoring of wound status [126]. The adoption of *smart wound dressings* can enhance wound management strategies and minimise the need for frequent dressing changes, which can decrease wound bed temperature and potentially inhibit healing. By minimising unnecessary disturbances to the sensitive regranulating wound bed, wound therapy can be optimised, and patient comfort can be significantly improved. Bodo *et al.* emphasised that non-invasive methods for measuring and quantifying wound healing markers align with the objectives and principles of evidence-based medicine [127]. Ideally, an electroceutical bandage would be capable of sensing the healing status of the wound and administering appropriate treatment. There is a critical demand for smart technologies that can treat wounds, report their status to healthcare providers, and thus positively influence wound management.

The ability to monitor wound status in real time and quantify it would enable the standardisation of wound healing reports and enhance the efficiency and efficacy of wound treatment. The future of wound care is set to be transformed by the application of smart technology in preventing, treating and monitoring PUs.

9.4   Conclusion

The integration of ES for the prevention and treatment of PUs represents a promising advancement in healthcare for individuals with reduced mobility and/or conditions such as SCI. The evidence suggests that ES can improve tissue health by enhancing muscle size, blood circulation and local oxygenation, thereby reducing the risk of PU development. While current preventive measures are essential, ES offers a complementary approach that can potentially lower PU incidence. However, the clinical implementation of ES remains limited due to a lack of large-scale, well-controlled studies confirming its effectiveness in reducing PU recurrence and its cost-effectiveness. Furthermore, ES shows significant potential in the treatment of chronic wounds, promoting healing through enhanced cellular activity and improved blood flow. The development of smart wound care technologies, incorporating real-time monitoring and targeted treatment delivery, could revolutionise PU management, offering more effective, efficient and patient-friendly solutions. Future research should focus on optimising ES protocols, integrating machine learning for wound analysis and assessing the long-term impacts and cost-effectiveness of these innovative approaches. Overall, the application of ES in PU prevention and treatment holds substantial promise, necessitating further exploration and refinement to maximise its clinical benefits.

9.5   Suggestions for future research

- Large clinical trials to determine optimal ES parameters for both prevention and treatment of PUs.
- Development of advanced smart dressings: Future research could focus on the development and refinement of smart wound dressings that not only monitor wound status in real time but also deliver targeted treatments based on the

wound's healing stage. This includes integrating sensors for detecting infection, pH levels and moisture content, as well as delivering therapeutics like antibiotics or growth factors.

- Machine learning algorithms for wound analysis: There's a potential to develop and train machine learning algorithms to analyse data collected from smart wound dressings. These algorithms could predict healing outcomes, suggest adjustments in treatment protocols and identify early signs of complications like infections or delayed healing.
- Telehealth integration: Future work could explore the integration of smart wound care technologies with telehealth platforms. This would allow for remote monitoring of wounds by healthcare providers, reducing the need for in-person visits and enabling timely interventions.
- Longitudinal studies on patient outcomes: Conducting longitudinal studies to assess the impact of smart wound dressings and electroceutical bandages on patient outcomes, including healing times, incidence of infections and overall patient satisfaction and comfort, would provide valuable insights into their effectiveness and areas for improvement.
- Cost-effectiveness analysis: Research into the cost-effectiveness of using smart technologies in wound care is crucial. This includes analysing the potential savings from reduced clinic visits, shortened healing times and decreased incidence of complications against the cost of these technologies.
- Customisable and biodegradable options: Investigating the development of customisable smart dressings that can be tailored to the specific needs and contours of different wound types and locations. Additionally, exploring biodegradable options that minimise environmental impact could be a promising direction.

References

[1] Redelings MD, Lee NE, and Sorvillo F. 'Pressure ulcers: more lethal than we thought?'. *Adv Skin Wound Care*. 2005;18(7):367–372.

[2] Tsokos M, Heinemann A, and Püschel K. 'Pressure sores: epidemiology, medico-legal implications and forensic argumentation concerning causality'. *Int J Legal Med*. 2000;113(5):283–287.

[3] Kaya PK, Kaya M, Girgin NK, Kahveci FŞ, Akalın EH, and İşçimen R. 'Sepsis episodes caused by pressure injuries in critical illness: a retrospective observational cohort study'. *Wound Manag Prev*. 2023;69(4):4–9.

[4] Dolanová D, Bůřilová P, Krupová L, *et al.* 'Mortality related to pressure ulcers in Czech Republic - analyses of national health registries'. *J Tissue Viability*. 2023;32(3):365–370.

[5] Carter MJ, DaVanzo J, Haught R, Nusgart M, Cartwright D, and Fife CE. 'Chronic wound prevalence and the associated cost of treatment in Medicare beneficiaries: changes between 2014 and 2019'. *J Med Econ*. 2023;26 (1):894–901.

[6]   Research and Markets the world's largest market research store. *Global Wound Care Market Report and Forecast 2023-2031* [online], 2023. Available from: https://www.researchandmarkets.com/reports/5797982/global-wound-care-market-report-forecast#cat-pos-. [Accessed 12 June 2023].

[7]   Queen D, and Harding K. 'Estimating the cost of wounds both nationally and regionally within the top 10 highest spenders'. *Int Wound J.* 2024;21(2): e14709.

[8]   Powers JG, Higham C, Broussard K, and Phillips TJ. 'Wound healing and treating wounds: Chronic wound care and management'. *J Am Acad Dermatol.* 2016;74(4):607–625.

[9]   Brown TS, Hawksworth JS, Sheppard FR, Tadaki DK, and Elster E. 'Inflammatory response is associated with critical colonization in combat wounds'. 2011;12(5):351–7.

[10] Malone M, Radzieta M, Peters TJ, *et al.* 'Host-microbe metatranscriptome reveals differences between acute and chronic infections in diabetes-related foot ulcers'. *APMIS.* 2022;130(12):751–762.

[11] Puca V, Marulli RZ, Grande R, *et al.* 'Microbial species isolated from infected wounds and antimicrobial resistance analysis: data emerging from a three-years retrospective study'. *Antibiotics.* 2021;10(10):1162.

[12] Percival SL, Hill KE, Williams DW, Hooper SJ, Thomas DW, and Costerton JW. 'A review of the scientific evidence for biofilms in wounds'. *Wound Repair Regen.* 2012;20(5):647–657.

[13] White RJ, Cutting K, and Kingsley A. 'Topical antimicrobials in the control of wound bioburden'. *Ostomy Wound Manage.* 2006;52(8):26–58.

[14] Ferguson AC, Keating JF, Delargy MA, and Andrews BJ. 'Reduction of seating pressure using FES in patients with spinal cord injury'. A preliminary report. *Paraplegia.* 1992;30(7):474–8.

[15] Bogie KM, and Triolo RJ. 'Effects of regular use of neuromuscular electrical stimulation on tissue health'. *J Rehabil Res Dev.* 2003;40(6):469–75.

[16] Bogie KM, Wang X, and Triolo RJ. 'Long-term prevention of pressure ulcers in high-risk patients: a single case study of the use of gluteal neuromuscular electric stimulation'. *Arch Phys Med Rehabil.* 2006;87(4):585–91.

[17] van Londen A, Herwegh M, van der Zee CH, *et al.* 'The effect of surface electric stimulation of the gluteal muscles on the interface pressure in seated people with spinal cord injury'. *Arch Phys Med Rehabil.* 2008;89(9):1724–32.

[18] Smit CA, Haverkamp GL, de Groot S, Stolwijk-Swuste JM, and Janssen TW. 'Effects of electrical stimulation-induced gluteal versus gluteal and hamstring muscles activation on sitting pressure distribution in persons with a spinal cord injury'. *Spinal Cord.* 2012;50(8):590–4.

[19] Olive JL, Slade JM, Dudley GA, and McCully KK. 'Blood flow and muscle fatigue in SCI individuals during electrical stimulation'. *J Appl Physiol (1985).* 2003;94(2):701–8.

[20] Kim J, Chester HH, Wang X, and Bogie K. 'The use of sensory electrical stimulation for pressure ulcer prevention'. *Physiother Theory Pract.* 2010;26 (8):528–536.

[21]   Barton TJ, Low DA, Janssen TWJ, Sloots M, Smit CAJ, and Thijssen, DHJ. 'Femoral artery blood flow and microcirculatory perfusion during acute, low-level functional electrical stimulation in spinal cord injury'. *Am J Phys Med Rehabil.* 2018;97:721–726.

[22]   Smit, CAJ, de Groot, S, Stolwijk-Swuste, JM, and Janssen TWJ. 'Effects of electrical stimulation on risk factors for developing pressure ulcers in people with a spinal cord injury: a focused review of literature'. *Am J Phys Med Rehabil.* 2016;95(7):535–552.

[23]   Baldi JC, Jackson RD, Moraille R, and Mysiw WJ. 'Muscle atrophy is prevented in patients with acute spinal cord injury using functional electrical stimulation'. *Spinal Cord.* 1998;36(7):463–9.

[24]   Chilibeck PD, Jeon J, Weiss C, Bell G, and Burnham R. 'Histochemical changes in muscles of individuals with spinal cord injury following functional electrical stimulated exercise training'. *Spinal Cord.* 1999;37:264–268.

[25]   Rochester L, Barron MJ, Chandler CS, Sutton RA, Miller S, and Johnson MA. 'Influence of electrical stimulation of the tibialis anterior muscle in paraplegic subjects. 2. Morphological and histochemical properties'. *Paraplegia.* 1995;33(5):4–522.

[26]   Gerrits HL, de Haan A, Sargeant AJ, van Langen H, and Hopman MT. 'Peripheral vascular changes after electrically stimulated cycle training in people with spinal cord injury.' *Arch Phys Med Rehabil.* 2001;82:832–9.

[27]   Wijker BJ, de Groot S, van Dongen JM, *et al.* 'Electrical stimulation to prevent recurring pressure ulcers in individuals with a spinal cord injury compared to usual care: the Spinal Cord Injury PREssure VOLTage (SCI PREVOLT) study protocol'. *Trials.* 2022;23(1):156.

[28]   Kern H, and Carraro U. 'Home-based functional electrical stimulation of human permanent denervated muscles: a narrative review on diagnostics, managements, results and byproducts revisited 2020'. *Diagnostics (Basel).* 2020;10(8):529.

[29]   Liu LQ, Moody J, Traynor M, Dyson S, and Gall A. 'A systematic review of electrical stimulation for pressure ulcer prevention and treatment in people with spinal cord injuries'. *J Spinal Cord Med.* 2014;37(6):703–18.

[30]   Liu LQ, and Ferguson-Pell M. 'Blood perfusion changes during sacral nerve root stimulation versus surface gluteus electrical stimulation on in seated spinal cord injury'. *Assist Technol.* 2019;31(1):1–8.

[31]   Smit CA, de Groot S, Stolwijk-Swuste JM, and Janssen TW. 'Effects of electrical stimulation on risk factors for developing pressure ulcers in people with a spinal cord injury: a focused review of literature'. *Am J Phys Med Rehabil.* 2016;95(7):535–52.

[32]   Müller-Putz, G. *Neuroprosthetics and Brain-Computer Interfaces in Spinal Cord Injury; A Guide for Clinicians and End Users.* 1st edn. Switzerland: Springer; 2021. pp. 274–278.

[33]   Bekhet AH, Jahan AM, Bochkezanian V, Musselman KE, Elsareih AA, and Gorgey AS. 'Effects of electrical stimulation training on body composition

parameters after spinal cord injury: a systematic review'. *Arch Phys Med Rehabil.* 2022;103(6):1168–1178.

[34]   Kern H, Gargiulo P, Pond A, Albertin G, Marcante A, and Carraro U. 'To reverse atrophy of human muscles in complete SCI lower motor neuron denervation by home-based functional electrical stimulation'. *Adv Exp Med Biol.* 2018;1088:585–591.

[35]   Baron MV, Silva PE, Koepp J, *et al.* 'Efficacy and safety of neuromuscular electrical stimulation in the prevention of pressure injuries in critically ill patients: a randomized controlled trial'. *Ann Intensive Care.* 2022;12(1):53.

[36]   Smit CAJ, Berenpas F, de Groot S, Stolwijk-Swuste JM, and Janssen TWJ. 'Feasibility of overnight electrical stimulation-induced muscle activation in people with a spinal cord injury. A pilot study'. *Spinal Cord Ser Cases.* 2020;6:5.

[37]   Olive JL, Slade JM, Dudley GA, and McCully KK. 'Blood flow and muscle fatigue in SCI individuals during electrical stimulation'. *J Appl Physiol.* 2003;94(2):701–8.

[38]   Smit CA, Legemate KJ, de Koning A, de Groot S, Stolwijk-Swuste JM, and Janssen TW. 'Prolonged electrical stimulation-induced gluteal and hamstring muscle activation and sitting pressure in spinal cord injury: effect of duty cycle'. *J Rehabil Res Dev.* 2013;50(7):1035–46.

[39]   Kane A, Warwaruk-Rogers R, Ho C, *et al.* 'A feasibility study of intermittent electrical stimulation to prevent deep tissue injury in the intensive care unit'. *Adv Wound Care (New Rochelle).* 2017;6(4):115–124.

[40]   Bersch I. Webinar titled *'Patient Perspective: Pressure Ulcer Prevention, Wound Healing and Electrical Stimulation for People With Spinal Cord Injury'.* Webinar host was a collaboration of Spinal Injuries Association and IFESS, 8 March 2022.

[41]   Mohr T, Andersen JL, Biering-Sorensen F, *et al.* 'Long term adaptation to electrically induced cycle training in severe spinal cord injured individuals'. *Spinal Cord.* 1997;35:1–16.

[42]   Sloan KE, Bremmer LA, Byrne J, Day RE, and Scull ER. 'Musculoskeletal effects of an electrical stimulation induced cycling programme in the spinal injured'. *Paraplegia.* 1994;(32):407–415.

[43]   Bickel CS, Slade JM, Haddad F, Adams GR, and Dudley GA. 'Acute molecular responses of skeletal muscle to resistance exercise in able-bodied and spinal cord-injured subjects'. *J Appl Physiol.* 2003;(94):2255–2262.

[44]   Petrofsky JS, and Phillips CA. 'The use of functional electrical stimulation for rehabilitation of spinal cord injured patients'. *Cent Nerv Syst Trauma.* 1984;1(1):57–74.

[45]   Thijssen DHJ, Bruno RM, Van Mil ACCM, *et al.* 'Expert consensus and evidence-based recommendations for the assessment of flow-mediated dilation in humans'. *Eur Heart J.* 2019;40(30):2534–47.

[46]   Neubauer-Geryk J, Hoffmann M, Wielicka M, *et al.* 'Current methods for the assessment of skin microcirculation: Part 1'. *Postepy Dermatol Alergol.* 2019;36(3):247–254.

[47] Gorgey AS, Khalil RE, Davis JC, *et al*. 'Skeletal muscle hypertrophy and attenuation of cardio-metabolic risk factors (SHARC) using functional electrical stimulation-lower extremity cycling in persons with spinal cord injury: study protocol for a randomized clinical trial'. *Trials*. 2019;20(1):526.

[48] Arora M, Harvey LA, Glinsky JV, *et al*. 'Electrical stimulation for treating pressure ulcers'. *Cochrane Database Syst Rev*. 2020;1(1):CD012196.

[49] Bakkum AJ, de Groot S, Stolwijk-Swüste JM, *et al*. 'Effects of hybrid cycling versus handcycling on wheelchair-specific fitness and physical activity in people with long-term spinal cord injury: a 16-week randomized controlled trial'. *Spinal Cord*. 2015;53(5):395–401.

[50] Joseph A, Wanono R, Flamant M, and Vidal-Petiot E. 'Orthostatic hypotension: a review'. *Nephrol Ther*. 2017;13(Suppl 1):S55–S67.

[51] Electrophysical Agents. 'Contraindications and precautions: an evidence-based approach to clinical decision making in physical therapy'. *Physiother Can*. 2010;62(5):1–80.

[52] Keskin EA, Onur O, Keskin HL, Gumus II, Kafali H, and Turhan N. 'Transcutaneous electrical nerve stimulation improves low back pain during pregnancy'. *Gynecol Obstet Invest*. 2012;74(1):76–83.

[53] Katsura D, Tsuji S, Hayashi K, *et al*. 'Maternal and foetal physiological response of sacral surface electrical stimulation during pregnancy: a preliminary study'. *Exp Physiol*. 2022;107(5):508–514.

[54] Dowswell T, Bedwell C, Lavender T, and Neilson JP. 'Transcutaneous electrical nerve stimulation (TENS) for pain relief in labour'. *Cochrane Database Syst Rev*. 2009;(2):CD007214.

[55] Guo WC, and Wang F. 'Effect of nerve electrical stimulation for treating chemotherapy-induced nausea and vomiting in patients with advanced gastric cancer: a randomized controlled trial'. *Medicine (Baltimore)*. 2018;97 (51):e13620.

[56] Bennett MI, Johnson MI, Brown SR, Radford H, Brown JM, and Searle RD. 'Feasibility study of transcutaneous electrical nerve stimulation (TENS) for cancer bone pain'. *J Pain*. 2010;11(4):351–9.

[57] Janigro D, Perju C, Fazio V, *et al*. 'Alternating current electrical stimulation enhanced chemotherapy: a novel strategy to bypass multidrug resistance in tumor cells'. *BMC Cancer*. 2006;6:72.

[58] O'Connor D, Caulfield B, and Lennon O. 'The efficacy and prescription of neuromuscular electrical stimulation (NMES) in adult cancer survivors: a systematic review and meta-analysis'. *Support Care Cancer*. 2018;26 (12):3985–400.

[59] O'Connor D, Lennon O, Minogue C, and Caulfield B. 'Design considerations for the development of neuromuscular electrical stimulation (NMES) exercise in cancer rehabilitation'. *Disabil Rehabil*. 2021;43(21):3117–3126.

[60] Broderick BJ, Kennedy C, Breen PP, Kearns SR, and ÓLaighin G. 'Patient tolerance of neuromuscular electrical stimulation (NMES) in the presence of orthopaedic implants'. *Med Eng Phys*. 2011;33(1):56–61.

[61] Badger J, Taylor P, and Swain I. 'The safety of electrical stimulation in patients with pacemakers and implantable cardioverter defibrillators: a systematic review'. *J Rehabil Assist Technol Eng.* 2017;4:2055668317745498.

[62] Egger F, Hofer C, Hammerle FP, *et al.* 'Influence of electrical stimulation therapy on permanent pacemaker function'. *Wien Klin Wochenschr.* 2019; 131(13–14):313–320. Epub 2019 Apr 25.

[63] Bois-Reymond, E.H. Vorläufiger Abriss einer Untersuchung über den sogenannten Froschstrom und über die elektromotorischen Fische. *Annalen der Physik.* 1843;134(1):1–30. DOI:10.1002/ANDP.18431340102.

[64] Nuccitelli R. 'A role for endogenous electric fields in wound healing'. *Curr Top Dev Biol.* 2003;58:1–26.

[65] Jaffe LF, and Vanable JW Jr. 'Electric fields and wound healing'. *Clin Dermatol.* 1984;2(3):34–44.

[66] Zhao M. 'Electrical fields in wound healing-an overriding signal that directs cell migration'. *Semin Cell Dev Biol.* 2009;20(6):674–82. Epub 2008 Dec 25.

[67] Ennis WJ, Lee C, Gellada K, Corbiere TF, and Koh TJ. 'Advanced technologies to improve wound healing: electrical stimulation, vibration therapy, and ultrasound-what is the evidence?'. *Plast Reconstr Surg.* 2016;138 (3):94S–104S.

[68] Bai H, McCaig CD, Forrester JV, and Zhao M. 'DC electric fields induce distinct preangiogenic responses in microvascular and macrovascular cells'. *Arterioscler Thromb Vasc Biol.* 2004;24(7):1234–9.

[69] Asadi MR, and Torkaman G. 'Bacterial inhibition by electrical stimulation'. *Adv Wound Care (New Rochelle).* 2014;3(2):91–97.

[70] Ashrafi M, Novak-Frazer L, Morris J, Baguneid M, Rautemaa-Richardson R, and Bayat A. 'Electrical stimulation disrupts biofilms in a human wound model and reveals the potential for monitoring treatment response with volatile biomarkers'. *Wound Repair Regen.* 2019;27(1):5–18.

[71] Garber SL, Bryce TN, Gregorio-Torres, TL, Ho CH, and Rader C. *Pressure Ulcer Prevention and Treatment Following Spinal Cord Injury: A Clinical Practice Guideline for Health-Care Professionals.* 2nd edn. Washington, DC, USA: Paralyzed Veterans of America; 2014. Available at: https://pva.org/research-resources/publications/clinical-practice-guidelines/. Accessed 24 July 2024.

[72] European Pressure Ulcer Advisory Panel, National Pressure Injury Advisory Panel and Pan Pacific Pressure Injury Alliance. *Prevention and Treatment of Pressure Ulcers/Injuries: Quick Reference Guide.* In: Haesler E (ed.). EPUAP/NPIAP/PPPIA; 2019.

[73] Gould LJ, Alderden J, Aslam R, *et al.* 'WHS guidelines for the treatment of pressure ulcers-2023 update'. *Wound Repair Regen.* 2024;32(1):6–33.

[74] Girgis B, and Duarte JA. 'High voltage monophasic pulsed current (HVMPC) for stage II-IV pressure ulcer healing. A systematic review and meta-analysis'. *J Tissue Viability.* 2018;27(4):274–284.

[75] Harding K. 'Electrical stimulation for wound healing needs serious consideration'. *Wounds Int.* 2013;4(eSuppl):3.

[76] Clark M. 'Electrical stimulation and wound healing'. *Wounds Int.* 2013;4 (eSuppl):4–6.

[77] Aziz Z, and Bell-Syer SE. 'Electromagnetic therapy for treating pressure ulcers'. *Cochrane Database Syst Rev.* 2015;(9):CD002930.

[78] Houghton PE, Campbell KE, Fraser CH, *et al.* 'Electrical stimulation therapy increases rate of healing of pressure ulcers in community-dwelling people with spinal cord injury'. *Arch Phys Med Rehabil.* 2010;91(5):669–78.

[79] Lala D, Spaulding SJ, Burke SM, and Houghton PE. 'Electrical stimulation therapy for the treatment of pressure ulcers in individuals with spinal cord injury: a systematic review and meta-analysis'. *Int Wound J.* 2016;13 (6):1214–1226. Epub 2015 Apr 13.

[80] Khouri C, Kotzki S, Roustit M, Blaise S, Gueyffier F, and Cracowski JL. 'Hierarchical evaluation of electrical stimulation protocols for chronic wound healing: an effect size meta-analysis'. *Wound Repair Regen.* 2017;25 (5):883–891.

[81] Driver VR, Gould LJ, Dotson P, *et al.* 'Identification and content validation of wound therapy clinical endpoints relevant to clinical practice and patient values for FDA approval. Part 1. Survey of the wound care community'. *Wound Repair Regen.* 2017;25(3):454–465.

[82] Chen L, Ruan Y, Ma Y, Ge L, and Han L. 'Effectiveness and safety of electrical stimulation for treating pressure ulcers: a systematic review and meta-analysis'. *Int J Nurs Pract.* 2023;29(2):e13041.

[83] Gomes RC, Brandino HE, de Sousa NT, Santos MF, Martinez R, and Guirro RR. 'Polarized currents inhibit in vitro growth of bacteria colonizing cutaneous ulcers'. *Wound Repair Regen.* 2015;23(3):403–11.

[84] Tomic-Canc M, Agren MS, and Alvarez OM. 'Epidermal repair and the chronic wound' In: Rovee DT and Maibach HI (eds), *The Epidermics in Wound Healing*. Boca Raton, FL: CRC Press; 2004. pp. 25–57.

[85] Goldberg SR, and Diegelmann RF. 'Wound healing primer'. *Surg Clin North Am.* 2010;90(6):1133–46.

[86] Brown TS, Hawksworth JS, Sheppard FR, Tadaki DK, and Elster E. 'Inflammatory response is associated with critical colonization in combat wounds'. *Surg Infect (Larchmt).* 2011;12(5):351–7.

[87] White RJ, Cutting K, and Kingsley A. 'Topical antimicrobials in the control of wound bioburden'. *Ostomy Wound Manage.* 2006;52(8):26–58.

[88] Percival SL, Hill KE, Williams DW, Hooper SJ, Thomas DW, and Costerton JW. 'A review of the scientific evidence for biofilms in wounds'. *Wound Repair Regen.* 2012;20(5):647–57.

[89] Dowd SE, Wolcott RD, Sun Y, McKeehan T, Smith E, and Rhoads D. 'Polymicrobial nature of chronic diabetic foot ulcer biofilm infections determined using bacterial tag encoded FLX amplicon pyrosequencing (bTEFAP)'. *PLoS One.* 2008;3(10):e3326.

[90] Scales BS, and Huffnagle GB. 'The microbiome in wound repair and tissue fibrosis'. *J Pathol.* 2013;229(2):323–31.

[91]   Bertesteanu S, Triaridis S, Stankovic M, *et al.* 'Polymicrobial wound infections: pathophysiology and current therapeutic approaches'. *Int J Pharm*. 2014;463(2):119–26.

[92]   Costerton JW, Stewart PS, and Greenberg EP. 'Bacterial biofilms: a common cause of persistent infections'. *Science*. 1999;284(5418):1318–22.

[93]   Dalton T, Dowd SE, Wolcott RD, *et al.* 'An in vivo polymicrobial biofilm wound infection model to study interspecies interactions'. *PLoS One*. 2011; 6(11):e27317.

[94]   Wolcott RD, Rhoads DD, and Dowd SE. 'Biofilms and chronic wound inflammation'. *J Wound Care*. 2008;17(8):333–41.

[95]   Mihai MM, Holban AM, Giurcăneanu C, *et al.* 'Identification and phenotypic characterization of the most frequent bacterial etiologies in chronic skin ulcers'. *Rom J Morphol Embryol*. 2014;55(4):1401–8.

[96]   Lavigne JP, Sotto A, Dunyach-Remy C, and Lipsky BA. 'New molecular techniques to study the skin microbiota of diabetic foot ulcers'. *Adv Wound Care (New Rochelle)*. 2015;4(1):38–49.

[97]   Dana AN, and Bauman WA. 'Bacteriology of pressure ulcers in individuals with spinal cord injury: what we know and what we should know'. *J Spinal Cord Med*. 2015;38(2):147–60.

[98]   Holmes CJ, Plichta JK, Gamelli RL, and Radek KA. 'Dynamic role of host stress responses in modulating the cutaneous microbiome: implications for wound healing and infection'. *Adv Wound Care (New Rochelle)*. 2015;4 (1):24–37.

[99]   Sprockett DD, Ammons CG, and Tuttle MS. 'Use of 16S rRNA sequencing and quantitative PCR to correlate venous leg ulcer bacterial bioburden dynamics with wound expansion, antibiotic therapy, and healing'. *Wound Repair Regen*. 2015;23(5):765–71.

[100] Kunimitsu M, Nakagami G, Kitamura A, *et al.* 'Relationship between healing status and microbial dissimilarity in wound and peri-wound skin in pressure injuries'. *J Tissue Viability*. 2023;32(1):144–150.

[101] Verbanic S, Deacon JM, and Chen IA. 'The chronic wound phageome: phage diversity and associations with wounds and healing outcomes'. *Microbiol Spectr*. 2022;10(3):e0277721.

[102] Gomes F, Furtado GE, Henriques M, *et al.* 'The skin microbiome of infected pressure ulcers: a review and implications for health professionals'. *Eur J Clin Invest*. 2022;52(1):e13688.

[103] Dunyach-Remy C, Salipante F, Lavigne JP, *et al.* 'Pressure ulcers microbiota dynamics and wound evolution'. *Sci Rep*. 2021;11(1):18506.

[104] Rowley BA, McKenna JM, Chase GR, and Wolcott LE. 'The influence of electrical current on an infecting microorganism in wounds'. *Ann N Y Acad Sci*. 1974;238:543–51.

[105] Kincaid CB, and Lavoie KH. 'Inhibition of bacterial growth in vitro following stimulation with high voltage, monophasic, pulsed current'. *Phys Ther*. 1989;69(8):651–5.

[106] Ehrlich GD, Stoodley P, Kathju S, *et al.* 'Engineering approaches for the detection and control of orthopaedic biofilm infections'. *Clin Orthop Relat Res.* 2005;(437):59–66.

[107] Perez-Roa RE, Tompkins DT, Paulose M, Grimes CA, Anderson MA, and Noguera DR. 'Effects of localised, low-voltage pulsed electric fields on the development and inhibition of Pseudomonas aeruginosa biofilms'. *Biofouling.* 2006;22(5–6):383–90.

[108] Gomes Md, Teixeira CA, Barbosa GM, *et al.* 'Effects of direct electric current on Herpetomonas samuelpessoai: an ultrastructural study. *Bioelectromagnetics.* 2012;33(4):334–45.

[109] García D, Gómez N, Mañas P, Raso J, and Pagán R. 'Pulsed electric fields cause bacterial envelopes permeabilization depending on the treatment intensity, the treatment medium pH and the microorganism investigated'. *Int J Food Microbiol.* 2007;113(2):219–27.

[110] Bolton L, Foleno B, Means B, and Petrucelli S. 'Direct-current bactericidal effect on intact skin'. *Antimicrob Agents Chemother.* 1980;18(1):137–41.

[111] Goldman R, Rosen M, Brewley B, and Golden M. 'Electrotherapy promotes healing and microcirculation of infrapopliteal ischemic wounds: a prospective pilot study'. *Adv Skin Wound Care.* 2004;17(6):284–94.

[112] Korzendorfer H, and Hettrick H. 'Biophysical technologies for management of wound bioburden'. *Adv Wound Care (New Rochelle).* 2014;3 (12):733–741.

[113] Kawasaki L, Mushahwar VK, Ho C, Dukelow SP, Chan LL, and Chan KM. 'The mechanisms and evidence of efficacy of electrical stimulation for healing of pressure ulcer: a systematic review'. *Wound Repair Regen.* 2014; 22(2):161–73.

[114] Rippon M, Davies P, and White R. 'Taking the trauma out of wound care: the importance of undisturbed healing'. *J Wound Care.* 2012;21(8):359–60, 362, 364–8.

[115] Kannon GA, and Garrett AB. 'Moist wound healing with occlusive dressings. A clinical review'. *Dermatol Surg.* 1995;21(7):583–90.

[116] Metzger S. 'Clinical and financial advantages of moist wound management'. *Home Healthc Nurse.* 2004;22(9):586–90.

[117] McGuiness W, Vella E, and Harrison D. 'Influence of dressing changes on wound temperature'. *J Wound Care.* 2004;13(9):383–5.

[118] Kloth LC. 'Electrical stimulation for wound healing: a review of evidence from in vitro studies, animal experiments, and clinical trials'. *Int J Low Extrem Wounds.* 2005;4(1):23–44.

[119] Bogie KM, Schwartz K, Henzel MK *et al.* 'Health electroceutical bandage to decrease infection and accelerate healing of chronic ischemic wounds'. In: *Proceedings of the European Wound Management Association, 34th Conference*, London, England, 2024.

[120] Howe DS, Dunning J, Zorman C, Garverick SL, and Bogie KM. 'Development of an integrated surface stimulation device for systematic evaluation of wound electrotherapy'. *Ann Biomed Eng.* 2015;43(2):306–313.

[121]   Bogie KM. 'The Modular Adaptive Electrotherapy Delivery System (MAEDS): an electroceutical approach for effective treatment of wound infection and promotion of healing'. *Mil Med*. 2019;184(1):92–96.

[122]   Seshadri DR, Bianco ND, Radwan AZ, Zorman CA, and Bogie KM. 'An absorbent, flexible, transparent, and scalable substrate for wound dressings'. *IEEE J Transl Eng Health Med*. 2022;10:4900909.

[123]   Scalamandré A, and Bogie KM. 'Smart technologies in wound prevention and care'. In: Gefen A (ed), *Innovations and Emerging Technologies in Wound Prevention and Care*. London: Academic Press; 2020. pp. 225–244.

[124]   National coverage Determination (NCD) *Electrical Stimulation (ES) and Electromagnetic Therapy for the Treatment of Wounds 270.1* [online]. Available from: https://www.cms.gov/Medicare-Coverage-Database/view/ncd.aspx?ncdid=131&ver=3 [Accessed 12/18/2023]

[125]   Premera. *Electrical Stimulation and Electromagnetic Therapy for treating wounds*. Corporate Last revised 6 March 2023. [online]. Available from: https://www.premera.com/medicalpolicies/2.01.57.pdf [Accessed 12/18/23].

[126]   Mehmood N, Hariz A, Fitridge R, and Voelcker NH. 'Applications of modern sensors and wireless technology in effective wound management'. *J Biomed Mater Res B Appl Biomater*. 2014;102(4):885–95.

[127]   Bodo M, Settle T, Royal J, Lombardini E, Sawyer E, and Rothwell SW. 'Multimodal noninvasive monitoring of soft tissue wound healing'. *J Clin Monit Comput*. 2013;27(6):677–88.

Chapter 10

Diaphragm stimulation and respiratory function

Raymond P. Onders[1]

10.1   Introduction

The diaphragm is the primary muscle involved in respiration. The diaphragm atrophies faster than any other muscle when it is not being utilised. Diaphragm dysfunction arises from diseases affecting diaphragm innervation or its contractile properties with resulting dyspnea, decreased exercise performance, sleep disordered breathing, atelectasis with increased risk of pneumonia and respiratory failure. Bilateral diaphragm failure can lead to positive pressure ventilation with tracheostomy mechanical ventilation (MV) which not only significantly affects quality of life but drastically increases a patient's mortality. Diaphragm muscle electrical stimulation can free patients from ventilators in spinal cord injury (SCI), help phrenic nerve recovery from injuries and overcome loss of central control of respiration. Decreasing the need for abnormal positive pressure ventilation with electrical stimulation giving patients more freedom. This chapter will outline the development of intramuscular diaphragm pacing (DP) and review the present clinical indications for this functional electrical stimulation of the diaphragm in SCI, phrenic nerve/diaphragm dysfunction including unilateral abnormalities and temporary utilisation in the intensive care unit (ICU).

10.2   Background and development of diaphragm pacing

The diaphragm is the primary muscle of respiration responsible for approximately 80% of inspiratory lung volume during normal rest breathing. The diaphragm is primarily composed of fatigue-resistant slow-twitch type I myofibres and fast twitch type IIa myofibres. The diaphragm is innervated by the right and left phrenic nerves which originate with the phrenic lower motor neurons (LMN) in the spinal cord at the cervical levels 3–5. These cervical phrenic motor neurons are controlled involuntarily by upper motor neurons (UMN) in the respiratory control centre in the brain stem or voluntarily by the cerebral cortex UMN. There are numerous

[1]Margaret and Walter Remen Chair of Surgical Innovation, University Hospitals Cleveland Medical Centre, Case Western Reserve University School of Medicine, USA

aetiologies that can affect respiratory control leading to diaphragm contraction abnormalities resulting in respiratory compromise ranging from mild shortness of breath to invasive MV. The pre-Bötzinger complex, part of the ventral respiratory group (VRG), is responsible for respiratory drive and is the only drive during sleep. Injury or loss of pre-Bötzinger complex neurons may affect respiratory control and lead to central apneas [1]. Congenital Central Hypoventilation Syndrome (CCHS or Ondine's Curse) is an isolated genetic loss of this centre leading to the need for MV during sleep.

MV is the primary therapy for respiratory failure and a fundamental treatment in ICUs. MV is usually a time limited therapy that when discontinued with extubation, has no untoward sequelae. A decrease in diaphragm contractions leads to a decrease in ventilation that adversely affects normal respiratory physiology leading to hypercarbia from decreased minute ventilation or hypoxia from increased dead space. Numerous studies has shown that short exposure to MV leads to decreases in protein synthesis and increased proteolysis, which is histopathologically manifested as diaphragm muscle atrophy, with up to 50% of the diaphragm muscle atrophying and conversion to the non-functional, fast-twitch, type-IIb muscle fibres in less than one day [2]. The severity of this muscle atrophy worsens with increased time of MV exposure. This condition is called ventilator-induced diaphragm dysfunction (VIDD).

Electrical activation of the diaphragm muscle, by way of phrenic nerve stimulation (PNS) or through direct DP at the motor point was developed to improve ventilation in a number of diseases. The concept of PNS to provide ventilatory support was first reported in the 18th century. In the 1940s, a group first demonstrated that ventilation could be maintained with percutaneous electrodes in patients with poliomyelitis [3]. In the 1960s, significant technological advances were made that led to the development of traditional PNS systems (Avery Mark IV Breathing Pacemaker System, Avery Biomedical Devices, Commack, NY, USA). There was development of an implantable electrode/receiver system which could be activated by radiofrequency waves generated by a power source external to the body. The group at Yale and others also accumulated significant clinical experience which defined patient evaluation methods, surgical techniques and safe parameters of stimulation which resulted in diaphragm conditioning via stimulation of the phrenic nerve [4].

In the 1980s, the group at Case Western Reserve University in Cleveland showed the diaphragm could be directly stimulated at the motor point to provide ventilation [5–7]. In the 1990s, the device was refined for the initial human studies at University Hospitals Cleveland Medical Centre and a standard laparoscopic implantation was utilised for SCI patients in the 2000s [8–11]. Because muscle motor point electrodes can be removed and used for short periods of time, Onders and colleagues began investigating its' use in other patients including patients with injured phrenic nerves and for temporary use in the ICU.

To be effective in recruiting diaphragm muscle to provide ventilatory support, the phrenic nerve must be able to provide conduction pathways to the muscle. Therefore, the LMN in the spinal cord and the phrenic nerve must be intact to prevent muscle denervation and atrophy of the muscle. Previously an assessment of phrenic nerve function was performed in all patients contemplating DP. Unfortunately, some patients with SCI have sustained injury to the phrenic motor neurons in the spinal cord and/or phrenic rootlets. If phrenic nerve function is absent or significantly reduced, DP will not work and should not be done. Historically, phrenic nerve function was assessed both by measurements of phrenic nerve conduction times and/or by fluoroscopic evaluation of diaphragm movement during PNS as has been described [12].

Unfortunately, this method of detecting an intact phrenic nerve was plagued by technical difficulties and inherent false positive and false negative results. A recent publication describes that the "gold standard" phrenic nerve study had 50% false negative or false positive test results when compared to direct surgical stimulation of the diaphragm [13]. The ultimate decision on if a DP will work is direct surgical stimulation of the diaphragm muscle which is easily done with the laparoscopy. Patients are willing to accept a low risk diagnostic laparoscopy to see if DP can allow them to be removed from the adverse effects of a MV. If the diaphragm is stimulable during the diagnostic laparoscopy, then the DP system is implanted.

The present DP system has been implanted in over 2500 patients (NeuRx RA/ 4 System, Synapse Biomedical, Oberlin, Ohio, USA). This is a distinct method compared to phrenic nerve pacing and that was likewise originally developed to provide natural negative pressure ventilation and replace MV in SCI patients. The DP surgery has become standardised involving minimally invasive laparoscopic surgery that implants two electrodes in each hemi-diaphragm at the mapped phrenic motor points where maximal diaphragm contraction can occur. The first step involves exposure of the diaphragm with four-port laparoscopy. No neuromuscular blocking agents are used during the surgery. Simultaneous gastrostomy tubes can be done at the end of the procedure. The falciform ligament is divided, which allows easier visualisation of the medial aspect of the right diaphragm and easier exit of the pacing electrodes through the epigastric port.

The key aspect is mapping the diaphragm to identify the point on the abdominal side of the diaphragm at which stimulation causes the greatest diaphragm excursion- diaphragm motor point [9,10]. A standard laparoscopic dissector is used to deliver an electrical stimulus from an external customised electrical clinical station. Stimulation is applied in either a twitch or a burst mode from the clinical station. Mapping allows qualitative and quantitative data to be obtained. Quantitatively, changes in abdominal pressures are measured. Qualitatively, the diaphragm contraction is directly observed. The stronger the stimulated contraction, the closer to the motor point of the phrenic nerve (Figures 10.1 and 10.2).

During mapping, the entire diaphragm is assessed in grid pattern to be sure the point of maximal contraction is not missed. The primary electrode site is identified as the location of maximal pressure change in each hemidiaphragm. A secondary electrode site is identified as either a backup to the primary site or at a location in each hemidiaphragm that recruits another region of the diaphragm (e.g., anterior, lateral or posterior) at a similar magnitude.

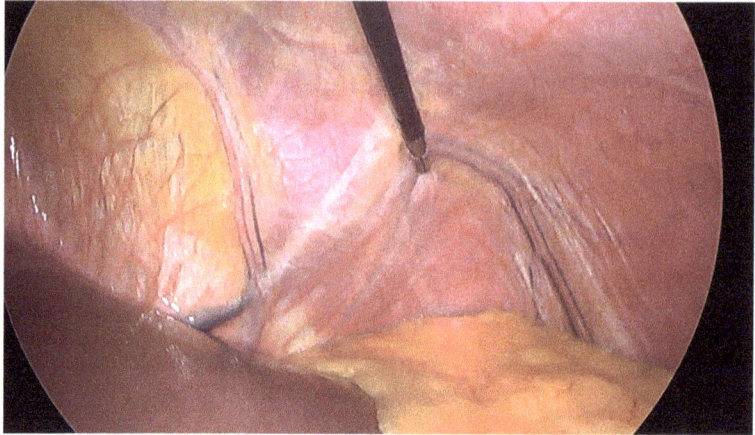

Figure 10.1   A standard laparoscopic dissector is used to map a patient's left diaphragm. A cable connects it to a specialised external clinical station to allow stimulation

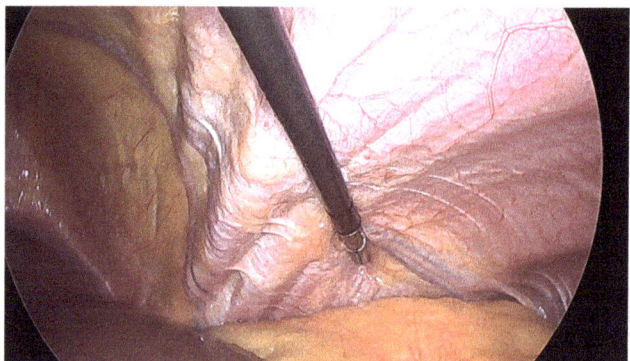

Figure 10.2   A burst stimulation was provided through the dissector and complete contraction of the diaphragm is seen confirming the correct location to implant an electrode which is lateral to the pericardium on the left diaphragm

The DP intramuscular electrode is a double helix wound lead with exposed 316LVM stainless steel stimulating surface and polypropylene reinforced core. The electrode also has a barb at the implanted end composed of 14 pieces of polypropylene suture fused together. The body of the lead is insulated with PFA (perfluoroalkoxy) fluoropolymer coating and terminated in a 316L stainless steel pin with a silicon reinforcing sleeve. An intramuscular electrode is introduced into the abdominal cavity with the electrode delivery instrument. The electrode is inserted into the diaphragm at an angle, so that the electrode lead travels parallel to the plane of the diaphragm before exit, and the delivery instrument is withdrawn (Figure 10.3).

The electrode is then tested to ensure the desired response to twitch stimuli is achieved, and the procedure is repeated for the remaining electrodes. A second electrode is implanted at the site previously marked during mapping (Figure 10.4).

These electrode from each hemi-diaphragm exit from the abdominal cavity through the epigastric port site and are tunnelled subcutaneously to a common exit site in the chest. An additional indifferent electrode is placed subcutaneously in the lower chest or upper abdomen through a separate percutaneous exit site. An external pulse generator is connected to the leads at the percutaneous exit site and delivers the stimulus pulses with a respiratory rate (Figure 10.5).

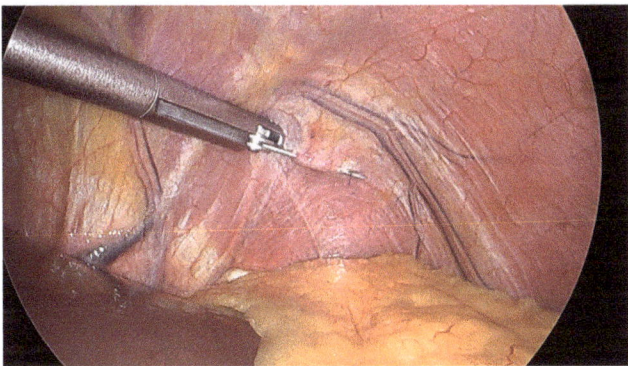

Figure 10.3   The laparoscopic electrode instrument with implantation needle, housing the electrode, is placed in the left diaphragm muscle. Countertraction applied with another laparoscopic instrument helps to deploy the electrode, which has a small barb, into the diaphragm muscles

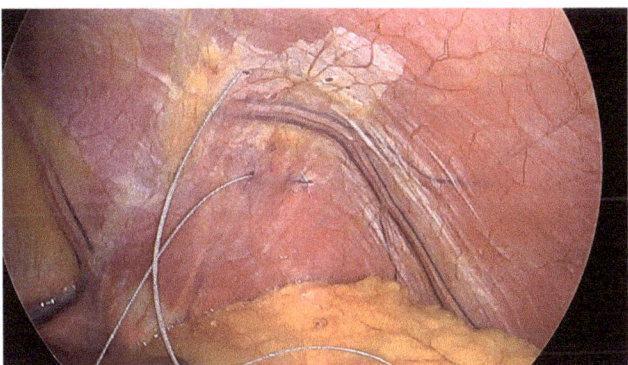

Figure 10.4   A second electrode has been implanted on the left diaphragm and will exit the abdomen through the epigastric port

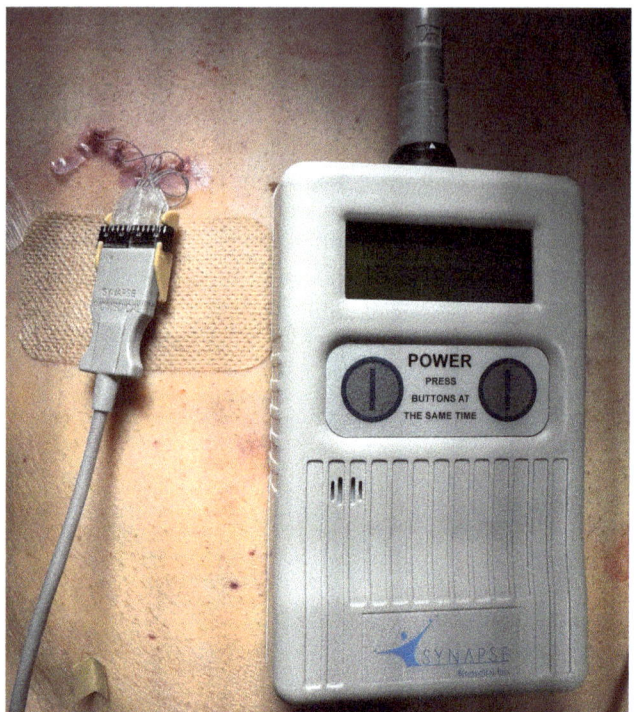

Figure 10.5   *The four implanted electrodes are tunnelled from the subxiphod port to*
an exit site where a subcutaneously inserted ground electrode is placed.
The electrodes are connected to a block which a cable then connects to
the external pulse generator. This is programmed to maximise
ventilation and provide diaphragm conditioning. This is an acutely
injured spinal cord injured patient in which stimulation starts in the
operating room to facilitate ventilator weaning post-operatively

Electrode evaluation is performed by adjusting individual stimulus parameters
(pulse amplitude, width, rate and frequency) for each electrode so that a
comfortable level of stimulation is identified for the initial diaphragm conditioning
sessions. The DP external pulse generator will be programmed to provide a tidal
volume that provides 15% above the basal needs (5–7cc/Kg) which is also tolerated
by the patient. The settings will always be below 25 in amplitude, below 20 in
frequency and below 200 in pulse width. The pulse generator provides a capaci-
tively coupled, charge balanced, biphasic stimulation to each electrode with a
common indifferent electrode that was placed in the subcutaneous tissue.
Essentially, DP electrically stimulates the intact diaphragm lower motor units
causing contraction for ventilation and rehabilitation of the diaphragm. DP over-
comes atrophy of the diaphragm and converts the diaphragm to slow twitch muscle
fibre type.

With DP, there are two weaning modalities. For acute, in-hospital patients, weaning is immediately initiated with continuous utilisation of DP with pressure support weaning. This is the usual fashion when DP is used soon after SCI which is the recommendation presently. For other patients who may have been chronically on a ventilator at home, a home-based conditioning program is employed wherein the DP is turned on and the ventilator is turned off. The ventilator is turned back on if/when the patient feels uncomfortable, or the pulse oximeter starts to drop. Due to disuse atrophy and the conversion of muscle fibres to fast-fatigueable type during periods of inactivity, patients who have been on long term MV will require conditioning of the diaphragm muscle to sustain ventilation. The diaphragm can recover quite rapidly from training, so that patients and their caregivers can repeat a session every hour. The length of time required to achieve more than 4 continuous hours on DPS stimulation depends on the amount of time the patient and caregivers devote to this process. The goal initially is daytime ventilation and then eventually full-time natural ventilation with DP.

To address the growing problem of prolonged MV in the ICU, the group in Cleveland designed temporary DP electrode with a temporary external impulse pulse generator and a human trial was completed. This was a prospective FDA study (IDE #G150040), IRB-approved, and listed on clinicaltrials.gov (NCT 02410798) that evaluated the feasibility of temporary diaphragm electrodes to provide ventilation through stimulation. At the end of the subject's primary surgery, two temporary DP electrodes were placed intramuscularly in each hemidiaphragm at the expected motor point where, with stimulation, diffuse diaphragm contraction would occur because of proximity to the phrenic nerve. No diaphragm mapping was done which is a component of the permanent electrode implantation. The electrodes exited the chest or peritoneal spaces laterally without tunnelling to a central location which allows easier eventual removal. The electrodes are attached to the external pulse generator at the bedside immediately to begin diaphragm conditioning.

This study included 8 males and 4 females with multiple comorbidities underwent three different surgical approaches: 4 median sternotomy, 4 laparoscopy and 4 laparotomy. Subjects had multiple comorbidities. Electrode stimulation in all patients exceeded ideal tidal volumes by an average of 37% (0–95%). This study confirmed that mapping the diaphragm in non SCI patients would not be necessary to adequately provide ventilation. A daily electromyogram was obtained to analyse respiratory function and confirm stability of electrode placement until removal of the temporary DP electrodes.

This study confirmed that temporary DP electrodes could be used throughout a patient's hospital course maintaining diaphragm strength and preventing diaphragm atrophy. There were no complications with electrode placement and all 48 study electrodes remained in place until complete removal of all 48 electrodes at bedside prior to discharge. This trial demonstrates the ease of placement, functionality, removal, and safety of temporary DP electrodes [14].

The diaphragm electrodes are also utilised to monitor and guide therapy through diaphragm electromyography (dEMG). The implanted electrodes do not change position allowing consistency in the recordings regardless of when the recordings are obtained. This allows dEMG to be evaluated during normal automatic breathing, with maximum inspirations, during sleep, and with positive pressure ventilation regardless of mode (invasive and or non-invasive MV). These recording allow the identification of the significant positive effects of DP on respiratory neuroplasticity.

This remainder of this chapter will discuss the clinical utility of intramuscular DP in three critical areas: spinal cord injured patients, injured or dysfunctional phrenic nerves and temporary DP in the ICU.

10.3   Current experience in spinal cord injured patients

In the cervical SCI population, 51% of all patients are discharged on MV with 4% of cervical SCI patients needing long-term MV. Respiratory complications are the leading cause of death in SCI, with pneumonia being the leading cause of death in those on MV [15]. Patients on chronic MV have significantly shorter life spans compared to those with the same level of injury who are not on MV. Long term MV is associated leads to increased anxiety for patients and their caregiver. It alters speech patterns, decreases sense of smell, adds bulk and weight to wheelchairs, impedes mobility, and noise and tubing attracts unwanted attention. The need for a tracheostomy increases secretions and may cause tracheomalacia.

DP was first implanted in a SCI patient in 2000. A 2022 publication of the results of the multi-centre initial FDA IDE study showed 96.2% of SCI patients dependent on tracheostomy MV could use DP to support ventilation at least four hours a day [16]. They also report a meta-analysis cohort of 196 patients exhibited a similar result of 92.2% of patients successfully using DP a minimum of four hours. Fifty percent were able to replace MV full time. The patients ranged in age from 18 years to 74 years (36 years old average). The report included 37 males with most injuries resulting from motor vehicle accidents followed by sports injuries. Patients were on PMV from 3 months to 27 years prior to DP implant with the average time of injury to implant being 5.6 years. There is no upper time limit for successful DP as long as there is an intact phrenic motor neuron, phrenic nerve and diaphragm muscle. The longer the time from SCI to surgery the greater amount of time it will take to rehabilitate the diaphragm and the lungs.

When comparing the monthly cost of maintaining a patient at home with a portable ventilator including the cost of long-term equipment replacement/rental, medical, need for continuous electrical supply with a generator and nursing care, DP is cost effective. Onders *et al.* describes the cost savings of $13,000 monthly for one SCI patient who was successfully weaned off the ventilator to full-time pacing [11]. As shown during multiple natural disasters, the lack of electricity negatively affects patients on MV, because the batteries of a MV last only several hours, whereas the loss of electricity and availability of reliable generators can last months, while the battery of DP external unit lasts a month and can be easily replaced [17]. DP provides natural negative pressure ventilation which aerates the posterior lobes of the lungs and increases respiratory compliance and therefore decreases pneumonia rate in the SCI patient population. Hirshfield *et al.* analysed 64 spinal cord patients with chronic respiratory insufficiency in whom 32 were able to receive a pacer and 32 who did not [18]. Pacing the diaphragm and allowing negative pressure ventilation decreased respiratory infections from 2 per 100 days to 0 with pacing (p<0.001). In an additional report, an analysis looked at the quality of life of patients with pacing compared to when they were on MV and all patients would recommend pacing over MV to other patients [19]. This report stated the diaphragm stimulation as opposed to MV improved patients' ability to go outside of their home, participate in leisure activities and relationships with others. They also report a significant improvement in olfaction and taste with the use of pacing. Patients who use DP as their primary mode of respiration report improved sense of smell, decrease in secretions and suctioning needs, and increased sense of freedom and independence.

SCI patients also have a significant risk for cardiac bradycardia requiring cardiac pacemakers, since there is a close proximity of the diaphragm muscle to the heart, the question of cardiac capture through an intramuscular electrode placed in the diaphragm needed to be addressed. A multicentre report focused on 20 SCI patients who had a permanent internal cardiac pacemaker and also had DP was completed in 2010 [20]. All patients in this study were able to achieve tidal volumes to meet their basic metabolic needs with 71% able to replace MV with DP full time. The internal cardiac pacemakers were interrogated at the time of DP implant with DP being set at maximal stimulations settings and the cardiac pacemakers being set at their most sensitive. No device interactions were noted. More recently there has been no interactions reported with cardiac defibrillators or left ventricular assist devices (LVAD) [21,22].

Recently there has been more focus on early DP after SCI with an initial report in 2014 [23]. Their analysis included 29 patients: 22 of whom were implanted and 7 patients had denervated non-stimulatable diaphragms at surgery. These diaphragms could not be stimulated because of complete destruction of the LMN from the trauma insult, which occurs more commonly in gunshot wounds (GSW) because of the blast injury. The average time frame of injury to implant was 3–112 days with a median of 33 days. Seventeen percent of patients were weaned completely off MV in an average of 13.1 days. The earlier DP was done allowed a faster wean as shown in a subset of patients implanted within 11 days of injury weaning off MV in 5.7 days. Some patients (36%) implanted early after injury had recovery of respiration and were able to wean off DP. The ability to record dEMG in this SCI population highlighted the potential of electrical stimulation from DP and neuroplasticity of the spinal cord allowing recovery of control of respiration. Also noteworthy was the fact that early identification of those patients with denervated non salvageable diaphragms would save significant amounts of time, frustration and money on unsuccessful futile ventilator weaning. The knowledge of a complete loss of phrenic nerve function would allow early consideration of the growing use of nerve transfer techniques to allow eventual recovery of diaphragm function.

During the COVID-19 pandemic there was a significant fear among SCI patients of being denied access to MV if hospitalised [24]. There were also reports of increased risks of COVID-19 infections because of decreases pulmonary reserve in SCI patients along with a higher mortality rate [25,26]. The management of tracheostomies during the COVID-19 pandemic both during insertion since it is an aerosol generating procedure and long term care is a risk for health care workers requiring increased personal protective equipment [27]. The group from Cleveland addressed this by aggressively implanting DP early in SCI patients on MV to prevent even the need for tracheostomy [28]. In patients with intact bilateral diaphragms 44% were able to be weaned from MV without a tracheostomy. Subsequent decannulation of tracheostomy occurred in further patients so that only 22% required long term tracheostomy. Hospital length of stay was significantly shorter for patient not requiring tracheostomy at 19 days versus 31 days. Being free from the ventilator also allowed earlier and more access to further SCI rehabilitation.

The increased publications outlining the success of DP in SCI recently lead the American College of Surgeons trauma quality program in March of 2022 to update their best practices guidelines for spine injury and listed, as a key point to consider, stimulation of the diaphragm to become ventilator free [29]. The largest single site study to date assessing DP for SCI of any aetiology is a retrospective study in 2018 in which 92 patients' outcomes were reviewed [30]. The mechanism of injury included: motor vehicle accident (MVA) 44, sports 22, falls 12, GSW 7, crush injury 3, traumatic forceps delivery 2, assault 1 and electrocution 1. The age at time of injury ranged from birth to 74 years old with the average of 27.3 years and median age of 23 years. 74 of the injured were males (80%). Time on MV prior to DP was an average of 47.5 months (6 days to 25 years with median of 1.58 years). A total of 88% of patients achieved four consecutive hours of pacing (81 patients). Fifty-six patients (60.8%) utilised DP full time 24 hours a day with no MV. A subgroup of 33 patients that were implanted in the first year after their injury had 72.7% success in being removed from the ventilator 24 hours a day (24 out of 33). Patients implanted after two years only had a 51% success in being able to be free of the ventilator 24 hours a day with DP (22 out of 43).

A total of nine patients in this report had ventilator dependence with SCI from GSW. Two were excluded from implantation on initial laparoscopy since their diaphragms were not stimulatable. Two never weaned and one was only able to go off the ventilator less than four hours a day. Four were able to use DP 24 hours a day. Overall for GSW there was only a 44% success rate. With acts of violence primarily from GSW increasing in the USA up to 13.5% of all SCI, it is unfortunate that the result of DP only approaches 44%. This appears to be due to blast affect and the infarction of the phrenic motor neurons causing loss of the diaphragm motor units so there cannot be stimulation.

A 2018 report showed that early mortality rate decreased from 15% to 3% with the use of DP early in the hospital course for SCI [31]. The length of hospitalisation was also decreased in the DP group (65+/−61 vs 43+/−24 days for the control and DP groups, respectively, p=0.03). Kerwin *et al.* reported statistically significant saving of $144,444 (P=.003) in average hospital charges with the use of DP compared to a propensity matched group. They also report a more rapid wean from MV of 10.1 days as opposed to 29.2 days (p<0.001) [32]. A multicentre European registry encompassing 34 patients concluded that DP is a good alternative to MV with 77% achieving at least 4 hours of freedom from MV [33].

DP has been implanted successfully in paediatric SCI patients. An initial report of six patients ranging in age 3–17 (average 9 years old) with an average weight of 32.6 kg were successfully implanted with no technical difficulties [34]. In a more recent report, 14 paediatric patients were successful with greater than four hours of DP without MV [30]. Their average time on MV was 35.5 weeks (range of 0–90 weeks). Average age at implant was 9.4 years (range 1–17 years). Eight of them able to completely wean from all MV (57%). Three patients either had no tracheostomy or were decannulated. Two patients had completer recovery of breathing and had DP electrodes removed. Unique to the paediatric population is growth that may require DP reprogramming and scoliosis development from SCI. Scoliosis can affect respiration and ventilator weaning and may need to be addressed prior to DP implantation. Overall, paediatric patients experienced the same success with being liberated from MV with DP as their adult counterparts.

10.4   Current experience in phrenic nerve injury/ diaphragm dysfunction

Phrenic nerve injury/diaphragm dysfunction is an under recognised source of dyspnoea in patients. Even with unilateral diaphragm paralysis, patients can be extremely symptomatic to the point of requiring MV. These patients may have phrenic nerve paresis or complete paralysis from numerous aetiologies including open heart surgeries, shoulder surgery, manipulation of cervical spine, lung transplants, mediastinal tumours, COVID-19 pandemic and idiopathic. During cardiac surgery, damage to the phrenic nerve can be from transection, stretching, crushing or hypothermia occurring from 7.6% to 38% of cases. When there is phrenic nerve injury after cardiac surgery there is higher rate of difficulty weaning from MV(32% versus 5%, p<0.001), lower rate of extubation at 24 hours (50% versus 92%, p<0.001), a higher post-operative rate of pneumonias (23.9% versus 8.7%, p<0.0001), higher reintubation rate (8.8% versus 2.9%, p<0.0001) and higher tracheostomy rate (3.3% versus 0.3%, p<0.0001) [35,36]. Phrenic nerve injury leading to diaphragm dysfunction also occurs commonly in lung transplants. When this occurs, there is a much higher rate of delay in extubation, higher re-intubations rates and higher rate of pneumonias [37–40]. Phrenic nerve injury is a notable complication associated with heart transplant and LVAD, and are susceptible to damage due to their anatomical proximity to the heart and major blood vessels [31,41]. Abnormalities of diagnostic tests to identify diaphragm dysfunction include decreased vital capacity, decline in supine vital capacity, decreased maximal inspiratory pressure, decreased movement on fluoroscopy and no thickening of diaphragm during inspiration when assessed with ultrasound [42].

DP has been utilised in patients with unilateral or bilateral diaphragm dysfunction from phrenic nerve injury or dysfunction. The initial report was in 2014 on the extended use of DP in patients with diaphragm dysfunction leading to symptomatic hypoventilation [43]. In this study 21 patients with a mean of 36 months of respiratory symptoms were implanted with DP. The most common cause of diaphragm dysfunction in this report was a result of phrenic nerve injury from thoracic surgery (4), shoulder surgery or neck trauma (5), idiopathic (9), spinal muscle atrophy (SMA) (1), Charcot Marie Tooth disease (1) and one diaphragm flutter (1). Thirteen patients (62%) had clinically relevant respiratory improvements, and four had partial improvement. Four patients were able to be completely weaned from MV. Hejal *et al.* in 2020 summarised the use of DP in unilateral diaphragm paralysis in 62 patients [44]. Aetiologies of the paralysis codified into 7 groups: idiopathic (27), chest/thoracic procedure (13), neck/shoulder surgery/injection (8), trauma (4), infection (3), chemo/radiation (3), Parsonage Turner's syndrome (3) and post liver transplant (1). The average age was 60 ± 10 years. All patients had significant respiratory symptoms and abnormal chest radiographs. Symptom duration was greater than six months in 49 patients (79%). Post implantation, 44 (70.9%) of patients reported improvement in symptom and 46 (74%) had improvement in dEMG. The earlier the implantation the greater the chance of success. In patients with success the DP electrodes can be easily removed. The conclusion is that DP should be offered to patients to allow complete recovery of a normal functioning diaphragm before being considered for a diaphragm plication which was the only previous therapy.

DP pacing has shown to help in recovery of phrenic nerve injuries after cardiac surgery with the initial report in at the Society of Thoracic Surgery annual meeting in 2018. Ten cardiac surgery patients with injured phrenic nerves with prolonged ventilator support were implanted. Ages ranged from 57 to 81 years old. Mean duration of positive pressure ventilation prior to DP was 53 days (range 5−148) (7 patients received invasive MV). All patients were successfully weaned with DP in an average of 16 days (1−35). All six tracheostomy patients decannulated. The average survival is 32 months (2−77 months). Serial dEMGs showed diaphragm recovery in nine out of ten patients. This same group then reported on a larger group of 31 patients who had injured phrenic nerves with diaphragm dysfunction from 2011 to 2022 [45]. All 31 patients had evidence of diaphragm dysfunction identified by chest X-ray or fluoroscopy that was not present pre-operatively confirming operative phrenic nerve injury. Of this, 15 were still on MV (48%) with 13 having tracheostomies, and 8 were dependent on NIV (26%). Indications for index cardiac procedure include 20 CABG or valve or combined cases, six heart transplants, three LVAD and two recurrent atrial myxoma resections. Time from initial cardiac surgery to DP implantation averaged 208 days with range of 5–755 days. Post-operative dEMG data confirmed minimal to no activity on the affected side and was basis for follow up recovery along with radiologic imaging. Twenty-four patients had demonstrated complete long-term diaphragm and phrenic nerve improvement (77.4%). There were no implantation complications and device related morbidity.

DP has a role in injured phrenic nerves during heart transplants or LVAD implantation. Eight patients had DP system implanted for identified phrenic nerve injuries (five after heart transplant and three from LVAD) [22]. All patients were dependent on NIV or MV. Average time from initial event to DP was 231 days (19–738). 87.5% (7/8) had recovery of phrenic nerve dysfunction. Two patients who were tracheostomy dependent on MV were weaned with tracheostomy decannulation.

A recent prospective publication identified 4.1% patients with an elevated diaphragm post COVID-19 pneumonia secondary to neuropathy [46]. Phrenic nerve dysfunction and diaphragm weakness may contribute to post-acute sequela of COVID-19 leading to breathlessness. DP was safely used in five patients and improved recovery of the involved phrenic nerves and diaphragms [47]. Phrenic nerve dysfunction should be identified prior to lung transplant for possible rehabilitation prior to transplant which is important in that 7% of lung transplants were performed for respiratory failure secondary to COVID-19 in 2021 [48]. Functioning diaphragms are critical post lung transplant.

10.5 Temporary diaphragm pacing in intensive care units overcoming ventilator induced diaphragm dysfunction

The previous sections showed how DP can significantly improve ventilation in SCI patients and in patients with phrenic nerve injury. There is a much larger group of patients who require prolonged temporary use of MV. These patients, often referred to as 'failure to wean' (FTW), present a significant physiologic burden to themselves and economic burden to the health care system. Up to 50% of ICU patients require invasive MV with 20% still requiring a ventilator for over seven days. Over 40% of this time is spent rehabilitating the diaphragm and weaning a patient after the initial intubation to be placed on the ventilator. There are multiple aetiologies contributing to FTW resulting in long-term MV which includes heart failure, primary pulmonary disease and critical illness neuropathy. In addition to medical aetiologies, MV has its own deleterious effects to the diaphragm. Positive pressure MV leads to inactivity of the diaphragm muscle which leads to the significant atrophy of the muscle. Since the diaphragm is the primary inspiratory muscle, VIDD is widely recognised as a major contributing factor to FTW.

Diaphragm weakness is a common occurrence among critically ill individuals, often manifesting both prior to admission to the ICU and during their stay therein. Numerous factors contribute to this weakness, with notable ones including sepsis and the duration and methods of MV [49]. This critical illness-induced diaphragm weakness is consistently linked to unfavourable outcomes, including increased ICU mortality, challenges in the weaning process, and the necessity for prolonged MV [50,51]. These individuals consume a significantly higher share of healthcare resources and tend to experience suboptimal long-term results [52–54]. They often describe a reduced quality of life, face significant functional and cognitive constraints, necessitate prolonged informal caregiving, and exhibit elevated one-year mortality rates [55–57]. It is anticipated that the annual count of patients requiring prolonged MV will surpass 600,000 within the next ten years, with corresponding hospital expenditures ranging between $50 billion and $60 billion [52].

In 2016 was one of the first reports of the DP system in a series of FTW patients that were not SCI patients. This was a retrospective review of compassionate, off-label use of an FDA-approved device under IRB approval [21]. Immediately after implantation the DP system was used to drive ventilation, with subsequent weaning from MV. Ten patients were implanted laparoscopically with no complications. Mean duration of positive pressure MV prior to intervention was 44 days (range 4–148 days). All ten were successfully weaned. Mean time to completely wean from MV was 15 days (range 1–35). All tracheostomy patients were decannulated. The conclusion of this initial report is that DP could be a possible therapy to treat FTW allowing earlier extubation.

The FDA recognised the need to decrease the burden of MV during the COVID-19 pandemic and authorised under an emergency use authorisation (EUA) a temporary DP system since DP may decrease weaning times by 64% in other patient groups. At the 19th Annual STS Perioperative and Critical Care Conference in 2022 the group from Cleveland reported on 21 patients utilising these temporary electrodes that were placed at the time of high risk of respiratory failure patients during cardiac surgery. The risk factors that predict prolonged MV included: prior open cardiac surgery, LVEF less than 30%, COPD, history of CVA, elevated diaphragm on radiologic evaluation or pre-operative intra-aortic balloon pump. Prior to sternotomy closure, the pleural space is opened, two electrodes are placed in each hemi- diaphragm, and the wires are tunnelled percutaneously. Once in the ICU DP ensued using a disposable multi-channel stimulator generating a fused diaphragm contraction preventing diaphragm atrophy and VIDD. 90% (19/21) of these patients were extubated in less than 24 hours with 67% extubated in less than 12 hours. They also analysed patients with pre-operative elevated unilateral diaphragm problems and 93% (13/14) of the patients were rapidly extubated with DP overcoming the known paradoxical movement of the diaphragm and associated sleep dysfunction. Immediate DP post high risk cardiac surgery can prevent VIDD.

In 2023, Chung *et al.* aimed to assess the applicability of DP technology in reducing prolonged MV and other respiratory complications following extensive aortic reconstructive surgery [58]. They conducted a retrospective review of patients who underwent temporary DP post-surgery between 2019 and 2022 at their single centre. Fourteen patients, identified as high-risk for prolonged MV based on their smoking and respiratory history, underwent intra-muscular temporary DP following extensive aortic repair with immediate diaphragm stimulation on admission to the ICU. The primary endpoint was the incidence of prolonged MV, with secondary endpoints including improvements in dEMG, length of hospitalisation, duration of ICU stays, and reintubation rates. Results indicated a mean total duration of postoperative MV of 31.9 hours, with an average pacing duration of 4.4 days. Two patients required prolonged MV, with an average duration of 75.4 hours, while two patients required reintubation. Notably, no complications related to DP wire placement or removal were reported. The study concluded that DP is both safe and feasible for patients at high risk of pulmonary insufficiency following extensive aortic reconstructive surgery.

Onders *et al.* conducted an initial retrospective analysis examining the application of DP in lung transplantation, aiming to address diaphragm dysfunction and mitigate VIDD alongside phrenic neuropathy-associated morbidity [59]. Utilising both chronic and temporary DP systems, eight patients underwent DP without device-related adverse events. Among them, three recipients with DD post-transplant received chronic DP and one pre lung transplant, exhibiting subsequent diaphragm function recovery. Additionally, temporary DP electrodes were implanted at the time of lung transplant in four recipients. Notably, DP facilitated recovery of phrenic nerve injury and prevented VIDD. This group recently reported on their much larger group of 30 lung transplant patients implanted with the temporary DP system. They identified 70% of lung transplant patients with phrenic/diaphragm dysfunction. The DP stimulation allowed phrenic nerve rehabilitation with an 80% diaphragm/phrenic nerve recovery rate [60]. The findings suggest that DP warrants consideration in lung transplantation to optimise outcomes by identifying and improving phrenic nerve function.

The Cleveland group has focused on addressing the prolonged MV (MV greater than 24 hours) after cardiac surgery since that significantly increases patient morbidity and mortality [61]. In the initial pilot trial in 32 high risk cardiac patients the diaphragm electrodes were safely placed in all patients. Diaphragm stimulation was started on the ten patients who required MV at 24 hours. The median percent of time on MV in the first 120 hours post op was 35.7% in stimulated group versus 80% in a non-stimulated group [62]. This led to a recently completed randomised trial of 40 high risk cardiac patients who all received implantation of diaphragm electrodes. The patients were then randomised to immediate diaphragm stimulation (Tx) on arrival to the ICU or standard of care (SOC). The Tx group had a 71% relative risk reduction of requiring MV at 24 hours post-operatively. Predictive enrichment strategy was utilised to identify patients most likely to respond to DP therapy. In this analysis, the median time of MV for the 15 SOC patients was 17.7 hours (IQR 8.3–23.4), while the MV time for the 13 Tx patients was 9.4 hours (IQR 7.14–12.5). This showed an improvement in MV time of eight hours with the utilisation of DP ($p<0.05$). In another subgroup analysis of 13 patients with COPD, 7 SOC patients had a median MV time of 23.4 hours (IQR 8–41) hours in comparison to the 6 Tx patients who had a median MV time of 8.2 hours (IQR 4–13), showing a 15-hour improvement in extubation times utilising DP [63].

10.6   Conclusion

DP has a profound positive effect in helping SCI patients replace their MV with natural negative pressure ventilation. Early implantation of DP has substantial benefits and no known drawbacks. All SCI patients should be offered DP. Phrenic nerve dysfunction can significantly affect a patient's respiratory function and DP helps in the recovery of the phrenic nerve injury with significant improvement in the respiratory status of patients. Prolonged MV in the ICU is a significant and growing healthcare problem. The use of temporary DP can have a significant positive effect in decreasing the length of time on MV that would change the paradigm of therapy in ICUs.

References

[1] Onders RP Stimulation for inspiration. In: Krames S, Peckham H, and Resai AR (eds.) *Neuromodulation. 2nd edn.* Amsterdam: Elsevier; 2018. pp. 1347–1354.

[2] Levine S, Nguyen T, Taylor N, *et al.* Rapid disuse atrophy of diaphragm fibers in mechanically ventilated humans. *N Engl J Med* 2008;35:1327–35.

[3] Whittenberger JL, Sarnoff SJ, and Hardenbergh E Electrophrenic respiration: its use in man. *J Clin Invest* 1949;28:124–28.

[4] Glenn WW Diaphragm pacing: present status. *Pacing-Clin-Electrophysiol* 1978;1:357–70.

[5] Nochomovitz ML, Hopkins LM, Broadkey J, Montenegro H, Mortimer JT, and Cherniak, NS Conditioning of the diaphragm with phrenic nerve stimulation after prolonged disuse. *Am-Rev-Respir-Dis* 1984;130(4):685–8.

[6] Nochomovitz M, Peterson DK, and Stellato TA Electrical activation of the diaphragm. *Clin-Chest-Med* 1988;9:349–58.

[7] Peterson DK, Nochomovitz ML, Stellato TA, and Mortimer JT Long-term intramuscular electrical activation of phrenic nerve: efficacy as a ventilatory prosthesis. *IEEE Trans Biomed Eng* 1994;41:1127–1135.

[8] Dimarco AF, Onders RP, Kowalski KE, Miller ME, Ferek S, and Mortimer JT Phrenic nerve pacing in a tetraplegic patient via intramuscular diaphragm electrodes. *Am J Respir Crit Care Med* 2002;166:1604–1606.

[9] Onders RP, Aiyar H, and Mortimer JT Characterization of the human diaphragm muscle with respect to the phrenic nerve motor points for diaphragmatic pacing. *Am Surg* 2004;70:241–247.

[10] Onders RP, Ignagni AI, Aiyer H, and Mortimer JT Mapping the phrenic nerve motor point: the key to a successful laparoscopic diaphragm pacing system in the first human series. *Surgery* 2004;136:819–826.

[11] Onders RP, Ignagni AI, DiMarco AF, and Mortimer JT The learning curve of investigational surgery: lessons learned from the first series of laparoscopic diaphragm pacing for chronic ventilator dependence. *Surg Endosc* 2005; 19:633–7.

[12] Alshekhlee A, Onders R, Syed T, Elmo M, and Katirji B Phrenic nerve conduction studies in spinal cord injury: applications for diaphragmatic pacing. *Muscle Nerve* 2008;36:1546–52.

[13] Onders RP, Elmo M, Stepien C, and Katirji B Spinal cord injury level and phrenic nerve conduction studies do not predict diaphragm pacing success or failure - all patients should undergo diagnostic laparoscopy. *Am J Surg* 2021; 221(3):585–588.

[14] Onders RP, Markowitz A, Ho VP, *et al.* Completed FDA feasibility trial of surgically placed temporary diaphragm pacing electrodes: a promising option to prevent and treat respiratory failure. *Am J Surg* 2018;215(3):518–521.

[15] Onders R. Recovery of diaphragm function through functional electrical stimulation: diaphragm pacing. In: Elkwood A, Kaufman M, and Schneider L (eds.) *Rehabilitative Surgery* Berlin: Springer; 2017

[16] Onders RP, Khansarinia S, Ingvarsson PE, *et al.* Diaphragm pacing in spinal cord injury can significantly decrease mechanical ventilation in multicenter prospective evaluation. *Artif Organs* 2022;46(10):1980–1987.

[17] Onders RP, Elmo MJ, and Ignagni AR. The hurricanes of 2005 and ventilators: the benefits of diaphragm pacing stimulation (DPS) system. *J Spinal Cord Med* 2006;29:315.

[18] Hirchfield S, Exner G, Luukkaala T, and Baer GA Mechanic ventilation or phrenic nerve stimulation for treatment of spinal cord-induced respiratory insufficiency. *Spinal Cord* 2008;46:738–742.

[19] Adler D, Gonzalez-Bermejo J, Duguet A, *et al.* Diaphragm pacing restores olfaction in tetraplegia. *Eur Respir J* 2009;34:365–370.

[20] Onders RP, Khansarinia S, Weiser T, *et al.* Multi-center analysis of diaphragm pacing in tetraplegic with cardiac pacemakers: positive implications for ventilator weaning in intensive care units. *Surgery* 2010;148:893–7.

[21] Onders R, Elmo M, Kaplan C, Nearman H, and Schilz R Diaphragm pacing as successful salvage therapy in failure to wean patients: should this novel technology be used earlier? *Am J Respir Crit Care Med* 2016;193:7653.

[22] Onders R, Elmo M, Carl N, *et al.* Diaphragm dysfunction from phrenic nerve injuries during LVAD or heart transplants: positive role of diaphragm pacing. *J Heart Lung Transplant* 2024;43(4):S254.

[23] Posluszny JA, Onders R, Kerwin AJ, *et al.* Multicenter review of diaphragm pacing in spinal cord injury: successful not only in weaning from ventilators but also in bridging to independent respiration. *J Trauma Acute Care Surg* 2014;76:303–310.

[24] Monden KR, Andrews E, Pilarksi C, Hearn J, Wudlick R, and Morse LR COVID-19 and the spinal cord injury community: concerns about medical rationing and social isolation. *Rehabil Psychol* 2021;66:373–379.

[25] Burns SP, Eberhart AC, Sippel JL, Wilson GM, and Evans CT Case-fatality with coronovirus disease 2019 (COVID 19) in United States veterans with spinal cord injuries and disorders. *Spinal Cord* 2020;58:1–2.

[26] Hoogenes B, Queree M, Townson A, Willms R, and Eng JJ COVID-19 and spinal cord injury: clinical presentation, clinical course and clinical outcomes: a rapid systematic review. *J Neurotrauma* 2021;38:1242–1250.

[27] Onders RP, Elmo M, Young B, and Tinkoff G Observational study of early diaphragm pacing in cervical spinal cord injured patients to decrease mechanical ventilation during the COVID-19 pandemic. *Surgery* 2023;173(3):870–875.

[28] Meister KD, Pandian V, Hillel AT, *et al.* Multidisciplinary safety recommendations after tracheostomy during COVID-19 pandemic: state of the art review. *Otolaryngol Head Neck Surg* 2021;164(5):984–1000.

[29]　American College of Surgeons *Trauma Quality Programs Best Practice Guidelines Spine Injury*. Available from: https://www.facs.org/-/media/files/quality-programs/trauma/tqip/spine_injury_guidelines.ashx [Accessed May 2024].

[30]　Onders RP, Elmo MJ, Kaplan C, Schilz R, Katirji B, and Tinkoff G Long-term experience with diaphragm pacing for traumatic spinal cord injury: early implantation should be considered. *Surgery* 2018;164(4):705–711.

[31]　Kerwin A, Yorkgitis B, Ebler D, Madbak F, Hsu A, and Crandall M Use of diaphragm pacing in the management of acute cervical spinal cord injury. *J Trauma Acute Care Surg* 2018;85(5):928–31.

[32]　Kerwin AJ, Diaz Zuniga Y, Yorkgitis BK, *et al.* Diaphragm pacing decreases hospital charges for patients with acute cervical spinal cord injury. *Trauma Surg Acute Care Open* 2020;5(1):e000528.

[33]　Wijkstra PJ, van der Aa H, Hofker HS, *et al.* Diaphragm pacing in patients with spinal cord injury: a European experience. *Respiration* 2022;101(1):18–24.

[34]　Onders RP, Ponsky TA, Elmo MJ, Lidsky K, and Barksdale E First reported experience with intramuscular diaphragm pacing in replacing positive pressure mechanical ventilators in children. *J Pediatr Surg* 2011;46:72–6.

[35]　Laghlam D, Lê MP, Srour A, *et al.* Diaphragm dysfunction after cardiac surgery: reappraisal. *J Cardiothorac Vasc Anesth* 2021;35(11):3241–3247.

[36]　Bruni A, Garofalo E, Pasin L, *et al.* Diaphragmatic dysfunction after elective cardiac surgery: a prospective observational study. *J Cardiothorac Vasc Anesth* 2020;34(12):3336–3344.

[37]　LoMauro A, Righi I, Privitera E, *et al.* The impaired diaphragmatic function after bilateral lung transplantation: a multifactorial longitudinal study. *J Heart Lung Transplant* 2020;39(8):795–804.

[38]　Crothers E, Kennedy DS, Emmanuel S, *et al.* Incidence of early diaphragmatic dysfunction after lung transplantation: results of a prospective observational study. *Clin Transplant* 2021;35(9):e14409.

[39]　Hernández-Hernández MA, Sánchez-Moreno L, Orizaola P, *et al.* A prospective evaluation of phrenic nerve injury after lung transplantation: incidence, risk factors, and analysis of the surgical procedure. *J Heart Lung Transplant* 2022;41(1):50–60.

[40]　Boscolo A, Sella N, Pettenuzzo T, *et al.* Diaphragm dysfunction predicts weaning outcome after bilateral lung transplant. *Anesthesiology* 2024;140(1):126–136.

[41]　Zaleska-Kociecka M, Celińska Spodar M, Paszyń K, *et al.* Diaphragmatic palsy - a common complication after LVAD implantation. *J Heart Lung Transplant* 2024;43:S348.

[42]　McCool FD, and Tzelepis GE Dysfunction of the diaphragm. *N Engl J Med* 2012;366(10):932–42.

[43]　Onders R, Elmo MJ, Kaplan C, Katirji B, and Schilz R Extended use of diaphragm pacing in patients with unilateral of bilateral diaphragm dysfunction: a new therapeutic option. *Surgery* 2014;156:772–86.

[44] Hejal RB, Alonz A, Elmo M, and Onders R Unilateral diaphragm paralysis: the evolving role of diaphragm pacing. *Am J Respir Crit Care Med* 2020;201: A3261.

[45] Onders R, Elmo M, Carl N, *et al. Diaphragm Dysfunction From Phrenic Nerve Injuries During Cardiac Procedures: Positive Role of Diaphragm Pacing to Improve Nerve and Diaphragm Dysfunction.* Beverly, MA: American Association for Thoracic Surgery Program Book; 2023. Available from: https://www.aats.org/resources/abstract.pdf?abstract=1250825 (Accessed May 2024).

[46] Law SM, Scott K, Alkarn A, *et al.* COVID-19 associated phrenic nerve mononeuritis: a case series. *Thorax* 2022;77(8):834–838.

[47] Onders R, Elmo M, Carl N, Hejal R, and Schilz R Diaphragm dysfunction from COVID-19 neuropathy: utilization of diaphragm pacing for recovery. *J Heart Lung Transplant* 2023;42:S316–317.

[48] Roach A, Chikwe J, Catarino P, *et al.* Lung transplantation for Covid-19-related respiratory failure in the United States. *N Engl J Med* 2022;386 (12):1187–1188.

[49] Ambrosino N, and Vitacca M The patient needing prolonged mechanical ventilation: a narrative review. *Multidiscip Respir Med* 2018;13:6.

[50] Dres M, Goligher EC, Heunks LMA, and Brochard LJ Critical illness-associated diaphragm weakness. *Intensive Care Med* 2017;43:1441–52.

[51] Goligher EC, Dres M, Fan E, *et al.* Mechanical ventilation-induced diaphragm atrophy strongly impacts clinical outcomes. *Am J Respir Crit Care Med* 2018;197:204–13.

[52] Zilberberg MD, and Shorr AF Prolonged acute mechanical ventilation and hospital bed utilization in 2020 in the United States: implications for budgets, plant and personnel planning. *BMC Health Serv Res* 2008;8:242.

[53] Cox CE, Carson SS, Govert JA, Chelluri L, and Sanders GD An economic evaluation of prolonged mechanical ventilation. *Crit Care Med* 2007; 35:1918–27.

[54] Nelson JE, Meier DE, Litke A, Natale DA, Siegel RE, and Morrison RS The symptom burden of chronic critical illness. *Crit Care Med* 2004;32:1527–34.

[55] Cox CE, Martinu T, Sathy SJ, *et al.* Expectations and outcomes of prolonged mechanical ventilation. *Crit Care Med* 2009;37:2888–94.

[56] Nelson JE, Tandon N, Mercado AF, Camhi SL, Ely EW, and Morrison RS Brain dysfunction: another burden for the chronically critically ill. *Arch Intern Med* 2006;166:1993–9.

[57] Van Pelt DC, Milbrandt EB, Qin L, *et al.* Informal caregiver burden among survivors of prolonged mechanical ventilation. *Am J Respir Crit Care Med* 2007;175:167–73.

[58] Chung JM, Wogsland AA, Bose S, Schilz R, Onders RP, and Cho JS Temporary diaphragm pacing for patients at risk of prolonged mechanical ventilation after extensive aortic repair. *J Vasc Surg Cases Innov Tech* 2023;9(4):101319.

[59] Onders R, Elgudin Y, Abu-Omar Y, *et al.* Diaphragm pacing in lung transplant patients: to identify and treat diaphragm function abnormalities. *J Heart Lung Transplant* 2021;40(4):S316–S317.

[60] Onders R, Elmo M, Carl N, *et al.* Current utilization of diaphragm pacing in lung transplant patients: identifying and treating phrenic and diaphragm function abnormalities. *J Heart Lung Transplant* 2024;43(4):S344–S345.

[61] Sharma V, Rao V, Manlhiot C, Boruvka A, Fremes S, and Wąsowicz M A derived and validated score to predict prolonged mechanical ventilation in patients undergoing cardiac surgery. *J Thorac Cardiovasc Surg* 2017;153 (1):108–115.

[62] Jarrett C, Onders R, Pelletier M, *et al.* Pilot study of percutaneous temporary diaphragm pacing wires in cardiac surgery patients at risk of prolonged mechanical ventilation during the covid-19 pandemic. *Can J Cardiol* 2021; 37(10):S112–3.

[63] Hungate J, Onders R, El Diasty M, *et al. Randomized Study of Temporary Diaphragm Pacing for Enhanced Recovery After Surgery in Cardiac Surgery Patients at Risk of Prolonged Mechanical Ventilation.* Beverly, MA: American Association for Thoracic Surgery Program Book; 2024. Available from: https://www.aats.org/resources/randomized-study-of-temporary-7360#abstract (Accessed May 2024).

Chapter 11

Practical considerations in running a clinical service

Ian Swain[1], Paul Taylor[2], Christine Singleton[3],
Ines Bersch-Porada[4] and Lisa Brown[5]

11.1 Introduction

The aim of this chapter is to highlight the issues that need to be considered and the facilities that need to be put in place to run a clinical electrical stimulation (ES) service. It is not a definitive, prescriptive list, as it will vary depending upon the country's healthcare system, funding mechanisms, resources available and the patient groups being treated. Rather, it provides guidelines and suggestions and poses a series of questions that need to be addressed to set up a clinical service and ensure its continued success. A checklist of the various factors that need to be considered is provided in Appendix A.

The chapter is based on the authors' experience of establishing clinical services and we seek to show what we learnt and the approaches we used in the research setting, and how we applied research to service provision.

Currently, there is an increasing interest in using ES and functional electrical stimulation (FES) (where the stimulation is applied to facilitate function) in particular, to address the perceived deficit in rehabilitation, as, particularly in the UK, the clinical workforce struggles to implement the recommendations for more rehabilitation post stroke [1,2]. While the new guidelines are specific to adult stroke rehabilitation, it is clear that the principles are equally applicable to the rehabilitation of people with other neurological and musculoskeletal conditions. There is increasing recognition that ES can play an important role in allowing the therapy

[1]Orthopaedic Research Institute, Faculties of Science and Engineering and, Health and Social Care, Bournemouth University, UK
[2]Bournemouth University, Odstock Medical Limited and The National Clinical FES Centre, Salisbury District Hospital, UK
[3]Functional Electrical Stimulation (FES) Clinic, Birmingham Community Healthcare NHS Foundation Trust, UK
[4]International FES Centre®, Swiss Paraplegic Centre Nottwil, Switzerland
[5]Sargent College of Health and Rehabilitation Science, Boston University, USA

team to do more with its resources, enabling their patients to achieve better outcomes, greater independence and quality of life.

While the wider use of ES and FES is to be welcomed, the need to maintain high standards of clinical practice is a concern. To address this, ACPIN (Association of Charted Physiotherapists In Neurology) in the UK has published a set of Clinical Practice Guidelines (CPG) that make recommendations for each stage of the patients journey and how the clinical service should operate [3]. CPGs for the use of AFO and FES post stroke were also developed and endorsed by the American Physical Therapy Association (APTA) and the Academy of Neurologic Physical Therapy (ANPT) to further support clinical decision making and the evidenced-based application of FES across outcomes spanning participation, activity and body structure and function [4].

11.2   Identifying the clinical need

An important parameter to determine is the size of the clinical service that will be provided. In many case the service will be set up as an adjunct to an existing neurorehabilitation service and may be intended to provide the needs of those using the service. However, experience indicates that there may be many patients who are not existing clients of the service, who could benefit from the new service. It is therefore important to consider this too. It is relatively easy to find out the prevalence (the total number of people) and incidence (the number of new cases per year) for the condition you are working with, usually expressed as numbers per 100,000 or 1,000,000 of the general population [5]. However, you will need to ensure that you use numbers applicable to your geographical area as number vary for different populations. For example, stroke has lower prevalence in Mediterranean areas, multiple sclerosis (MS) is more prevalent closer to the poles and spinal cord injury is more prevalent in countries with greater gun crime.

The next step is to find out the total population in your catchment area and hence calculate the likely numbers of people with your chosen condition. However, this will not be the number of people who could benefit from your service; other factors need to be considered. For example, most strokes occur over the age of 75 and older people may be less likely to participate in active rehabilitation. Some people will have co-morbidities that prevent the use of FES. For example, in spinal cord injury it is common for there to be mix of upper and lower motoneuron damage, particularly at the level of the spinal cord injury. Lesions at C4 will often cause denervation of the deltoid muscle, while lesions at C5 will usually cause denervation of the wrist extensors. The extent of denervation varies and can cross several spinal levels. Also, potential service users may prefer other interventions, or may not be interested in or able to afford what you have to offer or be unable to travel to your clinic. The final factor to consider is the proportion of the population, or their gate-keeper referrers, who know about your service and agree that it is the best option. It can take a long time for a new service to become standard care.

11.3   Worked examples for possible dropped foot stimulator users who have multiple sclerosis

The prevalence of MS in the UK is 1568 per one million population [6]. To estimate the number of people who have MS who may choose to use a dropped foot stimulator, the proportion of the MS population who have an expanded disability scale score (EDSS) of between 4 (significant disability but you can walk without an aid for 500 m) and 7 (unable to walk more than 5 m even with an aid) can be calculated [7]. A study by Jacobs, found that 31% of people who have MS have EDSS score in this range, of whom about 50% have a dropped foot that can be treated with FES and may choose to use FES. Hence, we estimate approximately 15% of the MS population may use FES. By taking the total population of your catchment area the likely maximum size of the service can be estimated, although it may take a long time to reach this size. Appendix B attempts to give some idea of the number of people, with a wide range of conditions, who could be suitable for treatment using ES or FES, based on UK figures.

One factor to consider is whether the proposed clinical service will provide short- or long-term treatment. FES used in the acute or subacute medical setting may be used for short-term therapeutic goals while in other clinical settings FES may be a long-term assistive device. This may also be determined by the condition being treated. For example, neurological or musculoskeletal conditions. There can be a fundamental difference between these two populations and the desired affects when using ES and FES. People with neurological conditions often require long-term use of ES in the form of an active orthosis, whereas people with musculoskeletal conditions usually use it for a limited time for a training or therapeutic effect to improve muscle strength and endurance following injury, or before or after surgery. This is an oversimplification, and some people with musculoskeletal conditions will require long-term use, while some people following a stroke will only use it for a short time. Therefore, each person must be individually assessed, and an appropriate treatment plan made. It is also necessary to decide whether the service you are planning will primarily provide long-term care, short-term care or both; as not only does this change the type of equipment required, but also the structure of the service. Although you might be able to operate an outpatient facility with people coming to the clinic, if you have musculoskeletal patients, people with neurological conditions will need to be provided with equipment to take home. This means that factors such as training patients to use the equipment, a repair service, long-term funding, etc, must be considered. For example, if your service is based in the acute / subacute setting, will you be able to refer patients on to another service for long-term FES care, if that is required? All these issues, and many others, will be addressed in more detail later in this chapter.

Next to be considered is how will you access this pool of potential referrals and how and from whom you will receive payment. This will obviously vary from country to country depending on how healthcare services are provided and funded.

However, there are some common issues that are independent of the healthcare system. These include.

How do you make people aware of the service?
Who do you take referrals from?
Do the public have open access to the service?
Is the service directly funded by the patients, is it funded by the local healthcare system, is it funded by insurance companies or is it a mixture of sources?

All will be considered in more detail in this chapter.

11.4   Building a business case

Once you have decided who the main cohort of patients will be, it is necessary to build a case to demonstrate that the necessary expertise exists among the clinical staff to deliver the service and that proposed treatments are based on sound scientific evidence. This is particularly true if you need to convince a local or national healthcare system to fund it, but should also be undertaken to ensure that self-funded patients also receive optimal, appropriate care.

A team-based approach ensures best clinical practice and outcomes for patients, especially when considering the use of FES for individuals with a neurologic diagnosis. In the United States, a common team model that supports effective clinical decision-making and improves the likelihood of insurance coverage includes a physical therapist, a device vendor for the FES units, a physician, and of course the patient. The physical therapist provides expertise in movement analysis and functional retraining supported by FES. A device vendor provides knowledge of the application of the FES unit itself, but also supports justification to insurance providers for coverage of the device. Many insurance companies also require a prescription or the support of a physician for insurance coverage. The composition of the clinical team will vary from country to country. In the UK, Clinical Scientist are State Registered and can provide a lot of clinical as well as the technical input that must be provided by companies in other parts of the world. Lastly, the most important member of the team is the patient. Studies have shown that FES benefits increase and abandonment of devices decreases when the decision-making process centres on the patient and on how FES can help them achieve their goals [8].

It is important to understand how your local healthcare system is structured and funded. This will determine whether the clinical service is nationally, or state funded, funded through insurance or is totally self-funded by the patient.

11.5   Providing evidence of efficacy

Funding the service is considered in more detail in the next section, but irrespective of how the system is funded, it is essential to be clear about how the planned service will be structured and the types of patients will be treated. This requires a good knowledge of the scientific literature and evidence-based national or international clinical guidelines. Although there are many papers published on ES, it is surprisingly difficult to determine which will be most relevant to your proposed service. The significant problem in producing the evidence for your business case is that there is a wide range of treatment and assessment protocols presented in the literature and therefore meta-analysis of clinical trials is often difficult or impossible. Where such reviews are undertaken, they often combine papers quoting different techniques which may not be representative of your chosen intervention. In addition, there is great variation in the condition and its stage of progression being treated. As a result, healthcare funders may conclude that there is no standard ES treatment and hence there is insufficient evidence of clinical efficacy. Therefore, when building your business case, you will need to focus on the specific services you are providing, what type of patients you plan to treat and the treatment regimens you will follow. This case should include data and provide objective comparison with alternative treatments to justify funding. Therefore, be clear what your service is trying to achieve and find the evidence necessary to justify it, using the strongest available evidence. This includes relying on evidence-based CPGs, endorsed by national organisations such as ACPIN [3] or the APTA [4] clinical decision-making tools [9] and randomised control trials whenever possible. An example of some of the information that needs to be included in a business case is given below. The provision of a dropped foot stimulation service is given as an example. In addition, a list of useful guidelines and sources of evidence is provided at the end of this chapter.

11.6 Considering health economics: a brief introduction using the model of a dropped foot service

To develop a successful business case, you will need to have an appreciation of health economics, especially if you are targeting a local/national healthcare provider or an insurance company as your main source of funding. You will also need to consider cost effectiveness and potential savings that your service might generate compared to alternative treatments. Examples of these could be reduction in falls, improvement in quality of life and improved therapist/patient ratios.

Funders of health care will use health economics to decide if an intervention provides sufficient benefit for the cost of providing that service (cost/utility benefit). To compare different interventions used for different conditions a technique known as QALY (Quality Adjusted Life Years) analysis is used [10]. The technique attempts to calculate the gain in quality of life and the period that it is maintained for by using a non-condition specific health related quality of life measure and relates this to the cost of the intervention and related cost savings. A gain of one QALY is equal to one year of perfect health. So, for example, if a treatment prevented death and resulted in one extra year's life with perfect health, the QALY gain would be 1. More likely, an intervention may not extend life expectancy but may improve quality of life, say from a QALY of 0.50–0.65 for the five years giving a gain of 0.15 for each year and total gain of 0.75. However, because the benefit from the treatment is spread over an extended time, a system of discounting is used to slightly reduce the overall gain, typically 3% per year. In this example 3% discounting would reduce the QALY gain over five years to 0.706. If the cost of providing the intervention for one year is £1,000, the cost per QALY in the first year would be £1000 / 0.15 = £6,667 per QALY. The cost of maintaining the use of the device for the next four years may, for example, be £500 a year giving a total cost of £3000 and a cost per QALY of £4247. In the UK, the National Institute for Clinical Excellence (NICE) has used a threshold of £20,000 per QALY for accepting treatments for use in the NHS with an upper threshold for consideration for acceptance of £30,000 per QALY. These thresholds have been in place for some time and in practice are a guide rather than an absolute limit for acceptable cost per QALY.

Various health-related quality of life measures have been used for calculating QALYs but the one most used is the Euro Quol 5 domain, five-level (EQ-5D-5L) or it's earlier three-level version, the EQ-5D-3L [11]. The questionnaire rates the current quality of life in five domains (mobility, self-care, usual activities, pain and discomfort, anxiety and depression) using a five-point scale (no problem, slight, moderate, severe, extreme). From the 5 scores a utility index can be calculated from look-up tables of between 1 (perfect health) to 0 (death). In fact, scores can be derived that are below 0 for health states considered 'worse than death'. The scores were derived from thousands of questionnaires administered to a cross section of the general population and there are different data sets for different populations around the world. It is important to use the data set that relates to the geographical area you are in.

While QALY analysis attempts to provide a framework for comparing different treatments and conditions, it is not an exact science and is highly dependent on estimates and assumptions. Below is an example of how the cost utility analysis of FES for correction of dropped foot was first estimated and further developed as new data and methods became available.

11.7   Evidence from the literature on cost effectiveness that could be used to support a business plan for FES for dropped foot

There are six reports that estimate the QALY gain associated with the use of FES for correction of dropped foot [12]. The first report was from the Development and Evaluation Committee of the South and West Regional Health authority 1996 [13,14]. It was this report that was submitted to the NHS to justify the establishment of the first clinical service in the UK for FES drop foot. The report was reviewed and accepted by the Health Authority. The report used data from the randomised controlled trial of the Odstock Drop Foot Stimulator (ODFS®) performed between 1993 and 1995 with 32 people who had had a stroke. The trial compared the effect of using the device with a standard treatment consisting of physiotherapy. The QALY gain was calculated using a combination of data including change in walking speed and physiological cost index, change in Hospital Anxiety and Depression Index (HAD) and change in a mobility score derived from a custom designed questionnaire, closely aligned with the Health-Related Quality of Life (IHQL). After 12 weeks of intervention, it was calculated that the FES group received a QALY gain of 0.065 while the physiotherapy group had a gain of 0.023, a difference of 0.042. At 1996 prices this gave a cost per QALY of £19,821 for one year's FES use and £10,037 over five years. In 2007 the report was re-examined and costs per QALY calculated for current prices [14]. This gave a cost per QALY of £39,047 at one year and between £13,524 and £19,237 at five years depending on the number of follow-up clinic appointments received. However, this analysis assumes that a comparison is made with an individual who receives physiotherapy. In clinical practice the ODFS® is used as a long-term aid while physiotherapy is rarely received for more than a few weeks. It may therefore be fair to attribute the whole of the QALY gain seen by FES users rather that the difference between FES and Physiotherapy interventions. This gives a cost per QALY gain of £25,230 at one year and between £8,738 and £12,431 at five years.

From an audit of patients who began FES use in 1999, it is now known that the average length of time FES was used for was 4.9 years and that the average cost per patient was £2,965 (based on an average of 10.9 hospital appointments per patient) [16]. It can therefore be calculated that for this cohort of 127 patients and assuming the same QALY gain calculated above, the mean cost per QALY was £9,658, well within the willingness to pay threshold of £30,000 used by NICE. It was not considered appropriate to apply discounting to the QALY gain as FES is a continuing intervention. This is supported by records of the difference in walking speed recorded, with and without FES, and Visual Analogue Scale (VAS) assessments of the impact of FES on various aspects of walking and quality of life. These results showing that these benefits were maintained over the whole period that FES was used [15,16].

A further economic report was produced by the Purchasing and Supply Agency of the NHS in the UK in February 2010 [17]. It took a different approach to calculating QALY gain. Its main indicator of effect was walking speed. The mean gain in walking speed due to FES was calculated by averaging the results from four published studies, two of which used the ODFS®. It was found that the mean increase in walking speed was 0.18 ms^{-1}. The change in walking speed was compared to Perry's criteria for mobility based on walking speed. Perry calculated that the mean threshold for becoming a moderate community walker was 0.58 ms^{-1} and for becoming a functionally independent walker was 0.80 ms^{-1}. By examining the range of walking speeds, it was possible to calculate the proportion of FES users who would cross these thresholds, and this could be corresponded to changes in the HUI3 (Health Utility Index v3) scale. The other input to the model was the number of FES users who received dis-benefit due to skin reaction to the electrodes. This was the only reported adverse effect of FES. 22% of FES users were reported as having minor skin irritation while 3% received a major skin reaction sufficient to cause discontinued use of FES. Using this technique an overall QALY gain of 0.041 was calculated. This compares with a QALY gain of 0.042 in the earlier study. A cost per QALY was found at one year of £52,336 and at five years of £19,238.

The Purchasing and Supply Agency report which examined data on skin irritation due to electrodes from the 1999 clinical rehab paper on patient's perceptions of use of the ODFS® may have been exaggerated [17]. As described above in the section on adverse effects, the types of electrodes used, and clinical procedures have since been improved since 1999 and this means the prevalence of new cases in the clinic significantly reduced to around 1–1.5%. Further, in the randomised controlled trial of the ODFS® with people who have secondary progressive MS, there were no reports of skin irritation in the period of the trial [18]. Also, in the audit of patients who began use of FES in 1999, only one FES user discontinued FES due to skin reactions in the whole ten-year follow-up period [15]. These results suggest that the dis-benefit effect of skin irritation has been significantly exaggerated in the Purchasing and Supply Agency report, resulting in a smaller QALY gain than might otherwise have been expected.

Street *et al.* examined the cost – utility of FES using the EQ-5D-5L questionnaire, the standard health economics instrument, to estimate the health utility index from using the ODFS® [19]. 45 pwMS and 27 pwCVA completed the questionnaire before beginning FES and again after 20 weeks use. The study showed a QALY gain of 0.114 (p=0.02) in both groups. Justified by the observation that the mean increase in walking speed due to FES remains steady throughout the time FES was used, the QALY gain was extrapolated over 4.9 years giving a total gain of 0.542 after discounting at 3% per year. From the long-term audit, the mean cost was £3095, giving a mean cost per QALY of £5,705.

In a similar study design, Juckes *et al.* recorded the walking speed, EQ-5D-5L and the Psychosocial Impact of Assistive Devices Scale (PIADS) of 82 consecutive pwMS who received the ODFS® Pace over six months [20]. An increase in walking speed from 0.670 m/s without FES at the start of treatment to 0.768m/s with FES at six months was found (p<0.001). The Utility index changed from 0.486 to 0.596 (p<0.001) over the same period giving a QALY gain of 0.110 and an estimated cost per QALY over five years of £6137. Statistically significant changes were also recorded in all 3 domains of the PIADS indicating improved device related quality of life.

Renfrew *et al.* compared the effect of using the ODFS® Pace with a custom moulded ankle foot orthosis [21]. 85 pwMS who had not used either FES or AFO for dropped foot correction were randomly allocated to each group and used the interventions for 12 months. While both interventions improved walking speed over 12 months, twice as many participants in the AFO group (n=21) discontinued the intervention than the FES group (n=11), primarily due to discomfort from wearing the AFO. FES users reported higher PIADS score indicating a greater device related quality of life (p=0.001). Despite higher costs for the FES group, because of a greater QALY gain, there was an incremental cost-effectiveness ratio of £14,285, modelled over two years, indicating the FES gave better value for money than an AFO.

11.8 Evidence from the literature on cost savings

Two studies have shown a 72% reduction in the incidence of falls when FES has been used [18,22]. No published data on the incidence of falls requiring medical treatment for people with MS could be found. However, data does exist for a general elderly population. Nurmi and Luthje (2002) performed an audit of falls amongst the elderly in institutional care [23]. They reported an incidence of falls of 1398 falls per 1000 person years and that one third of falls resulted in injury. The average cost per injury was €944. The average cost per fall per year was therefore €440. If falls that resulted in injury were reduced by the same proportion as in the ODFS® trial, there would be an annual saving of €329 or €1650 over five years. Allowing for an inflation rate of 44% (retail price index) between 2002 and 2014 the annual saving would be €474 (£374) or €2376 (£1877) over five years at 2014 prices (exchange rate 14 July 2014). From an individual perspective, the mean time between injuries would increase from 2.15 years to over seven years.

11.9 Financial considerations – ways of funding and costing of the service

Whoever provides the funding for the service, the costs and savings need to be considered and accurately calculated. Costs will depend upon the type of service you are planning to provide and the type of patients you are planning to treat. Many factors need to be considered to determine these costs. Not only do the costs need to cover the running of the service, it is also essential to include the service set up costs and the equipment necessary to provide that service. This might be comparatively simple if the service is to provide ES solely in the clinic alongside other treatment modalities, but is far more complex if the service plans to provide orthotic treatment for people with neurological disabilities where they will use the equipment for many years.

Given that as the lead clinician you are experienced in ES and you know what patient cohort you are going to treat, let us first consider the primary way the service is funded. Are you just going to see the occasional person for ES/FES treatment or are you going to put in a business case to a state, national or insurance-based healthcare provider? The first scenario is quite straight forward, and it is possible to grow the service incrementally. Therefore, we will concentrate on the steps needed to set up a full clinical service. However, if you are planning to only see individual patients there are still several issues that need to be considered.

11.10 Issues to be considered in a business case to treat individual patients

Referrals
  Patients self-refer?
  Referred by health professional, if so who? Doctor only, physio?

Funding
  Initial assessment, full cost or reduced cost to encourage referrals?
  Pay per visit?
  Pay for course of treatment?
  Is the cost of equipment and consumables included in the treatment package?

Equipment
  Loaned by clinic? If so, how is equipment maintained?
  What initial stock levels are required?
  Bought by patient through the clinic or directly purchased?
  Vendor relationships and contracts

Consumables
  Included in cost of treatment package?
  Any limit on amount provided?
  Bought by patient, if so, does the clinic keep items in stock?

Follow-up
  Outcome measures
  How to decide when treatment ends
  Collecting evidence on efficacy of the service

If you decide to go down the route of setting up a larger, externally funded clinical service, you will need to produce a robust business case. The content of this business case will depend upon the requirements of the proposed funder and therefore it is essential that you have a thorough, detailed knowledge of how healthcare is provided in your area.

11.11   Issues to be considered in a business case for an externally funded service

Referrals
  What geographical area or you planning to serve?
  How many people with the condition you plan to treat are there in your chosen area?
  How many people per year do you plan to see?
  Is there a 'gate keeper' to referrals? Some systems will only accept a doctor's referral.
  Is there any competition in your area?

Funding
  Per individual patient?
  A 'block' contract for 'x' patients per year?
  Funding only for initial assessment?
  Are treatment and equipment funded jointly, or separately?
  How do you indicate to the funders that a given person is suitable for treatment?

Set up cost
How much equipment/consumables do you need?
Who is going to fund this stock? Your institution?
Do you require any additional clinical staff?
Admin support for appointments, enquires etc.?
Have you included overheads in your business plan?
Have you produced and provided information about your service for referrers?

Facilities and resources
Time allocated for treatment, clinic letters, equipment management etc.
Space for clinical assessment, walking tests etc. 10m walk test
Follow-up appointments
Long-term stability of the service
Possible use of clinical assistants.
Succession planning and training of new staff

Equipment
What level of stock do you require?
Who owns the equipment? The funder? Your institution? The patient?
Is equipment leased, bought or provided as part of the treatment package.
What is the lifetime of the equipment
Does it comply with Medical Device Regulations

Consumables
What level and range of consumables do you require and how often do you restock?
What is the average number of consumables that a given person will use in a year?

Follow-up
Are follow-up appointments funded in the treatment package or separately?
Have you secured long-term funding for long-term users (neuro patients especially)?
What evidence does the funder require for them to continue funding?

Monitoring and Audit
How will monitor the quality of your service and what standards will you compare it against[3]?

11.12   Equipment and consumables

The choice of equipment will be dependent upon the type of service you are running and the conditions of the patients you are planning to treat. Whatever you plan to do, it is essential that the equipment you use is fit for purpose and will provide the stimulation parameters that you require. More details on equipment and the essential requirements, as well as regulations that need to be met are provided in Chapter 3. However, it is worth pointing out here that there is some confusion when using neuromuscular stimulators in that not all are built to the standards of medical devices. Many of the cheaper ones are sold as sports devices and as such, do not meet the requirements of standards such as the Medical Devices Regulations (MDR), CE marking or FDA approval. Obtaining such standards is an expensive process for manufacturers, and therefore stimulators that meet these strict demands are often significantly more expensive than those that do not, even though such stimulators are still available on the internet. However, using stimulators that are classified as Medical Devices will ensure that you are able to provide a safe clinical service.

It is often a good idea to standardise on a range of equipment as this make it easier for both clinical staff and service users. This is particularly true if a stimulator breaks, as a replacement can be taken off the shelf without any need for retraining hence ensuring that there is no gap in treatment. This replacement process can also be improved if details of stimulation parameters are recorded, thus potentially enabling replacement of the device by post. It might also be possible to negotiate discounts when purchasing the initial stock as well as building a relationship with the manufacturer.

In the last section on funding, the issue was raised as to who owns the stimulator. Not only is this important in the initial purchase, it is also important if something goes wrong with the stimulator. This is simple if the stimulator fails under warrantee, but what happens if there is accidental damage? Who pays for it then? Are you able to supply as long term/temporary replacement? Do you have a repair service so that it can be mended? There is no right or wrong answer to these questions, but it essential when setting up a service to consider how such problems will be dealt with when they undoubtedly occur. Also, if you are planning to run a service for many years, you must have an equipment replacement programme. If the manufacturers quote the service life of the device, this can be used to plan a replacement programme. You will also need to consider any staff and patient retraining that may be required for the introduction of updated devices.

Consumables are another area that require thought, especially for long-term users such as people with chronic neurological conditions. As well as electrodes, wires, footswitches, sensors etc only have a limited lifespan and often come in a range of sizes. The lifespan of these accessories should be available from the manufacturer and should be used to determine your stock levels. Therefore, a stock of such items is required and details readily available as to exactly what set up each individual patient has, length of leads, shoe size, etc. Maintaining this stock can be a time-consuming task, especially as patient numbers grow and is yet another task for a busy clinician. Hence consideration should be given to having sufficient administrative and technical support to enable someone to answer these requests and dispatch the relevant items. One answer could be to determine if the equipment manufacturer is able to provide the required support.

Finally, it is important to decide who funds the consumables and determine whether the cost of them is built into your treatment plan and funding model or whether you will charge for them on an individual basis. If the latter, is it the funder or the user who pays for them. Like many of the issues raised, there is no simple answer. You therefore need to determine what your service is going to provide and that depends on your local healthcare system and the way funding is provided in your area.

11.13   Staff training

ES is rather unusual in that it can be, and is, provided by a variety of different disciplines. It is usually provided by physiotherapist, but can be provided through a standard medical model, by occupational therapist, particularly in upper limb stimulation, by orthotists in some countries, and in the United Kingdom by Clinical Scientists trained in Rehabilitation Engineering. Irrespective of the primary discipline of the clinician, it is essential that they have a thorough appreciation of the skills and expertise of the other professions listed above, as well specific training in ES. This is because for ES to be optimised, it is often reliant on good medical management, the physical condition of the user, orthotic provision, the underlying pathology, whether the person has spasticity, as well as fitting in with activities of daily living and the use of the equipment itself. This is particularly true for people with long-term neurological conditions. Therefore, it is essential that training is provided to meet this multidisciplinary need, taking account of the primary disciplines of the staff involved.

Although some basic knowledge of ES is provided in some professional courses; physiotherapy and Clinical Scientist training in the UK, it is not always comprehensively taught and is of variable quality. The ACPIN Clinical Practice Guidelines recommend that specific training is obtained in the devices used [3]. We would strongly encourage anyone considering setting up an ES service to look at undertaking further training courses and to consult other professionals in the field who are already providing successful clinical services who might be willing to provide advice, support and shadowing opportunities. There is not an abundance of such courses available, with many being device specific provided by equipment manufacturers.

Once a service has been established, the need for ongoing training still remains. New staff, again possibly from a variety of disciplines, might start working in the service and in addition, new devices and techniques might become available. As a result, all staff must undertake Continual Professional Development (CPD) with the senior clinician leading by keeping up to date with current developments to develop suitable mentoring systems for junior staff and to provide relevant training materials. A supervised, in-house training scheme could be beneficial in this, with new staff having to undertake a series of 'competencies', as well as their initial training, whilst being supervised by senior staff.

Training in undertaking outcome measures is also required to support the use of the device and how well it is used to achieve the patient's goals. Particularly if several staff are undertaking these outcome measures, as it is essential that all measurements are made in the same way in order to enable that data to be combined and used in reports. These reports can then be used for research and audit purposes and to provide evidence of the efficacy of the service to funders.

11.14   Specific requirements in the management of electrical stimulation patients

For ES to be successful, it inherently depends upon 'buy-in' from the patient, for unless the patient is included in the decision-making process and the link between the device and how it will help the patient achieve their goals is explicitly made, is unlikely to be successful. This is apparent right from the initial assessment. The sensation of ES can be unusual, and some people find it uncomfortable and a few even painful. Therefore, everything must be done to explain carefully to the patient what to expect and to minimise the sensation of stimulation. It is often a good idea to practice all the techniques that you plan to use on yourself and if necessary, demonstrate on yourself in front of the patient. The main factor in minimising the sensation is good quality electrodes with even contact on well hydrated skin and they should be as large possible without causing overflow to muscle or nerves you don't want to recruit. Use long ramp up and down times, as sudden contractions are more uncomfortable. Make sure the patient is relaxed and fully understands what is going on.

The stimulation level needs to be turned up gradually, but not too gradually, with confidence whilst reassuring the patient the whole time. Overcoming the problem of the initial sensation is particularly difficult with children where one possibility is to demonstrate on their parents, or alternatively to let them turn the amplitude up themselves so that they feel in control. The subject of using ES with children and is covered in depth in Chapter 16.

To ensure that people can effectively participate in the treatment regime, they do need to fully understand the rationale behind the treatment and how to use the equipment when away from clinic. Although it might seem simple to set up a drop foot stimulator in the clinic for a professional, it is another matter for a stroke patient to set the system up at home, using only one hand without professional supervision, where problems such as finding the optimal electrode position soon become apparent. This subject will be dealt with in more detail in the following section on Patient Education.

11.15   Treatment protocols

There are a range of stimulation protocols used in the literature, which can make drawing conclusions from research data challenging. Therefore, clear protocols, based on scientific evidence, are necessary and need to be adhered to throughout treatment. One example is the FES clinical decision-making tool which was developed to assist clinicians when selecting ES parameters when applying FES in neurorehabilitation [9]. It is also essential that both stimulation parameters and outcome measures are accurately recorded. It can then be made clear to patients what the treatment is trying to achieve, what is expected of them and how you will determine, and demonstrate, that the treatment has been successful. The inclusion of PROMS (patient reported outcome measures) and objective data is often beneficial in achieving this, as it will enable them to determine what is important to them and the effect that ES will have on their quality of life [24]. A clinical practice guideline outlining a core set of outcome measures for individuals with neurologic diagnosis has been recommended. This core set of outcomes are standardised and include participation and activity level outcomes [25]. The results of the data collected can then be used to demonstrate to them the success of treatment, or if not successful, why treatment should be changed or stopped. This feedback can then be considered as 'soft advertising' to those referring into the service as well as educating them on what type of patients will be most likely to benefit from your service. Clinical data that is collected in a consistent manner, can then be pooled and used to enable an audit of the service, comparing outcomes to original research studies or, better still, published to increase the evidence base. While cases series studies such as these are often dismissed by academia as lower quality evidence, bodies such as NICE (National Institute for Health and Clinical Excellence in the UK) do use such studies in their guidelines. They may also give a truer picture of the real effect of an intervention in real life than formal research studies, as long as a consistent method for recording outcomes are used and all patients accounted for [26]. It is rarely possible with formal research studies, which are often restricted in size and time, to study the whole period an intervention may be used, and this is only really possible with long-term case series data. Recently introduced changes in the European Medical Device Regulations require manufactures to collect post-market clinical follow-up to demonstrate that their device continue to perform as they did in the original evaluations, both in terms of safety and clinical performance. Case series studied are therefore invaluable to maintain a clinical device in service.

To ensure ongoing funding you need to collect consistent outcome measures to provide objective data on the effectiveness of your service. This is particularly necessary if you need to make a presentation to a healthcare funder (insurance company, state/local health service etc) where their limited research will often show a diversity of treatment regimens and hence their conclusion is that the treatment is still at the research stage and hence not suitable for funding. Unless you know exactly what your clinical service will provide and have evidence to support why you are doing exactly what you are doing, this will be very difficult to counter.

One way in facilitating the provision of this data and standardised treatment pathways is to develop a standard set of clinic paperwork. This will make the service more efficient and ensure accurate record keeping which can then be used to produce the necessary evidence.

11.16 Patient education and information

ES is a technique that is very dependent upon the patient having a good understanding of both the equipment to be used and the basics of anatomy so that they can both locate the correct electrode positions and to adjust the stimulator to achieve a good, appropriate muscle contraction. This might appear to be a simple task for an experienced clinician, but it will be far more difficult for a patient to be able to undertake at home without any medical supervision. It is all too easy for them to appear to be confident in the clinic when asked if they understand what to do, only to later have great problems in replicating a good muscle contraction when trying to do so when alone in their own home. Therefore, to increase patient inclusion and compliance, patient education is an essential component of any clinical stimulation service.

This education of the patient and their carers and family should start right from the initial assessment by carefully explaining what you are going to do and how ES can help the patient achieve their desired goals. Explain how muscle contraction normally occurs in the body and how this is replicated by ES. Explain why you are putting the electrodes where you are, what structures you are stimulating under the skin and how you locate the appropriate point to place the electrodes. This might be the location of motor points when stimulating the quadriceps, or the location and structure of the peroneal nerve when setting up a dropped foot stimulator. Then adjust the stimulation intensity until the required response is achieved, whilst explaining to the patient why you require a given level of stimulation to achieve that response. Given that the initial assessment is successful and ES is determined to be appropriate, the patient should then also have a good understanding why ES is to be recommended for them.

Good patient education takes a considerable time, but it is critical in ensuring benefit and compliance, particularly in the more complex neurological cases. Therefore, a variety of approaches and sources of information are required in order to provide for the needs of a wide spectrum of patients and conditions. These include.

- The way you structure the clinical appointments to optimise patient education.
- What printed material should be given to the patient?
  - Instruction manual for the stimulator
  - Specific instructions for the individual patient
  - Detailed photographs and diagrams showing the correct electrode positions

- The use of the patients' or the carers' mobile phones can also be very useful and can be used to capture photographs. Short videos can also be used to show the movement that needs to be achieved.
- On-line information they can access
- Helpline and ability to talk to a clinician within a given time frame – rapid response to problems is needed to give patients confidence in the technology and the service providers.

The initial assessment can often be long, an hour or more, as there is a lot to fit in; taking a medical history, physical examination, explaining ES and trying stimulation to see if it gives the desired response. Also, at that appointment the patient might be rather nervous and apprehensive, or over optimistic about what ES might achieve. Although there is no 'correct' answer or 'gold standard' on how to structure clinics, it is essential to think about services from the patients' perspective and how clinics can be arranged to maximise the benefits that patients experience from the treatment. This relies upon the patient fully understanding the treatment. The level of medical and technical detail that each patient receives must be judged by the clinician. Some people are made more anxious by an overload of information, others are curious to understand the technique in detail. In Salisbury we have found it best for neurological patients to keep the initial assessment appointment to just assessing the patient and seeing if the technique works. If it does, two further appointments are made to set the system up, separated by seven days at which considerable time can be given to patient education and the clinician can observe the patient setting up the system themselves. On the second day we also ask the patient to attend the clinic having set the system themselves without supervision. The Birmingham FES clinic has a slightly different structure with a longer initial assessment to enable assessment and set up on the same day, as this reduces the need for the person to travel. Whatever the structure of the initial assessment appointment, if the patient has a partner or carer, we strongly encourage them to be fully involved in this education process. We have also found that if problems do occur, or the treatment needs modification, such as slight change in electrode positions, they usually do so during the first few weeks. Therefore, a third appointment is made at six weeks and the time between subsequent appointments increased as the patient becomes more familiar with the technique. As we say, this is not a prescriptive clinic pattern, just what we have found works for us. It is the principles of ensuring that the patient fully understands the treatment and can use the equipment effectively that is important.

Good, printed material is also an essential part of patient education, particularly as a lot of people attending ES clinics are likely to be elderly and might not want to access all information electronically. This information should include; a patient or user version of the stimulator instruction manual, specific instructions for the specific person, such as how long to stimulate each day, and what settings to have the stimulator on etc. In addition, diagrams or photographs of electrode positions are often very useful (Figure 11.1). These can either be taken by the clinician and printed out or could be photographed using the patient/partner's phone. Close ups showing how the electrode placement related to marks on the skin, moles etc, can also be useful. Alternatively, simple hand-drawn diagrams can make it easy for patients to see how to place electrodes.

The book, Anatomical Guide for the Electromyographer has some clear diagrams of motor points which can be particularly useful for clinicians if their location is what you are trying to demonstrate to the patient [27]. In this book motor points are clearly marked and can be simply related to bony landmarks using fingers breadths etc.

Although this information must be available in a printed form to prevent digital exclusion, some people will prefer electronic versions. Therefore, every attempt should be made to have the information available in a variety of media.

Finally, there will be times when despite all the education and information provided problems occur to which the patient requires answers. These might be technical problems with the equipment or clinical problems resulting from stimulation. In such cases there is no substitute for patients being able to talk to someone,

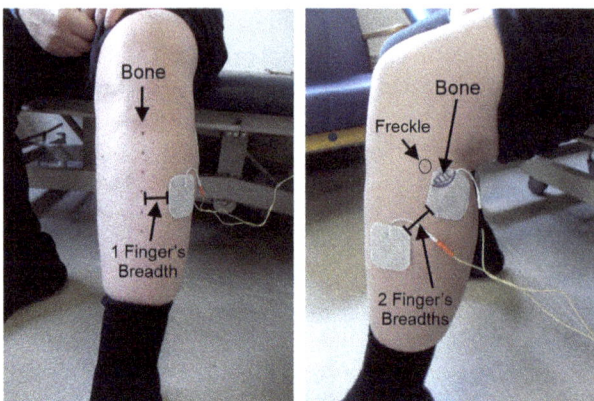

Figure 11.1   An example of an annotated photograph issued to an FES user to teach them how to use their anatomical landmarks to correctly place electrodes. The tibial crest is marked on the persons skin using a non-permanent marker pen. The head of the fibular bone is marked on the electrode. The FES user is instructed to find the head of fibula by feal and to place the shaded part of the electrode over the bone. Distances are marked in finger breadths. Other distinguishing marks such as freckles can also be marked.

especially as those with chronic conditions such as stroke or MS, who can become heavily reliant on their stimulation systems, and it is essential that problems are resolved swiftly. Problems with the equipment may require technical expertise and if this is not available, replacement parts or systems should be provided. Many cases, however, require a clinician to respond to ensure that optimal treatment is always maintained. To minimise problems occurring, regular, planned follow-up is required so that concerns and potential problems can be addressed before they start affecting the efficacy of the treatment.

11.17   Safety – precautions and contraindications

Looking at the instruction manual of any ES device will show a long list under the heading of 'Contraindication, Warnings, and Precautions'.

Whilst a lot of these are simply common sense, some do need further consideration as anyone running a clinical service will come across certain cases where you still want to use ES in a person whose underlying condition or other medical conditions require special consideration. The most common are; patients with pacemakers, a history of epilepsy, implanted metalwork and high level spinal cord injuries that are associated with autonomic dysreflexia. Therefore, in your clinical paperwork it is a good idea to have specific questions relating to these so that they are not overlooked when undertaking the initial assessment to determine if a given patient is suitable for ES. In addition to these specific patient groups, which will be considered in more detail below, it is essential to always consider the current path when applying ES and if possible, to ensure that a direct line between the electrodes does not pass through any structures to be avoided such as the heart, cancerous tissue etc. However, the decision on whether to proceed with ES should be considered on a patient-by-patient basis following a holistic assessment.

11.17.1   Pacemakers and implanted defibrillators

In our clinical experience this is the most commonly occurring area for potential contraindications. This is because many of the people seen in ES/FES clinic are elderly, such as those post stroke, after joint replacement etc. and hence will often have multiple pathologies and hence likely to be fitted with an implanted device. As a result, there have been a number of studies and literature reviews in order to determine whether or not problems are likely to occur [28,29]. Again, it is the current path that is important and such studies have shown that with modern pacemakers and implanted cardiac defibrillators (ICDs), stimulation of the limbs when both electrodes are on the same limb are safe and hence FES for drop foot could be considered low risk. Several short-term safety studies investigating lower limb NMES have concluded that it is safe to use in patients with pacemakers and ICDs; and there have been no case studies describing electromagnetic interference (EMI) when ES is applied to the lower limb. However, a definitive safety study that is intervention specific, long term, and high powered is needed to identify the exact risk. Until such a study is realised, ES should be used with caution in patients with pacemakers and ICDs. We recommend that pacemaker and ICD function should be monitored for EMI by a Cardiac Technician when ES is first used. This is more important when stimulation is applied to the upper limb or torso as the risk of interference increases the closer the electrodes are to the active device [30].

11.17.2   Uncontrolled epilepsy

Taking a detailed medical history is an essential component to operating any clinical service and as such can be used to identify if a given patient has a history of epilepsy, and if so, if that epilepsy is well controlled. It is recommended that people need to have been seizure free for at least three months. However, it is worth checking whether your local health authority has any set guidelines concerning this. Given this proviso, a history of epilepsy itself is not a contraindication, but such people will require more guidance. Asking them if they have any advanced warning of a coming seizure might also be useful and they should be carefully observed at the initial assessment when first trying stimulation.

11.17.3   Implanted metalwork

Any implanted metalwork for fracture fixation etc may alter the current path if that metalwork is between the two electrodes, as the metal will be more conductive than the surrounding tissues. In such cases alterative electrode placements will be necessary if the FES user experience unusual sensation from the stimulation. However, implanted metalwork should not be considered a contraindication.

11.17.4   Autonomic dysreflexia

Autonomic dysreflexia is a common complication of spinal cord injury above the spinal level of T6 [31]. It manifests as a sudden rise in blood pressure in response to a noxious stimulus below the level of the lesion and, if the stimulus is not removed, can lead to headache, cutis anserina (goose flesh), paraesthesia, shivering, flushing and other symptoms. The rise in blood pressure is a consequence of reflex arterial spasm in response to the stimulation of the sympathetic nervous system. In the intact nervous system baroreceptors in the cerebral vessels, carotid sinuses and aorta detect the hypertension and stimulate a parasympathetic response, the principal consequence being vasodilatation and a fall in heart rate. However, in this group, the signals are blocked by the lesion, the resulting vasodilatation only occurring above the lesion. ES can act as such a stimulus [32]. Therefore, when using ES for the first time in such a person it is good clinical practice to measure their blood pressure before starting and to monitor it throughout the clinical session. People with autonomic dysreflexia are usually well informed about this condition and are aware of when it is occurring. Stimulation should be stopped if the blood pressure reaches dangerous levels. This is often defined as an increase of over 20 mmHg in systolic blood pressure over baseline, or a resting level as more than 180 mmHg systolic and 120mmHg diastolic [33]. It should also be noted that AD can occur in people with MS if they have spinal lesions [32]. Clinical experience shows that blood pressure quickly falls after turning the stimulator off [34,35].

11.18   Clinical goals and outcome measures

There are many reasons why clinical goals and outcome measures are an integral part of any clinical ES service. Unlike many medical treatments, ES is often not well known and as such needs more justification and evidence of its efficacy than more routine treatments, such as the use of antibiotics or mechanical joint replacement. Therefore, it is important to provide this evidence, not only to the patients so that you can show that the treatment is working for them, but also to build up a body of evidence to justify the use of ES to healthcare funders. It is often useful to make this evidence specific to your service as for reasons stated above in the section on 'Evidence of Efficacy – Building a Case'.

Therefore, the outcome measures that you choose must be appropriate to the service that you are providing. They should be validated and used correctly, see the Rehabilitation Measures website [24]. As will be seen from looking at this website, there are many outcome measures that can be used and hence it will be necessary to only choose those that are relevant to your service and that can be done correctly in the clinic in the time available. In the US The ANPT have also developed free training tools for core outcome administration and comprehensive reference materials that include key psychometric data to determine clinical meaningful changes [25]. Time also has to be allocated to entering that information into databases and to analysing that data.

It is important to consider for how long the patient is going to use stimulation and hence how long, and how often you are going to follow them up. In some musculoskeletal cases this might only be for a few months, or even weeks, whereas for neurological patients it might be for ten years or more. Therefore, both long-term and short-term goals are required and again outcome measures chosen to show when these goals are met. Meeting these goals in the short term might be used as a reason for changing the treatment regime, such as increasing stimulation duration or intensity, or as a reason for stopping treatment because the treatment has been shown to be effective and the initial treatment goals met. If the goals are not met, it might be necessary to change the treatment regime, or indeed to stop treatment entirely if it is shown that the treatment is not being effective. It is particularly important to realise when treatment is not being effective, and when it is time to stop. When that decision is made, objective evidence is often useful to justify to the patient why it is being made. It is all too easy to continue ES when it is no longer being effective which is both a waste of your clinic resources and inappropriate for the funder, whether that is the healthcare provider or the patient themselves.

It is important when choosing your goals and outcome measures to consider the whole person and again relate the goals and outcome measures to what you are trying to achieve by your treatment regime. Objective data such as walking speed, energy expenditure, muscle volume etc can be very useful and can be free from risk of bias, whereas more care needs to be taken with more qualitative measures. However more qualitative outcome measures are important as they do provide a lot of information on the effect of your intervention on the quality of life of the patient. It might not matter to your patient if they are now walking at 0.9 m/s compared to 0.8 m/s before. However, it does matter to them whether they are able to walk down to the corner shop whereas previously they could not. Therefore, PROMs are an integral component of any clinical service.

At the outset of setting up a clinical stimulation service, consideration needs to be given in determining what goals and outcome measures are to be used, how they are used, who does them and how is the data collected recorded and analysed. Talking to the funders of the service at the outset, particularly state funders, insurance companies etc. can also be beneficial as they might have specific data they want to be collected. An example of this could be in reducing the number of falls by the use of a dropped foot stimulator in people with neurological problems such as stroke and MS. Therefore, if this is known at the outset, appropriate outcome measures can be built into the system and the data collected used to justify further funding.

Goals may change over time, particularly with long-term neurological conditions such as MS. Additional co-morbidities, or even aging, may compromise their mobility and therefore FES is no longer appropriate for functional walking, but NMES might still be useful for maintenance of joint range, muscle strength or in spasticity management. These can be evidenced with appropriate outcome measures including patient reported testimonials. Whatever outcome measures are used, it is essential that they are validated and used correctly. Staff training might be needed so that if several people are undertaking these measures, they are all done in the same way so that the data collected can be pooled for subsequent analysis. It is also important to ensure that such outcome measures are practical in your clinical setting and that they can be done within the clinic appointment, by all staff, at every appointment. Good quality data can be extremely useful, but it does take time to collect and analyse. Poor quality data is of no use to anyone.

11.19   Long-term support and follow-up

In general, there are two categories of people who might be referred to an ES service; those with a long-term disability or condition who require ES as a long-term orthotic device, such as for foot drop, or those recovering from an injury or surgery, or those recovering from a milder stroke or an incomplete SCI who require it as a method of muscle strengthening to enable them to complete their rehabilitation. In the second case people will normally only use stimulation for a period of months, whereas in the first case some people with neurological conditions might well use an FES device for more than ten years.

If you are seeing a significant number of people with long-term neuro-disability such as stroke, SCI, MS, then provision for these patients does need to be included in your service. This turn raises a number of issues that need to be addressed, such as

○ How long has funded been approved for?
○ Are you going to have an annual follow-up?
○ How are you going to supply consumables?
○ Who pays for these consumables?
○ How long will the equipment last?
○ How will you replace it if it breaks?
○ Will you provide replacement equipment if it is being mended?
○ What happens if the patient's condition changes? It can improve or get worse.
○ Do you need to provide data to the funder every year to secure funding?

It is also important here to consider plans for the continuation of the service in the event of loss of staff, either due to staff leaving or long-term illness, problems with supply of equipment or change in regulations which might have an impact on the clinical service. Each of the possible scenarios needs to be considered and a risk analysis undertaken to minimise the effects of any disruption.

The answers to these questions are not easy and there is no 'right' answer in many cases, as they will depend upon your local situation and how your service is funded. However, many people with long-term neurological conditions can be very dependent upon their FES device if it is making a significant difference to their function and hence quality of life. Therefore, they do need to be addressed to ensure continual high-quality care, sometimes over many years.

11.20   Research and audit

The basic question to be answered here is 'How are you going to show that you are providing a good clinical service'. Although carefully chosen and well conducted clinical goal setting and outcome measures can provide good feedback to an individual patient, unless it is undertaken in a structured, systematic way as specified above (section on outcome measures) they will not be able to be analysed and used for research and audit purposes. This analysis might be used to provide feedback to funders or for publication in the scientific literature and presentation at conferences. To do so, any statistical methods used should be appropriate and the results obtained analysed correctly. If you do not have these skills yourself, you will need to consult a qualified statistician. This could also be a good way into collaboration with your local academic institutions.

In addition to using the data collected as part of the clinical service, it is also worth collecting other sources of data that can be useful in determining whether you are providing a good service or not. These are less formal than traditional research methods but can still provide useful information which might also be useful to funders of the service. These are user questionnaires and audit. There are guidelines developed by ACPIN which you can audit your service against [3].

Although user questionnaires are less formal than standard qualitative research methods, their development does require experience to ask the correct questions in the correct way. If you are running a service and do decide to undertake one, we strongly suggest you consult an experienced qualitative researcher first. It is also useful to contact the funder as well as they might have standard information that they require, such as 'Would you recommend the service to a friend or member of your family' It is obviously best to know what they want to know at the start of the process rather than at the end. Your local Patient Experience Team in your hospital might be able to help with this, and it could provide evidence of patient satisfaction within your institution. It should also be decided how such a questionnaire is to be administered.

○ Is it going to be sent through the post?
○ Available only via the internet?
○ Questions asked by the clinician in clinic?
○ Completed by the patient in clinic?
○ Are you going to have a separate questionnaire for carers?

Obviously, you want to obtain as much information as possible about their perception of your service but remember not to make the questionnaire too long or complex or people will simply not complete it. If you do decide to undertake a questionnaire, how will you decide to implement their suggestions for improvement and how long will it be before you repeat it?

Audit will show if your clinical service is functioning well and efficiently. Audits can be designed to provide various useful pieces of information such as; time from referral to first appointment, percentage of initial assessments leading to people starting treatment, how complete is your database, average consumables used per patient per year, etc. These will be mainly a paper exercise and the information required should be available through the patients notes.

11.21   Providing feedback to funders

The main question to ask here is 'What information do the funders want? It is far better to talk to them to ascertain this before writing the report and hence save wasting considerable time. It should then simply be the case of directly answering their questions, making sure your answers are clear and presented in a simple way that they can understand. Do not waste your time in providing lots of data and analysis that they have not asked for. For example, in Germany and Switzerland reports in a written form are required on regular basis (every three month in Switzerland) and must be sent to the insurance companies regarding Efficiency Effectiveness and Expediency Criteria.

The use of clear graphs and/or diagrams is also important, as the information contained can be more easily understood and its relevance apparent, compared to the reader having to wade through many pages of text to acquire the same information. It can be useful in comparing the presentation of this data in a similar way to what you would do for a talk or a conference poster. In such cases the message needs to be clear and simply presented so that the reader, or listener at a conference, is in no doubt what you are trying to say and why your results are important. Think about what is most important in what you are trying to say, and make sure that your message is clear.

Appendix A   Checklist

Essentials of a clinical service
- Introduction
- Identifying the clinical need
- Building a business case
- Financial considerations – ways of funding and costing the service
- Equipment and consumables
- Staff training
- Specific requirements in the management of electrical stimulation patients
- Treatment protocols
- Patients education and information
- Safety – precautions and contraindications
- Clinical goals and outcome measures
- Long-term support and follow-up
- Research and audit
- Providing feedback to funders
    At the end do we provide a checklist so that people can tick off if they have considered all the various factors required to set up a clinical service.

- Introduction
  - This is not a recipe to follow but is an attempt to pose a series of questions and issues that you need to consider before trying to take what might be an interesting research /development finding into the real world of providing services for a wider community. It is not exhaustive but does contain the key issues that do need to be addressed if you are to successfully run a service. Between the authors of this chapter we have treated over 20,000 people with a wide range of ES/FES treatments and modalities for many neurological and MSK conditions.

- Identifying the clinical need
  - Which patient group Neuro &/or MSK. Difference between the two groups.
  - Patient numbers. Will your service be able to cope
  - Therapy – orthotic provision.
  - MSK often short term whereas Neuro long term.
  - How are you going to access your potential patients
  - Who will you take referrals from

- Building a business case
  - Know your healthcare system
  - Know the literature
  - Be aware of any national or international guidelines
  - Make the evidence appropriate to your local need.
  - Be treatment specific
  - Be aware of changing patterns of disease. SCI today is not the same as it was 30 years ago.
  - Be clear what your service is trying to achieve and find evidence to justify this

- Financial considerations – ways of funding and costing the service
  - Who is going to fund your service – patients or your healthcare system
  - Who is going to enable the setting up of your service. Purchase of equipment, consumables, staff.
  - Who needs to refer patients so you can get the funding.
  - Funded by visit or by course of treatment?
  - Who is going to own the equipment
    - Bought by patient
    - Leased to patient
    - Covered by the cost of the treatment package.
  - Funded by individual patient or by block contract.
  - Long-term funding required for patients if long-term orthotic use.
  - Have you included overhead costs in your funding model.
  - What other services are nearby, is there any competition
  - Be clear what your service is trying to achieve

- Equipment and consumables
  - Does the equipment meet the appropriate standards. Medical or Sports use.

- ○ Can you standardise on equipment
- ○ What stock of consumables are you going to carry
- ○ Who provides the consumables
- ○ Who pays for the consumables
- ○ How much is provided.
- ○ Who repairs any stimulators
- ○ Do you carry spare stock so replacements can be provided while equipment is being mended.
- ○ Accidental damage? Who pays.
- ○ Replacing equipment if worn out or introducing new versions.

- Staff training
  - ○ What level of lead/supervision do the funders require. Consultant led?
  - ○ Are you proving as an enhancement to therapy – physio/OT led or as part of equipment provision – orthotist led.
  - ○ How do you determine staff are well trained.
    - – Initial training
    - – Set competences
    - – Updates on new equipment and techniques. – user days?
- Specific requirements in the management of electrical stimulation patients
  - ○ How do you explain everything to your patients.
  - ○ Minimise the unpleasant sensation of ES
  - ○ Be familiar which how stimulation feels in whatever treatment you are providing
  - ○ Dealing with children

- Treatment protocols
  - ○ Be clear to the patient what you are trying to achieve
  - ○ Demonstrate that the proposed treatment is relevant to their needs.
  - ○ Keep your treatment consistent with your protocols so that you can audit them and undertake research.

- Patients education and information
  - ○ How do you train the person to use the system
  - ○ Will they be able to do so without professional supervision
  - ○ Do they have a helper/prime carer.
    - ○ If so, have you trained them .
  - ○ Do you check that they are competent to use it once they leave the clinic.
  - ○ Have you provided clear written information for them to take away.
  - ○ Can they get help remotely once they get home.

- Safety - contraindications and precautions
  - ○ Pregnancy
  - ○ Uncontrolled epilepsy
  - ○ Pacemakers and defibs
  - ○ Local cancers

- o Stimulating head/neck/chest areas
- o How do you ensure safety in complex cases

- Clinical goals and outcome measures
  - o Do the goals and outcome measure set meet the need of what your service is trying to achieve.
  - o Set clear goals – short term and long term
  - o How long are you going to provide treatment for.
    - – Long-term orthotic use
    - – Short-term therapeutic
    - – Don't continue treatment when not achieving anything.
    - – Have objective data to determine when goals are met
    - – Discharge people when ES/FES no longer beneficial
  - o Are your outcome measures relevant and validated.
  - o Can they be done simply in a clinical environment
  - o How long do they take to perform

- Long-term support and follow-up
  - o Undertake a risk analysis to determine how you will cope with disruptions to the service
  - o People with long-term orthotic use might require follow up for many years >10
  - o In such cases, as people age their needs might change
    - – MS patients will deteriorate
    - – Some stroke patients might improve

- Research and audit
  - o How are you going to show that you are providing a good service
    - – Objective data
    - – PROMS
    - – User satisfaction questionnaires
    - – Audit

  - o How good/complete is your data
  - o Are any statistics meaningful and statistically significant
  - o Does it meet the requirements of funders

- Providing feedback to funders
  - o What information do they want.
  - o Does your feedback answer their questions.
  - o Don't waste time providing information they don't ask for.
  - o Do they require objective data or PROMS or both?
  - o Is the feedback presented in a way they can understand.

Appendix B Number of people who might benefit from electrical stimulation

Condition	Incidence per year in England [per 1M] [5]	Prevalence in England [per 1M] [5]	Number who may use FES for walking [per 1M]	Number who may use FES for upper limb [per 1M]	Number of other application users [per 1M]	Notes
Bell's palsy	8,000 [142]		n/a	n/a	4,000 [71]	Half recover without treatment. Half may benefit from ES to aid recovery if spontaneous recovery has not occurred.
Bladder dysfunction-neurogenic	22,081 [393]	623,410 [11,093]	n/a	n/a	Not known	FES to the posterior tibial nerve has been documented to help support or improve bladder control in neurogenic bladder conditions, reducing the symptoms of urge incontinence. Research continues into stimulation of the genital nerve to increase bladder capacity in SCI.
Brain and spinal tumours	7,820 [139]	85,927 [1,529]	Not known	Not known	n/a	FES may be appropriate if the tumour has affected the part of the brain or spinal cord associated with movement, or the affected by the surgery/radiation treatment used to treat the tumour
Charcot-Marie-Tooth	Not Known	19,376 [345]	Not known	n/a	Not known	This is an under investigated area. People with slowed nerve conduction may benefit from FES to assist walking.
Chronic obstructive pulmonary disease	115,000 [2,046]	Estimated at 3 million	n/a	n/a	Not known	Estimated that 2% of the population and 4.5% of over 40's are living with COPD. FES for strengthening Respiratory muscles has some research evidence to suggest improvements in lung capacity and strength of COPD patients.
Critical care patients	170,900 [3,041]	n/a	n/a	n/a	Not known	New Research is indicating that FES can reduce a patient's time on ICU ventilation and aid in quicker Respiratory recovery.

(Continues)

Condition					
Dystonia	Not Known	58,970 [1,049]	Not known	Not known	n/a	Dystonia affecting the ankle can be treated with FES
Functional neurological disorder	Not known	n/a	Not known	n/a	n/a	There has been reports of success with some people with this condition. There is no information on what proportion may be assisted with FES
Guillain-Barre syndrome	1,200 [21]	Not known	>100	n/a	n/a	Normally recovers spontaneously after many months. Does not normal respond to FES. Some evidence that some people may benefit from FES late on if spontaneous recovery has plateaued.
Hemiplegia (at birth)	Not known	55,600 [989]	20,000? [356]	Not known	n/a	A subgroup of cerebral palsy. Will respond to FES
Hereditary spastic paraparesis	Not known	1,000 [18]	200+[3.6]	n/a	n/a	Does respond to FES. About 10% of people use FES for walking
Irritable bowel syndrome – constipation	Not known	2,750,000 [48,932]	n/a	n/a	n/a	A common digestive disorder affecting 7-21% of the general population. Symptoms; abdominal pain, bloating or discomfort, changes in the appearance or frequency of bowel movements. As FES increases transit time of Peristalsis, IBS-C should be helped by FES.
Motor neurone disease	1,124 [20]	5,000 [89]	Not known	Not known	n/a	Usually does not respond to FES. Some reports of short-term use of FES to improve ankle movement
Multiple sclerosis	4,120 [73]	90,590 [1,612]	30,000 [534]	3,000 53]	n/a	FES is beneficial for walking over the middle part of the progression of FES. This means up to 1/3rd of people may benefit from FES for dropped foot. Constipation effects about 50%
Multiple system atrophy	321 [5.7]	2,354 [42]	Not known	Not known	n/a	Similar to PD. It is not known if FES is useful but is a possibility.

(Continues)

Appendix B (Continued)

Condition	Incidence per year in England [per 1M] [5]	Prevalence in England [per 1M] [5]	Number who may use FES for walking [per 1M]	Number who may use FES for upper limb [per 1M]	Number of other application users [per 1M]	Notes
Myelopathy	Not Known	Not Known	Not Known	Not Known	n/a	Muscle weakness due to a central cause. Can respond to FES
Paediatric constipation Idiopathic 0–19yrs	15,560 [277]	Not known	n/a	n/a	Not known	1 in every 3 children will need medication for constipation at some point in their childhood. There is evidence that FES does help children suffering and is offered at many private Physio practices.
Parkinson's disease	19,461 [346]	121,927 [2,170]	40,000 [712]	Not known	n/a	FES may be appropriate for moderately effected people with PD. This is about 40,000 of the UK population. It is not known what proportion will respond to FES.
Pressure ulcers Gen	412,000 [7,331]		n/a	n/a	Not known	FES may be appropriate to help distribute pressure on high-risk contact areas and could assist pressure sore prevention on older people with low mobility, critical care patients or disabled patients when sitting or lying in one position for prolonged periods are unavoidable.
Pressure ulcers Sci	1,500 [27]	17,500 [311]	n/a	n/a	Not known	FES has been applied to muscles for *pressure sore prevention*; *FES* induces a change in the *pressure* distribution on the contact area during sitting or lying, so could prove beneficial.
Post-polio syndrome		100,800 [1,794]	Not Known	n/a	n/a	Acute polio does not respond to FES. There is some evidence that FES may be beneficial to some people in long-term recovery.

(Continues)

Progressive supranuclear palsy	2,040 [36]	6,500 [116]	Not Known	n/a	n/a	Similar to PD but more progressive. One report of a person with PSP who responded well for a few months, reducing the frequency of falls.
Respiratory function In Sci	100 [1.8]	2,000 [36]	n/a	n/a	Not known	The ability to breathe deeply and cough forcefully is impaired to varying degrees depending on the level and completeness of the SCI, with greater dysfunction seen at higher injury levels. Evidence to date suggests that FES can improve respiratory ability in this group of patients.
Respiratory function – ICU acquired weakness	Approx. 100,000 [1,779]	100,000 [1,779]	n/a	n/a	Not known	Approx. 40% of ICU patients may suffer with acquired respiratory weakness. 258,000 ICU patients were seen in NHS hospitals 2016–2017, this number will be increasing, especially with COVID-19 patients requiring lung support. A recent study by Dr McCaughey investigated whether abdominal FES is a safe and effective to reduce mechanical ventilation duration for critically ill mechanically ventilated patients. The results of two studies that show abdominal FES may reduce mechanical ventilation duration and ICU length of stay for the critically ill, thus reducing incidents of ICU related weakness or muscle wasting.
Sexual dysfunction neurogenic	750 [13]	Not known	n/a	n/a	Not known	*FES* can be a beneficial treatment option when the cause of Erectile Dysfunction is related to degeneration of cavernous smooth muscle. Little research has been conducted to see if FES could be beneficial for female sexual functioning.

(Continues)

Appendix B (*Continued*)

Condition	Incidence per year in England [per 1M] [5]	Prevalence in England [per 1M] [5]	Number who may use FES for walking [per 1M]	Number who may use FES for upper limb [per 1M]	Number of other application users [per 1M]	Notes
Spina bifida	200 [3.6]	Not Known	Not Known	n/a	n/a	Very little experience but lesions in the thoracic or cervical cord may allow FES to be used.
Spinal cord injury	2,500 [44]	50,000 [890]	6000? [107]	Not known	n/a	FES applicable to patients with lesions above T12. Incomplete injury patients do benefit from dropped foot stimulators. Upper limb FES is commonly used in recovery phase. Long-term orthotic use is possible.
Stroke	80,000 [1,423]	1,000,000 [17,793]	100,000 [1,779]	100,000 [1,779]	n/a	About 65% after the age of 75 years and older stroke survivors are less likely to become FES users. About 10% may use FES for walking or hand function.
Transverse myelitis	253 [4.5]	3,820 [70]	Not known	Not known	n/a	Successful treated with FES. Proportion of patients for whom FES is suitable is unknown
Traumatic brain injury	140,000 [2,491]	1,095,152 [19,487]	Not known	Not known	n/a	While FES can be used for some people with TBI, the proportion for which it is appropriate for is unknown.

References

[1] *Stroke Rehabilitation in Adults*. NICE Guideline [NG236]. London: National Institute for Health and Care Excellence; 2023. Available at: https://www.nice.org.uk/guidance/ng236. Accessed 12 July 2024.

[2] *National Clinical Guideline for Stroke for the UK and Ireland*. London: Intercollegiate Stroke Working Party; 2023. Available at: www.strokeguideline.org. Accessed 12 July 2024.

[3] Bulley C, Adonis A, Burridge J, *et al. Evidence Based Clinical Practice Guidelines for the Use of Functional Electric Stimulation to Improve Mobility in Adults With Lower Limb Impairment Due to an Upper Motor Neuron Lesion.* Queen Margaret University and ACPIN; 2022. Available at: https://www.acpin.net/pdfs/2210%20QMU%20Report.pdf. Accessed 12 July 2024.

[4] Johnston TE, Keller S, Denzer-Weiler C, and Brown L. A clinical practice guideline for the use of ankle-foot orthoses and functional electrical stimulation post-stroke. *J Neurol Phys Ther*. 2021;45(2):112–196.

[5] *Neuro Numbers* 2019. A report by the Neurological Alliance; 2019. Available at: https://www.neural.org.uk/wp-content/uploads/2021/04/neuro-numbers-2019-1.pdf. Accessed 12 July 2024.

[6] GBD 2016 Multiple Sclerosis Collaborators. Global, regional, and national burden of multiple sclerosis 1990–2016: a systematic analysis for the global burden of disease study 2016. *Lancet*. 2019;18(3):269–285. doi:https://doi.org/10.1016/S1474-4422(18)30443-5.

[7] Jacobs LD, Wende KE, Brownscheidle CM, *et al*. A profile of multiple sclerosis: the New York State Multiple Sclerosis Consortium. *Mult Scler*. 1999;5(5):369–376.

[8] Abouzakhm N, Choy S, Feld R, *et al*. Evaluating the validity of a functional electrical stimulation clinical decision making tool: a qualitative study. *Front Neurol*. 2022;13:1001123. Accessed 12 July 2024.

[9] Abouzakhm N, Choy S, Feld R, *et al*. Evaluating the validity of a functional electrical stimulation clinical decision making tool: a qualitative study. *Front Neurol*. 2022;13:1001123. doi:10.3389/fneur.2022.1001123.

[10] Manca A, Hawkins N and Sculpher MJ. Estimating mean QALYs in trial-based cost-effectiveness analysis: the importance of controlling for baseline utility. *Health Econ*. 2005;14:487–496. doi:10.1002/hec.944.

[11] EQ-5D-5L. Rotterdam: EuroQol. Available at: https://euroqol.org/information-and-support/euroqol-instruments/eq-5d-5l/. Accessed 12 July 2024.

[12] Taylor P, and Street T. *The Case for the Odstock Dropped Foot Stimulator (ODFS®). A Summary of the Published Evidence for the Odstock Dropped Foot Stimulator*. National Clinical FES Centre; 2022. Available at: https://odstockmedical.com/wp-content/uploads/The-Case-for-the-Odstock-Dropped-Foot-Stimulator-Dec-2022.pdf. Accessed 12 July 2024.

[13] Swain ID, Taylor PN, Burridge JH, Hagan SA, and Wood DE. Report to the development evaluation committee: Common peroneal stimulation for the correction of drop-foot. National Clinical FES Centre; 1996. Available at: https://odstockmedical.com/wp-content/uploads/the_dec_report.pdf. Accessed 12 July 2024.

[14] Taylor P, Mann G, Jolley C, and Swain I. Economic Justification for the Odstock Dropped Foot Stimulator (ODFS®). *International Society for Prosthetics and Orthotics annual scientific meeting*, 3 Nov 2007.

[15] Taylor P, Humphreys L, and Swain I. The long-term cost-effectiveness of the use of functional electrical stimulation for the correction of dropped foot due to upper motor neuron lesion. *J Rehabil Med.* 2013;45:154–160.

[16] Singleton C and Street T. The effect of dropped foot stimulation on walking speed for people with multiple sclerosis – a longitudinal study. In: *Proceedings of the International FES Society UK and Ireland Chapter Scientific Meeting*, 8–9 May 2015, Sheffield, UK.

[17] Taft R, Lovell A, and Cook R. *Economic Report: Functional Electrical Stimulation for Dropped Foot of Central Neurological Origin.* CEP10012. Centre for Evidence-based Purchasing; 2010. Available at: https://odstock-medical.com/wp-content/uploads/econ-report.pdf. Accessed 12 July 2024.

[18] Barrett CL, Mann GE, Taylor PN, and Strike P. A randomized trial to investigate the effects of functional electrical stimulation and therapeutic exercise on walking performance for people with multiple sclerosis. *Mult Scler.* 2009;15(4):493–504.

[19] Street T, Taylor P, and Swain I. Quality of Life following the use of functional electrical stimulation for multiple sclerosis. In: *Proceedings of the International FES Society UK and Ireland Chapter Scientific Meeting*, 8–9 May 2015, Sheffield, UK.

[20] Juckes FM, Marceniuk G, Seary C, and Stevenson VL. A cohort study of functional electrical stimulation in people with multiple sclerosis demonstrating improvements in quality of life and cost-effectiveness. *Clin Rehabil.* 2019;33(7):1163–1170.

[21] Renfrew LM, Paul L, McFadyen A, *et al.* The clinical- and cost-effectiveness of functional electrical stimulation and ankle-foot orthoses for foot drop in multiple sclerosis: a multicentre randomized trial. *Clin Rehabil.* 2019;33(7): 1150–1162. doi:10.1177/0269215519842254.

[22] Esnouf JE, Taylor PN, Mann GE, and Barrett CL. Impact on falls and activities of daily living of use of a functional electrical stimulation (FES) device for correction dropped foot in people with multiple sclerosis. *Mult Scler.* 2010;16:1141–1147.

[23] Nurmi I, and Lüthje P. Incidence and costs of falls and fall injuries among elderly in institutional care. *Scand J Prim Health Care.* 2002;20(2):118–22.

[24] Shirley Ryan AbilityLab. *Rehabilitation Measures Database.* Available at: https://www.sralab.org/rehabilitation-measures. Accessed 12 July 2024.

[25] Moore JL, Potter K, Blankshain K, Kaplan S, O'Dwyer L, and Sullivan J. A core set of outcome measures for adults with neurologic conditions

undergoing rehabilitation: a clinical practice guideline. *J Neurol Phys Ther.* 2018;42(3):174–220. doi:10.1097/NPT.0000000000000229.

[26]   Benson K and Hartz AJ. A comparison of observational studies and randomized controlled trials. *N Engl J Med.* 2000;342:1878–86.

[27]   Perotto AO. *Anatomical Guide for the Electromyographer.* Springfield, IL: Charles C Thomas; 2011.

[28]   Badger J, Taylor P, and Swain I. The safety of electrical stimulation in patients with pacemakers and implantable cardioverter defibrillators: a systematic review. *J Rehabil Assist Technol Eng.* 2017;4:2055668317745498.

[29]   Egger F, Hofer C, Hammerle FP, *et al.* Influence of electrical stimulation therapy on permanent pacemaker function. *Wien Klin Wochenschr.* 2019;131 (13–14):313–320. https://doi.org/10.1007/s00508-019-1494-5.

[30]   Crevenna R., Wolzt M, Fialka-Moser V, *et al.* Long-term transcutaneous neuromuscular electrical stimulation in patients with bipolar sensing implantable cardioverter defibrillators: a pilot safety study. *Artif Organs.* 2004;28:99–102.

[31]   Krassioukov A, Linsenmeyer TA, Beck LA, *et al.* Evaluation and management of autonomic dysreflexia and other autonomic dysfunctions: preventing the highs and lows. *J Spinal Cord Med.* 2021;44(4):631–683. doi:10.1080/10790268.2021.1925058.

[32]   Asley EA, Laskin JJ, Olenik LM, *et al.* Evidence of autonomic dysreflexia during functional electrical stimulation in individuals with spinal cord injuries. *Paraplegia.* 1993;31:593–605.

[33]   Medine Plus. *The National Library of Medicine.* Available at: https://medlineplus.gov/highbloodpressure.html. Accessed 12 July 2024.

[34]   Sampson EE, Roberts MD, Burnham RS, and Andrews BJ. Functional electrical stimulation effect on orthostatic hypotension after spinal cord injury. *Arch Phys Med Rehabil.* 2000;81:139–143.

[35]   Taylor PN, Tromans AM, Harris KR, and Swain ID. Electrical stimulation of abdominal muscles for control of blood pressure and augmentation of cough in a C3/4 level tetraplegic. *Spinal Cord.* 2002;40:34–36.

Chapter 12

Neurogenic bladder, bowel and sexual management

Sarah Knight[1,2], Sean Doherty[2] and Tamsyn Street[3]

12.1 Introduction – basic anatomy and physiology with clinical considerations

Maintaining control of the bladder and bowel is normally achieved during the early years whilst control of sexual function is more variable and is usually achieved in later years. In adults, the ability to defer bladder or bowel emptying and maintain continence are taken for granted, therefore, loss of this control following neurological damage or disease, can have a devastating effect on quality of life. Control of bladder, bowel and sexual function relies on a complex interaction between the voluntary and autonomic nervous systems, along with a number of reflexes. Any neurological damage through either disease or trauma can have a profound effect on these functions, leading to incontinence, infections, impotence and reduced quality of life [1]. Bladder, bowel and sexual functioning are high priorities for people living with neurological conditions such as spinal cord injury (SCI) and multiple sclerosis (MS) [2,3]. Developments in electrical stimulation technologies, have led to a variety of techniques which aim to manage the symptoms of the neurological damage, as well as correct the underlying neurological deficit. This has been achieved through both implanted and wearable devices using neurostimulation and neuromodulation [3]. This chapter aims to outline some of these developments and their practical applications.

12.1.1 Bladder function

The lower urinary tract comprises the bladder, urethra and associated sphincters including the bladder neck, external urethral sphincter and pelvic floor. The bladder has two main functions, to act as a safe storage reservoir for the output of the kidneys, and to empty this reservoir safely and efficiently at socially convenient times. Neurological innervation (Figure 12.1) of the bladder and

[1]London Spinal Cord Injury Centre, Royal National Orthopaedic Hospital, Stanmore, UK
[2]ASPIRE CREATE, Department of Medical Physics and Biomedical Engineering, University College London, UK
[3]Clinical Sciences and Engineering, Salisbury NHS Foundation Trust, Salisbury, UK

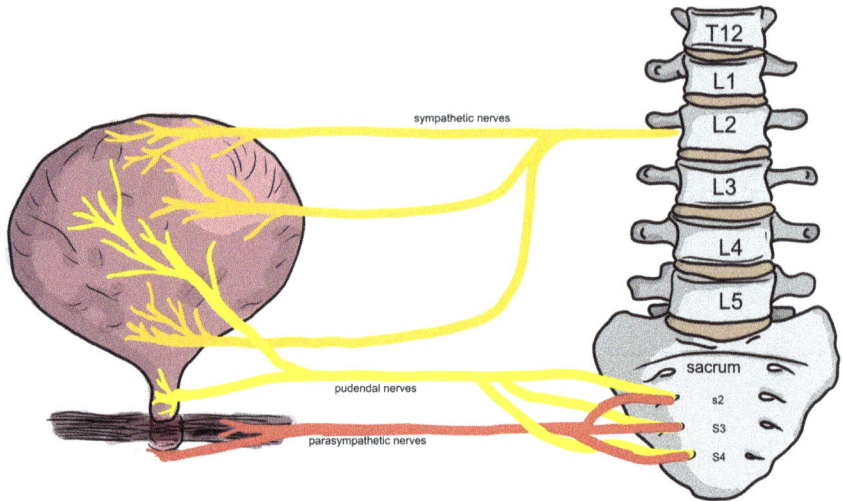

Figure 12.1   Innervation of bladder, sphincter and pelvic floor. © Sarah Knight 2024.

associated sphincters is predominantly through the autonomic nervous system (parasympathetic and sympathetic) and reflexes but there is also a degree of voluntary control which is gained through the period known as 'potty training' which enables continence and social convenient bladder emptying. The para-sympathetic innervation of the bladder and sphincters arises from the sacral segments (S2-S4) and runs through the pelvic nerve; parasympathetic innerva-tion causes contraction of the bladder (detrusor muscle) and relaxation of the urethra. Sympathetic innervation arises in T12-L2 levels of the spinal cord and runs via the hypogastric nerve; it leads to relaxation of the bladder wall and contraction of the sphincters. The somatic (voluntary) control is via the pudendal nerve, again originating in S2-S4. The main higher centres of control lie in the cerebrum where initiation of bladder emptying in response to bladder sensations occurs. An area of the pons, known as the pontine micturition centre, is responsible for the co-ordination of bladder and sphincter through both sto-rage and voiding phases.

Disruption to the neurological innervation of the lower urinary tract leads to development of neurogenic lower urinary tract dysfunction (NLUTD). In supra-sacral spinal cord lesions, in addition to loss of control of voluntary voiding and sensation of bladder fullness, neurogenic detrusor over-activity (NDO) and detrusor-sphincter dys-synergia (DSD) will develop in response to the loss of descending control from the pontine micturition centre [1]. This can lead to reduced bladder capacity, high bladder storage pressures, incontinence and potential damage to the upper tracts [4]. 70–84% of patients with SCI will develop lower urinary tract symptoms that require management in order to protect the upper tracts [5]. The basis of good management of the neurogenic bladder is restoring safe storage pressures and providing an efficient means of complete emptying of the bladder. Safe storage pressures are most commonly achieved through pharmaceutical interventions. These may include anticholinergic medication which can block contraction of the detrusor muscle, or injections of botulinum toxin into the detrusor muscle to temporarily cause paralysis. However, these techniques can have side effects including dry mouth, constipation and they have recently been implicated in a potential decline in cognitive function following long term use. Bladder emptying is most commonly achieved through intermittent or permanent catheterisation which can lead to increased risk of infection. These management options do not restore normal function, but simply address the resulting dysfunction. Developments in electrical stimulation technologies, using neurostimulation and neuromodulation, may offer the potential to restore bladder function by attempting to utilise neuroplasticity of the underlying circuits, without the side effects of medications.

12.1.2 Bowel function

The lower gastro-intestinal tract consists of the large intestine (ascending, transverse, descending and sigmoid colon), the rectum and the anal sphincters. Similarly to the bladder, bowel function relies on an interaction of the voluntary and autonomic nervous systems, but there is also a level of intrinsic control (Figure 12.2). The parasympathetic and sympathetic innervation of the bowel arises in the same areas as that of the bladder. Parasympathetic activity increases motility and peristalsis and maintains tone, whilst sympathetic activity contracts the sphincters. Defecation occurs when stool enters the rectum and the sensation of urge is experienced. If it is socially convenient, the internal and external anal sphincters will relax, allowing stool to pass.

Neurogenic bowel dysfunction (NBD) will occur when there is disease or trauma of the neurological circuits that control bowel function. Neurogenic bowel following a SCI is described as being either reflexic, flaccid or mixed depending on the level of the lesion [6]. A reflex bowel is associated with a lesion above T12, which leaves the reflex arc between the spinal cord, the colon and the anorectum intact. In contrast, a flaccid bowel (injuries to L1 and below) results in a lax sphincter and pelvic floor muscles with damage to the reflex arc between the spinal cord, the colon and the anorectum. A lesion between T12 and L2 may show aspects of both a reflex and flaccid bowel.

The majority of individuals with a SCI will suffer from NBD including constipation, faecal incontinence or abdominal pain [7]. For patients with a reflex bowel, bowel emptying can be facilitated through the use of digital stimulation alongside a suppository or alone. These interventions make use of the intact reflex arc enabling the facilitation of a bowel movement. The use of laxatives and transanal irrigation may also be required to ensure complete bowel emptying and to reduce the risk of faecal incontinence [8]. The duration of time required for bowel emptying is strongly associated with impact on quality of life. Faecal incontinence is estimated to be present in around 75% of people with a SCI. Faecal incontinence may or may not be associated with incomplete bowel

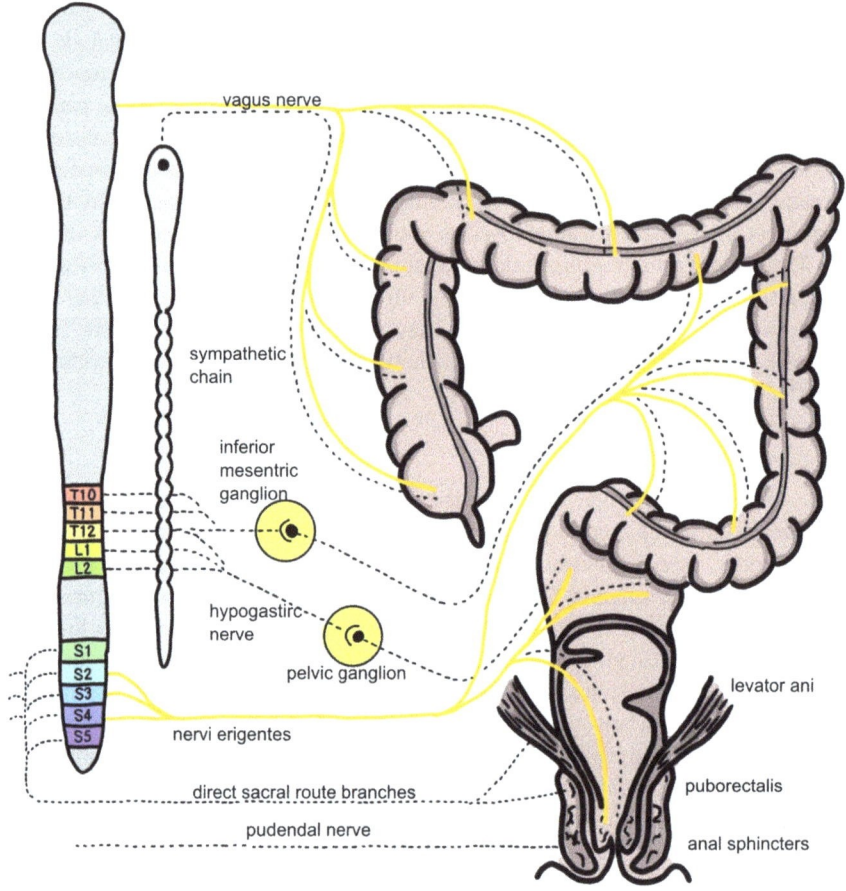

Figure 12.2   Innervation of bowel and sphincter. © Sarah Knight 2024.

emptying and is associated with having a profound effect on quality of life and may cause severe restriction on social activities. Inadequate bowel management is also associated with a life-threatening condition called autonomic dysreflexia (AD) which is an uninhibited sympathetic nervous system response occurring in patients with SCI at the thoracic six (T6) level and above. AD can in rare cases occur below the T6 level.

A systematic review of the evidence for the effectiveness of treatments targeted at neurogenic bowel specifically for SCI included 52 studies [9]. Although multi-faceted programs incorporating a wide variety of treatments are the first line approach to treating neurogenic bowel, they were associated with lower levels of evidence in the review suggesting the need for further research in the area. Of the non-pharmacological (conservative and non-surgical) interventions, there was good evidence for the use of transanal irrigation to reduce constipation and fecal incontinence.

However, a reduced compliance with irrigation techniques was also found over time. The reduced compliance may be due to a number of factors such as usability and the experience of the treatment. There was strong evidence for using prokinetic agents when conservative management was not effective. However, prokinetic agents may be associated with a number of side effects such as abdominal pain and diarrhoea [10]. The systematic review [9] further found there were lower levels of evidence supporting the use of surgical interventions, when conservative and pharmacological treatments had failed, suggesting the need to focus on interventions and good bowel management strategies at an early stage. The review also found some promising initial research in the area of electrical stimulation for bowel management.

12.1.3   Sexual function

Sexual dysfunction is a commonly occurring physical syndrome, which affects both men and women [11]. Sexual dysfunction encompasses a multitude of different individual physical characteristics including sexual desire dysfunction, sexual arousal disorders, erectile dysfunction (ED), persistent sexual arousal dysfunction, early ejaculation, delayed ejaculation, orgasmic dysfunction, anejaculation, dyspareunia, vaginismus and sexual aversion [12]. Sexual dysfunction for women is often characterised by reduced sensation in the genital areas and reduced lubrication which may make intercourse difficult. Many women with a SCI are often able to enjoy an active sex life when provided with the right support and knowledge [13]. Fertility is not usually affected by SCI, though patients should be carefully monitored during pregnancy and birth for complications such as urinary tract infections and AD. Interventions for sexual dysfunction for women are beyond the scope of this chapter. See [13] for further information.

ED, defined as the "inability to attain or maintain a penile erection sufficient for successful vaginal intercourse" [14], is a common condition affecting 42.6% of males in the United Kingdom [15]. ED is a multi-factorial condition with a complex pathophysiology. Male sexual function, erection and ejaculation, relies on a combination of psychogenic factors and lumbosacral reflexes. The ejaculatory reflex (ER) is a complex set of events culminating in projectile expulsion of semen from the urethra. The reflex is usually a combination of cerebral and genital input, co-ordinated in the thoracolumbar area of the spinal cord. The afferent input from tactile sensation is via dorsal penile nerves lying in the pudendal nerves (S2-S4). The efferent reflex controlling seminal emission and bladder neck closure arises from sympathetic fibres T11-L2 lying in the hypogastric nerve (Figure 12.3).

Following SCI or other neurological damage or disease to these pathways, sexual dysfunction may occur. Sensation in the genital regions may be lost or impaired. An inability to achieve a sustained erection (reflex erections may remain but may be insufficient for intercourse) is common. Anejaculation or retrograde ejaculation may also occur. Male fertility can also be affected through degradation of sperm quality, Table 12.1 provides an overview of interventions used for ED,

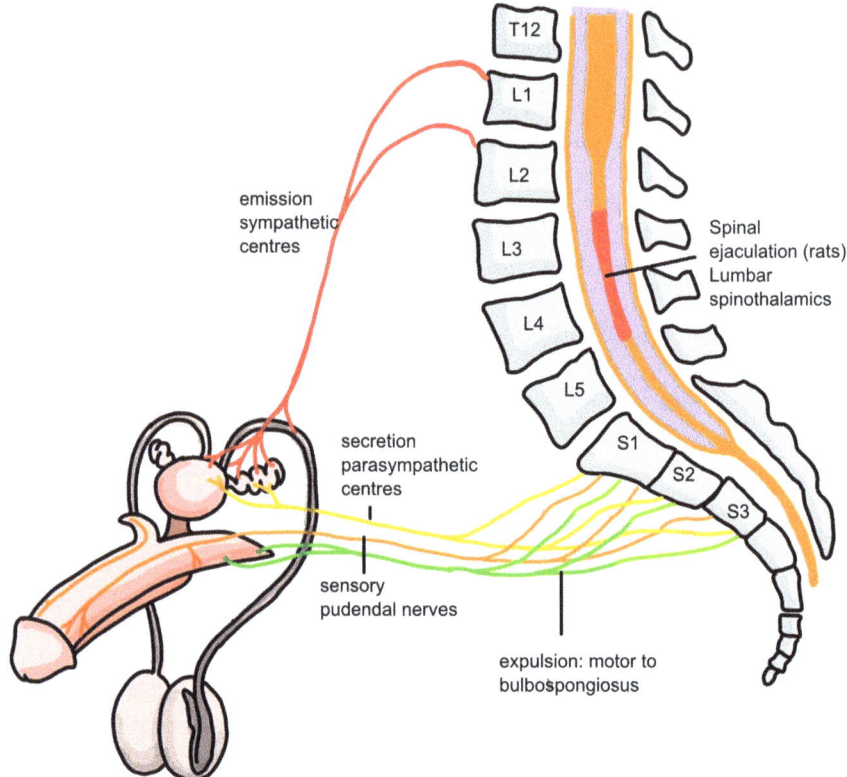

Figure 12.3   Innervation of male sexual organs. © Sarah Knight 2024

Table 12.1   Interventions used for erectile dysfunction, proposed mechanism and associated risks and adverse events

Intervention	Mechanism	Risks, adverse events and issues
Lifestyle modification	Physical activity inducing the reversal of causative conditions including diabetes, vascular disease.	Challenging amongst patients with reduced mobility.
PDE-5 inhibitors	Relaxation of smooth muscle, resulting in increased blood flow.	Side effects include tachycardia, headaches, internal haemorrhage, and allergic reactions. Patients may also experience tachyphylaxis (resistance to treatment)
Intracavernosal injection	Direct relaxation, and dilation or Corpus Cavernosum	Pain at injection site, infection, fibrosis and priapism.
Vacuum erectile device	Increased blood flow into corpus cavernosum through negative pressure.	Ischaemic injury, subcutaneous haemorrhage, sensation of 'Cold-Penis'.

mechanism and associated issues. Currently, ED in males is treated with phos-phodiesterase type -5 inhibitors such as oral sildenafil (Viagra) injectables an aprostadil injection which is injected directly into the penile tissue. Vacuum erec-tile devices and lifestyle modifications may also be used to improve erectile functioning. Penile vibratory stimulation (PVS) can be used to elicit reflex ejacu-lation in people with an intact reflex. Sperm can also be retrieved surgically directly from the epididymis or testicular tissue.

12.2   Techniques for treatment of neurogenic bladder, bowel and sexual dysfunction using electrical stimulation technologies

12.2.1   Overview

Many of the pharmaceutical and surgical solutions for bladder, bowel and sexual dysfunction after SCI have unwanted side effects, which may limit their com-pliance or take up. Electrical stimulation through both implanted and non-invasive (transcutaneous) electrodes have been investigated extensively for many years for its role in the restoration of lower urinary tract, bowel and sexual function after neurological lesions. Electrical stimulation is normally classified as neurostimula-tion if it is stimulating the organ directly via the efferent nerve, or neuromodulation if the stimulation uses reflex pathways to modulate function usually through the afferent nerves. A number of different anatomical and neurological targets have been investigated, these have included stimulation of the skin, bladder/bowel, peripheral nerves, sacral roots and spinal cord [16,17]. Two of the most important developments in the restoration of bladder, bowel and sexual function are the Finetech Brindley Sacral Anterior Root Stimulator (SARS) Implant and the Sacral Nerve Neuromodulation implants which are now commercially available. There are on-going research studies in other areas, including genital nerve stimulation and spinal cord stimulation which will be discussed, but which have not yet produced commercially available devices. Table 12.2 provides an overview of different electrical stimulation sites and effect on bladder, bowel and sexual functioning.

12.2.2   Treatment of neurogenic bladder function using electrical stimulation

The principle aims of using electrical stimulation to restore bladder function after neurological lesion are to provide a good capacity, low pressure storage reservoir that enables continence, and a means of emptying the bladder efficiently and safely. Early investigations by Nashold *et al.* [18] of the effect of direct stimulation of the detrusor muscle were unsuccessful mainly due to the high voltage required which caused pro-blems with pain, thermal damage and high bladder pressures which were potentially damaging to the upper tracts. Subsequent research investigated the use of electrical stimulation of the sacral roots, (anterior/motor/ventral and posterior/sensory/dorsal) which led to the development of two commercially available implantable devices, the Finetech Brindley SARS and the Sacral Neuromodulation (SNM) devices.

Table 12.2   Different stimulation sites and effect on bladder, bowel and sexual function

Anatomical site	Type of electrode	Bladder	Bowel	Sexual function
Sacral roots (S2-S4)	Implanted book electrodes	Bladder emptying	Bowel emptying	Erection
Sacral nerves (S3)	Tined electrodes	Improving continence and capacity	Improving continence	
Dorsal genital nerve (pudendal nerve)	Surface electrodes on penis or clitoris	Improving continence and capacity	Improving continence	
Genital nerve (pudendal nerve)	Implanted cuff electrodes	Improving continence and capacity	Improving continence	
Tibial nerve	Percutaneous, implanted or transcutaneous electrodes	Improving continence and capacity	Improving continence	
Spinal cord	Transcutaneous or epidural electrodes	Potentially emptying and continence	Potentially emptying and continence	Potentially erection and ejaculation
Rectum	Anal probe electrode			Ejaculation
Bladder wall	Catheter mounted electrode	Bladder emptying		
Abdominal wall	Surface electrodes		Bowel emptying	Ejaculation

12.2.2.1   Finetech-Brindley sacral anterior root stimulator

Brindley [19] developed an implantable electrical stimulation device in the 1970s which comprised electrodes implanted directly onto the sacral anterior nerve roots (S2-4) which when stimulated caused bladder contractions which were sufficient to completely empty the bladder (Figure 12.4). Stimulation of the anterior roots causes a concomitant contraction of both the detrusor muscle and the external urethral sphincter. This can lead to an implant driven dys-synergia which prevents bladder emptying. However, Brindley discovered that he could utilise the different physiological responses of smooth (detrusor) and striated (sphincter) muscle to electrical stimulation to overcome this problem, and by using a burst of stimulation, efficient post-stimulus voiding could be achieved (Figure 12.1). The initial implants were very successful at bladder emptying, but did not overcome the NDO, which still lead to reduced bladder capacity and incontinence. Therefore, later implants were combined with a sacral deafferentation or posterior rhizotomy of the

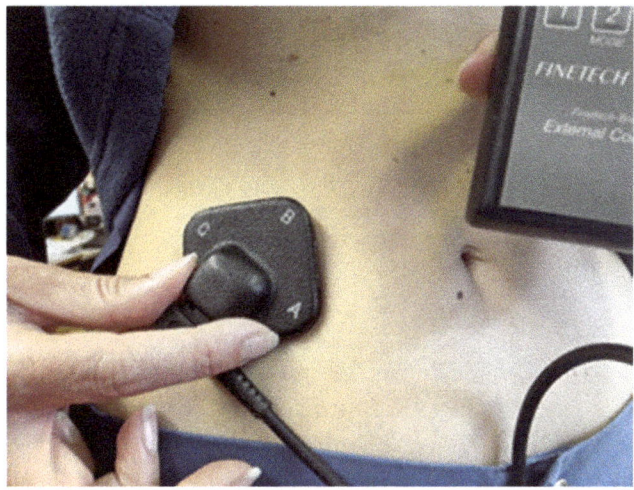

Figure 12.4   External components of Finetech-Brindley sacral anterior stimulator implant. © Sarah Knight 2024

S2-4 nerve roots to abolish the reflex arc responsible for NDO and DSD. The implanted components include the electrodes (either intrathecal or extradural) and cables to a radio receiver block which is normally located on the chest wall or abdomen (Figure 12.2). The power supply and stimulation programming are located in an external control unit. Sacral anterior root stimulation also enables bowel emptying and production of implant driven erections. This device has been used successfully world-wide over the last 50 years to restore bladder, bowel and sexual function to people following a clinically complete SCI.

Implantation of the Finetech Brindley SARS is only performed in specialist centres, where surgeons have been trained in the technique. The operation has been carried out by both Neurosurgeons and Urologists. Intra-operative identification of appropriate nerve roots requires specialist equipment and skills and is normally combined with intra-operative urodynamics to identify which sacral roots provide the best bladder contraction. Following implantation, the device is set up with different programmes for bladder and bowel evacuation and implant driven erections. Stimulation of the S3 or S4 anterior sacral roots at 30 Hz, pulse width 200–300 microseconds, causes an immediate contraction of the sphincters, but a slower response in the detrusor muscle as described above. By delivering the electrical stimulation in bursts (typically 3 s on 5 s off) urine can be expelled from the bladder when the sphincters relax (during off phase) but whilst the bladder is still contracting. Whilst unphysiological, the post-stimulation burst voiding is incredibly efficient. (Figure 12.5). Although the Finetech-Brindley SARS is a very effective management technique for bladder and bowel management following SCI [20], popularity has reduced due to a number of factors including loss of surgical expertise, and reluctance amongst patients to undergo sacral deafferentation which

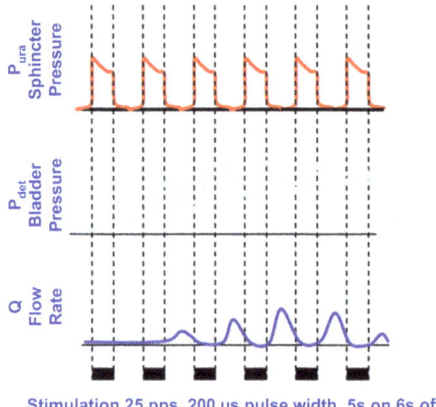

Stimulation 25 pps, 200 μs pulse width, 5s on 6s off

Figure 12.5   Post stimulus voiding using the Finetech-Brindley sacral anterior root stimulator implant. © Sarah Knight 2024

is perceived as destructive, especially in the light of the possibility of spinal cord regeneration, and the popularity of intra-detrusor botulinum toxin injections. These factors have led to the search for alternative methods to control NDO which would eliminate the need for sacral deafferentation, but still enable the use of sacral anterior root stimulation for bladder and bowel management. At present, there is no solution to this issue and further research is required into the role of combined neurostimulation, pharmaceuticals and nerve blocking.

12.2.2.2   Sacral nerve neuromodulation

At a similar time to Brindley developing the sacral anterior nerve root stimulator, Tanagho and Schmidt [21] developed a technique for stimulation of the extradural sacral nerve roots to target afferent fibres entering the spinal cord to modulate the micturition reflex thereby reducing bladder over-activity. This was commercialised as the Medtronic Interstim which comprised an implanted pulse generator and tined electrodes (with small forks to reduce displacement) which were placed adjacent to the sacral nerve roots (normally unilateral or bilateral S3) accessed via the sacral foramina (Figure 12.6) Patients normally undergo an initial trial period with percutaneous temporary electrodes to evaluate efficacy before implantation of the sacral nerve electrodes and implanted pulse generator. The electrode implantation does not involve a laminectomy, making it much less invasive than the Finetech-Brindley SARS, however it does not enable bladder emptying. This form of stimulation is termed SNM and has been used successfully to control over-active bladder (OAB) in people without neurological damage, and to a limited extent in people with NDO [22]. However, due to a lack of randomised controlled trials in people with a neurogenic bladder it is not widely used in this group. It has also been used to control faecal incontinence and urinary retention on people without neurological damage. Stimulation parameters for neuromodulation are usually lower than those of neurostimulation. Typical stimulation frequencies are 10–15 Hz and

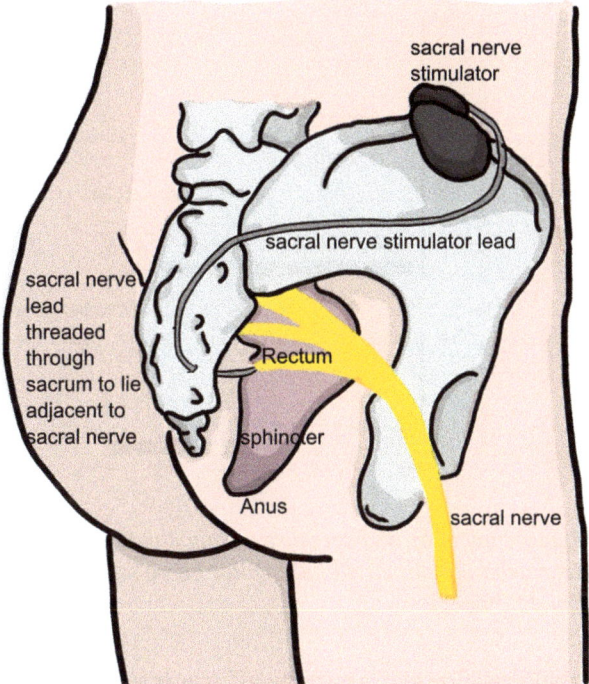

Figure 12.6 Sacral nerve stimulation through commercially available implanted pulse generator. © Sarah Knight 2024

amplitude is normally sub-motor threshold. The original Medtronic Interstim device was not MRI compatible however, Axionics developed a similar device which is MRI compatible.

Kirkham *et al.* [23] investigated whether sacral nerve neuromodulation could be combined with sacral anterior root stimulation in a single implant that could empty the bladder without the need for rhizotomy. Extradural electrode placement (where motor and sensory nerve fibres are combined, enabled neuromodulation of NDO, but caused a reflex dys-synergia of the sphincters which prevented complete bladder emptying, even with burst stimulation. However, intrathecal electrodes placed on the separated motor and sensory nerve roots enabled both neuro-stimulation and neuromodulation in a single case study [23]. Although this technique was successful in producing both bladder emptying and increased capacity with neuromodulation, further work to develop and implement the technique into clinical practice has not been completed.

12.2.2.3 Pudendal nerve (genital) stimulation

The pudendal nerve originates from the S2-4 segments of the sacral spinal cord and electrical stimulation has been shown to both elicit and suppress voiding reflexes

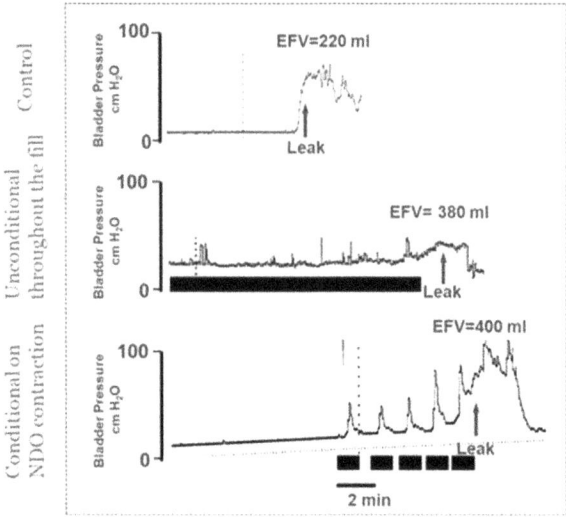

Figure 12.7   Bladder capacity and pressure during conditional and unconditional
neuromodulation via dorsal genital nerve stimulation © Sarah
Knight 2024

[24,25]. The dorsal genital nerve is a purely afferent branch of the pudendal nerve which due to its superficial path, can be easily stimulated through non-invasive surface electrodes placed on the dorsum of the penis or adjacent to the clitoris. Stimulation of the DGNS activates a reflex contraction of the external urethral and anal sphincter which in turn is thought to inhibit detrusor contractions. Surface electrical stimulation of the dorsal genital nerve has been extensively investigated and can successfully suppress NDO and increase bladder capacity in people with SCI [26,27]. A meta-analysis of previous studies [28] found stimulation parameters are usually in the range 10–30 Hz with lower frequencies having the advantage of lower power consumption. Pulse widths are generally in the region 200–300 μs. The optimised stimulation amplitude has been reported to be twice the threshold for the pudendo-anal reflex [26]. Stimulation can be applied continuously or conditionally (Figure 12.7). Conditionally neuromodulation requires the recognition of a trigger threshold; various triggers have been used including patients own sensation of urge, anal sphincter EMG [29] and bladder pressure [30]. Most of the studies that have been performed have been in the laboratory, with relatively few studies investigating the home use of DGNS [31,32]. Algorithms for closed loop systems, which identify bladder contractions and trigger stimulation to suppress NDO are currently being developed [33]. More recently, implantable genital nerve stimulators are being trialled.

12.2.2.4   Tibial nerve stimulation

The tibial nerve is a mixed nerve branch of the sciatic nerve which contains fibre projections from the L4-S3 spinal segment. Although not explicitly involved in the control of the lower urinary tract, it originates in the same spinal segments that do control bladder function. McGuire [34] first showed that tibial nerve stimulation could suppress detrusor over-activity in the 1970's. Commercially, tibial nerve (common peroneal) stimulation for control of over-activity of the micturition reflex was first introduced by Stoller in the 1990s [35] in non-neurogenic patients [36]. Percutaneous electrodes are normally inserted 4–5 cm cephalod to the medial malleolus and sensory and motor threshold (flexion of toe) are observed. Stimulation is generally delivered percutaneously in weekly 30 mins sessions with typical stimulation parameters of 20 Hz, 200 ms pulse width and 0.5–9 mA currents. Clinically, TNS is still most widely used for the treatment of OAB in non-neurogenic patients, and a number of RCTs against placebo, sham and pharmaceutical interventions have been reported in non-neurogenic patient groups [37,38]. The Urgent PC is a commercially available percutaneous tibial nerve stimulation device which is normally used in percutaneous stimulation (PTNS) clinics. Transcutaneous (TTNS) and PTNS of the tibial nerve has been successful in controlling OAB in non-neurogenic patients but the evidence for success in NLUTD is less compelling due to lack of high-powered trials. TTNS may be more acceptable in the home environment, whilst implantable devices for tibial nerve stimulation are currently being developed (Yamashiro *et al.*, 2019).

12.2.2.5 Intravesical electrical stimulation

Electrical stimulation using electrodes that are placed directly into the bladder via an electrocatheter have been used to increase bladder capacity and improve sensations of bladder filling in patients with myelomeningocele (spina bifida) and incomplete SCI. 15–20 weekly sessions of 90 minutes are given to retrain the bladder. Stimulation parameters are reported to be 1–10 mA, 2–6 ms pulse width and 60–90 Hz [39].

12.2.2.6 Spinal cord stimulation

Epidural spinal cord stimulation (eSCS) was originally developed for the treatment of pain. However, the technique has recently been used to investigate its potential for restoration of locomotor control in people with SCI (see Chapter 4). These studies have led to promising results in restoration of autonomic functions such as bladder and bowel function although research is still in its infancy [40–42]. Although the exact mechanisms of action are not properly understood, spinal cord stimulation is thought to modulate both afferent and efferent spinal pathways and alter the responsiveness of spinal networks and facilitate neuroplasticity [18]. Transcutaneous and epidural spinal cord stimulation systems have been investigated with respect to restoration of bladder function. Epidural stimulation requires the implantation of paddle or lead electrodes into the epidural space along with the implantation of a pulse generator. Transcutaneous spinal cord stimulation requires the application of surface electrodes over the relevant area of the spinal cord. Epidural stimulation may be associated with high complication rates due to the necessity for surgery, however the stimulation may be more targeted. Transcutaneous stimulation would require the patient to apply electrodes and may be more painful. In 2011, Harkema *et al.* [43] reported the first use of eSCS to enable standing in a person with SCI, however they also noted off-target functional improvements in bladder, bowel and sexual function and carried out further studies to optimise the effects. Further studies [41] reported spontaneous voiding and improved voiding efficiency with stimulation at the lower end of the stimulating electrode corresponding to L5-S1 using 30 Hz, however they did concede that the results were very patient specific. Transcutaneous electrical stimulation over the T11-L1 spinal cord has also demonstrated improvements in bladder capacity, and voiding efficiency [40]. Stimulation parameters include the use of high frequency (10 kHz) carrier frequencies and long pulse widths (1 ms). This is an exciting area of research and one to watch for future developments.

12.2.3   Treatment of neurogenic bowel function using functional electrical stimulation

12.2.3.1   Finetech-Brindley sacral anterior root stimulation

As described in the bladder section, the Finetech Brindley SARS device can also facilitate bowel emptying. Similar stimulation parameters are used for bowel management, but a continuous stimulation regimen is normally used as the issue of sphincter dys-synergia is not a problem for bowel evacuation. Typically, stimulation frequencies of 25–30 Hz with a 200–300 microseconds pulse width and above motor threshold amplitude are required. Patients normally stimulate for 20 s and then stool is evacuated or moved into the rectum where digital removal may be required. About 50% of patients with a SARS report bowel evacuation, whilst those who still have to use manual evacuation report significant decreases in the time spent on bowel management [8]. This intervention is only available to people with a complete SCI.

12.2.3.2   Dorsal genital nerve stimulation

There are far fewer studies investigating the role of dorsal genital nerve stimulation on neurogenic bowel compared to bladder. However, Worsoe *et al.* [44] investigated its role in reducing faecal incontinence in people with SCI. Using similar parameters to those for neurogenic bladder dysfunction (20 Hz, 200 μs pulse width and twice pudendo-anal threshold). Researchers reported a significant reduction in rectal cross-sectional area and postulated that this may prevent faecal incontinence in this patient group.

12.2.3.3   Transrectal stimulation

Transrectal stimulation has been reported as a technique for the treatment of NBD in children with myelomeningocele, but not in adults. In one study of 55 children with NBD due to spina bifida [45], a rectal probe was inserted 2–3 cm into the rectum and electrical stimulation of 15–20 Hz, 2–3 mA and 5 s ON and 3–4 s OFF performed in daily 30 min sessions for 2–3 weeks. Significant improvements in faecal continence, sensation, and ability to defer call to stool were reported. More recently, researchers conducted a SCI case study providing the participant with burst-pattern electrical rectal stimulation (ERS) (50 mA, 20 pulses/s at 100 Hz), via a rectal probe electrode until bowel emptying was achieved. This was provided over a six-week period at randomly selected bowel management sessions. Out of 17 session, 16 provided a bowel movement. Out of 17 sessions, 13 provided a complete bowel emptying [46]. As an implant solution this technique may provide an effective alternative to using digital stimulation to elicit the recto colic reflex to promote bowel emptying.

12.2.3.4   Sacral nerve neuromodulation

Medtronic and Axionics SNM devices have been used extensively for bowel management in idiopathic faecal incontinence. There are fewer studies investigating its use in people with NBD. However, Jarrett *et al*, [47] investigated the use of SNM in 13 patients with an incomplete SCI to assess the efficacy for faecal incontinence. He reported improvements in the number of episodes of faecal incontinence per week, the number of days per week with staining or pad use, the ability to empty the bowel completely. The exact mechanism of action of SNM in the treatment of NBD remains unclear. It may include enhancement in striated muscular activity and neuromodulation of sacral reflexes. However, sacral nerve stimulation, overall, remains a relatively safe and effective technique for patients with NBD and SCI despite the minimally invasive procedure.

12.2.3.5   Transcutaneous spinal cord stimulation

Spinal cord stimulation for NBD is an emerging area of interest. Kreydin et al. [48] reported on a single SCI patient case study (AIS A C5) in time required for bowel emptying was reduced. Stimulation was delivered using transcutaneous electrodes over T11 and L1 and reference electrodes were placed over the iliac crests. A therapeutic waveform consisted of two alternating pulses of opposite polarities separated by a 1 μS delay forming a delayed biphasic waveform. The pulses consisted of a high frequency biphasic carrier pulse (10 kHz) combined with a low frequency (30 Hz) burst pulse each with a pulse width of 1 ms. The participant received 18 days of stimulation at therapeutic levels followed by 18 days of sham stimulation and was blinded as to which was being provided. The participant used the stimulation for 1 hour 5 days a week. The average time for bowel emptying reduced from 75 minutes down to 15 minutes. When sham stimulation was applied bowel management time increased. Well-designed randomised controlled studies are required to examine efficacy and effectiveness incorporating different lesion levels, usability for the patient and impact on health-related quality of life.

12.2.3.6   Abdominal electrical stimulation

Transcutaneous electrical stimulation of the abdominal area is a developing area with a growing number of clinical studies being conducted. Some clinics are now offering the treatment in clinical practice. There are two main types of electrical stimulation that have been used. Initial studies using transabdominal interferential electrical stimulation focused predominantly on non-neurological paediatric slow transit constipation [49–53]. Significant improvements in symptom reduction such

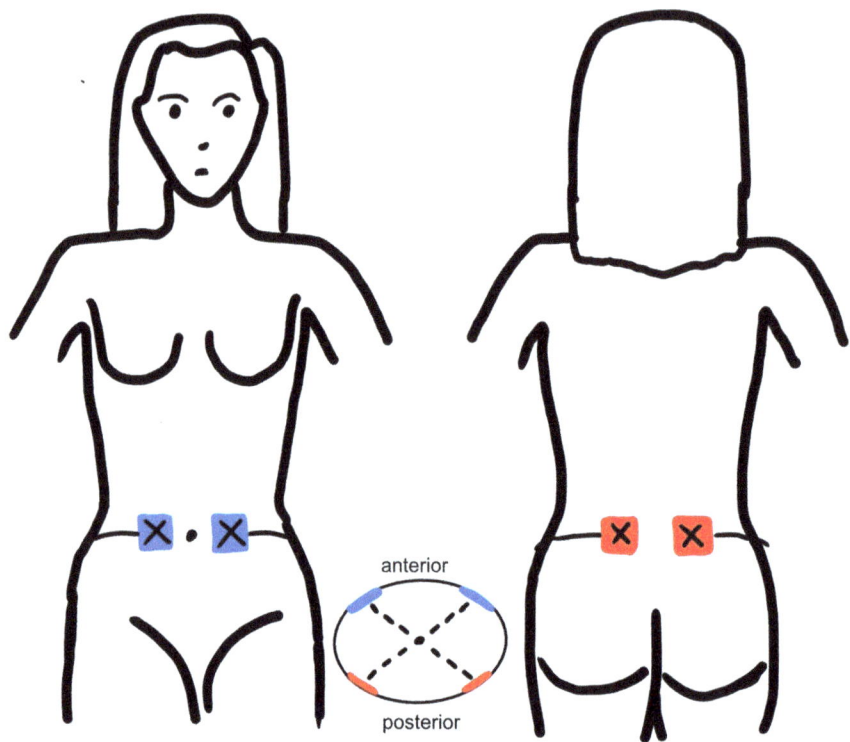

Figure 12.8   Transabdominal interferential electrical stimulation proposed mechanism © Sarah Knight 2024

as reduced overall bowel management time have been reported in clinical trials. Long term therapeutic benefits have also been observed. It is proposed that interferential current stimulation is produced through crossing two medium-frequency alternating currents that are slightly out of phase (2 paraspinal and 2 abdominal electrodes are applied at a comfortable intensity (<40 mA, carrier frequency 4 kHz, varying beat frequency 80–150 Hz [54]. A new modulated current is presumed to be created where the two diagonally opposed currents meet (Figure 12.8). It is hypothesized that interferential electrical stimulation is more effective than the more commonly used neuromuscular stimulation through reaching 'deeper structures' in the body. There is little to support the proposed in the literature and few studies have been conducted [54].

Initial studies using neuromuscular electrical stimulation of the abdominal area included people with neurogenic bowel due to SCI. Korsten *et al.* [55] hypothesised that increasing intra-abdominal pressure and activating muscles that are usually used for defecation would lead to improved bowel movements. Researchers included eight people with SCI. Electrical stimulation was applied using an abdominal belt with electrodes placed on the transverse and abdominal obliques and participants received stimulation of either 8 Hz or 20 Hz for 30 minutes each day for two weeks. The stimulation current was turned up to 80 mA or until an abdominal contraction was observed. Researchers found improvements in time to first stool, total bowel care time and colonic transit time using radiopaque markers. A further small study [56] assigned seven people with complete SCI to either a sham control or intervention group. Participants were provided with 25 minutes of electrical stimulation of the abdominal muscles, five days a week for eight weeks. Using radiopaque markers researchers reported a decrease in colonic transit times in the ascending, descending and transverse colon but not the rectosigmoid compared to the sham control condition. Participants also reported an improved aesthetic appearance in the intervention group.

Abdominal electrical stimulation studies for other types of neurogenic bowel have also provided encouraging initial results. Singleton *et al.* [57] included four people with MS and neurogenic bowel and administered abdominal electrical stimulation for 30 minutes twice a day for six weeks. Electrical stimulation was administered to the external oblique and transverse abdominis muscles at 40 Hz, 330 μ pulse width, and 40–50 mA (Figure 12.9). Using wireless motility capsules whole gut transit time and colonic transit time were reduced over the six- week period. Constipation related quality of life improved in all participants and the use of laxatives was reduced. A further observational study with the same methodology including 20 people with MS over an 8 week period reported similar findings in terms of constipation related quality of life and reduction in laxative use [58]. Interestingly, a subsequent study which also found an improvement in constipation related quality of life did not find an improvement in whole gut transit time and colonic transit time using wireless motility capsules [59].

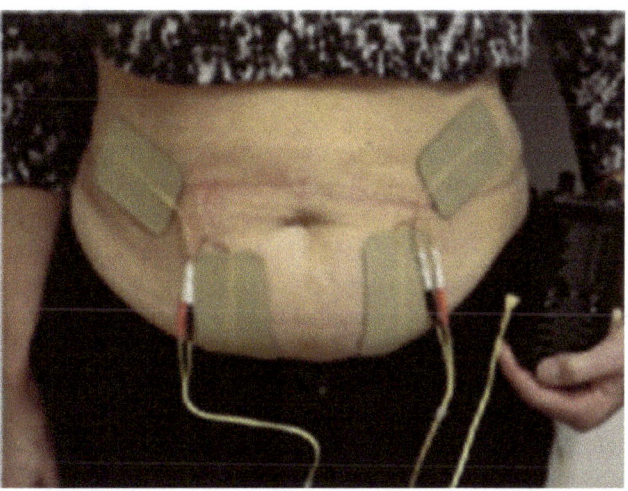

Figure 12.9   Abdominal electrical stimulation electrode placement copyright
© 2016 Christine Singleton et al. [1]

Well-designed randomised controlled studies with adequate sample sizes are required to examine efficacy and effectiveness incorporating different lesion levels, usability for the patient and impact on health-related quality of life. Comparisons between different types of electrical stimulation parameters within the same patient group may provide useful information about the relative benefits and usability of the different techniques. Researchers using transabdominal interferential stimulation with paediatric populations where electrodes are placed on the paraspinal area may not encounter usability issues as parents will be available to support administration of the intervention. In contrast, adults with MS or SCI who have neurogenic bowel may experience usability issues administering this technique independently in contrast to application of the electrodes to the abdominal area due to upper limb dexterity.

12.2.4   Treatment of sexual function using functional electrical stimulation

12.2.4.1   Finetech-Brindley sacral anterior root stimulator implant

The SARS implant, if the S2 anterior nerve roots are placed on an electrode, they can be used to produce an implant driven erection in male patients. The stimulation parameters are slightly different to those used for bladder and bowel emptying, with a frequency of 2 pps and continuous stimulation used. The stimulation needs to be applied throughout the duration of required erection, which can cause some issues with maintaining contact between the transmitter and receiver. In addition, the lower limb movements caused by the stimulation of the nerve roots, can cause a hindrance. As a result, the implant is less commonly used for erection than it is for bladder and bowel emptying.

12.2.4.2   Electro-ejaculation

Patients who have failed an attempt to produce antegrade ejaculation with PVS can be trialled with electroejaculation [60]. The electro-ejaculation technique was originally developed by veterinary surgeons in order to obtain semen samples from endangered species [61]. A commercially available device is available through Dalzell Medical (Saegar Electroejaculation). This involves electrical stimulation of the prostate and seminal vesicles via a rectal probe electrode. In patients with no sensation in sacral segments S4-5, the technique can be used without anaesthesia, however, if sensation is preserved, it is recommended that the procedure is performed under anaesthesia. Patients with a SCI above T6, who are at risk of developing AD, are given a pre-stimulation dose of nifedipine, as the stimulation is likely to trigger an episode of AD. Stimulation parameters for producing antegrade electroejaculation are 50 Hz, 100–250 mA and 5–10 V. Stimulation is usually given in 5–15 three second bursts of stimulation before ejaculation occurs and then a further ten or more bursts to ensure complete ejaculate. The rectal mucosa is very sensitive, so thorough inspection of mucosal membranes should be performed during and after the procedure. Patients with known rectal disease are contra-indicated.

12.2.4.3   Abdominal electrical stimulation and penile vibratory stimulation

PVS is commonly used to achieve antegrade ejaculation in people with a supra-sacral SCI. It is successful in approximately 80% of SCI patients who have an intact pudendal reflex (Sønksen and Ohl, 2002). Commercially available vibrators (Multicept and Vibronox) have vibrating plates (100–120 Hz and 1–3 mm amplitude) which are held against the glans of the penis and stimulate the ER. A small single subject study showed that a combination of abdominal electrical stimulation and PVS was more successful at producing antegrade ejaculation than PVS alone [62].

12.3   Practical clinical considerations for application of electrical stimulation to bladder, bowel and sexual function

12.3.1   Overview

All the techniques described in this chapter should only be attempted by clinicians with the required expertise and training in neurogenic bladder and bowel management, including ability to perform and interpret urodynamic investigations. Commercially available devices should only be used in the context of a fully funded clinical service. All other techniques should only be investigated in the confines of an ethically approved clinical investigation under the clinical governance of a consultant urologist or gastroenterologist with a special interest in neurogenic bladder and bowel. Full urodynamic assessments of bladder function should be performed prior to the use of any electrical stimulation technique to ensure that bladder function has been well characterised and that techniques do not cause any potential harm to upper tracts.

Due to the proximity of the pelvic organs, it is recommended that any research conducted examining one pelvic function should include consideration and monitoring of other functions, to determine benefits as well as adverse effects of the intervention. This can be achieved through using a validated patient reported questionnaire. Impact on sexual functioning is often overlooked during studies examining bladder and bowel functioning with any changes reported anecdotally after the study without use of any validated outcome measures.

12.3.2   Specific risks

12.3.2.1   Autonomic dysreflexia

People with a SCI above T6 are at potential risk from AD; an abnormal over-reaction of the involuntary (autonomic) nervous system to noxious stimulus below the level of the lesion. This can cause a sudden and severe rise in blood pressure leading to the risk of cardiac arrest or cerebrovascular accident. The most effective treatment is to remove the stimulus and give sublingual nifedipine. Bladder and bowel distension is one of the most common triggers for development of AD. Noxious skin stimulation such as electrical stimulation may also trigger AD in some people. Therefore, it is extremely important that when using FES techniques for bladder, bowel or sexual function in people with SCI above T6, blood pressure is monitored carefully, and nifedipine is available at all times. An assessment for severity of AD prior to using any intervention is recommended as well as gradual introduction of the stimulation at a low level for a short duration.

12.3.2.2   Electrode related risks

Surface electrical stimulation has the potential risk for allergic reaction to adhesives and skin damage. Precautions should be taken to prevent these complications during an initial assessment to determine suitability for the treatment. Stimulation should not be applied if the patient has a major skin condition or allergy in the area proposed for stimulation. The patient should be advised that stimulation should not be applied where there is an open wound or irritated skin, e.g. where hair has been removed. Patients should be provided with instructions on how to apply and care for their electrodes and skin to minimise any chances of skin irritation. Patients should further be advised on how to determine the difference between increased blood flow and skin irritation. Increased blood flow may lead to a change in the appearance of the skin for around an hour. In contrast, skin irritation may be present several hours later. Patients should be advised to stop using stimulation immediately if skin irritation is suspected and contact their clinician for further advice on management which may be dependent on the type of stimulation used. Any episodes of skin irritation should be documented and monitored.

For implant technology the most serious complications include lead migration, implant site pain and infection, foreign body reactions and the development of scar tissue around the implant site which may require additional surgical intervention. Other issues to consider include the therapeutic benefits of the treatment may be lost over time or the device may stop working or become less effective. Medical device technology parts used to support and update the device may no longer be manufactured and therefore available to support continued use. Due to the specialist nature of the area, there might not be viable alternatives to provide for replacement. With the potential life changing benefits of the treatment these are serious considerations that patients are made aware of prior to starting treatment.

12.3.2.3   Concomitant medication

Any changes to medications for bladder, bowel or sexual function should be implemented gradually and only on the advice of the consultant responsible for the care of the patient.

12.4   Conclusions

Effective control of bowel and bladder are frequently ranked as the highest priorities for people with SCI and other neurological conditions. Sexual functioning is also frequently rated as a high priority. We have outlined in this chapter various uses of electrical stimulation techniques to manage and restore bladder, bowel and sexual function particularly after neurological damage or disease. At present there are a limited number of commercially available devices using electrical stimulation to target bowel, bladder and sexual function, however these areas are being actively researched with many promising areas for development in the future. Clinical trials conducted in this area are often small and underpowered, influenced by a variety of challenges. The diverse heterogeneity of patients who may benefit from interventions in this area will often make it necessary to tailor interventions to individual specific needs, contributing to the challenge to conduct well designed randomised controlled trials. The burden on participants to travel to take part in research who often have mobility issues is an additional challenge. Conducting studies remotely or including remotely conducted components should be considered to support participation where possible. Potential research participants often present with a variety of comorbidities that may lead to a deterioration in health. Researchers should factor in the potential for significant delays, reduced retention of participants and subsequent impact on data collection and analysis.

Future developments particularly for bladder functioning which may impact on other areas include overcoming the need for posterior rhizotomy, selective nerve stimulation and techniques for nerve blocking. For sexual functioning further well-designed studies examining hybrid approaches may provide greater effectiveness and introduce further choice to patients. Similarly for bowel management further well-designed clinical trials are required to determine effectiveness of initial encouraging results from observational and case study date. Future comparisons between different types of electrical stimulation technologies may be used to optimise treatment for patient benefit.

Glossary

NDO	Neurogenic detrusor over-activity – unwanted bladder contractions caused by neurological damage to the nerves controlling bladder function leading to incontinence and high bladder pressures
NBD	Neurogenic bowel dysfunction – functional disorder of the bowel caused by neurological injury or disease.
SCI	Spinal cord injury – damage to the spinal cord caused by trauma or disease which interrupts messages between the brain and end organs.
DSD	Detrusor sphincter dys-synergia – co-contraction of bladder muscle and sphincter which makes it difficult to empty the bladder efficiently
PVS	Penile vibratory stimulation – a handheld device which is used to deliver mechanical vibratory stimulation to the glans of the penis to elicit a reflex ejaculation
SCS	Spinal cord stimulation
Upper urinary tracts	the kidneys and ureters
Lower urinary tract	bladder, sphincters and pelvic floor

References

[1]   Craggs MD, Balasubramaniam AV, Chung EAL, and Emmanuel AV. Aberrant reflexes and function of the pelvic organs following spinal cord injury in man. *Auton Neurosci Basic Clin*. 2006;126–127:355–70.

[2]   Anderson KD. Targeting recovery: priorities of the spinal cord-injured population. *J Neurotrauma*. 2004;21(10):1371–83.

[3]   Wheeler TL, de Groat W, Eisner K, *et al.* Translating promising strategies for bowel and bladder management in spinal cord injury. *Exp Neurol*. 2018; 306:169–76.

[4]   Selzman AA, and Hampel N. Urologic complications of spinal cord injury. *Urol Clin North Am*. 1993;20(3):453–64.

[5]   Hamid R, Averbeck MA, Chiang H, *et al.* Epidemiology and pathophysiology of neurogenic bladder after spinal cord injury. *World J Urol*. 2018;36 (10):1517–27.

[6]   Krassioukov A, Eng JJ, Claxton G, Sakakibara BM, and Shum S. Neurogenic bowel management after spinal cord injury: a systematic review of the evidence. *Spinal Cord*. 2010;48(10):718–33.

[7]   Krogh K, Emmanuel A, Perrouin-Verbe B, Korsten MA, Mulcahey MJ, and Biering-Sørensen F. International spinal cord injury bowel function basic data set (Version 2.0). *Spinal Cord*. 2017;55(7):692–8.

[8]   Coggrave M, Norton C, and Wilson-Barnett J. Management of neurogenic bowel dysfunction in the community after spinal cord injury: a postal survey in the United Kingdom. *Spinal Cord*. 2009;47(4):323.

[9]   Krassioukov A, Eng JJ, Claxton G, Sakakibara BM, and Shum S. Neurogenic bowel management after spinal cord injury: a systematic review of the evidence. *Spinal Cord*. 2010;48:718.

[10] Quigley EMM. Prokinetics in the management of functional gastrointestinal disorders. *Curr Gastroenterol Rep*. 2017;19(10):53.

[11] Christensen BS, Grønbaek M, Osler M, Pedersen BV, Graugaard C, and Frisch M. Associations between physical and mental health problems and sexual dysfunctions in sexually active Danes. *J Sex Med*. 2011;8(7):1890–902.

[12] Lewis RW, Fugl-Meyer KS, Bosch R, *et al.* Epidemiology/risk factors of sexual dysfunction. *J Sex Med*. 2004;1(1):35–9.

[13] Courtois F, Alexander M, and McLain ABJ. Women's sexual health and reproductive function after SCI. *Top Spinal Cord Inj Rehabil*. 2017;23(1):20–30.

[14] Shamloul R, and Ghanem H. Erectile dysfunction. *Lancet*. 2013;381 (9861):153–65.

[15] Goldstein I, Goren A, Li VW, Tang WY, and Hassan TA. Epidemiology update of erectile dysfunction in eight countries with high burden. *Sex Med Rev*. 2020;8(1):48–58.

[16] Gross T, Schneider MP, Bachmann LM, *et al.* Transcutaneous electrical nerve stimulation for treating neurogenic lower urinary tract dysfunction: a systematic review. *Eur Urol*. 2016;69(6):1102–11.

[17] Parittotokkaporn S, Varghese C, O'Grady G, Svirskis D, Subramanian S, and O'Carroll SJ. Non-invasive neuromodulation for bowel, bladder and sexual restoration following spinal cord injury: a systematic review. *Clin Neurol Neurosurg*. 2020;194:105822.

[18] Nashold BSJ, Friedman H, Glenn JF, Grimes JH, Barry WF, and Avery R. Electromicturition in paraplegia. Implantation of a spinal neuroprosthesis. *Arch Surg Chic Ill 1960*. 1972;104(2):195–202.

[19] Brindley GS, Polkey CE, and Rushton DN. Sacral anterior root stimulators for bladder control in paraplegia. *Paraplegia*. 1982;20(6):365–81.

[20] Van Kerrebroeck PE, Koldewijn EL, and Debruyne FM. Worldwide experience with the Finetech-Brindley sacral anterior root stimulator. *Neurourol Urodyn*. 1993;12(5):497–503.

[21] Tanagho EA, and Schmidt RA. Electrical stimulation in the clinical management of the neurogenic bladder. *J Urol*. 1988;140(6):1331–9.

[22] Kessler TM, La Framboise D, Trelle S, *et al.* Sacral neuromodulation for neurogenic lower urinary tract dysfunction: systematic review and meta-analysis. *Eur Urol*. 2010;58(6):865–74.

[23] Kirkham APS, Knight SL, Craggs MD, Casey ATM, and Shah PJR. Neuromodulation through sacral nerve roots 2 to 4 with a Finetech- Brindley sacral posterior and anterior root stimulator. *Spinal Cord*. 2002;40(6):272–81.

[24] McGuire EJ, Zhang SC, Horwinski ER, and Lytton B. Treatment of motor and sensory detrusor instability by electrical stimulation. *J Urol*. 1983;129 (1):78–9.

[25] Yoo PB, Klein SM, Grafstein NH, *et al.* Pudendal nerve stimulation evokes reflex bladder contractions in persons with chronic spinal cord injury. *Neurourol Urodyn*. 2007;26(7):1020–3.

[26] Kirkham AP, Shah NC, Knight SL, Shah PJ, and Craggs MD. The acute effects of continuous and conditional neuromodulation on the bladder in spinal cord injury. *Spinal Cord*. 2001;39(8):420–8.

[27] Wheeler JSJ, Walter JS, and Zaszczurynski PJ. Bladder inhibition by penile nerve stimulation in spinal cord injury patients. *J Urol*. 1992;147(1):100–3.

[28] Bourbeau DJ, Creasey GH, Sidik S, Brose SW, and Gustafson KJ. Genital nerve stimulation increases bladder capacity after SCI: a meta-analysis. *J Spinal Cord Med*. 2018;41(4):426–34.

[29] Doherty S, Vanhoestenberghe A, Duffell L, Hamid R, and Knight S. A urodynamic comparison of neural targets for transcutaneous electrical stimulation to acutely suppress detrusor contractions following spinal cord injury. *Front Neurosci*. 2019;13:1360.

[30] Fjorback MV, Rijkhoff N, Petersen T, Nohr M, and Sinkjaer T. Event driven electrical stimulation of the dorsal penile/clitoral nerve for management of neurogenic detrusor overactivity in multiple sclerosis. *Neurourol Urodyn*. 2006;25(4):349–55.

[31] Bourbeau DJ, Gustafson KJ, and Brose SW. At-home genital nerve stimulation for individuals with SCI and neurogenic detrusor overactivity: a pilot feasibility study. *J Spinal Cord Med*. 2019;42(3):360–70.

[32] Opisso E, Borau A, Rodríguez A, Hansen J, and Rijkhoff NJM. Patient controlled versus automatic stimulation of pudendal nerve afferents to treat neurogenic detrusor overactivity. *J Urol*. 2008;180(4):1403–8.

[33] Majerus S, Nguyen C, Brose S, Nemunaitis G, Damaser M, and Bourbeau DJ. Automated closed-loop stimulation to inhibit neurogenic bladder overactivity. *Proc Inst Mech Eng [H]*. 2024;238(6):619–627

[34] McGuire EJ. Experimental observations on the integration of bladder and urethral function. *Invest Urol*. 1978;15(4):303–7.

[35] Stoller ML. Afferent nerve stimulation for pelvic floor dysfunction. *Int Urogynecology J*. 1999;10:P99–P99.

[36] Schneider MP, Gross T, Bachmann LM, *et al.* Tibial nerve stimulation for treating neurogenic lower urinary tract dysfunction: a systematic review. *Eur Urol*. 2015;68(5):859–67.

[37] Peters KM, Carrico DJ, Wooldridge LS, Miller CJ, and MacDiarmid SA. Percutaneous tibial nerve stimulation for the long-term treatment of overactive bladder: 3-year results of the STEP study. *J Urol*. 2013;189(6):2194–201.

[38] Stalder SA, Gross O, Anderson CE, *et al.* bTUNED: transcutaneous tibial nerve stimulation for neurogenic lower urinary tract dysfunction. *BJU Int*. 2023;132(3):343–52.

[39] Kaplan WE, and Richards I. Intravesical transurethral electrotherapy for the neurogenic bladder. *J Urol*. 1986;136(1 Pt 2):243–6.

[40] Gad PN, Kreydin E, Zhong H, Latack K, and Edgerton VR. Non-invasive neuromodulation of spinal cord restores lower urinary tract function after paralysis. *Front Neurosci*. 2018;12:432.

[41] Herrity AN, Williams CS, Angeli CA, Harkema SJ, and Hubscher CH. Lumbosacral spinal cord epidural stimulation improves voiding function after human spinal cord injury. *Sci Rep*. 2018;8(1):8688.

[42] Kreydin E, Zhong H, Latack K, Ye S, Edgerton VR, and Gad P. Transcutaneous electrical spinal cord neuromodulator (TESCoN) improves symptoms of overactive bladder. *Front Syst Neurosci*. 2020;14:1.

[43] Harkema SJ, Wang S, Angeli CA, *et al.* Normalization of blood pressure with spinal cord epidural stimulation after severe spinal cord injury. *Front Hum Neurosci*. 2018;12:83.

[44] Worsoe J, Rasmussen M, Christensen P, and Krogh K. Neurostimulation for neurogenic bowel dysfunction. *Gastroenterol Res Pract*. 2013;2013:563294.

[45] Palmer LS, Richards I, and Kaplan WE. Transrectal electrostimulation therapy for neuropathic bowel dysfunction in children with myelomeningocele. *J Urol*. 1997;157(4):1449–52.

[46] Colasante C, Brose SW, Gustafson K, and Bourbeau D. Minimally invasive electrical rectal stimulation promotes bowel emptying in an individual with spinal cord injury. *J Spinal Cord Med*. 2023;46(6):975–9.

[47] Jarrett MED, Matzel KE, Christiansen J, *et al.* Sacral nerve stimulation for faecal incontinence in patients with previous partial spinal injury including disc prolapse. *Br J Surg*. 2005;92(6):734–9.

[48] Kreydin E, Zhong H, Lavrov I, Edgerton VR, and Gad P. The effect of non-invasive spinal cord stimulation on anorectal function in individuals with spinal cord injury: a case series. *Front Neurosci.* 2022;16:816106.

[49] Chase J, Robertson VJ, Southwell B, Hutson J, and Gibb S. Pilot study using transcutaneous electrical stimulation (interferential current) to treat chronic treatment-resistant constipation and soiling in children. *J Gastroenterol Hepatol.* 2005;20(7):1054–61.

[50] Clarke MCC, Catto-Smith AG, King SK, *et al.* Transabdominal electrical stimulation increases colonic propagating pressure waves in paediatric slow transit constipation. *J Pediatr Surg.* 2012;47(12):2279–84.

[51] Hutson JM, Dughetti L, Stathopoulos L, and Southwell BR. Transabdominal electrical stimulation (TES) for the treatment of slow-transit constipation (STC). *Pediatr Surg Int.* 2015;31(5):445–51.

[52] Yik YI, Hutson J, and Southwell B. Home-based transabdominal interferential electrical stimulation for six months improves paediatric slow transit constipation (STC). *Neuromodulation J Int Neuromodulation Soc.* 2018;21(7):676–81.

[53] Yik YI, Ismail KA, Hutson JM, and Southwell BR. Home transcutaneous electrical stimulation to treat children with slow-transit constipation. *J Pediatr Surg.* 2012;47(6):1285–90.

[54] Moore JS, Gibson PR, and Burgell RE. Neuromodulation via interferential electrical stimulation as a novel therapy in gastrointestinal motility disorders. *J Neurogastroenterol Motil.* 2018;24(1):19–29.

[55] Korsten MA, Fajardo NR, Rosman AS, Creasey GH, Spungen AM, and Bauman WA. Difficulty with evacuation after spinal cord injury: colonic motility during sleep and effects of abdominal wall stimulation. *J Rehabil Res Dev.* 2004;41(1):95–100.

[56] Hascakova-Bartova R, Dinant JF, Parent A, and Ventura M. Neuromuscular electrical stimulation of completely paralyzed abdominal muscles in spinal cord-injured patients: a pilot study. *Spinal Cord.* 2008;46(6):445–50.

[57] Singleton C, Bakheit AM, and Peace C. The efficacy of functional electrical stimulation of the abdominal muscles in the treatment of chronic constipation in patients with multiple sclerosis: a pilot study. *Mult Scler Int.* 2016;2016:4860315.

[58] Street T, Peace C, Padfield E, and Singleton C. Abdominal functional electrical stimulation for bowel management in multiple sclerosis. *Neurodegener Dis Manag.* 2019;9(2):83–9.

[59] Lin SD, Butler JE, Boswell-Ruys CL, *et al.* The effect of abdominal functional electrical stimulation on bowel function in multiple sclerosis: a cohort study. *Mult Scler J - Exp Transl Clin.* 2020;6(3):2055217320941530.

[60] Sønksen J, and Ohl DA. Penile vibratory stimulation and electroejaculation in the treatment of ejaculatory dysfunction. *Int J Androl.* 2002;25(6):324–32.

[61] Seager SW, and Halstead LS. Fertility options and success after spinal cord injury. *Urol Clin North Am.* 1993;20(3):543–8.

[62] Goetz LL, and Stiens SA. Abdominal electric stimulation facilitates penile vibratory stimulation for ejaculation after spinal cord injury: a single-subject trial. *Arch Phys Med Rehabil.* 2005;86(9):1879–83.

Chapter 13

Use of electrical stimulation in the intensive care unit (ICU)

Gad Alon[1], Paulo Eugênio Silva[2,3], Juliette Francovich[4] and Annemijn Jonkman[4]

13.1 The problem: challenges encountered in the ICU

The use of therapeutic electrical stimulation (ES) is currently only rarely used clinically in the intensive care unit (ICU), yet there is clear potential for its benefits. This chapter therefore focuses on the rationale for its use, the evidence from research studies and how it could be implemented.

Patients admitted to an ICU are likely to be immobilised in bed, sometimes for several weeks. They are likely to become weak and experience respiratory dysfunction. While some patients are in good health prior to admission, others may have a long and complex medical-physical history. The most common co-morbid conditions include diabetes mellitus (DM), cardio-pulmonary disease, chronic obstructive pulmonary disease (COPD), kidney disease, osteoarthritis, and obesity. Collectively, these diseases contribute to ICU-acquired weakness, which is characterised by a loss of muscle mass mostly affecting the limbs and respiratory muscles, and a diminished ability to contract the muscles due to impaired excitability. While in the ICU, many patients require life-support in the form of mechanical ventilation; while taking over the function of the respiratory muscle pump, mechanical ventilation is one of the key contributors to the development of respiratory muscle dysfunction. In addition, some patients develop sepsis, some are connected to a heart-lung machine (ECMO), and many are sedated, all further contributing to prolonged immobilisation and further deterioration of bodily functions. Delayed ventilator weaning and longer stay in the ICU or hospital increase morbidity and mortality.

The primary goal of rehabilitation is to mobilise the patients and activate the weakened and sarcopenic muscles and in doing so accelerate the recovery and

[1]School of Medicine, University of Maryland, USA
[2]Faculty of Medicine, University of São Paulo, Brazil
[3]Research Group FICIT (CNPq), Hospital de Base do Distrito Federal (IGESDF), Brazil
[4]Department of Adult Intensive Care, Erasmus Medical Center, The Netherlands

shorten the time in the ICU. Specifically, the goal of incorporating neuromuscular electrical stimulation (NMES) to the management of ICU patients is to minimise the loss of muscle force generation and augment the recovery of strength and mobility. Additional consideration highlights the importance of providing a continuum of care following discharge from the ICU to the hospital wards or into the community for ongoing management and recovery has rarely been addressed. This chapter will offer a discussion of the reasons for absence of continuum of care and why it is essential to promote it.

Five million patients are admitted annually to intensive care units (ICUs) in the USA (www.sccm.org/Communications/Critical-Care-Statistics). In the UK, approximately 200,000 admissions occur each year according to the intensive care national audit and research centre (Intensive Care National Audit & Research Centre). Acute respiratory failure is the most common reason for admission to the ICU and can be caused by, e.g. infection, trauma, heart failure, or complications during elective surgery. Patients with severely injured lungs and critically low levels of arterial oxygen require life-saving breathing support with mechanical ventilation. Other reasons for admission include the need for intensive and at times invasive monitoring; support of cardio- and peripheral vascular circulation; stabilisation of internal organs functions and comprehensive management of injury and/or illness. Ohsfeldt *et al.* [1] summarised data reporting that the overall median cost of an ICU stay of patients with COVID-19 was $13,443 with a median length of stay (LOS) of five days. Patients requiring mechanical ventilation had the highest hospital and ICU median costs ($47,454 and $41,510) and LOS (16 and 11 days), respectively [1].

Some patients are in good health prior to admission, including those who sustain multi-system injury [2], while others may have a long and complex medical-physical history [3–7]. One of the most common co-morbidities found in the ICU is DM. DM contributes to disorders of the circulatory, nervous, and immune systems, especially with the added complications of COVID-19 as a new comorbidity in the past few years [8–10]. Patients with chronic cardiovascular disease (CVD), particularly heart failure, are admitted to the ICU following complications of heart surgery [11]. Patients with a history of COPD are typically admitted to ICU when they require continuous support of their pulmonary function [3,7,12,13]. Even patients with chronic kidney disease, particularly after kidney transplant, may end up in the ICU. Comorbidities such as obesity or osteoarthritis may further delay the recovery and discharge from the ICU. Collectively, these diseases contribute to the terms ICU acquired weakness (ICUAW) that is characterised by loss of muscle mass, loss of muscle force generation, and diminished ability to contract the muscles due in part to impaired excitability [14–16]. Furthermore, although mechanical ventilation is the cornerstone of supportive therapy in the ICU temporarily taking over the function of the respiratory pump, respiratory muscle function may further deteriorate in ventilator-bound patients. Diaphragm dysfunction has been diagnosed in 64% of patients within the first 24h of mechanical ventilation [17] and a rapid decline in diaphragm function of 30% within the first 5-6 days of controlled mechanical ventilation was reported in a follow-up study [18]. Respiratory muscle weakness may also already exist prior to ICU admission and many factors beyond mechanical ventilation could contribute to worsening respiratory muscle function (e.g., sepsis, systemic inflammation and malnutrition) [19,20]. Respiratory muscle dysfunction was found twice as frequently as limb muscle weakness [21] and is associated with prolonged weaning and poor patient outcome.

It is well established in the research literature that structural changes and metabolic disorders, particularly those associated with diabetes, sedation, and sepsis, all appear to be influenced by the LOS in the ICU [22]. The longer a patient remains in the ICU, the more challenging his/her management becomes, specifically the challenge to minimise ICU-acquired weakness and loss of muscle mass, while helping to recover the excitability of the neuromuscular system. Using ultrasound imaging Silva *et al.* found a significant decrease in the thickness of limb muscles of patients admitted to the ICU [23]. In a further study, Vanhorebeek and colleagues [24] found that critically ill patients showed 34% to 80% lower mRNA expression of FNDC5, KYAT1, and amylases compared to a control group. Furthermore, the lower FNDC5 expression in patients was independently associated with a higher ICU mortality and ICU-acquired weakness, whereas the lower expression of amylase in ICU survivors was independently associated with a longer ICU stay. The authors concluded that expression of the studied myokines was affected by critical illness and associated with clinical outcomes. One exception was the finding that NMES increased FNDC5 expression compared with non-stimulated muscle [24]. Moreover, it was reported that the DNA methylation signature in muscles of critically ill patients differs from DNA methylation in muscles of healthy controls, further illuminating the deleterious consequences of being admitted to the ICU [25].

Several studies have quantified the deterioration of muscle mass and structural changes using either ultrasound or computed tomography (CT) [26]. Quantification includes muscle volume and cross-sectional area (CSA). Adding infiltrate of intermuscular adipose tissue is also a recognised biomarker associated with ICUAW and impaired excitability [27,28]. However, these changes are not uniform and vary from muscle to muscle and likewise depend on the LOS in the ICU. Most published studies have concluded that deterioration while in ICU is particularly impactful when comparing data between the first week and 13–14 days post admission [3,13,29,30]. The impaired excitability is of particular concern because it diminishes the ability to volitionally contract the muscles, making it even more difficult for the patient to move. Additional comorbidities that adversely affect excitability include sedation, and sepsis. Sekhniashvili *et al.* used the term critical illness neuromyopathy and reported that 'slow conditioning electrical pulses at motor nerves and directly at the muscles' resulted in a temporary yet remarkable improvement of excitability in sepsis-associated neuropathy [31]. Silva *et al.* used the term neuromuscular electrophysiological disorders (NED) and demonstrated an increase of 48% in NED from day 1 to day 14 [23].

Respiratory muscle dysfunction in the critically ill is a different entity as compared to ICUAW- of the extremities [21]. Clinically it has been shown to be as a reduced motion and thinning of the diaphragm or as reduced capacity to generate sufficient pressure/force [19]. It is a multifactorial problem and strong evidence suggests that diaphragm inactivity under mechanical ventilation is the main contributor to the development of diaphragm weakness [19]. In diaphragm biopsies from ventilated ICU patients, both slow-twitch and fast-twitch muscle fibre atrophy was demonstrated, as well as activation of proteolytic pathways and an increased number of neutrophils and macrophages [32]. More recently, extracellular matrix accumulation was found in such biopsies, partially replacing the lost areas by myofiber atrophy (replacement fibrosis) [33], and animal models revealed that titin, a component of striated muscle, elasticity may affect the response of the diaphragm to inactivity [34]. This replacement fibrosis could also explain why ultrasound studies found different patterns of diaphragm and expiratory muscle thickness over the course of mechanical ventilation: either increases, decreases or no changes in thickness were reported [35,36], without a clear association with atrophy on histology [37]. Furthermore, both decreases and increases in diaphragm thickness were associated with prolonged mechanical ventilation [35,36]. Besides muscle inactivity, prolonged sedation, high inflammatory state, and malnutrition are important factors contributing to the development of respiratory muscle dysfunction [19].

Malnutrition before and while in the ICU is an important comorbid factor that is known to adversely affect the brain, internal organs, and musculoskeletal functions. Compared to a control group that received a mean calorie delivery of 16.8 kcal/kg/day and mean protein delivery of 0.8 g/kg/day, an intervention group consumed a mean calorie delivery of 20.1 kcal/kg/day and mean protein delivery of 1.4 g/kg/day. As a result, by day 10 the experimental group was able to move to a sitting position significantly earlier and lost significantly less thigh muscle volume than the control group [38]. Other researchers have reported similar results and also noted that the nitrogen balance remained negative in a control group, while the nitrogen balance became positive after 3–4 days in the experimental group where adequate calories and proteins were supplemented [3,13]. A positive nitrogen balance is an indicator of the anabolic process of recovery of the contractile properties of the muscle [38]. In all three independent studies, the authors combined the higher doses of nutritional intake with NMES. Table 13.1 summarises the known comorbidities and their impact on the various body functions.

13.2   Rationale for the use of ES in the ICU to improve mobility and minimise muscle loss and overview of evidence from the literature

To date, there are over 100 peer-reviewed published articles where NMES and functional electrical stimulation (FES) were administered in the ICU. This section elaborates on the rationale and mechanism that guides the application of ES. The questions of when to apply ES, to which muscle groups, and how to program the stimulation parameters, as well as ways to calculate treatment dosing are discussed. This section also reviews the evidence supporting the principle that ES should not be applied in isolation but that it should rather complement other interventions typically

Table 13.1 Selected adverse interactions between comorbidities and body's functions.

	Skeletal muscles	Joints' motion	Pulmonary	Heart	Peripheral circulation	Kidney
Immobility	1. Loss of muscle mass 2. Decrease force capacity	1. Loss of range 2. Joint stiffness	1. Depressed respiratory function 2. Fluid accumulation in the lungs	1. Reduction of venous return, 2. Increased heart rate 3. Cardiac muscle deconditioning	1. Venous stasis 2. Venous dilation 3. Decreased blood pressure 4. Edema 5. Embolus formation 6. Thrombophlebitis 7. Orthostatic hypotension	1. Decreased creatinine clearance 2. Decreased urinary excretion of waste products 3. Risk of the formation of renal calculi 4. Risk of UTI
Diabetes	1. Suppression of growth & proliferation of muscle cells 2. Loss of mass 3. Loss of strength		1. Increase COPD 2. Pulmonary fibrosis 3. Chance of pneumonia	1. Damage blood vessels and the nerves 2. Cardio myopathy 3. Decrease cardiac output	1. Peripheral artery disease 2. Diabetic neuropathy	1. Damage to blood vessels inside the kidneys 2. Decrease filtration 3. High blood pressure
Osteo arthritis	1. Loss of mass 2. Loss of strength	1. Loss of range 2. Joint stiffness 3. Pain				
Sepsis	1. Decrease ATP/ADP ratio 2. Mitochondrial dysfunction 3. Inflammation 4. Loss of strength		1. Respiratory distress syndrome 2. Lung injury	1. Drop in blood pressure 2. Increase heart rate 3. Fever	1. Vasodilatation of peripheral vessels 2. Unstable autonomic control of circulation	1. Decrease renal blood flow 2. Decrease filtration
Brain damage	1. Loss of motor control 2. Loss of strength 3. Abnormal activation 4. Myofascial adhesion and shortening	1. Loss of range 2. Joint stiffness	1. Increased sympathetic activity 2. Pulmonary venous constriction 3. Higher capillary permeability	1. Arrhythmias 2. Release of biomarkers of cardiac injury 3. Left ventricle dysfunction (most are transient and reversible)	1. Decrease arterial flow 2. Decrease venous flow 3. Decrease lymphatic flow	

offered in the ICU. To ascertain effective communication among the disciplines of engineering, rehabilitation researchers, physicians/consultants, physical therapists, occupational therapists, and ICU nurses, a review of selected relevant terminology is necessary. A frequently asked question is what are the commonalities and differences between stimulators that are being used in the ICU and are labelled generically in peer-reviewed publications as NMES or FES? Common to both is a stimulator that generates medically safe electric current in the form of pulses. Both offer control over phase/pulse duration (in numerous publication the term pulse width is used instead of pulse duration), stimulus intensity, pulse frequency also reported as the number of pulses per second (pps) and are programmed to repeatedly induce muscle contractions known as ON time and muscle relaxation known as OFF time. When the stimulation occurs without volitional activation by the patient, the term NMES is appropriate. In contrast, when the electrically induced contractions coincide with task specific or functional activity, the preferred term is FES [39].

Patients who require mechanical ventilation and are sedated, or in a coma are particularly at risk of the complications associated with immobility. The longer they remain minimally mobile, the more pronounced the acquired weakness, loss of muscle mass, and diminished neuromuscular excitability. In addition to providing around the clock medical care, the primary rehabilitative goal is to mobilise the patients and activate the weakened and sarcopenic muscles and in doing so accelerate the recovery and shorten the time in the ICU. ES is one intervention that may enhance recovery, shorten the time on mechanical ventilation, decrease the time of immobility, and may shorten the LOS in the ICU. Several studies delineated the electrophysiological mechanisms by which ES minimises deterioration and helps the recovery of muscle strength while concurrently enhancing muscle perfusion because of improved micro and macro circulation [40,41]. Based on studies of patients hospitalised for general surgery or admitted to the ICU, NMES induces a cascade of responses beginning at the molecular level and include up-regulation of IGF-1, modulation of MuRF-1, up-regulation of relevant markers of differentiating satellite cells, remodelling of the extracellular matrix, maintenance of satellite cell function and reduction of fibrosis [42]. Other investigators have found enhancement of protein synthesis [43], and increased FNDC5 expression [24]. Collectively, these molecular and cellular responses provide a mechanistic foundation and explain the significant improvement in electrically-induced contraction of impaired-sarcopenic muscles while minimising muscle loss during the ICU stay [3,16,29,30,38,44–52]. Whereas the mechanistic foundation is thoroughly described in available clinical literature, the success of achieving clinically useful electrically-induced muscle contractions may be impeded by the reported decline in excitability the longer the patient remains immobile [23]. As a result, one study reported that only 50% of ICU patients tolerated clinically observed contraction induced by NMES [53]. Gronow *et al.* concluded that only 38% were considered responders to NMES [54]. Of note are the data that while in the ICU, not all muscles undergo decline in neuromuscular excitability. The less responsive patients also have lower limb oedema, are treated with vasopressors, are sedated, impaired by sepsis, and exhibit decreased metabolic capacity [16,53,54].

A literature search on PubMed to identify articles relating to the use of ES to mobilise patients in the ICU, resulted in 22 original studies ranging from a single case to one study that included 139 patients [3,12,13,22,29,38,40,42,44–51,53,55–59]. Six investigations stimulated the quadriceps [13,30,42,53,55,56], two studies applied the NMES to the quadriceps and dorsiflexors bilaterally, two studies applied stimulation to the quadriceps, hamstrings, dorsiflexors, and plantar flexors [12,45]. Baron *et al.* found significant reduction of incidence of pressure sores by applying NMES to gluteus maximus bilaterally [40], whereas Cebeci *et al.* stimulated the biceps brachii and quadriceps bilaterally [29]. Leite *et al.* stimulated the quadriceps in one group and the diaphragm in another group [49], while Bao *et al.* focused on the plantar and dorsiflexors bilaterally [47]. These studies varied not only in the selection of different muscle groups but also in outcome measures. None of the studies prevented loss of muscle mass, whether measured by muscle volume or CSA of muscle tissue. However, they were able to successfully and significantly minimise the loss of muscle mass compared to non-stimulated muscles of the control groups. A common feature of all clinical studies was the short time frame of NMES application lasting only between 5 and 14 days. A more detailed review of these studies will be presented in the next section.

13.2.1 Implementing electrical stimulation to improve mobility and minimise muscle loss in the ICU environment

This section reviews published data guiding the selection of stimulation parameters, electrode management, treatment dosage, combination with other technologies including motorised cycling and breathing technologies. It also explains the rationale and evidence supporting the utilisation of non-invasive ES as an integral part of the physical rehabilitation program offered to patients in the ICU.

The ICU environment is rich with electrical and electronic equipment, some devoted for continuous monitoring of the cardio-pulmonary functions and selected internal organ functions. The addition of intravenous (IV) infusion of fluids nutrients and medications, various non-electric tubing, suction systems, and in some cases heart-lung machines (ECMO), create a very crowded ICU environment. Of particular importance for a clinician who wishes to use NMES or FES in the ICU is to ascertain the compatibility of the stimulation system with all other electronic/ electrical systems so that there is no interference that affects the functions of any of the systems. Specific precautions are discussed later in the chapter. Unique to an ICU is the involvement of multiple medical specialists, ICU nurses, respiratory and physical therapists, nutritionists, all working in shifts 24/7 caring for and communicating with the patient and his/her immediate family. Being familiar with the ICU environment is critical for a clinician (typically a physician-consultant or a physical therapist) who wishes to implement an ES program. At the time of this writing, ES is not considered a common practice in the ICU and all available data have been published as research studies. Possible explanations for not becoming a common practice in the ICU will be offered later in this chapter. In several studies the stimulation program was applied in isolation if the patient was not responsive to command due to coma or heavy sedation and was not able to add volitional movement while stimulated [22,49,53,54]. In contrast, recent data support an advance practice model that initiate a multi-modal intervention consisting of nutritional supplements, electrical stimulation (NMES or FES), and mobility activities as soon as the patient begins to follow commands [12,30,38,46, 48,51,53,60–64].

As was stated earlier in the chapter, when the ES occurs without volitional activation by the patient, the term NMES is appropriate. In contrast, when an electrically induced contraction coincides with a task specific or functional activity, the preferred term is FES [65]. Principally, whenever the patient is responsive to instructions the stimulation should be combined with volitional effort to move. The ability to move while in the ICU depends on the medical status and mobility restrictions following surgical interventions. Support to implement exercise and mobility training is advocated in at least two publications [66,67]. A sample of typical exercises is presented in Table 13.2.

Data in support of using NMES or FES in the ICU are abundant. Angelopoulos *et al.* [59] stimulated the quadriceps bilaterally applying 75 Hz or 45 Hz NMES for a single 30 min session. The study included 31 patients admitted to ICU with systemic inflammatory response syndrome (SIRS) or sepsis. Peripheral micro-circulation was monitored at the thenar eminence using near-infrared spectroscopy (NIRS) to obtain tissue O2 saturation (StO2) and local microcirculation of the vastus lateralis. The authors concluded that a single NMES session of either

Table 13.2   Examples of exercises in the ICU

Activity	In bed	Transfer and in chair	Ambulation with/ without assistive devices
Strengthening and endurance exercises	Leg press Hip extension Hip abduction Push knees to slide up the bed Ankle dorsi/plantar flexion Shoulder girdle movements Upper extremities against resistance If able to sit on bedside 30–60 sec rhythmic stabilization	Bed to chair transfer In chair: Standing up and siting down (gradual increase repetition) In standing rhythmic stabilization Stepping, extending and flexing the knees while sitting Shoulder girdle movements Upper extremities against resistance, Push ups using the armrest	Walk in ICU then out in the corridor Increase distance each day Stationary bike Squat exercise against a wall Walk up and down Stairs/step

frequency enhanced local and systemic skeletal muscle microcirculation. In a study reported by Segers and colleagues [53], 50 patients were recruited on day 3–5 of their ICU stay and were treated with 25 min applying NMES of both quadriceps, five days per week. Only 50% of the patients had adequate quadriceps contraction in at least 75% of the NMES sessions. Univariate analysis showed that lower limb oedema (p < 0.001), sepsis (p = 0.008), admission to the medical ICU (p = 0.041), and treatment with vasopressors (p = 0.011) were associated with impaired quadriceps contraction. Patients responded better to NMES in the beginning of their ICU stay in comparison with after one week of ICU stay. Collectively, critically ill patients having sepsis, oedema, or receiving vasopressors were less likely to respond to NMES with an adequate quadriceps contraction [53].

The feasibility of applying NMES to fully sedated, comatose patients, being treated in the ICU provided novel data on the efficacy of minimising the loss of skeletal muscles force generation. The researchers investigated the efficacy of providing NMES to six patients. One leg was subjected to twice-daily NMES of the quadriceps muscle for a period of 6–8 days whereas the other leg acted as a non-stimulated control. Quadriceps muscle biopsies were collected from both legs to assess muscle fibre-type-specific CSA. Phosphorylation status, muscle protein synthesis was assessed, and mRNA expression of selected genes was measured. In the control leg type 1 and type 2 muscle fibres-CSA decreased by 16 +/− 9% and 24 +/− 7%, respectively, whereas no muscle atrophy was observed in the stimulated leg. Also, mRNA expression of key genes involved in muscle protein breakdown either declined or remained unchanged because of the stimulation. The authors concluded that NMES represents an effective and feasible interventional strategy to minimise the loss of skeletal muscle contraction force [42].

Three independent studies have utilised a multi-modal approach by combining mobility exercises, nutrition supplements and FES [3,13,38]. In one study conducted in the pulmonary ICU, the quadriceps and dorsiflexors of both lower extremities were stimulated for 30 min twice daily for up to 13–14 days. Twenty-three received usual critical care including standard nutrition and mobility training (control group) while 16 patients received high doses of Whey proteins and calorie delivery combined with FES and mobility exercises (experimental group). The FES parameters included symmetric biphasic waveform pulses of 300 μs phase duration and pulse rate of 30 pulses per second (30 Hz). Contraction and relaxation times were each set at 10 s and all four muscle groups were stimulated concurrently. Stimulation intensity during each session was adjusted by the research therapist to assure as strong as possible visible muscle contraction. For participants who were sedated, FES sessions were provided twice a day for 30 min at an intensity to elicit the highest tolerated overt intermittent contractions. The loss of volume and CSA of the thigh and leg muscles were much less evident in the experimental group compared to the control group [3]. A similar experimental paradigm was applied in the neuro ICU to patients with a subarachnoid haemorrhage. In this study, FES was applied only to the quadriceps bilaterally. Again, the loss of muscle volume and CSA were significantly less compared to the control group [13]. A highly complex intervention protocol was published by Nakano *et al.* using a combination of high dose nutritional supplements, structured mobility exercises, and NMES applied via belt-like electrodes. Unfortunately, the authors did not specify the groups of muscles they stimulated in the trunk and both lower extremities [68]. The stimulation lasted 20 min once a day but on average was delivered only 2.0 (1.8–4.0) times during the first ten days. The stimulation frequency was 20 Hz with a pulse width of 250 μs. The ON (contraction time) was set to 5 s and the OFF (relaxation time) was 2 s. Data on thigh muscle volume obtained from 56 patients in the experimental showed significant less loss compared to the control group (11.6 vs 14.5%, $p = 0.03$) [38].

Bao *et al.* [47] experimental design applied NMES over the plantar and dorsiflexors (group 1) or plantar flexors only (group 2) of patients admitted to the ICU following complicated pelvic or spinal fractures. NMES parameters were asymmetric biphasic pulsed current, 300 μs phase duration, 30 Hz frequency, on/off ratio of 1:4, adjusting the current intensity until palpable and visible muscle contraction was achieved within patient's tolerance. Treatment was delivered for 20 min, twice a day for an average ICU stay of 13.3 days, The NMES was combined with passive and active exercises of both ankles. Muscle strength of the plantar flexors increased significantly compared to a control group at ICU discharge. Whereas all three groups lost ankle active range of motion (AROM), the loss was significantly less particularly in the NMES group 1. Similar outcome was found from the CT-derived CSA data, where the loss was less evident for NMES group 1 [47].

Early mobility in the ICU is a recognised and recommended practice. Combining it with ES has been the objective of several recent clinical trials [30,46,60,69]. Campos and colleagues [46] enrolled 139 consecutive mechanically ventilated patients randomly assigned either to an early mobilisation (EM) group and EM+NMES group both received EM daily or a EM+NMES group. Both groups received NMES applied to the quadriceps and the tibialis anterior, once a day for 60 minutes, five days a week, starting in the first 48 hours of ICU admission and continuing until ICU discharge. The NMES was applied while the patient assumed a supine position using biphasic symmetrical rectangular pulse (80 Hz, 400 μs, 5 s on, 10 s off). The intensity of the stimulus was increased to the point that there was full, visible, or palpable muscle contraction. Among the interesting findings, patients in the EM+NMES group presented a significantly higher score of functional status score (FSS) for the ICU scale when compared with the EM group at ICU discharge and hospital discharge. They also had better functional status, took less days to stand up during the ICU stay, and had a significantly shorter hospital LOS, lower frequency of ICU-acquired weakness, and better global muscle strength [46]. Shorter LOS in the ICU was also documented by other investigators [40,44,70].

Leite and associates [49] divided 67 subjects requiring mechanical ventilation (MV) into three groups: (a) control group (CG), (b) stimulation of quadriceps (quadriceps group-QuadG), and (c) stimulation of diaphragm (diaphragm group-DiaG). The QuadG and DiaG patients received consecutive daily ES sessions at specific points from the first day of randomisation until ICU discharge. For the NMES of the quadriceps, the following parameters were used: 10 kHz carrier frequency modulated to 50 Hz, 8 s muscle contraction and 30 s relaxation. The modulated frequency of the diaphragm NMES was 30 Hz, the contraction was only 1 s and the relaxation 20 s. The size of the electrodes was not reported but appears to be about 2.5 × 5 cm each. Two channels (four electrodes) were secured over each quadriceps. Four channels (eight electrodes of the same size) were secured over the 7th and 8th intercostal spaces and assumed to stimulate the diaphragm; however, considering the deep structure of the diaphragm and lack of quantitative verification of its activation upon stimulation, it is questionable whether effective contractions were elicited (most likely the intercostal muscles were stimulated). Leite *et al.* [49] reported that muscle strength improved significantly in the QuadG and the functional independence at ICU discharge was significantly better in QuadG. The QuadG also presented a better Barthel Index compared to DiaG or CG and also presented better FSS compared to CG [49]. While Leite *et al.* [49] applied diaphragm stimulation to mechanically ventilated patients, the outcomes were not related to respiratory function but rather functional independence and hospital LOS [38]. Typical electrodes placements to induce muscle contraction is illustrated in Figures 13.1 and 13.2.

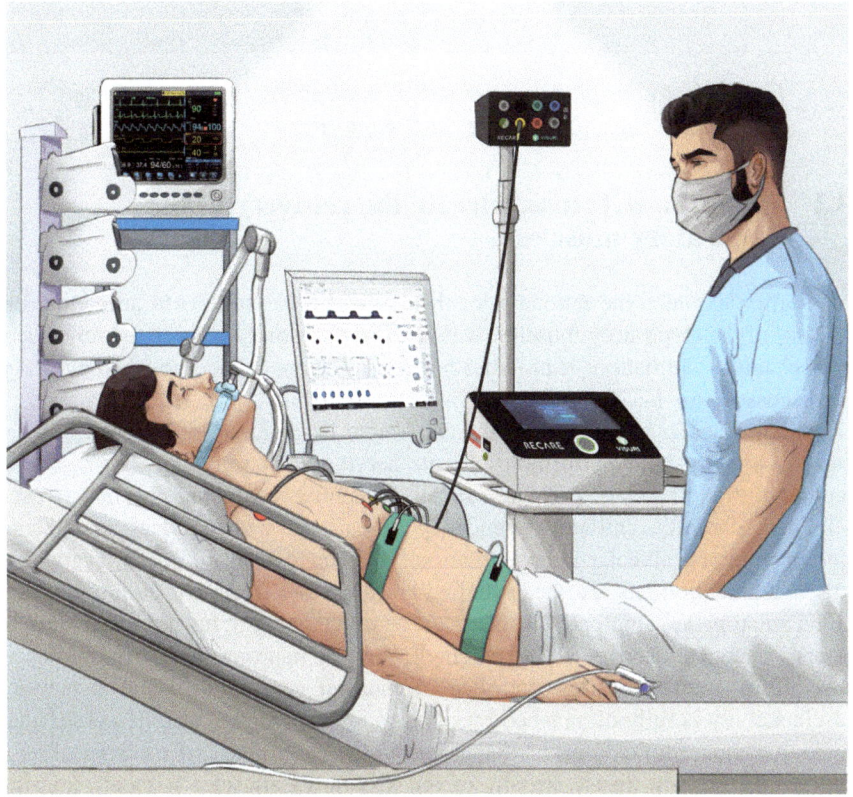

Figure 13.1   Electrodes placement configured to stimulate the abdominal to assist in exhalation

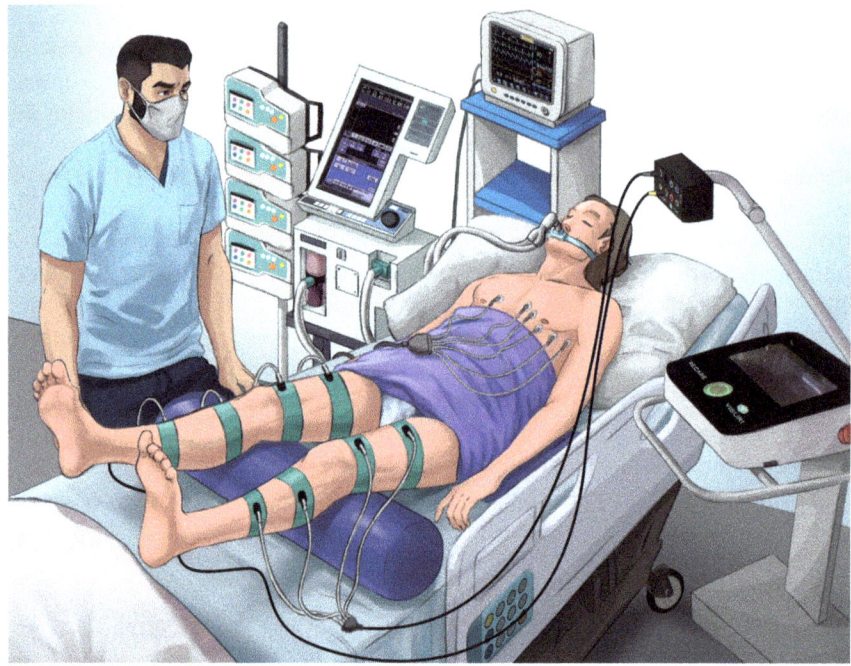

Figure 13.2   Electrodes placement to induce contractions of the quadriceps,
hamstrings, dorsiflexors, and plantarflexors.

13.3   ES in the ICU to accelerate the recovery of the respiratory muscles

This section explains the rationale for the use of ES in the ICU to accelerate the recovery of the respiratory muscles, which could potentially shorten the time spent on mechanical ventilation. It provides a brief overview of the published evidence and discusses how ES can be implemented in the ICU.

Neuromuscular stimulation of the diaphragm has been used diagnostically for years, as a method for performing phrenic nerve conduction studies or assessing diaphragm strength [71–75]. Therapeutically, diaphragm stimulation is mostly used to prevent chronic ventilator-dependency in patients with cervical spinal cord injuries or central alveolar hypoventilation [76,77]. However, in recent years, diaphragm stimulation has also been studied as adjunctive treatment during mechanical ventilation as a potential way to preserve diaphragm function and thus to improve patient outcomes [78]. This application of neuromuscular stimulation is especially of interest during controlled mechanical ventilation where respiratory muscle activity is fully diminished by sedation and sometimes also neuromuscular blocking agents. Stimulating the diaphragm in the early phase of mechanical ventilation may prevent disuse atrophy of the respiratory muscles without requiring patient cooperation. Since the diaphragm is innervated by the phrenic nerves, phrenic nerve stimulation (PNS) is a method to achieve independent activation of the diaphragm [79]. There are several methods for PNS with varying degrees of invasiveness, such as transcutaneous ES in the neck [80,81] or in the intercostal spaces over the diaphragm [82,83], percutaneous ES [78,84–86], or transvenous electrical PNS [87–90]. Considering the anatomy of this muscle, diaphragm stimulation using surface electrodes is not recommended. While effectiveness might increase with increased invasiveness, application might not always be feasible in ICU patients (see Table 13.3 for an overview of (dis)advantages for different routes of diaphragm stimulation in the ICU). Recently, Morris *et al.* [91] demonstrated the feasibility of continuous on-demand transvenous PNS to prevent diaphragm inactivity during controlled mechanical ventilation. Further research is needed to investigate feasibility of continuous PNS beyond 24 hours and to determine whether it can enhance clinical outcomes [91].

When applying diaphragm stimulation in the ICU, it is important to measure and titrate the diaphragm response to stimulation to achieve optimal results and prevent diaphragm fatigue or injury through overstimulation. A mapping process can be performed to identify the most suitable electrode combination and/or positioning and determine the threshold stimulation intensity [87,88]. Diaphragm contractions have been assessed by visual inspection or palpation in some studies [82,87,88,90]. However, it should be noted that it is hardly feasible to distinguish diaphragm activation from intercostal muscle stimulation when using surface electrodes for stimulation. Diaphragm ultrasound can also be used to observe the response to stimulation [81]. Moreover, diaphragm electromyography (EMG) and mechanomyography can be used to monitor the electrical and mechanical response to stimulation [81]. Finally, stimulated diaphragm contractions are associated with an increase in tidal volume, airway flow and transdiaphragmatic pressure (which can be measured as the difference between gastric and oesophageal pressure) as compared to unstimulated breaths [73,87,90,92]. These parameters can therefore be used to titrate and monitor the diaphragm's response to stimulation. Most importantly, the intensity of the stimulation must be adjusted to the level that is tolerated by the patient [88].

It is also important to select patients who are at risk for diaphragm atrophy and/ or prolonged weaning and therefore might benefit from the therapy. This patient population is not yet established and still a topic of research debate. Patients with an (expected) ventilation duration of at least 72 hours have been enrolled in few studies [87,88,93]. Dres *et al.* [88] showed in a randomised clinical trial in 112 patients with difficult weaning that PNS sessions 2–3 times daily could enhance diaphragm function (measured as maximum inspiratory pressure); however, no differences on weaning outcome were found. A larger follow-up study is now ongoing (clinicaltrials.gov NCT03783884). Regarding stimulation in the acute phase of mechanical ventilation to prevent atrophy development, only a few preclinical experimental studies exist and show the potential benefit of this approach [79,94,95]. Several clinical pilot studies of various quality have shown that electrical PNS may be associated with prevention of a decrease in diaphragm thickness or even increase thickness in patients on mechanical ventilation [84,86,93].

Table 13.3   Different routes for diaphragm stimulation

Technique	Stimulation location	Advantages	Disadvantages
Transcutaneous electrical PNS		<ul><li>Non-invasive.</li><li>Optimal electrode position does not change with changing head position [1].</li><li>Concentric electrode arrangement can constrain the area of activation, limiting off-target effects, and allowing more targeted stimulation of the phrenic nerve [1].</li></ul>	<ul><li>Possible complications: shoulder discomfort, pain from cutaneous afferent activation, contraction of neck muscles, and upper airway obstruction [1,2].</li><li>Technically challenging in ICU: Edema, beard growth, a jugular catheter or a tracheostomy can limit current delivery to target, or electrodes can block central venous access in the neck [1,3].</li></ul>
Percutaneous electrical diaphragm stimulation		<ul><li>Intramuscular stimulation is very targeted and proven effective in patients with chronic ventilator-dependency [4,5].</li><li>Temporary leads can be removed percutaneously [6,7].</li></ul>	<ul><li>Only applicable in patients who undergo surgical intervention before/ during ICU admission [6,7].</li><li>Risk of infection of percutaneous leads [8].</li></ul>
Percutaneous electrical PNS		<ul><li>Bedside lead insertion using ultrasound guidance [8,9].</li><li>Applicable in patients with a larger neck circumference [8].</li><li>Tolerable for conscious patients due to lower stimulation currents needed close to target [9].</li></ul>	<ul><li>More invasive than transcutaneous stimulation, risk of infection [8].</li><li>Risk of nerve injury [8].</li></ul>

(Continues)

Table 13.3   (*Continued*)

Technique	Stimulation location	Advantages	Disadvantages
Transvenous electrical PNS		• Safe and feasible bedside technique for generating diaphragm contractions [10–13]. • Lungpacer LIVE® Catheter allows for phrenic nerve pacing and delivery of intravenous fluids [12].	• Not able to place catheter or achieve effective stimulation in all patients due to individual differences [10,13]. • Risks associated with central venous catheter placement (e.g., pneumothorax, bleeding, infection, venous thrombosis) [1,11,14].
Transcutaneous electrical diaphragm stimulation		• Non-invasive. • Does not block central venous access in the neck.	• Placement of electrodes in the parasternal region beside the xiphoid process limits specificity of diaphragm activation [15]. • Risk of skin reactions to electrodes [16].
Transcutaneous electrical stimulation abdominal wall		• Non-invasive and well-tolerated [17–19]. • Safe and feasible in ICU patients [17,18]. • Optimal electrode position for force generation is posterolateral on the abdominal wall [20,21].	• Physiological effects and effects on clinical outcomes in ICU patients remain to be studied [17,18].

However, it should be noted that changes in diaphragm thickness and atrophy (and contractile function) are not necessarily correlated [37].

Several contraindications to diaphragm stimulation should be considered, such as phrenic nerve damage, diaphragm paralysis, pneumothorax, and the effect on cardiac pacemakers [93]. In patients with high adiposity or oedema, it can be difficult to achieve threshold stimulation and to determine the level of activation of the diaphragm [88]. In addition, factors such as sepsis, vasopressors or steroid treatment are associated with critical illness myopathy and may reduce the response to neuromuscular stimulation [53,79,96]. Finally, it is important to prevent patient-ventilator asynchronies during diaphragm stimulation. Thus, the stimulation should be synchronised to the inspiratory phase of the ventilator. To note, some suggest that diaphragm stimulation might even contribute to lung-protective ventilation as tidal volume can be increased during PNS without increasing ventilator driving pressure [84,90], and by reducing dorsal atelectasis it could improve the homogeneity of ventilation distribution [97]. On the other hand, by generating negative pleural pressure swings, diaphragm stimulation may be harmful in the context of heart failure or shock [97].

A review in 2023 by Etienne *et al.* [78] provided an overview of anatomical and technical aspects of diaphragm neurostimulation and reviewed the physiological rationale and clinical evidence of diaphragm neurostimulation in mechanically ventilated critically ill patients [78]. They concluded that diaphragm neurostimulation in mechanically ventilated patients may offer advantages at different stages of mechanical ventilation to prevent diaphragm atrophy and during weaning to improve diaphragm function [78]. However, further research is necessary to evaluate feasibility, optimal dosing, tolerance, physiological effects, and patient-centred outcomes in various ICU populations [78].

In contrast, studies on the impact of neuromuscular stimulation of the abdominal expiratory muscles during mechanical ventilation are limited [98]. These muscles are an important component of the respiratory muscle pump and are frequently recruited with high respiratory load or low inspiratory muscle (diaphragm) capacity [99]. Expiratory muscle weakness could occur in ICU patients as demonstrated by a reduction in maximum expiratory pressure (MEP) or cough strength and associations with weaning outcome have been reported [100–102]. In patients with spinal cord injury, neuromuscular stimulation of the expiratory muscles is applied to improve coughing and lung vital capacity [103]. In ICU patients, neuromuscular stimulation of the expiratory muscles has been applied in the context of maintaining core strength by stimulating the abdominal muscles and reducing ICUAW [70]. Breath-synchronised neuromuscular stimulation of the expiratory muscles has expanded the possibilities from purely maintaining core muscle strength to also improving respiratory function of patients on mechanical ventilation [98,104].

Expiratory muscle activity can be achieved safely by transcutaneous ES of the abdominal muscles [98,104]. To achieve highest force generation, the most optimal electrode position is posterolateral on the abdominal wall (see Table 13.3) [105,106]. When titrating the response of the expiratory muscles, it is important to note that the abdominal wall muscles consist of three layers that are, under normal physiological conditions, recruited in a fixed order [99]. First the transversus abdominis muscle is recruited, followed by the internal oblique muscle and the external oblique muscle, and finally the rectus abdominis muscle [99]. NMES activates these layers all at once, and reaching these deeper layers requires increasing the stimulation intensity. Therefore, it is important to assess

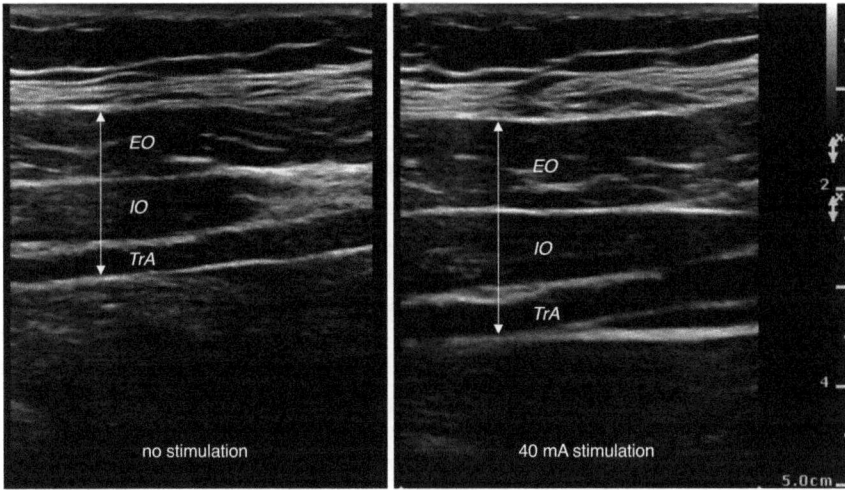

Figure 13.3   Ultrasound being used to determine stimulation threshold

the response to stimulation to ensure that all layers are activated. The response to stimulation can be assessed visually and by palpation. Ultrasound assessment is recommended to find the stimulation threshold to ensure that all different layers are activated (Figure 13.3). Functional consequences of expiratory muscle stimulation can be assessed as well. Activation of the expiratory muscles will increase intra-abdominal pressure during expiration, which can be quantified using a gastric balloon [99].

Ultrasound example of the expiratory abdominal wall muscles (EO = external oblique, IO = internal oblique, TrA = transversus abdominis) at rest and during abdominal FES with a stimulation intensity of 40 mA. Since the muscle layers have different fibre directions and more degrees of freedom to move as compared to e.g., the diaphragm, activation of one layer may deform the other muscle layers. Ultrasound should therefore mainly be used as a qualitative assessment to evaluate whether FES results in muscle activation and to determine the threshold intensity that is needed.

Similar as for diaphragm stimulation, factors such as adiposity, oedema, sepsis, steroids, and vasopressors will increase the stimulation threshold, making it difficult or sometimes impossible to achieve contractions with tolerable stimulation intensities [96,107]. Contraindications for neuromuscular stimulation of the abdominal muscles are, for instance, use of neuromuscular blockers, congenital myopathies, recent abdominal surgery, pregnancy, and cardiac pacemaker [98,108].

Neuromuscular stimulation of the abdominal muscles has been applied with the aim to prevent muscle atrophy and maintain/enhance strength during mechanical ventilation. In a first pilot study, McCaughey *et al.* [104] performed breath-synchronised ES of the abdominal muscles to demonstrate its feasibility in 20 ICU patients in Australia [104]. Using a double-blind, sham-controlled design the transversus abdominis and internal and external oblique muscles were stimulated using 5 cm ×10 cm rectangular electrodes, generating symmetric biphasic pulses with a frequency of 30 Hz and a pulse duration of 350 µs. FES was applied for 30 min, twice per day, for five days per week, until weaning from mechanical ventilation. Synchronised stimulation was performed using a flow sensor placed in the ventilator circuit: once expiratory flow was detected, the stimulator was triggered to apply NMES. Abdominal and diaphragm muscle thickness were measured using ultrasound three times in the first week, and weekly thereafter by a blinded assessor. There appeared to be no between-group thickness changes of the rectus abdominis or diaphragm on day 3. However, ICU LOC ($p = 0.011$) and ventilation duration ($p = 0.039$) appeared to be shorter in the intervention compared to the control group, although this study was not powered on patient outcomes. The patient population in this pilot study consisted mainly of neurocritically ill patients [104]. In a second pilot study, Jonkman *et al.* [98] performed a multicentre study to test the feasibility and safety of abdominal muscle stimulation in a more heterogeneous group of ICU patients, using a similar stimulation protocol [98]. A pooled analysis of the results of these two pilot studies was performed in that same report to analyse the difference in muscle thickness assessed by ultrasound between patients who received abdominal muscle stimulation and patients who received sham stimulation [98]. A significant difference was found between the groups in favour of active stimulation after three days but not after five days, probably owing to the large inter-observer variability in ultrasound assessment of abdominal muscle thickness and a limited sample size [98]. The main message of these studies are that abdominal NMES synchronised with the expiratory phase of mechanical ventilation is feasible and safe in the critically ill ICU patient.

Although significant advancements have been made through the development of breath-synchronised stimulation of the expiratory muscles, research of neuromuscular stimulation of the expiratory muscles in ICU patients is still very much in the early stages. A large international study powered on patient weaning outcomes is now being initiated to assess whether abdominal NMES could improve ventilator weaning (clinicaltrials.gov: NCT05759013).

13.4   Case study

This case report presents a critically ill patient undergoing mechanical ventilation and sedation, who was treated with whole-body NMES. The major outcomes encompassed neurophysiological studies, ultrasound imaging of the muscles, and evoked peak force measurements.

13.4.1 Patient information

A 30-year-old male patient was admitted to the Neurocritical Care Unit at the Base Hospital of the Federal District, Brazil, following a traumatic brain injury resulting from a motorcycle crash. Prior to the accident, he enjoyed good health and led an active lifestyle, engaging in regular exercise routines.

13.4.2 Clinical findings

He was admitted sedated and on mechanical ventilation, with an Injury Severity Score of 30 and a Simplified Acute Physiology Score III (SAPS III) of 40. His Glasgow Coma Scale score at the scene of the accident was 7. Throughout his ICU stay, no neurological surgical procedures were deemed necessary.

13.4.3 Timeline

He remained on mechanical ventilation for 13 days and spent a total of 15 days in the ICU, during which he received 12 days of sedation and five days of vasoactive drugs before being discharged from the ICU. Following his discharge from the ICU, the patient remained in the hospital ward for an additional 27 days, resulting in a total hospital stay of 42 days.

13.4.4 Diagnostic assessment

The patient underwent assessments within the first 24 hours of admission to the ICU, as well as on the 7th and 14th days of ICU stay.

13.4.4.1 Muscle architecture

Muscle architecture was assessed through muscle thickness and echogenicity using B-mode ultrasonography, with an ultraportable ultrasound device, Mobissom M2D Linear Model (Mobissom, São Paulo, Brazil), Figure 13.4A.

Muscle thickness and echogenicity were analysed utilising ImageJ software (http://imagej.nih.gov/ij/) [109]. Muscle echogenicity was measured through a quantitative grayscale analysis, where the most affected muscles had a white presentation (i.e., increased echogenicity). The echogenicity assessment area of analysis was selected in each muscle, including the maximum possible area (trace technique) [23] with an 8-bit image resolution, in values ranging from 0 (black) to 255 (white). The echogenicity and thickness were determined in each muscle, considering the mean value of the three replicate measurements [110].

Evaluation of the rectus femoris was conducted at the mean distance between the anterior superior iliac spine and the superior border of the patella. The tibialis anterior was evaluated at the proximal 1/4 of the distance between the inferior border of the patella and the lateral malleolus [16].

13.4.5 Neuromuscular electrophysiology testing

The presence of NED was assessed through two tests: Stimulus Electrodiagnosis Test (SET) and Strength-Duration Test (SDT). These tests were carried out

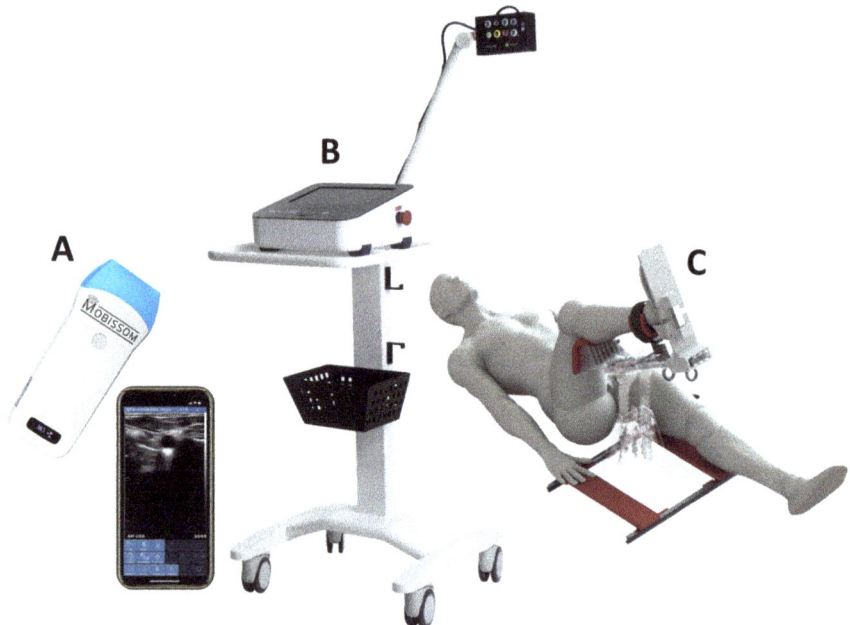

Figure 13.4   This figure presents all devices used to assess neuromuscular responses to neuromuscular electrical stimulation. A: An ultraportable linear ultrasound device, the Mobissom model MD2, with a frequency range of 7.5 to 10 MHz and a maximum depth of 5.5 cm. B: Recare electrical stimulator^{TM}, developed specifically for stimulating critically ill patients. It features 12 channels and a maximum pulse charge of 2400 µC (Pulse duration: 20,000 µs; Pulse amplitude: 120 mA). C: A 3D image of an adjustable strength assessment platform designed by our research group to measure electrically evoked peak force.

using the electrical stimulator Recare®, Visuri SA, Belo Horizonte, Brazil (Figure 13.4B). The tibialis anterior was evaluated at the proximal 1/4 of the distance between the inferior border of the patella and the lateral malleolus [16].

The SET determined the rheobase and chronaxie [16]. Rheobase is the minimal current intensity necessary to reach the neuromuscular excitability threshold applied with a rectangular pulse with an infinite duration (e.g., 1 s). Chronaxie is defined as the shortest pulse duration required to reach the neuromuscular excitability threshold by peak current twice the intensity of the rheobase. For rheobase assessment, the intensity was increased from 1 to 80 mA with 1 mA increments until eliciting a slight and visible muscle contraction. The evaluation was performed using a rectangular, biphasic, and symmetric pulse with duration of 1 s and intervals of 2 s between pulses [23]. For the evaluation of chronaxie, the pulse duration was increased from 50 µs with a current amplitude twice the value of the rheobase until eliciting a slight but visible muscle contraction [23] NED was recognised when chronaxie values reached ≥ 1000 µs [109]. In addition to providing evidence of NED diagnosis, the chronaxie value is an important variable for guiding NMES parameters prescription. The selection of pulse duration based on chronaxie when setting NMES parameters for muscle strengthening purposes is likely to produce better evoked muscle contraction [110].

Different from SET that assess two points in a continuous line (rheobase and chronaxie), the SDT assess the changes in pulse amplitude (mA) as a function of changes in pulse durations (s). In the present case report the SDT was performed using a rectangular, biphasic, and symmetric pulse with durations: 0.1 ms, 0.2 ms, 0.5 ms, 1 ms, 2 ms, 5 ms, 10 ms, 20 ms and 50 ms. For every pulse duration, the examiner needed to find a minimal muscle contraction using an intensity ranging from 1 to 120 mA [111]. After plotting the curve, the algorithm embedded in the device calculated the area under the curve (AUC). It was found that SDT AUC values exceeding 650 µC were associated with patients exhibiting NED.

Both, SET and SDT were conducted on the deep fibular nerve over the dorsum of the foot, near the ankle, utilising two surface electrodes measuring 3 cm in diameter each, Figure 13.5.

13.4.6   Muscle strength assessment of the quadriceps

Muscle strength was assessed isometrically using a strength assessment platform to measure the electrically evoked peak force that was developed by our research group (Figure 13.4C) [16,112]. For this evaluation, a calibrated load cell attached to a platform and an electrical stimulator (Recare®, Visuri SA, Belo Horizonte, Brazil) were utilised. The patient was positioned in a supine position with a 30° bed elevation, and the platform was adjusted to maintain the knee at a 30° flexion. Stimuli were administered via tetanic contraction with parameters set at 60 mA of amplitude, a 1 s on, and a pulse duration and frequency of 800 µs and 35 Hz, respectively. To evoke contraction in the quadriceps femoris, two auto-adhesive electrodes measuring 13×7.5 cm each were placed along the line between the anterior superior iliac spine and the superior border of the patella at the proximal and distal motor points [113]. Three stimuli were applied, with a 1-minute interval between each measurement. The highest detected value among the measures was recorded.

13.4.7   Therapeutic intervention

The patient underwent treatment following the Brazilian Guidelines for Early Mobilization in the Intensive Care Unit [114]. A standardised rehabilitation protocol, tailored to the patient's condition, was implemented for 10–30 minutes twice daily on weekdays by the physiotherapy team. The intensity (time) and the difficulty level of standard physiotherapy were adjusted according to the patient's cardiorespiratory status, level of sedation, cooperation, and functional capacity.

During the 12 days of sedation, the rehabilitation protocol primarily focused on body positioning and passive range of motion exercises. The NMES therapy was

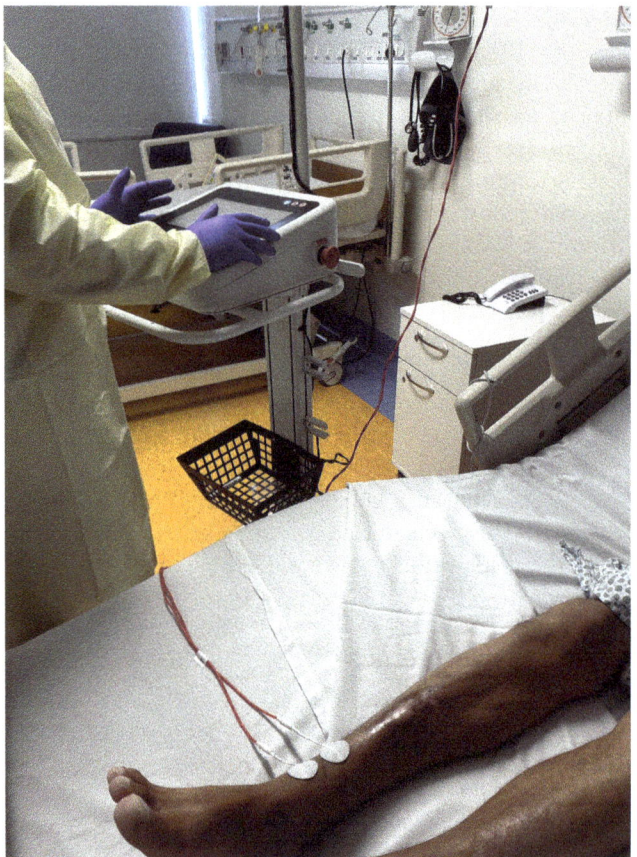

Figure 13.5   Stimulus electrodiagnosis test and strength-duration test were conducted on the deep fibular nerve over the dorsum of the foot, near the ankle, utilising two surface electrodes measuring 3 cm in diameter each.

initiated within 24 hours of ICU admission and continued until discharge from the ICU. Nutritional goals were established to achieve a daily intake of 30 kilocalories per kilogram of body weight, with a protein intake of up to 2.0 grams per kilogram per day.

The NMES therapy was administered using a 12-channel electrical stimulator, the Recare® device from Visuri SA, Belo Horizonte, Brazil. Electrodes were positioned according to the motor points, as previously described [113]. The muscles stimulated were the quadriceps femoris, hamstrings, dorsiflexors (tibialis anterior and extensor digitorum), and triceps surae, Figure 13.6.

Prior to initiating the NMES protocol, specific criteria were established for starting the therapy: hemodynamic stability (defined as mean arterial blood pressure > 60 or < 120 mmHg and/or heart frequency $< 70\%$ of the maximum predicted heart frequency), platelet count $> 20,000$, and body temperature maintained between 35°C and 37.5°C.

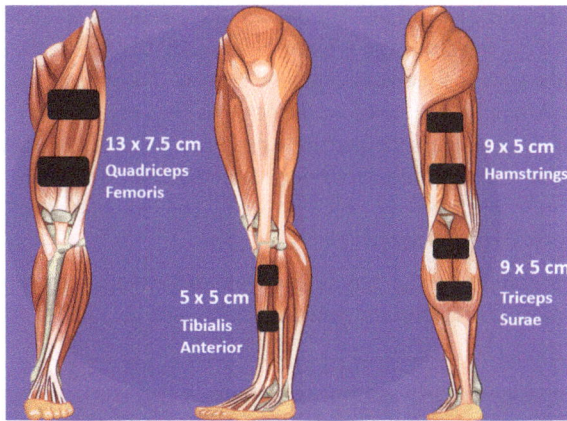

Figure 13.6 *The figure depicts the location and size of electrodes, which were positioned according to the motor points, as previously described [113]. The muscles stimulated were the quadriceps femoris, hamstrings, dorsiflexors (tibialis anterior and extensor digitorum), and triceps surae*

The NMES sessions were applied bilaterally twice a day, with a pulse duration ranging from 800 μs to 10,000 μs, adjusted based on neuromuscular excitability over the course of treatment. The pulse frequency was maintained at 35 Hz throughout. Each session consisted of 90 stimuli, totalling 180 stimuli per day, administered five days a week (900 stimuli per week). The Time On was set to 9 s (comprising 3 s of ramp up, 3 s of plateau, and 3 s of ramp down), while the Time Off was also adjusted to 9 s, resulting in a total session time of 27 minutes. The cumulative time of contractions, representing the duration during which muscles were under contraction, amounted to 810 s or 13.5 minutes per session.

The current amplitude was adjusted to the highest level possible to evoke maximally tolerated contractions in each muscle group, corresponding to type 5/5 according to the classification by Segers *et al.* [53]. NMES sessions were halted if the patient indicated pain perception exceeding a rating of 5 on the Behaviour Pain Scale [115], or in the event of hemodynamic instability, defined as mean arterial blood pressure falling below 60 mmHg or exceeding 120 mmHg, and/or heart frequency exceeding 70% of the maximum predicted heart frequency.

13.4.8 Follow-up and outcomes

The patient spent a total of 15 days in the ICU, during which seventeen out of twenty possible NMES sessions were conducted, resulting in a compliance rate of 85%. One session was interrupted due to pain, while two others were missed because of fever episodes. During the NMES sessions, the quadriceps femoris, hamstring, dorsiflexors, and triceps sural muscles were stimulated at a mean intensity of 105 $\pm$ 30 mA and a pulse duration of 1500 $\pm$ 800 μs each session. The mean pulse charge used to stimulate the muscles per session was 158 μC. By day 15, the patient's condition had improved significantly, allowing for discharge with an ICU mobility scale score of 8, indicating the ability to walk with the assistance of one person. The overall quality of evoked muscle contraction, as assessed based on the scale by Segers *et al.* [53], demonstrated a median score of 5, with an interquartile range of [5-5]. This indicates consistent palpable and visible tetanic contractions across the muscles evaluated.

13.4.9   Muscle architecture

Muscle ultrasonography of the rectus femoris and tibialis anterior revealed the preservation of muscle architecture over the 15 days of ICU bedrest. Further details are shown in Figure 13.7.

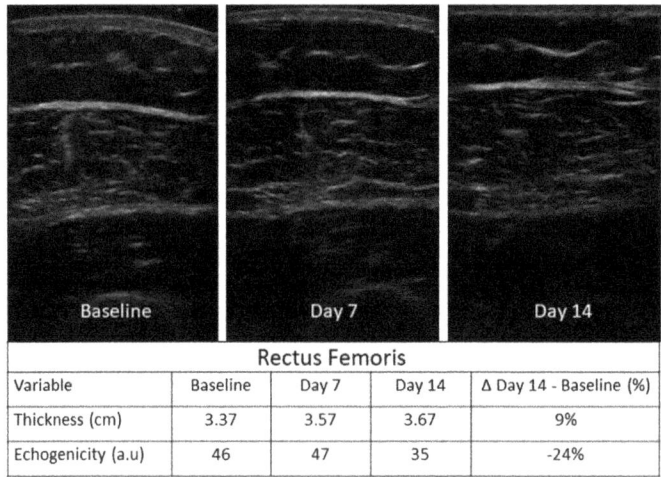

Rectus Femoris

Variable	Baseline	Day 7	Day 14	Δ Day 14 - Baseline (%)
Thickness (cm)	3.37	3.57	3.67	9%
Echogenicity (a.u)	46	47	35	-24%

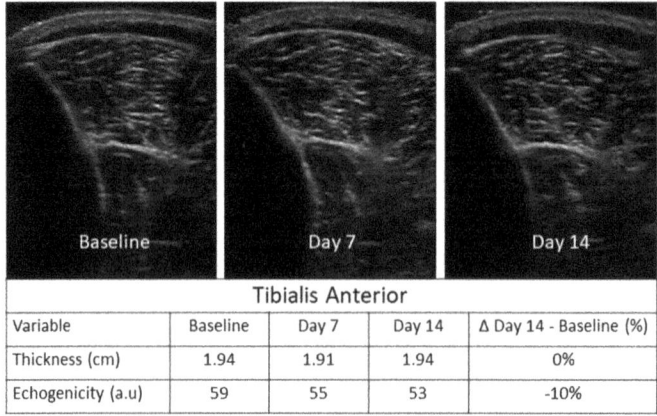

Tibialis Anterior

Variable	Baseline	Day 7	Day 14	Δ Day 14 - Baseline (%)
Thickness (cm)	1.94	1.91	1.94	0%
Echogenicity (a.u)	59	55	53	-10%

Figure 13.7   Muscle ultrasonography was used to measure muscle thickness and echogenicity. Evaluation of the rectus femoris was conducted at the mean distance between the anterior superior iliac spine and the superior border of the patella. The tibialis anterior was evaluated at the proximal 1/4 of the distance between the inferior border of the patella and the lateral malleolus

13.4.10 Neuromuscular excitability

The presence of a NED was confirmed at baseline, with a chronaxie of 3500 μs and strength-duration test's area under the curve of 1500 μC. However, over the course of treatment, there was a notable improvement in deep fibular nerve excitability, with values reaching normal levels by day 14. Specifically, the chronaxie decreased to 1500 μs by day 7 and further improved to 650 μs by day 14. Additional details are provided in Figure 13.8.

13.4.11 Muscle strength

Over the course of the treatment period, there was a significant increase in the evoked peak force of the quadriceps femoris muscle. Specifically, the force rose from 5 kgf at baseline to 24 kgf by day 14 (Figure 13.9).

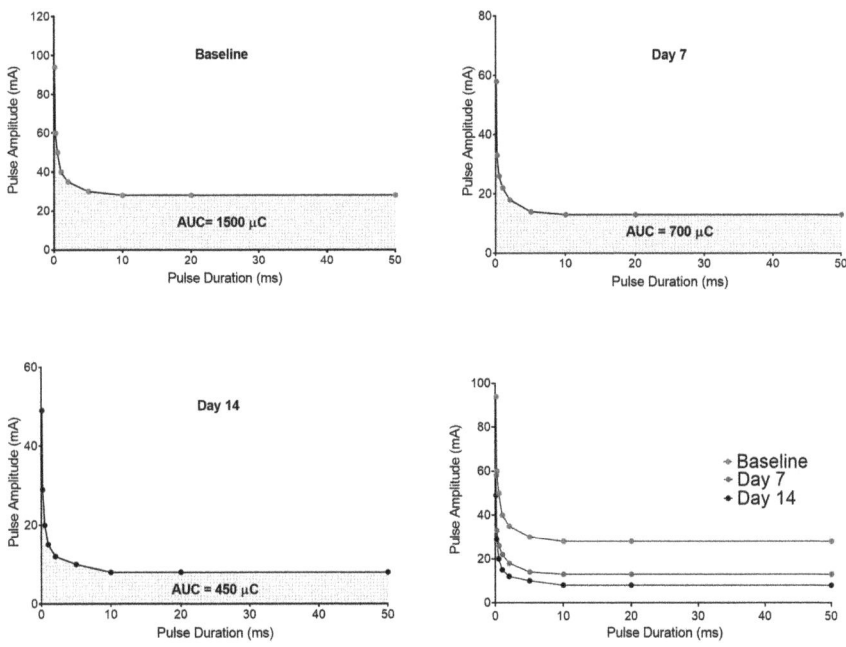

Figure 13.8 *The figure illustrates the strength-duration test's area under the curve over a span of 14 days. It demonstrates an improvement in nerve excitability, characterised by a reduction in the energy required to produce depolarisation (area under the curve)*

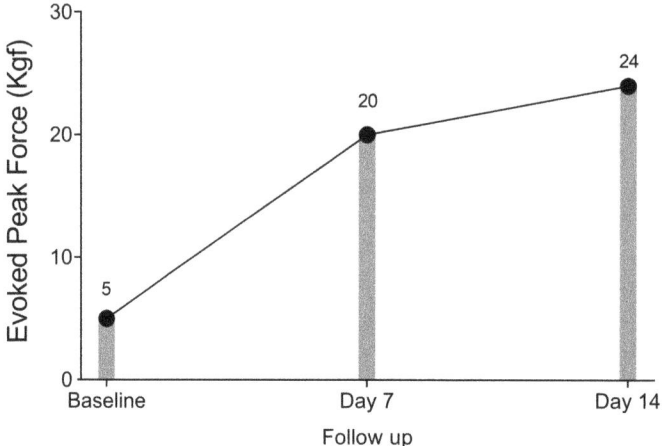

Figure 13.9   This figure presents the electrically evoked peak force of the quadriceps femoris muscle. The highest value after three bouts of electrical stimuli is reported. The contraction was elicited with a pulse duration and frequency of 800 μs and 35 Hz, respectively, with 60 mA amplitude and 1s contraction time. Two electrodes (13 × 7.5 cm) were placed over the quadriceps femoris motor points. kgf: kilogram force.

13.5   Discussion

The present case report highlights the successful treatment of a critically ill male patient utilising NMES during a 15-day ICU stay. It is widely acknowledged in the literature that passive exercise alone cannot effectively preserve muscle mass and neuromuscular excitability [23]. Therefore, for sedated patients, applying NMES to enhance muscle force generation and metabolism, are crucial components of their treatment regimen.

To achieve effectiveness, NMES must induce significant muscle contraction, as the magnitude of strength gain is directly correlated with evoked force [116]. Hence, it's essential to determine neuromuscular excitability through chronaxie and establish pulse duration and intensity to elicit high-quality muscle contraction (classified as 5/5) according to Segers *et al.* [53]. Additionally, the therapy prescription should consider the time of induced contraction, which should be at least nine minutes per session [117]. It's recommended to conduct at least one session per day, five days per week, with the option to increase frequency to two sessions per day, seven days per week, if necessary [117]. Finally, and equally important, is the concurrent stimulation of multiple muscle groups. NMES can elicit a systemic effect by releasing important anti-inflammatory, angiogenic, and anabolic myokines, and this effect appears to be correlated with the number of stimulated muscles. It is probable that stimulating multiple muscle groups (at least four) will be more effective than stimulating just the quadriceps [118].

In the case presented here, the maintenance of muscle architecture was demonstrated through assessments of muscle thickness and echogenicity. While an increase in muscle thickness can sometimes be associated with inflammation and oedema rather than hypertrophy, this distinction can be made using echogenicity. Inflammation, necrosis, and oedema typically increase echogenicity [23]. In this case study, there was a preservation of muscle mass alongside an improvement in muscle quality, as evidenced by a decrease in echogenicity.

Neuromuscular excitability was also improved, as evidenced by chronaxie values and Strength-Duration Test's area under the curve on the 14th day. Interestingly, the patient exhibited significant NED within the first 24 hours, which is consistent with findings from previous studies [16,119]. This demonstrates the importance of starting NMES as soon as possible, once the patient is admitted to the ICU, such as it was recommended in randomised clinical trials [16,46].

One hypothesis for this rapid deterioration in neuromuscular excitability is the reduction in nerve oxygenation. Nerve metabolism primarily relies on aerobic processes, and its nourishment occurs via active transport from the nucleus to the axon. In cases of hypoxia, adenosine triphosphate production is hindered, and nutrients fail to reach the axon, leading to axonopathy [119].

NMES plays a significant role in improving both microvascular and macro-vascular function, leading to enhanced tissue oxygenation and reduced oedema [120,121]. This could explain the observed improvement in neuromuscular excit-ability over the course of 14 days of NMES, as evidenced by the reduction in chronaxie values and Strength-Duration Test's area under the curve. The effect in microcirculation it is not restrict to the nerve, but also can preserve skin blood flow preventing pressure injury in critically ill patients [40].

After 15 days of ICU stay, the patient was discharged and capable of walking with the assistance of one person (ICU Mobility Scale = 8). This improvement in mobility is closely related to the maintenance of muscle strength. The observed increase in evoked force reinforces the notion of an improvement in muscle architecture and neuromuscular excitability, which aligns with findings from pre-vious studies [16]. Muscle strength is pivotal in enabling basic functional abilities such as walking. Therefore, strategies aimed at maintaining or improving muscle strength in the ICU setting are of utmost importance.

The case reported here represents just one among thousands of successful cases treated in the ICUs of the Base Hospital of the Federal District, Brazil, since 2014. NMES has been established as a standardised procedure in the ICU during this time, with a consensus among hospital staff that this therapy is essential for treating minimally mobile or non-collaborative critically ill patients. Despite the ample scientific evidence supporting the effectiveness of NMES in the ICU [16,40,46,122], its adoption and dissemination remain limited [123].

Over the last ten years, we have catalogued the main barriers to the adoption of NMES in the ICU. When the first randomised controlled trial was published in 2003 [124], the biggest barrier was the lack of scientific evidence. In just over a decade, dozens of RCTs have been published on this topic, and the lack of evidence no longer justifies the non-use of NMES in ICUs [122].

Currently, it seems that the shortage of human resources (due to the high demand for work in the ICU), the low level of knowledge among professionals, and the lack of specific stimulation equipment for critically ill patients are the primary obstacles to the widespread adoption of this treatment worldwide.

Improving the knowledge of ICU teams about NMES and promoting the use of appropriate NMES devices for critically ill patients could enhance staff commitment and facilitate the wider adoption of this therapy worldwide.

13.6   Practical clinical considerations

Until there is clinical evidence for effectiveness, agreement on objectives for its use and with which patients, how it should be applied (stimulation parameters, stimulation dosage, targeted muscles) and how benefits should be measured, ES is unlikely to become standard clinical practice in the ICU.

Of the stimulation parameters, the pulse frequency (Hz or PPS) and the ON-OFF times (s) are the most varied parameters. The pulse frequency determines the type of muscle contraction, which can range from a twitch, to incomplete tetanic, to complete tetanic contraction. To achieve the electro-physiologically preferred tetanic contraction, any pulse rate (frequency) of 20 Hz or higher (even 10 kHz) can be used. However, the higher the frequency, the faster electrically induced muscle fatigue.[59,117] Accordingly, selecting a pulse frequency between 20 and 40 Hz appears the clinically preferred choice. Selecting a higher frequency (50–75 Hz) is likely to yield the same tetanic contraction, but to minimise a probable muscle fatigue, the OFF (relaxation time) between successive contractions could be prolonged by a few seconds. Indeed, the ON-OFF ratio (seconds of contraction vs seconds of relaxation) is another parameter that can be modified to minimise muscle fatigue during the application of ES in the ICU.

Dosing the stimulation is a new concept that is not considered in available publications. However, it may be of considerable value to the interpretation of success vs. failure of implementing an ES training program. A guiding principle when formulating a clinically relevant dosing documentation is to determine how many contractions were induced each session and over multiple sessions. The formula does not consider the strength of electrically induced contraction because objective measurement of muscle force or moment (torque) is not feasible in most ICU environments. The proposed formula:

Training Dose = number of contractions per minute X length of

session (min) X number of sessions per day X number of days

per week X number of weeks.

For example, having ON time of 10 s and OFF time of 10 s yields three contractions per min. Having a session duration of 20 min would then yield

60 contractions. Assuming one session a day, repeated five times per week, increases the cumulative dose to 300 contractions. In this way, the cumulative dose can be documented and represents the total number of contractions while in the ICU. The offered formula can be used to compare the stimulation dosages between published studies and may offer insight into why the outcomes of one ES program are more effective than those of other programs. There is also a potential formula to calculate the dosage of electrical energy (in milli or microcoulombs) but it requires knowledge of the stimulator design being a constant current or constant voltage, the waveform used, the phase duration, the number of pulses per second (pps or Hz) the peak current intensity, and the treatment time. Such data are rarely disclosed in peer reviewed publications.

Consideration of which muscle groups should be the target of stimulation depends on the various comorbidities that each patient presents, which muscle groups are most likely to become atrophic and lose volitional force generation, and which muscle groups' loss of excitability. Collectively these factors adversely affect the responsiveness to ES because of sepsis, oedema, or heavily sedated patients receiving vasopressors [53]. In addition, when determining which muscle groups should be stimulated, it is essential to consider the primary clinical outcomes presented in published clinical studies. Is the primary outcome measure the time on mechanical ventilation? The LOS in the ICU? The time of immobility? Or minimising the loss of volitional force generation or muscle mass? Of note, data to date provide no evidence that ES can prevent muscle loss or increase muscle strength while in the ICU. It is therefore plausible to propose that at the time of this writing, published clinical studies in the ICU do not offer a clear road map to a predictive, effective treatment protocol that maximises the utilisation of ES in clinical practice.

Further elaboration on clinical considerations evolves around the principle of continuum of care. In most, if not all studies conducted in the ICU, ES was applied for only a few days, the longest for 13–14 days and was discontinued as soon as the patients were discharged from the ICU or even earlier when some patients were able to get out of bed (details are offered in earlier section of the chapter). The reasons for discontinuing the ES program are not disclosed in the peer-reviewed publications, but extrapolation from the available data using the dosage formula may provide an estimate, indicating that in these studies the NMES dosage was very low. A low dose NMES program can explain the less than desired recovery, specifically of muscle mass and volitional force generation. Such low treatment dosage is an inherent shortcoming of a legacy experimental design that allocates the same ES dosage, irrespective of the needs of each individual patient. The discontinuation of the ES as soon as the patients are discharged from the ICU is another major shortcoming inherent in a conventional experimental design. Moreover, the reports that only 50% [53] and only 38% [54] so-called 'responders', tolerated clinically meaningful muscle contraction while in the ICU diminish the likelihood of transferring the research finding into accepting ES as a valuable intervention to complement the standard treatment option in the ICU. These inherent deficiencies and ways to rectify them are discussed in the future studies section.

13.7   Precautions and contraindications

Other chapters in this book address contraindications to the application of non-invasive ES in the medical field in general, and the rehabilitation field specifically. When applied in the ICU some additional precautions must be added. One is to ensure compatibility of the ES with other electronic and mechanical technologies so any possible interference in the function of these technologies is avoided. Another precaution is to meet the requirements of a sterile environment and ensure compatibility with the patients' skin to minimise the chance of electrically induced skin irritation that may lead, if not careful, to skin lesion under the electrodes. In the reviewed studies in this chapter these adverse responses have not been reported. Additional precaution relates to the possibility of having a sudden heard attack, stroke or other unanticipated catastrophic events not associated with the ES application but are commonly reported in the ICU literature [125].

13.8   Suggestions for future research

Published studies conducted in the ICU, collectively offer evidence supportive of the efficacy of applying ES to the limb muscles. The application of ES to the respiratory muscle pump is yet in its infancy but has made a significant development in the last few years. Nonetheless, the magnitudes (the amount) of muscle force generation or patient improvement varied considerably among the studies suggesting that peer-reviewed published ES programs did not maximise the recovery of the patients' muscle strength, nor the time it took to discontinue mechanical ventilation, or to shorten the LOS in the ICU. As elaborated in this chapter, the patient population admitted to ICU is very diverse, with multiple aetiologies, medical histories, and circumstances that led to their admission. Applying NMES in an acutely and severely ill population could be challenging and should be done with caution, especially early during admission. For example, patients admitted to the neurological ICU because of a sudden subarachnoid hae-morrhage were fully independent prior to admission and many expected to recover independence in mobility following a short stay in the ICU [13]. In contrast, patients admitted to the pulmonary ICU typically have a prior comorbid medical history such as rheumatoid or osteoarthritis, liver, or kidney disease [14] that complicate the medical management, and may delay the recovery of unassisted mobility or recovery of muscle strength [3,12]. The two reasons for using ES in ICU are to prevent deterioration of muscle and to promote improved respiratory function. Importantly, ICU-acquired muscle weakness is different from diaphragm weakness and although both are associated with negative outcomes, the two types have only limited overlap [21].

   To generate evidence for the clinical use of ES in the ICU, the treatment program requires a working hypothesis and a priori specification of primary and secondary outcome measures. In accordance with the current consensus medical model that endorses the principles of personalised and precision medicine, future studies should be designed and constructed applying a more precise approach. It should include initial screening, increasing the dosage of ES application, and focusing on specific outcome measures that ES is likely to maximise. The physiological efficacy of the various PNS techniques to activate the diaphragm should be studied to define the optimal stimulation parameters and dosing. Future studies should first include a screening process to determine if any muscle group is not excitable. Extrapolating from the data that only 50% [53] or 38% [54] of patients in the ICU respond to the NMES, suggest that it would not be prudent to apply NMES to non-excitable muscles. Thus, the initial screening is critical to determine which muscle groups should be included in the training. Another component of the screening is to stratify groups allocation (control vs. experimental) based on mobility, once admitted to the ICU. Restoring independence in mobility so the patient can get out of bed and eventually out of the ICU without any assistance is the ultimate objective of all ICU admissions. The LOS has been a common outcome measure to quantify this objective. Using for example the functional status score for the intensive care unit (FSS-ICU) a score that reflects how much assistance is required to move, from 100% to no assistance, at the commencement of the study and assigning it equally between groups should minimise the between group variability and add power to the statistical analyses. Similar stratification should be applied when matching the groups for the length (time) of dependence on mechanical ventilation.

To adhere to the concept of personalised medicine, it is crucial to align the primary outcome measure in the study design with the muscle group(s) that are most likely to improve this outcome measure. For example, if shortening the time on mechanical ventilation is the primary outcome, the ES should be applied to induce contractions of the diaphragm or abdominal muscles and possibly the intercostal muscles [49,104]. Synchronisation with inhalation-exhalation cycle should be considered in future studies as it is essential to minimise the risk of patient-ventilator asynchrony and improve efficacy of stimulations. In contrast, if the primary objective is to accelerate the recovery of locomotion including standing up-sitting down and walking ability, the target of stimulation should be both lower extremities muscles.

Increasing the dosage of ES is a particular challenge for future studies. As shown in Chapter 6: Electrical Stimulation in Musculoskeletal Populations, most published studies concluded that the ES minimised significantly the loss of muscle volume or CSA, as well as loss of muscle strength, but failed to increase these variables. Alternatively, studies of patients with chronic conditions in outpatient or home environment reported strong evidence that ES is likely to strengthen atrophied-sarcopenic muscles after 4–6 weeks of training, implying a much higher dose of NMES application [126–129]. A research design question arises concerning how to adjust the stimulation dosage when approximately half of the patients are discharged between 5 and 7 days, while the remaining half is discharged at 13–14 days [3]. One option to increase the ES dose is to apply it 3–4 times daily. However, this is not a realistic option because of intense scheduling constraints in the ICU. The better option is to continue the ES program for 4–12 weeks regardless of whether the patient is being discharged to a non-ICU floor in the hospital, or to an outpatient clinic, or to his/her home environment [128,129]. This continuum of care model is hypothesised to maximise the recovery of muscle strength and may shorten the time it takes for the patients to regain mobility independence.

To make a continuum of care a successful pathway, patients' compliance-adherence with the ES training, especially after discharge from the ICU, is critical [104]. One way to improve compliance is to simplify the operations and application of the ES systems by further development of miniaturised electronic hardware, advanced software, and incorporation of new bio-compatible materials [39]. Having a 'patient friendly' multi-channel system that only takes 1–2 minutes to set up and operate is a challenge for today's biomedical engineers-researchers. The working hypothesis projects that these miniaturised-friendly ES systems will substantially increase patients' compliance. Furthermore, all members of the medical-rehabilitation personnel will be able to advise the patients when, how and why to utilise the ES during the prolonged recovery process.

In conclusion, there is a convincing rationale and pathophysiological evidence for the use of ES in the ICU to improve mobility, minimise muscle loss and accelerate the recovery of the respiratory muscles, all of which could reduce LOS and result in better long-term outcomes. Currently, however there is a paucity of clinical evidence to support standard clinical use. Challenges have been identified, many of which are common to other clinical applications of ES and include clearly stated objectives, agreed stimulation protocols and standardised outcome measures.

References

[1] Ohsfeldt RL, Choong CK, Mc Collam PL, Abedtash H, Kelton KA, and Burge R. Inpatient hospital costs for COVID-19 patients in the United States. *Adv Ther*. 2021;38(11):5557–95.

[2] Christmas AB, Freeman E, Chisolm A, *et al.* Trauma intensive care unit 'bouncebacks': identifying risk factors for unexpected return admission to the intensive care unit. *Am Surg*. 2014;80(8):778–82.

[3] Verceles AC, Serra M, Davis D, *et al.* Combining exercise, protein supplementation and electric stimulation to mitigate muscle wasting and improve outcomes for survivors of critical illness-the ExPrES study. *Heart Lung*. 2023;58:229–35.

[4] Silva-Gutierrez A, Artigas-Arias M, Alegria-Molina A, *et al.* Characterization of muscle mass, strength and mobility of critically ill patients with SARS-CoV-2 pneumonia: distribution by sex, age, days on mechanical ventilation, and muscle weakness. *Front Physiol*. 2023;14:1095228.

[5] Johnson BH, Johnston SS, Tewari P, Afolabi M, and Danker Iii W. Clinical and economic burden associated with prolonged air leaks among patients undergoing thoracic resection: a retrospective database analysis. *Clinicoecon Outcomes Res*. 2023;15:269–80.

[6] Thackeray M, Kotowicz MA, Pasco JA, Mohebbi M, and Orford N. Changes in body composition in the year following critical illness: a case-control study. *J Crit Care*. 2022;71:154043.

[7] Jiang T, Lin T, Shu X, *et al.* Prevalence and prognostic value of preexisting sarcopenia in patients with mechanical ventilation: a systematic review and meta-analysis. *Crit Care*. 2022;26(1):140.

[8] Nguyen C, Crowe CL, Kuti E, *et al.* Impact of pre-existing type 2 diabetes with and without cardiovascular disease on patients with COVID-19. *Diabetes Obes Metab*. 2023;25(9):2464–72.

[9] McDonnell ME, Garg R, Gopalakrishnan G, *et al.* Glycemic gap predicts mortality in a large multicenter cohort hospitalized with COVID-19. *J Clin Endocrinol Metab*. 2023;108(3):718–25.

[10] Lin S, Lai D, and He W. Association between hyperglycemia and adverse clinical outcomes of sepsis patients with diabetes. *Front Endocrinol (Lausanne)*. 2022;13:1046736.

[11] Nguyenhuy M, Chang J, Xu R, Virk S, and Saxena A. The fried frailty phenotype in patients undergoing cardiac surgery: a systematic review and meta-analysis. *Heart Surg Forum*. 2022;25(5):E652–E9.

[12] Okura K, Takahashi Y, Hasegawa K, *et al.* Early pulmonary rehabilitation with neuromuscular electrical stimulation in a patient with acute exacerbation of rheumatoid arthritis-associated interstitial lung disease: a case report. *Phys Ther Res*. 2022;25(3):156–61.

[13] Badjatia N, Sanchez S, Judd G, *et al.* Neuromuscular electrical stimulation and high-protein supplementation after subarachnoid hemorrhage: a single-center phase 2 randomized clinical trial. *Neurocrit Care*. 2021;35(1):46–55.

[14] Plaut T, and Weiss L. *Electrodiagnostic Evaluation of Critical Illness Neuropathy*. Treasure Island (FL): StatPearls; 2023.

[15] Fagoni N, Ferretti G, Piva S, *et al.* A reappraisal of the strength-duration test to assess neuromuscular impairment of critically ill patients. *J Electromyogr Kinesiol*. 2021;59:102555.

[16] Silva PE, de Cassia Marqueti R, Livino-de-Carvalho K, *et al.* Neuromuscular electrical stimulation in critically ill traumatic brain injury patients attenuates muscle atrophy, neurophysiological disorders, and weakness: a randomized controlled trial. *J Intensive Care*. 2019;7:59.

[17] Demoule A, Jung B, Prodanovic H, *et al.* Diaphragm dysfunction on admission to the intensive care unit. Prevalence, risk factors, and prognostic impact-a prospective study. *Am J Respir Crit Care Med*. 2013;188(2):213–9.

[18] Jaber S, Petrof BJ, Jung B, *et al.* Rapidly progressive diaphragmatic weakness and injury during mechanical ventilation in humans. *Am J Respir Crit Care Med*. 2011;183(3):364–71.

[19] Dres M, Goligher EC, Heunks LMA, and Brochard LJ. Critical illness-associated diaphragm weakness. *Intensive Care Med*. 2017;43(10):1441–52.

[20] Jonkman AH, Jansen D, and Heunks LMA. Novel insights in ICU-acquired respiratory muscle dysfunction: implications for clinical care. *Crit Care*. 2017;21(1):64.

[21]   Dres M, Dubé BP, Mayaux J, *et al.* Coexistence and impact of limb muscle and diaphragm weakness at time of liberation from mechanical ventilation in medical intensive care unit patients. *Am J Respir Crit Care Med.* 2017;195 (1):57–66.

[22]   Dall' Acqua AM, Sachetti A, Santos LJ, *et al.* Use of neuromuscular electrical stimulation to preserve the thickness of abdominal and chest muscles of critically ill patients: a randomized clinical trial. *J Rehabil Med.* 2017;49 (1):40–8.

[23]   Silva PE, Maldaner V, Vieira L, *et al.* Neuromuscular electrophysiological disorders and muscle atrophy in mechanically-ventilated traumatic brain injury patients: new insights from a prospective observational study. *J Crit Care.* 2018;44:87–94.

[24]   Vanhorebeek I, Gunst J, Casaer MP, *et al.* Skeletal muscle myokine expression in critical illness, association with outcome and impact of therapeutic interventions. *J Endocr Soc.* 2023;7(3):bvad001.

[25]   Van Dyck L, Guiza F, Derese I, *et al.* DNA methylation alterations in muscle of critically ill patients. *J Cachexia Sarcopenia Muscle.* 2022;13(3):1731–40.

[26]   Fazzini B, Markl T, Costas C, *et al.* The rate and assessment of muscle wasting during critical illness: a systematic review and meta-analysis. *Crit Care.* 2023;27(1):2.

[27]   Bunnell KM, Thaweethai T, Buckless C, *et al.* Body composition predictors of outcome in patients with COVID-19. *Int J Obes (Lond).* 2021;45 (10):2238–43.

[28]   Looijaard WG, Dekker IM, Stapel SN, *et al.* Skeletal muscle quality as assessed by CT-derived skeletal muscle density is associated with 6-month mortality in mechanically ventilated critically ill patients. *Crit Care.* 2016;20 (1):386.

[29]   Cebeci GC, Cebeci H, Kucuk MP, Kucuk AO, Bayrak IK, and Ulger F. Neuromuscular electrical stimulator as a protective treatment against intensive care unit muscle wasting in sepsis/septic shock patients. *J Coll Physicians Surg Pak.* 2022;32(10):1300–7.

[30]   Segers J, Vanhorebeek I, Langer D, *et al.* Early neuromuscular electrical stimulation reduces the loss of muscle mass in critically ill patients - a within subject randomized controlled trial. *J Crit Care.* 2021;62:65–71.

[31]   Sekhniashvili M, Bodechtel U, Toyka KV, and Baum P. Temporary reversal of nerve and muscle dysfunction by serial electrical stimulation in critical illness neuromyopathy. *Clin Neurophysiol.* 2022;142:244–53.

[32]   Hooijman PE, Beishuizen A, Witt CC, *et al.* Diaphragm muscle fiber weakness and ubiquitin proteasome activation in critically ill patients. *Am J Respir Crit Care Med.* 2015;191(10):1126–38.

[33]   Shi Z, van den Berg M, Bogaards S, *et al.* Replacement fibrosis in the diaphragm of mechanically ventilated critically ill patients. *Am J Respir Crit Care Med.* 2023;207(3):351–4.

[34] van den Berg M, Peters EL, van der Pijl RJ, *et al.* Rbm20$^{\Delta RRM}$ mice, expressing a titin isoform with lower stiffness, are protected from mechanical ventilation-induced diaphragm weakness. *Int J Mol Sci.* 2022;23(24):15689.

[35] Shi Z-H, de Vries H, de Grooth H-J, *et al.* Changes in respiratory muscle thickness during mechanical ventilation: focus on expiratory muscles. *Anesthesiology.* 2021;134(5):748–59.

[36] Goligher EC, Dres M, Fan E, *et al.* Mechanical ventilation-induced diaphragm atrophy strongly impacts clinical outcomes. *Am J Respir Crit Care Med.* 2018;197(2):204–13.

[37] Dot I, Pérez-Terán P, Francés A, *et al.* Association between histological diaphragm atrophy and ultrasound diaphragm expiratory thickness in ventilated patients. *J Intensive Care.* 2022;10(1):40.

[38] Nakano H, Naraba H, Hashimoto H, *et al.* Novel protocol combining physical and nutrition therapies, intensive goal-directed rehabilitation with electrical muscle stimulation and nutrition (IGREEN) care bundle. *Crit Care.* 2021;25(1):415.

[39] Alon G. Functional electrical stimulation (FES): clinical successes and failures to date. *J Nov Physiother Rehabil.* 2018;2:80–6.

[40] Baron MV, Silva PE, Koepp J, *et al.* Efficacy and safety of neuromuscular electrical stimulation in the prevention of pressure injuries in critically ill patients: a randomized controlled trial. *Ann Intensive Care.* 2022;12(1):53.

[41] Di Micco L, Marzocco S, Adesso S, *et al.* Muscle stimulation in elderly patients with CKD and sarcopenia. *G Ital Nefrol.* 2015;32(5).

[42] Dirks ML, Hansen D, Van Assche A, Dendale P, and Van Loon LJ. Neuromuscular electrical stimulation prevents muscle wasting in critically ill comatose patients. *Clin Sci (Lond).* 2015;128(6):357–65.

[43] Dirks ML, Groen BB, Franssen R, van Kranenburg J, and van Loon LJ. Neuromuscular electrical stimulation prior to presleep protein feeding stimulates the use of protein-derived amino acids for overnight muscle protein synthesis. *J Appl Physiol (1985).* 2017;122(1):20–7.

[44] Mahran GSK, Mehany MM, Abbas MS, *et al.* Short-term outcomes of neuromuscular electrical stimulation in critically ill patients. *Crit Care Nurs Q.* 2023;46(2):126–35.

[45] Nonoyama T, Shigemi H, Kubota M, Matsumine A, Shigemi K, and Ishizuka T. Neuromuscular electrical stimulation in the intensive care unit prevents muscle atrophy in critically ill older patients: a retrospective cohort study. *Medicine (Baltimore).* 2022;101(31):e29451.

[46] Campos DR, Bueno TBC, Anjos J, *et al.* Early neuromuscular electrical stimulation in addition to early mobilization improves functional status and decreases hospitalization days of critically ill patients. *Crit Care Med.* 2022; 50(7):1116–26.

[47] Bao W, Yang J, Li M, *et al.* Prevention of muscle atrophy in ICU patients without nerve injury by neuromuscular electrical stimulation: a randomized controlled study. *BMC Musculoskelet Disord.* 2022;23(1):780.

[48] Dos Santos FV, Cipriano G, Jr., Vieira L, *et al.* Neuromuscular electrical stimulation combined with exercise decreases duration of mechanical ventilation in ICU patients: a randomized controlled trial. *Physiother Theory Pract*. 2020;36(5):580–8.

[49] Leite MA, Osaku EF, Albert J, *et al.* Effects of neuromuscular electrical stimulation of the quadriceps and diaphragm in critically ill patients: a pilot study. *Crit Care Res Pract*. 2018;2018:4298583.

[50] Patsaki I, Gerovasili V, Sidiras G, *et al.* Effect of neuromuscular stimulation and individualized rehabilitation on muscle strength in intensive care unit survivors: a randomized trial. *J Crit Care*. 2017;40:76–82.

[51] Akar O, Gunay E, Sarinc Ulasli S, *et al.* Efficacy of neuromuscular electrical stimulation in patients with COPD followed in intensive care unit. *Clin Respir J*. 2017;11(6):743–50.

[52] Saitoh M, Dos Santos MR, Anker M, Anker SD, von Haehling S, and Springer J. Neuromuscular electrical stimulation for muscle wasting in heart failure patients. *Int J Cardiol*. 2016;225:200–5.

[53] Segers J, Hermans G, Bruyninckx F, Meyfroidt G, Langer D, and Gosselink R. Feasibility of neuromuscular electrical stimulation in critically ill patients. *J Crit Care*. 2014;29(6):1082–8.

[54] Grunow JJ, Goll M, Carbon NM, Liebl ME, Weber-Carstens S, and Wollersheim T. Differential contractile response of critically ill patients to neuromuscular electrical stimulation. *Crit Care*. 2019;23(1):308.

[55] Nozoe M, Kamo A, Shimada S, and Mase K. Neuromuscular electrical stimulation is ineffective for treating quadriceps muscle wasting with ruptured aneurysm: a case report. *Ann Med Surg (Lond)*. 2018;35:90–4.

[56] Koutsioumpa E, Makris D, Theochari A, *et al.* Effect of transcutaneous electrical neuromuscular stimulation on myopathy in intensive care patients. *Am J Crit Care*. 2018;27(6):495–503.

[57] Fischer A, Spiegl M, Altmann K, *et al.* Muscle mass, strength and functional outcomes in critically ill patients after cardiothoracic surgery: does neuromuscular electrical stimulation help? The Catastim 2 randomized controlled trial. *Crit Care*. 2016;20:30.

[58] Kho ME, Truong AD, Zanni JM, *et al.* Neuromuscular electrical stimulation in mechanically ventilated patients: a randomized, sham-controlled pilot trial with blinded outcome assessment. *J Crit Care*. 2015;30(1):32–9.

[59] Angelopoulos E, Karatzanos E, Dimopoulos S, *et al.* Acute microcirculatory effects of medium frequency versus high frequency neuromuscular electrical stimulation in critically ill patients - a pilot study. *Ann Intensive Care*. 2013;3 (1):39.

[60] Zhou J, Zhang C, Zhou JD, and Zhang CK. Effect of early progressive mobilization on intensive care unit-acquired weakness in mechanically ventilated patients: an observational study. *Medicine (Baltimore)*. 2022;101(44): e31528.

[61] Liu K, Shibata J, Fukuchi K, *et al.* Optimal timing of introducing mobilization therapy for ICU patients with sepsis. *J Intensive Care*. 2022;10(1):22.

[62] Lippi L, de Sire A, D'Abrosca F, *et al.* Efficacy of physiotherapy interventions on weaning in mechanically ventilated critically ill patients: a systematic review and meta-analysis. *Front Med (Lausanne).* 2022;9:889218.

[63] Wollersheim T, Grunow JJ, Carbon NM, *et al.* Muscle wasting and function after muscle activation and early protocol-based physiotherapy: an explorative trial. *J Cachexia Sarcopenia Muscle.* 2019;10(4):734–47.

[64] Stefanou C, Karatzanos E, Mitsiou G, *et al.* Neuromuscular electrical stimulation acutely mobilizes endothelial progenitor cells in critically ill patients with sepsis. *Ann Intensive Care.* 2016;6(1):21.

[65] Alon G. Functional electrical stimulation (FES): transforming clinical trials to neuro-rehabilitation clinical practice- a forward perspective. *J Nov Physiother.* 2013;3:176–85.

[66] Rousseau AF, Kellens I, Dardenne N, Misset B, and Croisier JL. Physical capacities assessment in critically ill patients: an exploratory study. *Aust Crit Care.* 2022;35(6):709–13.

[67] Moraes FDS, Marengo LL, Moura MDG, *et al.* ABCDE and ABCDEF care bundles: a systematic review of the implementation process in intensive care units. *Medicine (Baltimore).* 2022;101(25):e29499.

[68] Nakamura K, Kihata A, Naraba H, *et al.* Efficacy of belt electrode skeletal muscle electrical stimulation on reducing the rate of muscle volume loss in critically ill patients: a randomized controlled trial. *J Rehabil Med.* 2019;51 (9):705–11.

[69] Nankaku M, Ikeguchi R, Aoyama T, *et al.* A first view of the effect of a trial of early mobilization on the muscle strength and activities of daily living in mechanically ventilated patients with COVID-19. *Arch Rehabil Res Clin Transl.* 2022;4(3):100201.

[70] Liu M, Luo J, Zhou J, and Zhu X. Intervention effect of neuromuscular electrical stimulation on ICU acquired weakness: a meta-analysis. *Int J Nurs Sci.* 2020;7(2):228–37.

[71] Zifko U, Young BG, and Bolton CF. Electrophysiological monitoring in neurological respiratory insufficiency. *J Neurol Neurosurg Psychiatry.* 1997; 62(3):299.

[72] Watson AC, Hughes PD, Louise Harris M, *et al.* Measurement of twitch transdiaphragmatic, esophageal, and endotracheal tube pressure with bilateral anterolateral magnetic phrenic nerve stimulation in patients in the intensive care unit. *Crit Care Med.* 2001;29(7):1325–31.

[73] Buscher H, Valta P, Boie T, *et al.* Assessment of diaphragmatic function with cervical magnetic stimulation in critically ill patients. *Anaesth Intensive Care.* 2005;33(4):483–91.

[74] Hamnegård CH, Wragg SD, Mills GH, *et al.* Clinical assessment of diaphragm strength by cervical magnetic stimulation of the phrenic nerves. *Thorax.* 1996;51(12):1239–42.

[75] Mills GH, Ponte J, Hamnegard CH, *et al.* Tracheal tube pressure change during magnetic stimulation of the phrenic nerves as an indicator of diaphragm strength on the intensive care unit. *Br J Anaesth.* 2001;87(6):876–84.

[76]   Garara B, Wood A, Marcus HJ, Tsang K, Wilson MH, and Khan M. Intra-muscular diaphragmatic stimulation for patients with traumatic high cervical injuries and ventilator dependent respiratory failure: a systematic review of safety and effectiveness. *Injury.* 2016;47(3):539–44.

[77]   Glenn WW, Holcomb WG, Gee JB, and Rath R. Central hypoventilation; long-term ventilatory assistance by radiofrequency electrophrenic respira-tion. *Ann Surg.* 1970;172(4):755–73.

[78]   Etienne H, Morris IS, Hermans G, *et al.* Diaphragm neurostimulation assis-ted ventilation in critically ill patients. *Am J Respir Crit Care Med.* 2023;207 (10):1275–82.

[79]   Yang M, Wang H, Han G, *et al.* Phrenic nerve stimulation protects against mechanical ventilation-induced diaphragm dysfunction in rats. *Muscle Nerve.* 2013;48(6):958–62.

[80]   Demoule A, Morelot-Panzini C, Prodanovic H, *et al.* Identification of pro-longed phrenic nerve conduction time in the ICU: magnetic versus electrical stimulation. *Intensive Care Med.* 2011;37(12):1962–8.

[81]   Keogh C, Saavedra F, Dubo S, *et al.* Non-invasive phrenic nerve stimulation to avoid ventilator-induced diaphragm dysfunction in critical care. *Artif Organs.* 2022;46(10):1988–97.

[82]   Hsin YF, Chen SH, Yu TJ, Huang CC, and Chen YH. Effects of transcuta-neous electrical diaphragmatic stimulation on respiratory function in patients with prolonged mechanical ventilation. *Ann Thorac Med.* 2022;17(1):14–20.

[83]   Olimpio Junior H, Camilo GB, Fioritto AP, and Lopes AJ. Immediate hemodynamic responses to transcutaneous electrical diaphragmatic stimula-tion in critically ill elderly patients. *Rehabil Res Pract.* 2021;2021:9091278.

[84]   O'Rourke J, Sotak M, Curley GF, *et al.* Initial assessment of the percuta-neous electrical phrenic nerve stimulation system in patients on mechanical ventilation. *Crit Care Med.* 2020;48(5):e362–e70.

[85]   Onders R, Elmo M, Carl N, *et al.* Temporary diaphragm pacing wires in lung transplant patients: diagnostic and therapeutic utilization. *J Heart Lung Transplant.* 2023;42(Suppl 4):S101.

[86]   Sotak M, Roubik K, Henlin T, and Tyll T. Phrenic nerve stimulation prevents diaphragm atrophy in patients with respiratory failure on mechanical venti-lation. *BMC Polm Med.* 2021;21(1):314.

[87]   Ataya A, Silverman EP, Bagchi A, Sarwal A, Criner GJ, and McDonagh DL. Temporary transvenous diaphragmatic neurostimulation in prolonged mechanically ventilated patients: a feasibility trial (RESCUE 1). *Crit Care Explor.* 2020;2(4):e0106.

[88]   Dres M, de Λbreu MG, Merdji H, *et al.* Randomized clinical study of tem-porary transvenous phrenic nerve stimulation in difficult-to-wean patients. *Am J Respir Crit Care Med.* 2022;205(10):1169–78.

[89]   Evans D, Shure D, Clark L, *et al.* Temporary transvenous diaphragm pacing vs. standard of care for weaning from mechanical ventilation: study protocol for a randomized trial. *Trials.* 2019;20(1):60.

[90]   Reynolds S, Ebner A, Meffen T, *et al.* Diaphragm activation in ventilated patients using a novel transvenous phrenic nerve pacing catheter. *Crit Care Med.* 2017;45(7):e691–e4.

[91]   Morris IS, Bassi T, Bellissimo CA, *et al.* Proof of concept for continuous on-demand phrenic nerve stimulation to prevent diaphragm disuse during mechanical ventilation (STIMULUS): a phase 1 clinical trial. *Am J Respir Crit Care Med.* 2023;208(9):992–5.

[92]   Kaufmann EM, Krause S, Geisshuesler L, Scheidegger O, Haeberlin A, and Niederhauser T. Feasibility of transesophageal phrenic nerve stimulation. *Biomed Eng Online.* 2023;22(1):5.

[93]   Bao Q, Chen L, Chen X, *et al.* The effects of external diaphragmatic pacing on diaphragm function and weaning outcomes of critically ill patients with mechanical ventilation: a prospective randomized study. *Ann Transl Med.* 2022;10(20):1100.

[94]   Hicham M, Catherine C, Alexandre D, *et al.* Can phrenic stimulation protect the diaphragm from mechanical ventilation-induced damage? *Eur Respir J.* 2013;42(1):280.

[95]   Reynolds SC, Meyyappan R, Thakkar V, *et al.* Mitigation of ventilator-induced diaphragm atrophy by transvenous phrenic nerve stimulation. *Am J Respir Crit Care Med.* 2017;195(3):339–48.

[96]   Poulsen JB, Møller K, Jensen CV, Weisdorf S, Kehlet H, and Perner A. Effect of transcutaneous electrical muscle stimulation on muscle volume in patients with septic shock. *Crit Care Med.* 2011;39(3):456–61.

[97]   Morris IS, Dres M, and Goligher EC. Phrenic nerve stimulation to protect the diaphragm, lung, and brain during mechanical ventilation. *Intensive Care Med.* 2022;48(10):1299–301.

[98]   Jonkman AH, Frenzel T, McCaughey EJ, *et al.* Breath-synchronized electrical stimulation of the expiratory muscles in mechanically ventilated patients: a randomized controlled feasibility study and pooled analysis. *Crit Care.* 2020;24(1):628.

[99]   Shi ZH, Jonkman A, de Vries H, *et al.* Expiratory muscle dysfunction in critically ill patients: towards improved understanding. *Intensive Care Med.* 2019;45(8):1061–71.

[100] Su WL, Chen YH, Chen CW, *et al.* Involuntary cough strength and extubation outcomes for patients in an ICU. *Chest.* 2010;137(4):777–82.

[101] Savi A, Teixeira C, Silva JM, *et al.* Weaning predictors do not predict extubation failure in simple-to-wean patients. *J Crit Care.* 2012;27(2):221. e1–8.

[102] Kutchak FM, Debesaitys AM, Rieder Mde M, *et al.* Reflex cough PEF as a predictor of successful extubation in neurological patients. *J Bras Pneumol.* 2015;41(4):358–64.

[103] McCaughey EJ, Borotkanics RJ, Gollee H, Folz RJ, and McLachlan AJ. Abdominal functional electrical stimulation to improve respiratory function after spinal cord injury: a systematic review and meta-analysis. *Spinal Cord.* 2016;54(9):628–39.

[104]   McCaughey EJ, Jonkman AH, Boswell-Ruys CL, *et al.* Abdominal functional electrical stimulation to assist ventilator weaning in critical illness: a double-blinded, randomised, sham-controlled pilot study. *Crit Care.* 2019;23(1):261.

[105]   McCaughey EJ, Boswell-Ruys CL, Hudson AL, Gandevia SC, and Butler JE. Optimal electrode position for abdominal functional electrical stimulation. *J Appl Physiol.* 2018;125(4):1062–8.

[106]   Lim J, Gorman RB, Saboisky JP, Gandevia SC, and Butler JE. Optimal electrode placement for noninvasive electrical stimulation of human abdominal muscles. *J Appl Physiol.* 2007;102(4):1612–7.

[107]   Rodríguez-Blanco J, Rodríguez-Yanez T, Rodríguez-Blanco JD, *et al.* Neuromuscular blocking agents in the intensive care unit. *J Int Med Res.* 2022;50(9):3000605221128148.

[108]   Lago AF, de Oliveira AS, de Souza HCD, da Silva JS, Basile-Filho A, and Gastaldi AC. The effects of physical therapy with neuromuscular electrical stimulation in patients with septic shock: study protocol for a randomized cross-over design. *Medicine (Baltimore).* 2018;97(6):e9736.

[109]   Paternostro-Sluga T, Schuhfried O, Vacariu G, Lang T, and Fialka-Moser V. Chronaxie and accommodation index in the diagnosis of muscle denervation. *Am J Phys Med Rehabil.* 2002;81(4):253–60.

[110]   Silva PE, Babault N, Mazullo JB, *et al.* Safety and feasibility of a neuromuscular electrical stimulation chronaxie-based protocol in critical ill patients: a prospective observational study. *J Crit Care.* 2017;37:141–8.

[111]   Silva PE, Martins HR, de Deus Macedo JR, de Queiroz AV, Ultra de Aguiar A, and Martins EF. Towards innovative electrodiagnosis tests to investigate neuromuscular excitability dysfunction in critically ill patients: an agreement study. *Res Biomed Eng.* 2023;39(4):949–58.

[112]   Kennouche D, Luneau E, Lapole T, Morel J, Millet GY, and Gondin J. Bedside voluntary and evoked forces evaluation in intensive care unit patients: a narrative review. *Crit Care.* 2021;25(1):157.

[113]   Botter A, Oprandi G, Lanfranco F, Allasia S, Maffiuletti NA, and Minetto MA. Atlas of the muscle motor points for the lower limb: implications for electrical stimulation procedures and electrode positioning. *Eur J Appl Physiol.* 2011;111(10):2461–71.

[114]   Aquim EE, Bernardo WM, Buzzini RF, *et al.* Brazilian guidelines for early mobilization in intensive care unit. *Rev Bras Ter Intensiva.* 2019;31(4):434–43.

[115]   Ribeiro CJN, Araujo ACS, Brito SB, *et al.* Pain assessment of traumatic brain injury victims using the Brazilian version of the behavioral pain scale. *Rev Bras Ter Intensiva.* 2018;30(1):42–9.

[116]   Maffiuletti NA, Gondin J, Place N, Stevens-Lapsley J, Vivodtzev I, and Minetto MA. Clinical use of neuromuscular electrical stimulation for neuromuscular rehabilitation: what are we overlooking? *Arch Phys Med Rehabil.* 2018;99(4):806–12.

[117]   Balke M, Teschler M, Schafer H, Pape P, Mooren FC, and Schmitz B. Therapeutic potential of electromyostimulation (EMS) in critically ill patients-a systematic review. *Front Physiol.* 2022;13:865437.

[118] Sanchis-Gomar F, Lopez-Lopez S, Romero-Morales C, Maffulli N, Lippi G, and Pareja-Galeano H. Neuromuscular electrical stimulation: a new therapeutic option for chronic diseases based on contraction-induced myokine secretion. *Front Physiol*. 2019;10:1463.

[119] Latronico N, Bertolini G, Guarneri B, *et al*. Simplified electrophysiological evaluation of peripheral nerves in critically ill patients: the Italian multicentre CRIMYNE study. *Crit Care*. 2007;11(1):R11.

[120] Bahadori S, Immins T, and Wainwright TW. The effect of calf neuromuscular electrical stimulation and intermittent pneumatic compression on thigh microcirculation. *Microvasc Res*. 2017;111:37–41.

[121] Ojima M, Takegawa R, Hirose T, Ohnishi M, Shiozaki T, and Shimazu T. Hemodynamic effects of electrical muscle stimulation in the prophylaxis of deep vein thrombosis for intensive care unit patients: a randomized trial. *J Intensive Care*. 2017;5:9.

[122] Nakanishi N, Yoshihiro S, Kawamura Y, *et al*. Effect of neuromuscular electrical stimulation in patients with critical illness: an updated systematic review and meta-analysis of randomized controlled trials. *Crit Care Med*. 2023;51(10):1386–96.

[123] Fernandes JA, Silva MLd, Trancho AC, Macedo JRdD, Martins HR, and Silva PE. Assessment of neuromuscular electrical stimulation in critically ill patients: physical therapists' knowledge and barriers to its use. *Fisioterapia e Pesquisa*. 2022;29(3):245–51.

[124] Zanotti E, Felicetti G, Maini M, and Fracchia C. Peripheral muscle strength training in bed-bound patients with COPD receiving mechanical ventilation: effect of electrical stimulation. *Chest*. 2003;124(1):292–6.

[125] Parveez MQ, Yaddanapudi LN, Saini V, Kajal K, and Sharma A. Critical events during intra-hospital transport of critically ill patients to and from intensive care unit. *Turk J Emerg Med*. 2020;20(3):135–41.

[126] Teschler M, Heimer M, Schmitz B, Kemmler W, and Mooren FC. Four weeks of electromyostimulation improves muscle function and strength in sarcopenic patients: a three-arm parallel randomized trial. *J Cachexia Sarcopenia Muscle*. 2021;12(4):843–54.

[127] Boutry-Regard C, Vinyes-Pares G, Breuillé D, and Moritani T. Supplementation with whey protein, omega-3 fatty acids and polyphenols combined with electrical muscle stimulation increases muscle strength in elderly adults with limited mobility: a randomized controlled trial. *Nutrients*. 2020;12(6):1866.

[128] Sax OC, Gesheff MG, Mahajan A, *et al*. A novel mobile app-based neuromuscular electrical stimulation therapy for improvement of knee pain, stiffness, and function in knee osteoarthritis: a randomized trial. *Arthroplast Today*. 2022;15:125–31.

[129] Dasa V, Skrepnik NV, Petersen D, and Delanois RE. A novel mobile app-based neuromuscular electrical stimulation therapy for the management of knee osteoarthritis: results from an extension study of a randomized, double-blind, sham-controlled, multicenter trial. *J Am Acad Orthop Surg Glob Res Rev*. 2022;6(9):e22.

Chapter 14

Health, fitness, and physiological adaptations after functional electrical stimulation exercise

David R. Dolbow[1], Ashraf S. Gorgey[2] and Glen M. Davis[3]

14.1 Introduction

Over the past half-century, there has been a dramatic development in the importance and efficacy of electrical stimulation-evoked exercise to improve health and fitness. Electrically evoked exercise prescription parameters, including frequency, intensity, and training volume, can be manipulated to promote positive physiological adaptations affecting acute and chronic musculoskeletal, cardiovascular, and metabolic health outcomes. Such exercise has become a standard therapeutic tool to improve key health outcomes and functional mobility for individuals with neuromuscular disorders and other medical conditions. This chapter will describe some commonly deployed electrical stimulation exercises, including neuromuscular electrical stimulation (NMES) resistance training, functional electrical stimulation (FES) exercise and "hybrid" FES activities (i.e. a combination of electrical stimulation to the legs and voluntary exercises using the arms). The discussion will provide evidence regarding potential benefits, including muscle hypertrophy, physical fitness, and decreased risk of inactivity-based morbidities. Electrically evoked exercise is especially useful as a tool for conditioning the muscles of individuals with spinal cord injury (SCI) at various levels. However, there are also potential advantages for individuals who have sustained a cerebrovascular accident or have degenerative neuromuscular conditions. Exercise protocols that are effective during research or clinical use will be highlighted. Recent advancements in hybrid and high-intensity interval training (HIIT) will also be discussed. While electrical stimulation activities are safe when applied appropriately, some specific safety precautions and contraindications will be addressed. Finally, the chapter will include suggestions for advancing electrical stimulation as a therapeutic modality in clinical practice.

[1]Physical Therapy Program and College of Osteopathic Medicine, William Carey University, USA
[2]Hunter Holmes McGuire VA Medical Center and Virginia Commonwealth University, USA
[3]Clinical Exercise and Rehabilitation Unit, The Discipline of Exercise and Sport Science, Sydney School of Health Sciences, Faculty of Medicine and Health, University of Sydney, Australia

SCI can partially or wholly disrupt somatic and autonomic function below the level of injury. The loss of control over voluntary and involuntary functions below the level of SCI can limit physical activities and lead to a diminished systemic response to exercise [1,2]. The deleterious effects of SCI, coupled with decreased physical activity, can cause declines in body composition, cardiovascular fitness, and metabolic health. If left unaddressed, this scenario can lead to conditions such as obesity, diabetes mellitus, cardiovascular disease, and osteoporosis [1].

This chapter will discuss the advantages of electrically stimulated exercise, which can lead to increased muscle mass, reduced bone loss, and improved cardiovascular and metabolic health. We will also provide recommendations for exercise protocols and discuss safety factors associated with electrical stimulation activities.

Regular physical activity is crucial for maintaining health and mobility. Electrical stimulation of paralyzed muscles can induce muscle contractions and increase the range of physical activities performed [3]. This can be achieved through NMES, including resistance training or FES exercises, which include cycling and rowing activities [4]. For this book chapter, we will define FES as an electrical stimulation activity that is distinct from NMES. FES produces coordinated functional patterned activities such as cycling and rowing, whereas NMES is utilized for the muscle contraction of a single muscle or muscle group. Recent studies have shown that both NMES and FES can increase muscle volume and improve cardiovascular and metabolic outcomes [5,6].

14.1.1   *Muscle physiology*

After SCI, the body undergoes a rapid and significant atrophy of skeletal muscle below the level of injury [1]. The decrease in metabolically active skeletal muscle and the reduction in physical activity can decrease the total daily energy expenditure by half, a recipe for obesity and related conditions. One detrimental result is increased obesity associated with decreased physical function and greater dependence on others [7,8]. For nearly three decades, clinicians have utilized transcutaneous (skin surface electrodes) NMES and FES activities to help offset the loss of skeletal muscle after paralysis. NMES knee extension resistance training has been extensively studied and has shown promising results. For instance, Mahoney *et al.* [9] conducted a 12-week study on five individuals with chronic SCI, using NMES resistance training on their quadriceps muscles twice weekly. They found that the skeletal muscles' cross-sectional area increased by 33–39%. Similarly, Gorgey *et al.* [10] randomly assigned nine motor-complete SCI individuals to receive NMES resistance training plus dietary counselling versus dietary counselling alone. The once-weekly dietary counselling was designed to promote a 45% carbohydrate, 30% fat, and 25% protein diet. The authors found that the group receiving NMES resistance training plus dietary counselling increased their quadriceps cross-sectional area by 35%. In contrast, the dietary counseling-only group had no change over the same period. Although knee extension resistance training targeted the quadriceps muscle group, there were also significant gains in the cross-sectional area of the hamstrings (16%), indicating that the NMES-evoked muscle contractions can impact surrounding muscles. Moreover, insulin-like growth factor 1, associated with muscle hypertrophy, increased by 25%. A follow-up study to determine the effects of the stimulation overflow found that NMES resistance training for the quadriceps also significantly increased muscle cross-sectional area of the hamstrings, hip adductors, sartorius, and gracilis muscles, but not the trunk musculature [11]. In a case report using NMES resistance training, Gorgey and Shephard [12] demonstrated a 72% increase in skeletal muscle cross-sectional area of the thigh and a 53% decrease in intramuscular fat after twice weekly unilateral NMES resistance training over 12 weeks in a young man with tetraplegia. A summary of this case report is presented in Table 14.1.

Table 14.1   Case report one: NMES resistance training quadriceps muscles

Clinical indication	Twenty-two-year-old man five years post motorcycle accident with motor and sensory complete C5-C6 SCI.
Indication for NMES	Muscle atrophy with decreased cross-sectional area of the quadriceps muscle group and increased intra-muscular fat. No voluntary means to actively exercise the paralyzed muscles.
Electrode placement	Surface electrodes 8×10 cm in size were placed over the proximal quadriceps muscles approximately 30 cm above the patella and distally the electrodes were placed over the quadriceps muscles approximately 3 cm above the patella.
Stimulation settings	Frequency – 30 Hz, Pulse Width – 450 μs, Amplitude/Intensity – progressively increased until full knee extension was achieved. The current on/off time was set at five seconds on and five seconds off.
NMES protocol	The exercise protocol was four sets of ten repetitions with a two-minute rest between sets and a second hold at the terminal end of extension during each repetition. Resistance training sessions were two days per week for twelve weeks. During the first two weeks no weights were applied to the ankle. After full knee extension was reached for all 40 repetitions a one kg (two lbs) weight was applied to the ankles to increase resistance. This incremental process continued throughout the 12 weeks.

(Image adapted and redrawn from Noraxon MyoResearch EMG software tools (Noraxon USA Scottsdale, Arizona, USA) in Hamzaid NA. Development of an isokinetic functional electrical stimulation leg stepping trainer for individuals with neurological disabilities. PhD thesis, 2010.)

Studies have shown that FES cycling and rowing activities can increase muscle mass, although less significantly than NMES resistance training on specific muscle groups [13]. FES leg cycling exercise involves using surface electrodes on the quadriceps, hamstrings, and gluteal muscles to enable cycling. The intensity and duration of FES leg cycling protocols vary, but numerous studies have reported an increase in muscle mass ranging from 4% to 38%. In a recent randomized controlled trial, participants undertook FES leg cycling or performed arm crank exercises for 40 minutes five times a week for 16 weeks [14]. Resistance was increased to maintain the target heart rate at 75% of the maximal heart rate. After 16 weeks, those who performed FES leg cycling revealed a 7% increase in lean mass in their legs, a 4% increase in total body lean mass, and a 4% decrease in body fat percentage. The arm crank exercise group realized an increase in lean mass of 2% and a 4% decrease in body fat percentage. Another study found a near 7% increase in leg lean mass and a 2.5% decrease in body fat percentage after five individuals with chronic SCI cycled using resistance-guided HIIT FES for 30 minutes three times a week for eight weeks [15]. This protocol alternated cycling against $\geq 80\%$ of their predetermined maximal cycling resistance during high-intensity intervals interspersed with cycling against the lowest resistance the FES cycle would allow (0.5 Nm). Intervals were every 30 seconds, with ten seconds ramping up or down of the electrical current amplitude between the intervals. This group also received one-time per week dietary counselling. The five participants in the control group who received only dietary counselling showed no improvements. In a study that demonstrated an association between accumulated torque and muscle volume, Johnston and colleagues [16] compared individuals with SCI cycling at a low cadence but high torque to those with a high cadence at low torque. Participants who cycled at a low cadence (20 rev·min^{-1} with a maximal mean torque of 2.9 Nm) demonstrated a 19% increase in muscle mass. In comparison, those who cycled at a high cadence (50 rev·min^{-1} with a mean maximal torque of 0.8 Nm) experienced a 10% increase in muscle mass. In an earlier study, eight individuals with chronic SCI underwent six weeks of thrice weekly FES leg cycling with one leg training at 10 rev·min^{-1} and the other at 50 rev·min^{-1} [17]. The cycling sessions were 30 minutes in duration. The electrical stimulation current amplitude was increased linearly from 40 to 140 mA over ten minutes and then held constant. The frequency was 35 Hz, and the pulse width was 250 μs. The lower cadence provided more significant muscle hypertrophy, as demonstrated by an increase in thigh girth circumference nearly twice that of the high cadence leg (Figure 14.1).

An interesting recent randomized control trial compared 12 weeks of FES leg cycling to FES leg cycling plus concomitant progressive resistance training) in individuals with incomplete SCI [18]. Both isometric peak torque and muscle volumes were consistently higher in the FES leg cycling plus progressive resistance exercise group compared to the FES leg cycling only group. Their data suggested that adding high-load resistance training to FES cycling was more dose-potent for muscle size and strength than FES leg cycling alone. Two recent systematic review studies reported more evidence of the benefits of NMES and FES activities on muscle mass, with NMES and FES training activities increasing muscle mass from 6% to 75%, with

Figure 14.1   NMES resistance training illustration

NMES providing greater increases than FES activities [6,19]. A third review study focusing on FES leg cycling concluded that there was high certainty in the evidence for improving muscle mass and fatigue resistance [1]. In a case report on a 53 year old man 33 years post motor and sensory complete C4 SCI, six months of home-based FES leg cycling was completed to increase muscle mass and decrease body fat percentage [20]. The individual was first introduced to FES cycling in the medical center supervised by physical therapists and received training along with a family member on the safe set-up and operation of the FES cycling protocol. After it was shown that the individual could safely tolerate the activity and understood the safe operation of the FES cycle, the cycle was placed in the participants home and connected to the internet via Wi-Fi so that the rehabilitation staff could monitor the FES cycling performance of each session via internet connection. After six months of thrice weekly FES cycling sessions, total body lean mass increased by 3.3%, legs lean mass increased by 7.1% and body fat percentage decreased by 3.3%. Additionally, there were significant increases in the physical and psychological domains of 25% and 4.5%, respectively, on the World Health Organization Quality of Life Brief Questionnaire (Table 14.2).

FES rowing (FES for legs combined with voluntary arm exercise) has had mixed reports regarding muscle hypertrophy. Kim *et al.* [21] demonstrated a 6% increase in lean mass and a 14% decrease in body fat percentage after six weeks of FES rowing five times per week. Each session was 42 minutes long, with a five-minute warm-up and cool-down. The sessions consisted of five-minute bouts of FES rowing at 70% of their heart rate maximum, interspersed with 30 seconds of recovery. In contrast, both Jeon *et al.* [22] and Wilbanks *et al.* [23] observed no changes in lean mass after FES rowing programs of 12- and six-weeks duration.

Table 14.2 Case report two: FES cycling to increase muscle mass

Clinical indication	A 53-year-old man 33 years post motor vehicle accident with motor and sensory complete C4 SCI.
Indication for FES cycling	Muscle atrophy with decreased lean mass and increased body fat percentage 33.6%. No voluntary means to actively exercise the paralyzed muscles. For long term FES cycling the cycle was placed in the home and monitored by research and healthcare providers via internet connection.
Electrode placement	Surface electrodes 8 × 10 cm in size were placed over the proximal quadriceps approximately 30 cm above the patella and distally the electrodes were placed over the quadriceps muscle group approximately 3 cm above the patella. The proximal electrodes for the hamstrings were placed approximately 30 cm above the popliteal fossa while the distal electrodes were placed 2–3 cm above the popliteal fossa. Two electrodes were placed over the left and right gluteus maximus with at least 3 finger widths apart.
Stimulation settings	Frequency – 33 Hz, Pulse Width – 250 μs, Amplitude/Intensity – progressively increased until reaching a maximum of 140 mA.
NMES protocol	The exercise protocol started with a one-minute passive cycling warm-up. Then the electrical stimulation amplitude gradually increased until reaching the maximum intensity. The cycling speed was set at 50 rev·min^{-1} and the initial cycling resistance at the 0.5 Nm. The cycling resistance was increased as the patient tolerated with a maximum 1.7 Nm. The cycling duration progressed as the patient was able, starting at 14 minutes on the first cycling session to a maximum of 60 minutes.

However, Wilbanks and associates demonstrated an 8% increase in aerobic fitness and a significant decrease in shoulder pain. This is important because shoulder pain is reported in up to 83% of individuals with chronic SCI.

14.1.2 Clinical summary

NMES resistance training, FES cycling, and FES rowing are safe and effective ways to increase muscle mass and reduce body fat. Twice weekly sessions for

12 weeks of NMES resistance training can increase muscle strength and hypertrophy. FES leg cycling can also increase muscle mass and reduce body fat, but with lower gains in muscle mass than NMES resistance training. Hybrid FES cycling and FES rowing are natural forms of hybrid FES exercise. FES leg cycling and rowing have demonstrated improved muscle health with 30 minutes of exercise thrice weekly for eight weeks. Adding progressive resistance exercise to FES leg cycling has also been shown to elevate benefits. HIIT-FES leg cycling may benefit similarly to FES leg cycling with reduced exercise volume (Table 14.3).

Table 14.3   Effects of electrical stimulation exercise on muscle

Source	Participants*	Treatments	Results
Mahoney *et al.* [9]	*n* = 5 Chronic SCI	NMES-RT 2×/ week for 12 weeks	CSA Quadriceps +33–39%
Gorgey *et al.* [10]	*n* = 9 Chronic SCI	5 – NMES-RT and diet 2×/week for 12 weeks 4 – diet alone	NMES group CSA Quadriceps +35% Hamstrings +16% Insulin-like growth factor 1 +25% AUC Diet-only group no changes
Gorgey *et al.* [11]	*n* = 7 Chronic SCI	4 – NMES-RT 2×/week for 12 weeks and diet 3 – diet alone	NMES group CSA Quadriceps +31-39% Hip adductors +13–28% Sartorius +43–49% Gracilis +12–18% Diet-only group, no changes
Gorgey and Shepard [12]	Case report Chronic SCI	One leg NMES-RT 2×/week for 12 weeks One leg control	NMES leg CSA +72% thigh −53% intramuscular fat
Rosley *et al.* [18]	*n* = 23 chronic "incomplete" SCI	10 – FES-LCE +PRT 1 session PRT & 2 sessions FES-LCE weekly over 12 weeks 13 – FES-LCE three sessions weekly over 12 weeks	FES-LCE+PRT left hamstrings peak torque +45% change, consistently higher than FES-LCE FES-LCE+PRT right quadriceps peak torque +31% change, greater than the FES-LCE group. FES-LCE+PRT L muscle volume +7% increase.
Farkas *et al.* [14]	*n* = 13 Chronic SCI	6 – FES-LCE 5x/week for 16 weeks	FES +4% LM (ns) ACE +2% LM (ns) FES −4% BF% ACE −4.6% BF%

(Continues)

Table 14.3 Effects of electrical stimulation exercise on muscle (Continued)

Source	Participants*	Treatments	Results
Dolbow et al. [15]	n = 10 Chronic SCI	7 – ACE 5×/week for 16 weeks 5 – Interval FES cycling 3×/week for 8 weeks and diet 5 – diet alone	FES energy exp +38% ACE energy exp +85% Interval FES grp +5.7% legs lean mass −2.4% BF% Diet-only group no changes
Johnston et al. [16]	n = 17 Chronic SCI	Low cadence 20 rpm High cadence 50 rpm 9 – low cadence 8 – high cadence 3×/week for six months	Low cadence +19% LM High cadence +10% LM
Fornusek et al. [17]	n = 8 Chronic SCI	One leg low cadence 10 rpm One leg high cadence 50 rpm FES 3×/week for six weeks	Thigh of the low cadence leg increased nearly twice the circumference of the high cadence leg.

RT = resistance training; ns = not statistically significant; LM = lean mass; CSA = cross sectional area; LCE = leg cycling exercise; PRT = progressive resistance exercise; BF% = body fat percentage
*All participants had cervical or thoracic injuries and were wheelchair users.

14.1.3   Bone adaptation

While significant skeletal muscle hypertrophy may result from several weeks of NMES or FES cycling training, the slower metabolic processes within bone typically require at least six months or longer to produce smaller changes in bone mass compared to muscle mass. Early studies investigating bone mineral density (BMD) after electrical stimulation exercises lasting six months or less failed to increase bone mass. FES leg cycling three times per week for six months did not increase bone mass [24]. Thrice weekly NMES resistance training on the lower extremities of individuals with SCI found no improvements in BMD of the tibia after 36 sessions of training [25]. However, a pilot study by Holman and co-workers [26] noticed slight changes in the trabecular bone of the distal femur and medium changes in the proximal tibia in persons with SCI after NMES resistance training plus oral testosterone after four months. Adding testosterone treatments to NMES resistance training and using superior bone imaging for microarchitecture (magnetic resonance imaging versus dual-energy X-ray absorptiometry) may explain the observed changes in bone, while earlier studies lasting six months or less did not.

Studies that provided high-intensity stimulation for extended periods consistently demonstrated modest improvements in BMD. Mohr and co-workers [27] used FES leg cycling for 30 minutes thrice weekly over 12 months and found that the mean improvement in BMD was 10% in individuals with SCI who had lost 52% of their BMD after SCI before the FES cycling program commenced. However, the participants lost most of their improvement in BMD over the next six months while performing FES leg cycling only once weekly. Chen *et al.* [28] found an improvement in BMD in the distal femur and proximal tibia of individuals with SCI after FES leg cycling three times per week for six months. However, the increases faded after FES leg cycling was discontinued, suggesting bone quality regression after removing loading.

The comparison of individuals with SCI cycling at a low cadence with high torque to those cycling at a high cadence with low torque three times per week for six months demonstrated a non-significant 7% increase in trabecular number, and a 6% decrease in trabecular separation (increased density in the trabecular network), both beneficial signs regarding bone health and decreased risk of fracture [16]. Frotzler and colleagues [29] also investigated the effects of high-volume and high-intensity FES cycling on BMD. Individuals with SCI performed FES leg cycling at their highest power output three to four times per week for 12 months, improving 14% in the trabecular bone and 7% in the distal femur's BMD (total bone). Shields and Dudley-Javorski [30] further developed the approach of deploying high volume and intensity by applying NMES intensity at 1.5 times body weight five times per week for two years unilaterally in acutely spinal injured individuals, resulting in a 31% increase in BMD in the distal tibia. In another study on acute SCI individuals, electrical stimulation was applied to the plantar flexor muscles of one leg, with the other acting as the control. The stimulation was performed five days per week for three years with an intensity of 140% of body weight. The trained leg decreased BMD in the tibia 10% less than the untrained leg, indicating that electrical stimulation activities can slow bone loss if started early after SCI [31]. Contrary to earlier studies, Shields and Dudley-Javorski [32] investigated the effects of NMES training with the intensity at 110% of body weight five times per week for 6–11 months and found no changes in BMD.

In a case report of 60-year-old female two years post T6 motor complete SCI due to a motor vehicle accident, the individual performed FES leg cycling three times per week for 12 months to help restore a healthier body composition [33]. The individuals had severe osteopenia and nearing osteoporosis with a t-score of −2.4 as determined by dual-energy X-ray absorptiometry. After 12 months of FES leg cycling the individual's total body BMD improved from 0.934 to 1.023 and the t-score improved from −2.4 to −1.3. There was also a 7.7% increase in total body lean mass, 4.1% increase in legs lean mass, and a 2% decrease in body fat percentage (Table 14.4).

FES rowing for 30 minutes three times per week for 90 sessions was shown to improve trabecular BMD in the femur and tibia of individuals within two years after traumatic SCI [34]. Two individuals increased trabecular bone in the femur by 6% and 8%, respectively, while the other two participants decreased bone loss from 7% to 3% and 5% to 0%, respectively. Furthermore, after undergoing a conditioning program of NMES resistance training to the legs, combining FES rowing

Table 14.4   Case report three: FES cycling to increase bone mass

Clinical indication	A 60-year-old woman two years post motor vehicle accident with motor complete T6 SCI.
Indication for FES cycling	Bone loss with severe osteopenia (t-score −2.4) nearing osteoporosis. Also, a poor lean mass to fat mass ratio and a body fat percentage of 48.3%. No voluntary means to actively exercise the paralyzed muscles. For long term FES cycling the cycle was placed in the home and with the cycling performances monitored by research and healthcare providers via internet connection.
Electrode placement	Surface electrodes 8 × 10 cm in size were placed over the proximal quadriceps approximately 30 cm above the patella and distally the electrodes were placed over the quadriceps muscle group approximately 3 cm above the patella. The proximal electrodes for the hamstrings were placed approximately 30 cm above the popliteal fossa while the distal electrodes were placed 2–3 cm above the popliteal fossa. Two electrodes were placed over the left and right gluteus maximus with at least three finger widths apart.
Stimulation settings	Frequency – 33–50 Hz, Pulse Width – 250–300 μs, Amplitude/Intensity – progressively increased until reaching a maximum of 140 mA.
NMES protocol	The exercise protocol started with a one-minute passive cycling warm-up. Then the electrical stimulation amplitude gradually increased until reaching the maximum intensity. The cycling cadence ranged between 35 and 43 rev·min^{-1} and the cycling resistance ranged between 0.64 and 1.28 Nm. The cycling duration progressed as the patient was able to go from 20 minutes on the first cycling session to a maximum of 63 minutes. There was also a two-minute cooldown to finish.

(Image adapted and redrawn from Noraxon MyoResearch EMG software tools (Noraxon USA Scottsdale, Arizona, USA) in Hamzaid NA. Development of an isokinetic functional electrical stimulation leg stepping trainer for individuals with neurological disabilities. PhD thesis, 2010.)

Figure 14.2   FES rowing illustration

with oral zoledronic acid treatment for a year with thrice weekly exercise sessions at 75–85% of peak heart rate resulted in a reduction of bone loss. The results demonstrated that the combination of FES rowing and zoledronic acid, a bisphosphonate medication used to treat osteoporosis, decreased the loss of bone more than three times that of FES rowing only, from 8% to 2.5%, respectively (Figure 14.2) [35].

14.1.4   Clinical summary

The results of numerous studies indicate that FES and NMES activities can only provide a limited recovery or slow down the rate of bone loss after SCI. Based on the available evidence, it is not possible to draw any conclusions about the "best" dose-response protocol for bone health. However, it does suggest that high-volume and high-intensity exercise are required to demonstrate benefits in bone tissue. This would involve three to five times per week of high-power output exercise for at least six months to a year with low cadence and high-intensity cycling for FES leg cycling and FES rowing. Additionally, once bone loading stops, the benefits gradually go away. Testosterone or zoledronic acid treatments can be added to increase the results of FES activities (Table 14.5).

14.1.5   Cardiovascular and metabolic physiology

Living a sedentary lifestyle can weaken the cardiovascular system, especially for those with SCI and other neurological conditions, leading to an increased risk of cardiovascular disease and metabolic adverse sequelae, such as obesity, diabetes, dyslipidemia, hypertension, and systemic inflammation [36,37]. However, electrical stimulation exercises can help mitigate these risks by activating skeletal muscles and promoting beneficial physiological training adaptations [36]. Individuals with paralysis must produce sufficient muscle recruitment at intensity levels high enough to induce central and peripheral vascular changes (Table 14.6) [38].

The gold standard and most common measure for cardiovascular fitness is the highest oxygen consumption rate during incremental exercise testing (VO$_2$peak) [38]. Warburton *et al.* [39] completed a systematic review on cardiovascular health

Table 14.5   Effects of electrical stimulation exercise on bone

Source	Participants	Treatments	Results
Leeds *et al.* [24]	*n* = 6 Chronic SCI	FES-LCE 3×/week for six months	No change in BMD (proximal femur)
Rodgers *et al.* [25]	*n* = 12 Chronic SCI	NMES-RT 3×/week for 12 weeks	No change in BMD (tibia)
Holman *et al.* [26]	*n* = 10 Chronic SCI	NMES-RT 2×/week for	Distal femur – small trabecular increase Proximal tibia – medium trabecular increase
Mohr *et al.* [27]	*n* = 10 Chronic SCI	FES-LCE 3×/week for one year	+10% BMD (proximal tibia)
Chen *et al.* [28]	*n* = 15 Chronic SCI	FES-LCE 3×/week for six months	+11% BMD (distal femur) +13% BMD (proximal tibia)
Johnston *et al.* [16]	*n* = 17 Chronic SCI	Low cadence 20 rpm High cadence 50 rpm 9 – low cadence 8 – high cadence 3×/week for six months	Low cadence +7% trabecular number (ns) 6% decrease in trabecular separation (ns)
Frotzler *et al.* [29]	*n* = 11 Chronic SCI	FES-LCE 3–4×/week for one year	+14% BMD trabecular bone (distal femur) +7% BMD total bone (distal femur)
Shields and Dudley-Javorski, [30]	*n* = 7 (6 weeks post-SCI)	FES to plantar flexor muscles of one leg. The other leg was the control	+31% BMD (distal tibia)
Shields *et al.* [31]	*n* = 6 (4.5 months post SCI)	FES to plantar flexor muscles of one leg (daily, 1680 contractions per week up to 140% body weight). The other leg was the control.	Bone in both legs declined, but the trained leg's BMD declined 10% less.
Lambach *et al.* [34]	*n* = 4 <2 years SCI	Four FES rowing 3×/week for 90 sessions	
Morse *et al.* [35]	*n* = 20 Chronic SCI	Ten FES rowing 3×/week for one year Ten FES rowing 3×/week for one year + Zoledronic acid. Before FES rowing, 3–5 weeks of NMES-RT to quadriceps and hamstrings	FES rowing + Zoledronic had a greater reduced bone loss (8%) than FES rowing alone (2.5%)

ns=non-significant.

Table 14.6   Effects of electrical stimulation exercise on cardiovascular and metabolic physiology

Source	Participants	Treatment	Results
Allison *et al.* [40]	$n = 10$ Chronic SCI	FES cycling 3×/week for 12 weeks	+3.5% stroke Volume (ns) +2% cardiac output (ns) +17% resting femoral artery diameter +34% blood flow −46% CRP
Griffin *et al.* [42]	$n = 18$ Chronic SCI	FES cycling 2–3×/week for ten weeks	Insulin levels were lower at 60 and 120 min after dextrose consumption compared to the pre-test. −18.7% CRP −22.8% IL-6 −4.3% TNF-α
Mohr *et al.* [43]	$n = 10$ Chronic SCI	FES cycling 3×/week for one year. Then, FES cycling 1×/week for six months	+31.9% Insulin-stimulated glucose uptake +105% GLUT-4 Improvements dissipated after FES-LCE 1×/week for 6 months
Nash *et al.* [44]	$n = 8$ chronic tetraplegia	NMES-RT (four weeks) FES cycling six weeks	+35% left ventricular mass. +6.5% Internal dimension – end diastole +17.8% IVSTED +20% PWTED
Mercier *et al.* [45]	$n = 27$ Chronic SCI	Hybrid FES-rowing 3×/week for six months	+10% VO₂ peak Decreased risk of all-cause mortality
Brurok *et al.* [46]	$n = 6$ Chronic SCI	Hybrid HIIT-FES cycling 3×/week for eight weeks after a seven-week control period	+24.4% VO₂peak +33% stroke volume Decreased CVD risk
Gorgey *et al.* [51]	$n = 33$ Chronic SCI	20 NMES-RT 13 Passive Movement 2×/week for 12 weeks	NMES-RT +14% VO₂peak +35.7% power output Passive Movement No changes
Gorgey *et al.* [52]	$n = 14$ Chronic SCI	7 NMES-RT and T 7 T only	NMES-RT and T +34% MCS Increased SD T only No change in MCS & SD

(Continues)

Table 14.6 Effects of electrical stimulation exercise on cardiovascular and metabolic physiology (Continued)

Source	Participants	Treatment	Results
Gorgey *et al.* [10]	*n* = 9 Chronic SCI	5 – NMES-RT and diet 2×/week for 12 weeks 4 – diet alone	NMES-RT + diet −37.6% Triglycerides −33.8% AUC Plasma Insulin (ns) +10% HDL (ns)
Gorgey *et al.* [53]	*n* = 22 Chronic SCI	11 NMES-RT and T11 T only	NMES-RT and T −46% Il-6 Decreased TNF-α +14–18% BMR T only −37.8 Il-6

T = Testosterone; MCS = Mitochondrial citrate synthase; SD = succinate dehydrogenase; ns = non-significant; CRP = c-reactive protein; BMR = basal metabolic rate; AUC = area under the curve; CVD = cardiovascular disease; IVSTED = Interventricular septal thickness at end diastolic: PWTED = Posterior wall thickness at end diastolic.

and exercise after SCI and reported evidence that FES leg cycling training demonstrated VO_2peak and power output increases of 20 to 40%. They also reported evidence of improvement in the oxidative potential of muscle, improved glucose homeostasis, enhanced lipoprotein profiles, and overall cardiovascular fitness resulting from two months of thrice weekly FES leg cycling training.

A study by Allison *et al.* [40] observed that 12 weeks of thrice weekly FES leg cycling helped increase exercise capacity and reduce systemic inflammation in individuals with chronic SCI. Although there were only small nonsignificant central cardiovascular changes (stroke volume and cardiac output +3.5% and + 2%, respectively), peripheral vascular changes were significant. The resting femoral artery diameter increased by 17%, and blood flow increased by 34%. C-reactive protein decreased by 46%, suggesting reduced inflammation. However, cholesterol levels were unchanged. Electrical stimulation was provided via surface electrodes to the quadriceps, hamstrings, and gluteal muscle at 500 μs pulse width, at 50 Hz pulse frequency, and up to 140 mA for current amplitude, allowing the legs to cycle for up to 45 minutes at a cadence between 35 and 49 rev·min^{-1}. As the participants could tolerate, resistance was increased by 1.2 Nm increments. Exercise capacity was increased across participants in duration, distance, and speed.

Van der Scheer and co-authors [19] reported that although there was a lack of large cohort randomized control trials, 17 out of 21 lower-quality studies demonstrated significant increases in aerobic fitness, and 29 out of 30 studies reported significant improvements in power output after FES leg cycling training promoting the potential for cardiovascular and health benefits. In another systematic review, Farrow *et al.* [41] indicated that more extensive randomized control trials must be completed before a definitive cause-and-effect conclusion could be made regarding FES leg cycling and its effects on diabetic risk factors. Nevertheless, they found that five out of six FES cycling studies reported improved peripheral insulin sensitivity. Griffin and co-workers [42] demonstrated that two to three FES leg cycling sessions per week for ten weeks produced a reduction in both blood glucose and insulin at multiple time points using an oral glucose tolerance test and a reduction in the inflammatory markers like interleukin-6 (IL-6), tumor necrosis factor-alpha (TNC-α), and c-reactive protein. Similarly, thrice weekly FES leg cycling sessions of 30 minutes over a year in ten individuals with chronic SCI demonstrated an increased insulin sensitivity and insulin-regulated glucose transporter (GLUT-4) concentration. However, these improvements returned to pre-exercise levels after six months of FES cycling only once weekly, suggesting less than minimal dose-potency for health benefits [43].

An early study by Nash and colleagues [44] utilized NMES knee extension resistance training for the quadriceps muscles four weeks before FES leg cycling training for six months on eight individuals with chronic tetraplegia. The participants increased left ventricular wall thickness by 35%, septal wall thickness by 18%, and posterior wall thickness by 20%, reversing left ventricular atrophy. Mercier and co-authors [45] used hybrid FES rowing, which combined voluntary arm exercise with FES evoked leg exercise to increase the intensity of the activity. Twenty-seven individuals with chronic SCI completed FES rowing for 20 or more minutes three times per week for six months, significantly increasing their VO$_2$peak (10%). However, VO$_2$peak increased more over the first three months (7%) than over the second three months (4%), which indicated that most of the "training effect" appeared early in the program. The authors reported that increased aerobic capacity decreases the risk for all-cause mortality, demonstrating the importance of cardiovascular conditioning.

Brurok *et al.* [46] investigated the effects of hybrid HIIT-FES leg cycling, combining FES cycling with arm crank exercise with high- and low-intensity intervals. After seven weeks of no exercise for the control period, six individuals with chronic SCI completed eight weeks of thrice weekly hybrid HIIT-FES leg cycling with arm crank exercise. The high-intensity interval was 85–95% of their pre-determined peak watts. The mean increase of the group was 24.4% for VO$_2$peak and a 33% increase in stroke volume. The authors suggested that an increased peak aerobic power demonstrated a decrease in cardiovascular disease risk for the SCI population. Similarly, Mate and colleagues [47] completed a systematic review and meta-analysis study on hybrid FES cycling combining arm crank exercise with FES leg cycling and reported increased VO$_2$peak. Both central and peripheral adaptations were credited as instrumental for the increases in peak aerobic fitness. One study pointed to enhanced mitochondrial content, oxidative capacity, and improved blood flow in the paralyzed legs while highlighting the possibility of increased venous blood return to the heart [48].

While FES activities are a standard modality for individuals with SCI, they also demonstrate the ability to provide benefits to individuals with other neurological conditions. Scally and associates [49] completed a systematic review of the effects of FES leg cycling exercise on individuals with multiple sclerosis. The evidence supported the view that FES leg cycling exercise could reduce cardio-vascular disease risk, improve mobility through increased physical conditioning, and decrease muscle spasticity. Likewise, evidence from a randomized control trial found that FES leg cycling exercise significantly reduced physical impairments and activity limitations of individuals with hemiplegia after a stroke and traumatic brain injury [50]. Thirty individuals with a cerebral vascular accident resulting in hemi-plegia and three individuals with traumatic brain injury were randomly assigned to an FES leg cycling exercise group that performed 25 minutes of cycling five days per week for twenty training sessions or the non-FES leg cycling exercise control group. The FES leg cycling exercise group increased gait speed from 0.11 to 0.39 m·s^{-1}, improving power and function in the paretic leg and increasing potential overall activity levels (Figure 14.3).

Gorgey *et al.* [51] demonstrated that the conditioning of the legs with NMES resistance training could also impact VO$_2$peak performance when testing is com-pleted using FES leg cycling exercise. Twenty individuals with chronic SCI were randomly allocated to an NMES resistance training group, and thirteen were ran-domly assigned to a passive movement training group. Graded exercise testing was performed using FES cycling exercises before and after training twice weekly for 12 weeks in their respective groups. The NMES resistance training group increased their VO$_2$peak by 14% and power output by 36%, coinciding with a 30% increase in knee extensor muscle cross-sectional area. All changes in the NMES resistance training group were significantly higher than in the passive movement training group. NMES resistance training has also increased mitochondrial citrate synthase (34%) and succinate dehydrogenase in individuals with SCI, which may indicate improved skeletal muscle fiber metabolic function [52]. An earlier study of twelve

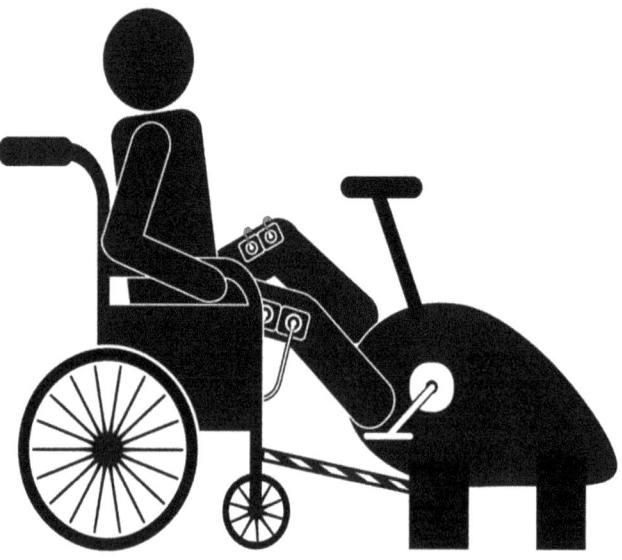

Figure 14.3   FES cycling illustration

weeks of twice-weekly NMES resistance training plus testosterone resulted in a 38% reduction in triglyceride blood level. A non-significant 34% reduction in plasma insulin was determined by the area under the insulin-time curve, and a non-significant 10% increase in HDL [10].

Sixteen weeks of twice-weekly NMES resistance training with daily testosterone patches (2-6 mg) demonstrated a 47% decrease in the inflammatory marker IL-6 and a decrease in TNF-α. In contrast, the testosterone-only group decreased IL-6 by 33.8%. NMES resistance training plus testosterone also produced a 10.5% increase in glucose effectiveness, which may indicate an improved ability of glucose to cause its cell uptake at basal insulin level and a 14–18% increase in basal metabolic rate [53].

14.1.6 Clinical summary

While large cohort randomized control trials are lacking, there is limited lower-level evidence of the effects of electrical stimulation exercise on cardiovascular and metabolic health. Hybrid exercise activities with FES leg exercise and voluntary arm exercise using higher cycling cadences at 35–50 (rev·min^{-1}) are more likely to produce positive changes. Studies have shown that hybrid FES cycling and rowing for 30–60 minutes three or more times per week for 8–24 weeks can produce improved VO$_2$peak, improved insulin sensitivity, and decreased systemic inflammation, all associated with cardiometabolic health. NMES resistance training may also improve VO$_2$peak via improved metabolic muscle function when combined with testosterone treatments. It may also help decrease inflammatory biomarkers and improve insulin sensitivity.

14.1.7 Exercise protocols

Medical referrals for electrical stimulation activities could be for improvement in muscle strength, muscular endurance, cardiovascular fitness, increased physical activity as part of a weight control program or possibly stimulation to maintain or decrease bone loss. Depending on the goal, the rehabilitation care plan is developed with input from the patient. After reviewing the medical history of the patient and the completion of a physical exam to determine the patients baseline physically, physiologically, and functionally. Most clinicians who aim to increase muscle size and strength follow a clinical practice similar to the "Dudley Protocol," described by Bickel *et al.* [54]. Using the quadriceps muscles as the example the NMES resistance training program is completed. The program consists of four sets of ten dynamic contractions of the quadriceps muscle group using two surface electrodes placed near the proximal and distal ends of the muscle group. The electrical stimulation parameters include a 30–35 Hz pulse frequency, 400–600 μs biphasic pulse width, and a current amplitude/intensity that is gradually increased until a vigorous contraction of the quadriceps muscles is reached with full knee extension. The amplitude maximum range is 100–200 mA. The training is performed from a seated position with the knee flexed at 90 degrees. Once the full extension of the knee is achieved, the amplitude is turned down, allowing the leg to return to the starting position safely. There are two to five minutes of rest between sets, and the legs are trained alternatively for each set. The resistance can be increased once a participant completes two consecutive training sessions with full knee extension. The protocol starts with zero resistance and then progresses with an increment of 1 kg (2 pounds) of ankle weights at each upgrade in resistance. A wide range of duration successfully increases muscle mass with NMES resistance training. However, the evidence suggests that 12 weeks of twice-weekly NMES is as beneficial as 16 weeks for increasing skeletal muscle [3].

For recommendations for general conditioning FES cycling is a logical choice. The protocol may include cycling at 30–50 rev·min^{-1} for 20–60 minutes three to five times per week for eight to sixteen weeks [19]. Resistance is set as the patient tolerates and increases by 0.5–1.0 Nm or 0.125 kp as the patient is able. The patient may also start at approximately six watts and increase according to tolerance [55]. If the goal is muscle hypertrophy of the legs, then a lower cycling cadence (20 rev·min^{-1}) allowing greater cycling torque is more advantageous [16,17]. For improved aerobic conditioning and cardiovascular health, applying greater muscle mass during exercise will increase intensity and enhance benefits [56]. This can be accomplished by incorporating voluntary arm exercise with the FES legs exercise as with hybrid FES cycling and FES rowing activities [38].

14.1.8   Electrical stimulation parameters and adjustments

The three electrical stimulation parameters usually manipulated for NMES and FES exercise are the pulse frequency (pulses per second), the pulse width (duration of each pulse), and the current amplitude (intensity of the current). According to prior research, NMES and FES activities are most successfully deployed at 30–50 Hz, while frequencies above 50 Hz can result in premature fatigue [55]. NMES protocols have been effective with pulse widths of 400–450 μs, increasing muscle size [55]. An investigation of FES leg cycling exercise showed that a pulse width of 350 μs resulted in higher energy expenditure than 200 μs. In comparison, a pulse width of 500 μs caused autonomic dysreflexia in some individuals with SCI [57]. These results have suggested that a pulse width of 350 μs is optimal for FES leg cycling exercise. Bochkezanian *et al.* [58] used NMES isometric contractions for high-intensity strength training, applying lower electrical stimulation frequency (30 Hz) and higher pulse width (1000 μs) on five participants with chronic SCI for 12 weeks. The amplitude of the electrical stimulation started at 30 mA and then increased by 10 mA every 20 seconds until muscle fatigue reached a maximum of 99 mA. This protocol increased the quadriceps cross-sectional area, improved strength, and decreased muscle spasticity.

Exercise-related health and fitness benefits are primarily dose-dependent, emphasizing the importance of maintaining a sufficient contraction force throughout the training sessions. Because the stimulation of motor units via NMES and FES activities is less efficient than the body's "natural" motor unit recruitment process due to the random and asynchronous stimulation of motor neurons through the surface electrodes, the muscles are at risk of early fatigue [59].

To offset the reduction in muscle contraction force due to fatigue during electrical stimulation exercise, it has been recommended that electrical stimulation frequency be reduced, and amplitude/intensity increased [3]. High subcutaneous fat content under the surface electrodes can also impact the amount of current that reaches the motor neurons and can reduce the strength of muscle contraction. A 50% cross-sectional area of subcutaneous fat in the thigh area impedes the electrical stimulus requiring 48–59% greater amplitude to complete knee extension [60].

14.1.9 Safety and contraindications

The safety of FES activities is exemplified by the increasingly common use in rehabilitation settings, research clinics, and home-based exercise programs [61–63]. The contraindications and precautions for electrical stimulation exercise can vary between countries, medical facilities, and institutions. So, it is always best to seek the participant's physician's advice regarding any safety questions. For example, many consider electrical stimulation activities contraindicated when the potential participant has a cardiac pacemaker or other implanted electrical device [64]. However, while some consider it a contraindication for all electrical stimulation activities, fearing that it may interfere with the electronics involved with the implanted device, others believe that electrical stimulation activities in other body regions are acceptable because of their distance away from the implant – for example, stimulation of the lower legs when the device is in the upper trunk [65]. As stated, seeking medical clarification before starting programs using electrical stimulation is always best. The following contraindications and precautions are commonly stated and should not be considered exhaustive. Common contraindications for NMES legs resistance training and FES cycling or rowing activities include pregnancy, cardiac failure or arrhythmias, uncontrolled autonomic dysreflexia, deep vein thrombosis or thrombophlebitis, hemorrhagic conditions, seizure disorders, severe osteoporosis or history of fragility bone fractures, recent or unhealed bone fractures in an area impacted by the stimulation, stimulation over or near electronic devices such as a cardiac pacemaker, stimulation over an area of damaged skin or recently radiated tissue, stimulation over an area of known or suspected malignancy, stimulation over an area of infection such as tuberculosis or osteomyelitis, or an area of impaired circulation, stimulation over eyes or genitalia, and stimulation on the neck over or near the carotid arteries [55,66].

Standard precautions regarding electrical stimulation activities include impaired cognition, impaired communication, fatigue, and skin disease. The skin should always be inspected before and after using surface electrodes because it is possible that the skin beneath the electrodes is susceptible to contact irritation and allergic reactions [67]. Redness may appear after use but should resolve within minutes. Figures 14.4–14.6 demonstrate the application of surface electrodes. The commonly used surface adhesive electrodes typically last 8–10 sessions before the adhesive wears away. Once the electrodes have difficulty sticking to the skin, they need to be replaced to allow good contact with the skin surface. Placing the

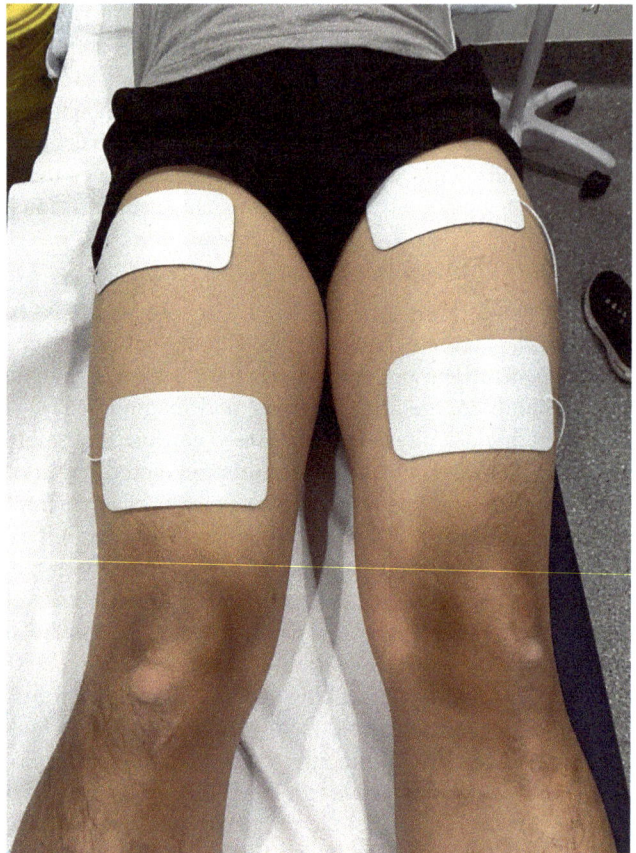

Figure 14.4   Placement of electrodes on quadriceps

electrodes over hairy areas of the skin or skin that has been covered with moisturizer can decrease the life of the surface electrodes.

Two medical sequelae should be closely monitored during NMES/FES exercise. These are occurrences of autonomic dysreflexia, especially in persons with sensorimotor complete spinal cord lesions above T6 [68] and orthostatic hypotension arising from lower limb blood pooling and the lack of a sub-lesional vascular muscle pump due to muscle paralysis [69]. Regarding potential autonomic dysreflexia reactions to electrical stimulation activities, the participant's blood pressure and symptoms should be monitored. A case report on a 36-year-old man with C6 American Spinal Cord Injury Association Impairment Scale (AIS) C nine years post injury performed ten FES interval cycling sessions even though he was prone to mild to moderate autonomic dysreflexia. The results of the study showed that increased systolic blood pressure of 20 mmHg or more and verbal expressions of symptoms by the cycler, both signs of autonomic dysreflexia coincided only 59%

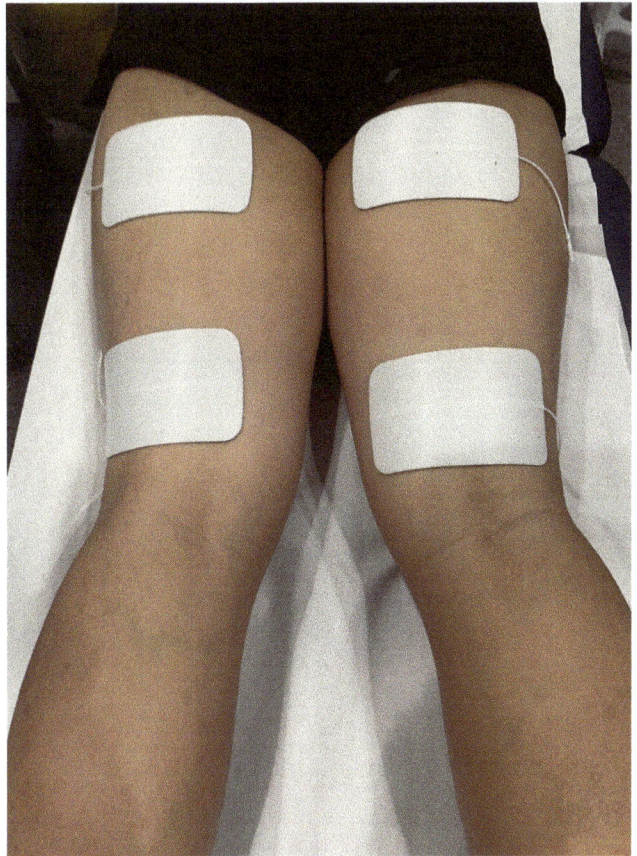

Figure 14.5   Placement of electrodes on the hamstrings

of the time, meaning that neither safety precaution is adequate if used alone [64]. In this case the symptoms were non-life-threatening resulting in chills, goosebumps, and sometimes a weird feeling as expressed by the cycler. Regardless of the non-severity of the symptoms, the non-synchronicity of the blood pressure elevations and the report of physical symptoms indicates that both blood pressure measures and symptom responses need to be monitored for individuals at risk of autonomic dysreflexia (Table 14.7). Once an autonomic dysreflexia response was recognized the electrical stimulation activity was stopped and the symptoms ceased. For safety reasons uncontrolled autonomic dysreflexia is considered a contraindication for electrical stimulation activities.

14.1.10   Conclusion and future research

NMES resistance training, FES cycling, and FES rowing have demonstrated the ability to improve physical function, body composition, and cardiometabolic health

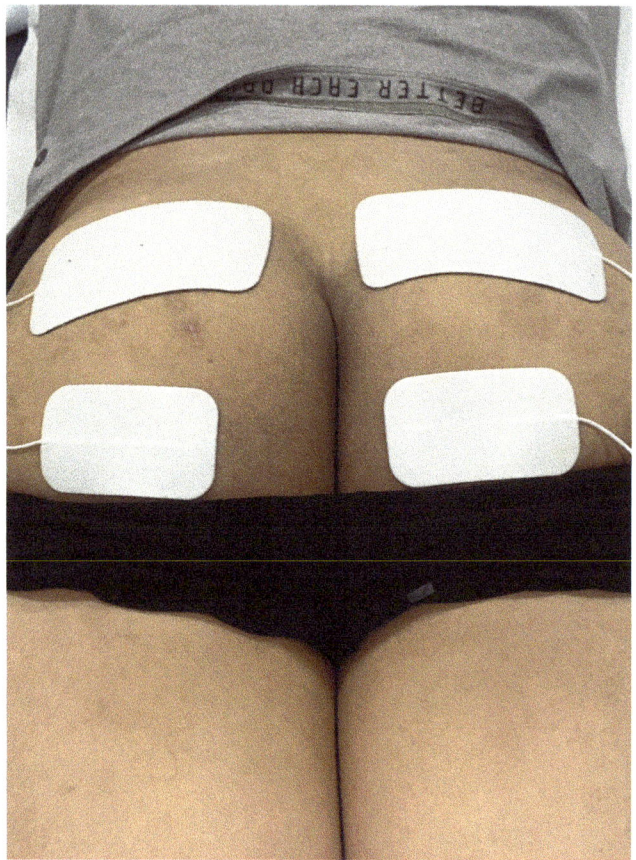

Figure 14.6   Placement of electrodes on the gluteal muscles

in individuals with paralytic and paretic conditions [70]. Recent advancements in hybrid FES cycling and rowing have shown the potential to provide greater benefits than FES leg cycling or arm crank exercise alone. HIIT-FES cycling and rowing protocols also have the potential to provide similar benefits with less training volume. Further study with large cohort randomized control trials is necessary to fully establish the efficacy of these activities.

To provide access to more well-being, future research should also focus on optimizing the dose-response relationship for different neurological conditions and reducing barriers to treatment. A recent survey revealed that inadequate time and staff training, financial cost, and a lack of availability of FES equipment are common complaints among individuals with SCI, healthcare providers, and researchers. To increase access, future FES research should concentrate on redu-cing the cost of clinical services and equipment [71].

Table 14.7   FES cycling with monitoring for autonomic dysreflexia reactions

Clinical indication	A 38-year-old man nine years post motor vehicle accident with incomplete C6 SCI AIS C and uncontrolled mild to moderate autonomic dysreflexia.
Indication for FES cycling	Chronic SCI, obese, 171 cm tall and 94.4 kg in weight with a BMI of 36.8 requires an exercise program but is prone to autonomic dysreflexia reactions.
Electrode placement	Surface electrodes 8 × 10 cm in size were placed over the proximal quadriceps approximately 30 cm above the patella and distally the electrodes were placed over the quadriceps muscle group approximately 3 cm above the patella. The proximal electrodes for the hamstrings were placed approximately 30 cm above the popliteal fossa while the distal electrodes were placed 2–3 cm above the popliteal fossa. Two electrodes were placed over the left and right gluteus maximus with at least three finger widths apart.
Stimulation settings	Frequency – 40 Hz, Pulse Width – 350 µs. During the higher intensity intervals, the Amplitude/Intensity – 48–50 mA for the quadriceps, 43–45 mA for the hamstrings, and 38–40 for the gluteal muscle groups. During the lower intensity intervals, the amplitude was reduced by 50%.
NMES protocol	The exercise protocol started with a one-minute passive cycling warm-up. Then the electrical stimulation amplitude gradually increased until reaching the maximum intensity. Then the high and lower intensity cycling intervals alternated every 30 seconds with a ten second ramp up and ramp down time between the intervals. The cycling cadence ranged between 35 rev·min^{-1} and the cycling resistance was 0.64 Nm during the high intensity interval and 0.5 during the low intensity interval. The cycling duration progressed to 20 minutes twice weekly. Blood pressures were monitored before, after and every five minutes during cycling. Symptoms monitored continuously.

(Image adapted and redrawn from Noraxon MyoResearch EMG software tools (Noraxon USA Scottsdale, Arizona, USA) in Hamzaid NA. Development of an isokinetic functional electrical stimulation leg stepping trainer for individuals with neurological disabilities. PhD thesis, 2010.)

As technology continues to advance, electrical stimulation activities will become more effective. Garment-based stimulation is a recent trend in FES training that simplifies the use of surface electrodes. These garments have embedded electrodes that stimulate the appropriate muscle groups. They are reportedly versatile, durable, cost-effective, and can be custom-made [72].

Electrically evoked exercise is currently the only way paralyzed muscles can receive a physiological training effect as provided by muscle contractions. As research continues to develop in this area, people with paralysis should have more access to these activities, which can potentially provide greater benefits.

References

[1]   Gorgey AS, Dolbow DR, Dolbow JD, Khalil RK, Castillo C, and Gater DR. Effects of spinal cord injury on body composition and metabolic profile – part I. *Journal of Spinal Cord Medicine*. 2014;37(6):693–702. doi:10.1179/2045772314Y.0000000245.

[2]   Deley G, Denuziller J, and Babault N. Functional electrical stimulation: cardiorespiratory adaptations and applications for training in paraplegia. *Sports Medicine*. 2015;45(1):71–82. doi:10.1007/s40279-014-0250-2.

[3]   Gorgey AS, Dolbow DR, Dolbow JD, Khalil RK, and Gater DR. The effects of electrical stimulation on body composition and metabolic profile after spinal cord injury–Part II. *Journal of Spinal Cord Medicine*. 2015;38(1):23–37. doi:10.1179/2045772314Y.0000000244.

[4]   Dolbow DR, Gorgey AS, Johnston TE, and Bersch I. Electrical stimulation exercise for people with spinal cord injury: a healthcare provider perspective. *Journal of Clinical Medicine*. 2023;12(9):3150. Published 2023 Apr 27. doi:10.3390/jcm12093150.

[5]   Atkins KD, and Bickel CS. Effects of functional electrical stimulation on muscle health after spinal cord injury. *Current Opinion in Pharmacology*. 2021;60:226–231. doi:10.1016/j.coph.2021.07.025.

[6]   Bekhet AH, Jahan AM, Bochkezanian V, Musselman KE, Elsareih AA, and Gorgey AS. Effects of electrical stimulation training on body composition parameters after spinal cord injury: a systematic review. *Archives of Physical Medicine & Rehabilitation*. 2022;103(6):1168–1178. doi:10.1016/j.apmr.2021.09.004.

[7]   Stenson KW, Deutsch A, Heineman AW, and Chen D. Obesity and inpatient rehabilitation outcomes for patients with a traumatic spinal cord injury. *Archives of Physical Medicine & Rehabilitation*. 2011;92:384–390.

[8]   Tian W, Hsieh CH, DeJung G, Backus D, Groah S, and Ballard PH. Role of body weight in therapy participation and rehabilitation outcomes among individuals with traumatic spinal cord injury. *Archives of Physical Medicine & Rehabilitation*. 2013;94(4 Suppl 2):S125–136.

[9]   Mahoney ET, Bickel CS, Elder C, *et al.* Changes in skeletal muscle size and glucose tolerance with electrically stimulated resistance training in subjects

with chronic spinal cord injury. *Archives of Physical Medicine & Rehabilitation*. 2005;86(7):1502–1504. doi:10.1016/j.apmr.2004.12.021.

[10] Gorgey AS, Mather KJ, Cupp HR, and Gater DR. Effects of resistance training on adiposity and metabolism after spinal cord injury. *Medicine & Science in Sports & Exercise*. 2012;44(1):165–174. | DOI:10.1249/MSS. 0b013e31822672aa.

[11] Gorgey AS, Dolbow DR, Cifu DX, and Gater DR. Neuromuscular electrical stimulation attenuates thigh skeletal muscles atrophy but not trunk muscles after spinal cord injury. *Journal of Electromyography & Kinesiology*. 2013; 23(4):977–984. doi:10.1016/j.jelekin.2013.04.007.

[12] Gorgey AS, and Shepherd C. Skeletal muscle hypertrophy and decreased intramuscular fat after unilateral resistance training in spinal cord injury: case report. *Journal of Spinal Cord Medicine*. 2010;33(1):90–95. doi:10.1080/ 10790268.2010.1168968113.

[13] Fenton JM, King JA, Hoekstra SP, Valentino SE, Phillips SM, and Goosey-Tolfrey VL. Protocols aiming to increase muscle mass in persons with motor complete spinal cord injury: a systematic review. *Disability Rehabilitation*. 2023;45(9):1433–1443.

[14] Farkas GJ, Gorgey AS, Dolbow DR, Berg AS, and Gater DR Jr. Energy expenditure, cardiorespiratory fitness, and body composition following arm cycling or functional electrical stimulation exercises in spinal cord injury: a 16-week randomized controlled trial. *Topics in Spinal Cord Injury Rehabilitation*. 2021;27(1):121–134. doi:10.46292/sci20-00065.

[15] Dolbow DR, Credeur DP, Lemacks JL, Stokic DS, Corbin GN, and Courtner AS. Electrically induced cycling and nutritional counselling for counteracting obesity after spinal cord injury: a pilot study. *Journal of Spinal Cord Medicine*. 2021;44(4):533–540. doi:10.1080/10790268.2019.1710939.

[16] Johnston TE, Marino RJ, Oleson CV, *et al.* Musculoskeletal effects of 2 functional electrical stimulation cycling paradigms conducted at different cadences for people with spinal cord injury: a pilot study. *Archives of Physical Medicine & Rehabilitation*. 2016;97(9):1413–1422. doi:10.1016/j. apmr.2015.11.014.

[17] Fornusek C, Davis GM, and Russold MF. Pilot study of the effect of low-cadence functional electrical stimulation cycling after spinal cord injury on thigh girth and strength. *Archives of Physical Medicine & Rehabilitation*. 2013;94(5):990–993. doi:10.1016/j.apmr.2012.10.010.

[18] Rosley N, Hasnan N, Hamzaid NA, Davis GM, and Haidzir M. Effects of a combined progressive resistance training and functional electrical stimulation-evoked cycling exercise on lower limb muscle strength of individuals with incomplete spinal cord injury: a randomized controlled study. *Turkish Journal of Physical Medicine & Rehabilitation*, 2023;69:23–30. doi: 10.5606/tftrd.2023.9418.

[19] Van der Scheer JW, Goosey-Tolfrey VL, Valentino SE, Davis GM, and Ho CH. Functional electrical stimulation cycling exercise after spinal cord injury: a systematic review of health and fitness-related outcomes. *Journal of*

NeuroEngineering and Rehabilitation. 2021;18(1):99. Available at: https://doi.org/10.1186/s12984-021-00882-8. Accessed 11 June 2024.

[20]   Dolbow DR, Gorgey AS, Moore JR, and Gater DR. A report of practicability of a six month home based functional electrical stimulation cycling program for an individual with tetraplegia. *Journal of Spinal Cord Medicine*. 2012;35 (3):182–186.

[21]   Kim DI, Park DS, Lee B, and Jeon JY. A six-week motor-driven functional electronic stimulation rowing program improves muscle strength and body composition in people with spinal cord injury: a pilot study. *Spinal Cord*. 2014;52; 621–624. Available at: https://doi.org/10.1038/sc.2014.76.Accessed 11 June 2024.

[22]   Jeon JY, Hettinga D, Steadward RD, Wheeler GD, Bell G, and Harber V. Reduced plasma glucose and leptin after 12 weeks of functional electrical stimulation-rowing exercise training in spinal cord injury patients. *Archives of Physical Medicine & Rehabilitation*. 2010;91(12):1957–1959. doi:10.1016/j.apmr.2010.08.024.

[23]   Wilbanks SR, Rogers R, Pool S, and Bickel CS. Effects of functional electrical stimulation assisted rowing on aerobic fitness and shoulder pain in manual wheelchair users with spinal cord injury. *Journal of Spinal Cord Medicine*. 2016;39(6):645–654. doi:10.1179/2045772315Y.0000000052.

[24]   Leeds EM, Klose KJ, Ganz W, Serafini A, and Green BA. Bone mineral density after bicycle ergometry training. *Archives of Physical Medicine & Rehabilitation*. 1990;71(3):207–209.

[25]   Rodgers MM, Glaser RM, Figoni SF, *et al*. Musculoskeletal responses of spinal cord injured individuals to functional neuromuscular stimulation-induced knee extension exercise training. *Journal of Rehabilitative Research & Development*. 1991;28(4):19–26. doi:10.1682/jrrd.1991.10.0019.

[26]   Holman ME, Chang G, Ghatas MP, *et al*. Bone and non-contractile soft tissue changes following open kinetic chain resistance training and testosterone treatment in spinal cord injury: an exploratory study. *Osteoporosis International*. 2021;32(7):1321–1332. doi:10.1007/s00198-020-05778-2.

[27]   Mohr T, Podenphant J, Biering-Sorensen F, Galbo H, Thamsborg G, and Kjaer M. Increased bone mineral density after prolonged electrically induced cycle training of paralyzed limbs in spinal cord injured man. *Calcified Tissue International*. 1997;61(1):22–25. doi:10.1007/s002239900286.

[28]   Chen SC, Lai CH, Chan WP, Huang MH, Tsai HW, and Chen JJ. Increases in bone mineral density after functional electrical stimulation cycling exercises in spinal cord injured patients. *Disability and Rehabilitation*, 2005;27 (22):1337–1341. doi:10.1080/09638280500164032.

[29]   Frotzler A, Coupaud S, Perret C, *et al*. High-volume FES-cycling partially reverses bone loss in people with chronic spinal cord injury. *Bone*. 2008;43 (1):169–176. doi:10.1016/j.bone.2008.03.004.

[30]   Shields RK, and Dudley-Javoroski S. Musculoskeletal plasticity after acute spinal cord injury: effects of long-term neuromuscular electrical stimulation training. *Journal of Neurophysiology*. 2006;95(4):2380–2390. doi:10.1152/jn.01181.2005.

[31]   Shields RK, Dudley-Javoroski S, and Law LA. Electrically induced muscle contractions influence bone density decline after spinal cord injury. *Spine (Phila Pa 1976)*. 2006;31(5):548–553. doi:10.1097/01.brs.0000201303. 49308.a8.

[32]   Shields RK, and Dudley-Javoroski S. Musculoskeletal adaptations in chronic spinal cord injury: effects of long-term soleus electrical stimulation training. *Neurorehabilitation & Neural Repair*. 2007;21(2):169–179. doi:10.1177/ 1545968306293447.

[33]   Dolbow DR, Gorgey AS, Gater DR, and Moore JR. Body composition changes after 12 months of FES cycling: case report of a 60 year old female with paraplegia. *Spinal Cord*. 2014;52(Suppl 1):S34.

[34]   Lambach RL, Stafford NE, Kolesar JA, *et al.* Bone changes in the lower limbs from participation in an FES rowing exercise program implemented within two years after traumatic spinal cord injury. *Journal of Spinal Cord Medicine*. 2020;43(3):306–314. doi:10.1080/10790268.2018.1544879.

[35]   Morse LR, Troy KL, Fang Y, *et al.* Combination therapy with zoledronic acid and FES-row training mitigates bone loss in paralyzed legs: results of a randomized comparative clinical trial. *American Society for Bone and Mineral Research Plus*. 2019;3(5):e10167. Published 2019 Feb 20. doi:10. 1002/jbm4.10167.

[36]   Garshick E, Kelley A, Cohen SA, *et al.* A prospective assessment of mortality in chronic spinal cord injury. *Spinal Cord*. 2005;43(7):408–416. doi:10. 1038/sj.sc.3101729.

[37]   Gater, DR, Farkas GJ, Dolbow DR, Berg A, and Gorgey AS. Body composition assessment after motor complete spinal cord injury: development of a clinically relevant equation to estimate body fat. *Topics in Spinal Cord Injury Rehabilitation*. 2021;27(1):11–22. doi:10.46292/sci20-00079.

[38]   Figoni SF, Dolbow DR, Crawford EC, White ML, and Pattanaik S. Does aerobic exercise benefit persons with tetraplegia from spinal cord injury? A systematic review. *Journal of Spinal Cord Medicine*. 2021;44(5):690–703. doi:10.1080/10790268.2020.1722935.

[39]   Warburton DE, Eng JJ, Krassioukov A, Sproule S; and the SCIRE Research Team. Cardiovascular health and exercise rehabilitation in spinal cord injury. *Topics in Spinal Cord Injury Rehabilitation*. 2007;13(1):98–122. doi:10. 1310/sci1301-98.

[40]   Allison DJ, Chapman B, Wolfe D, Sequeira K, Hayes K, and Ditor DS. Effects of a functional electrical stimulation-assisted cycling program on immune and cardiovascular health in persons with spinal cord Injury. *Topics in Spinal Cord Injury Rehabilitation*. 2016;22(1):71–78. doi:10.1310/ sci2201-71. PMID: 29398895; PMCID: PMC5790031.

[41]   Farrow M, Nightingale TE, Maher J, McKay CD, Thompson D, and Bilzon JLJ. Effect of exercise on cardiometabolic risk factors in adults with chronic spinal cord injury: a systematic review. *Archives of Physical Medicine & Rehabilitation*. 2020;101(12):2177–2205. doi:10.1016/j.apmr.2020.04.020.

[42] Griffin L, Decker MJ, Hwang JY, *et al.* Functional electrical stimulation cycling improves body composition, metabolic and neural factors in persons with spinal cord injury. *Journal of Electromyography & Kinesiology.* 2009; 19(4):614–622. doi:10.1016/j.jelekin.2008.03.002.

[43] Mohr T, Dela F, Handberg A, Biering-Sørensen F, Galbo H, and Kjaer M. Insulin action and long-term electrically induced training in individuals with spinal cord injuries. *Medicine & Science in Sports & Exercise.* 2001;33 (8):1247–1252. doi:10.1097/00005768-200108000-00001.

[44] Nash MS, Bilsker S, Marcillo AE, *et al.* Reversal of adaptive left ventricular atrophy following electrically-stimulated exercise training in human tetra-plegics. *Paraplegia.* 1991;29(9):590–599. doi:10.1038/sc.1991.87.

[45] Mercier HW, Picard G, Taylor JA, and Vivodtzev I. Gains in aerobic capacity with whole-body functional electrical stimulation row training and generalization to arms-only exercise after spinal cord injury. *Spinal Cord.* 2021;59(1):74–81. doi:10.1038/s41393-020-0527-2. Epub 2020 Jul 27. PMID: 32719528; PMCID: PMC7855132.

[46] Brurok B, Helgerud J, Karlsen T, Leivseth G, and Hoff J. Effect of aerobic high-intensity hybrid training on stroke volume and peak oxygen consumption in men with spinal cord injury. *American Journal of Physical Medicine & Rehabilitation.* 2011;90(5):407–414. | doi:10.1097/PHM.0b013e31820f960f.

[47] Máté S, Sinan-Fornusek C, Dhopte P, Singh MF, Hackett D, and Fornusek C. Effects of functional electrical stimulation cycling combined with arm cranking exercise on cardiorespiratory fitness in people with central nervous system disorders: A systematic review and meta-analysis. *Archives of Physical Medicine Rehabilitation.* 2023;104(11):1928–1940. S0003-9993(23) 00225-3. doi:10.1016/j.apmr.2023.03.026.

[48] Krauss JC, Robergs RA, Depaepe JL, *et al.* Effects of electrical stimulation and upper body training after spinal cord injury. *Medicine & Science in Sports & Exercise.* 1993;25(9):1054–1061.

[49] Scally JB, Baker JS, Rankin J, Renfrew L, and Sculthorpe N. Evaluating functional electrical stimulation (FES) cycling on cardiovascular, musculoskeletal and functional outcomes in adults with multiple sclerosis and mobility impairment: a systematic review. *Multiple Sclerosis & Related Disorders.* 2020;37:101485. doi:10.1016/j.msard.2019.101485.

[50] Ambrosini E, Ferrante S, Pedrocchi A, Ferrigno G, and Molteni F. Cycling induced by electrical stimulation improves motor recovery in postacute hemiparetic patients: a randomized controlled trial. *Stroke.* 2011;42 (4):1068–1073. doi:10.1161/STROKEAHA.110.599068.

[51] Gorgey AS, Lai RE, Khalil RE, *et al.* Neuromuscular electrical stimulation resistance training enhances oxygen uptake and ventilatory efficiency independent of mitochondrial complexes after spinal cord injury: a randomized clinical trial. *Journal of Applied Physiology (1985).* 2021;131(1):265–276. doi:10.1152/japplphysiol.01029.2020. Epub 2021 May 13. PMID: 33982590; PMCID: PMC8325607.

[52]   Gorgey AS, Graham ZA, Chen Q, *et al.* Sixteen weeks of testosterone with or without evoked resistance training on protein expression, fiber hypertrophy and mitochondrial health after spinal cord injury. *Journal of Applied Physiology (1985).* 2020;128(6):1487–1496. doi:10.1152/japplphysiol.00865.2019.

[53]   Gorgey AS, Khalil RE, Gill R, *et al.* Low-dose testosterone and evoked resistance exercise after spinal cord injury on cardio-metabolic risk factors: an open-label randomized clinical trial. *Journal of Neurotrauma.* 2019;36 (18):2631–2645. doi:10.1089/neu.2018.6136.

[54]   Bickel CS, Yarar-Fisher C, Mahoney ET, and McCully KK. Neuromuscular electrical stimulation-induced resistance training after SCI: a review of the Dudley protocol. *Topics in Spinal Cord Injury Rehabilitation.* 2015;21 (4):294–302. doi:10.1310/sci2104-294.

[55]   Dolbow DR, Gorgey AS, Sutor TW, Musselman K, Bochkezanian V, and Davis GM. Electrical stimulation exercise recommendations for individuals with spinal cord injury. *Archives of Physical Medicine & Rehabilitation.* 2023;104(5):847–851. doi:10.1016/j.apmr.2022.11.017.

[56]   Nightingale TE, and Gorgey A. Exercise testing and prescription for populations with other chronic diseases and health conditions (Spinal Cord). In: *American College of Sports Medicine's (ACSM's) Guidelines for Exercise Testing and Prescription.* Vol. 11th. 2020. Wolters Kluwer, Philadelphia, PA.

[57]   Gorgey AS, Poarch HJ, Dolbow DD, Castillo T, and Gater DR. Effect of adjusting pulse durations of functional electrical stimulation cycling on energy expenditure and fatigue after spinal cord injury. *Journal of Rehabilitative Research & Development.* 2014;51(9):1455–1468. doi:10.1682/JRRD.2014.02.0054.

[58]   Bochkezanian V, Newton RU, Trajano GS, Vieira A, Pulverenti TS, and Blazevich AJ. Effect of tendon vibration during wide-pulse neuromuscular electrical stimulation (NMES) on muscle force production in people with spinal cord injury (SCI). *Biomedical Central (BMC) Neurology.* 2018;18 (1):17. Published 2018 Feb 13. doi:10.1186/s12883-018-1020-9.

[59]   Gregory CM, and Bickel CS. Recruitment patterns in human skeletal muscle during electrical stimulation. *Physical Therapy.* 2015;85(4):358–364.

[60]   Gorgey AS, Cho GM, Dolbow DR, and Gater DR. Differences in current amplitude evoking leg extension in individuals with spinal cord injury. *NeuroRehabilitation.* 2013;33(1):161–70.61.

[61]   Dolbow DR, Gorgey AS, Khalil RK, and Gater DR. Effects of a fifty-six month electrical stimulation cycling program after tetraplegia: case report. *Journal of Spinal Cord Medicine.* 2017;40(4):485–488.

[62]   Kressler J, Ghersin H, and Nash MS, Use of functional electrical stimulation cycle ergometers by individuals with spinal cord injury. *Topics in Spinal Cord Injury Rehabilitation.* 2014;20(2):123–126.

[63]   Dolbow DR, Gorgey AS, Ketchum JM, Moore JR, Hackett LA, and Gater DR. Exercise adherence during home-based functional electrical stimulation

cycling by individuals with spinal cord injury. *American Journal of Physical Medicine & Rehabilitation.* 2012;91(11):922–30.

[64] Corbin GN, Weaver K, Dolbow DR, Credeur D, Pattanaik S, and Stokic DS. Safety and Preliminary Efficacy of Functional Electrical Stimulation Cycling in an Individual with Cervical Cord Injury, Autonomic Dysreflexia, and a Pacemaker: Case Report. *J Spinal Cord Med.* 2021;44(4):613–616. doi:10. 1080/10790268.2019.1692180.

[65] Badger J, Taylor P, and Swain I. The safety of electrical stimulation in patients with pacemakers and implantable cardioverter defibrillators: a systematic review. *Journal of Rehabilitation and Assistive Technologies Engineering.* 2017;4:2055668317745498. doi:10.1177/2055668317745498. PMID: 31186945; PMCID: PMC6453072.

[66] Rennie S. ELECTROPHYSICAL AGENTS – contraindications and precautions: an evidence-based approach to clinical decision making in physical therapy. *Physiotherapy Canada.* 2010;62(5):1–80. doi:10.3138/ptc.62.5. Epub 2011 Jan 5. PMID: 21886384; PMCID: PMC3031347.

[67] Almalty AR, Hamed SH, Jebril MY, and Abdelnour HM. The effect of electrical stimulation on skin vulnerability to irritants. *Skin Research and Technology.* 2024;30(2):e13591. doi:10.1111/srt.13591. PMID: 38279544; PMCID: PMC10818122.

[68] Ashley EA, Laskin JJ, Olenik LM, *et al.* Evidence of autonomic dysreflexia during functional electrical stimulation in individuals with spinal cord injuries. *Paraplegia.* 1993;31(9):593–605. doi:10.1038/sc.1993.95. PMID: 8247602.

[69] Sampson EE, Burnham RS, and Andrews BJ. Functional electrical stimulation effect on orthostatic hypotension after spinal cord injury. *Archives of Physical Medicine & Rehabilitation.* 2000;81(2):139–143. doi:10.1016/ s0003-9993(00)90131-x. PMID: 10668765.

[70] Dolbow DR, Gorgey AS, Sutor TW, Bochkezanian V, and Musselman K. Invasive and non-invasive approaches of electrical stimulation to improve physical functioning after spinal cord injury. *Journal of Clinical Medicine.* 2021;10:5356. Available at: https://doi.org/10.3390/jcm10225356. Accessed 11 June 2024.

[71] Triccas LT, Donovan-Hall M, Dibb B, and Burridge JH. A nation-wide survey exploring the views of current and future use of functional electrical stimulation in spinal cord injury, *Disability and Rehabilitation: Assistive Technology.* 2023;18(6):752–762. doi:10.1080/17483107.2021.1916631.

[72] Karamian BA, Siegel N, Nourie B, *et al.* The role of electrical stimulation for rehabilitation and regeneration after spinal cord injury. *Journal of Orthopaedics and Traumatology.* 2022;23(1):1–17. Available at: https://doi.org/ 10.1186/s10195-021-00623-6. Accessed 11 June 2024.

Chapter 15

Use of ES in spasticity management

Kathryn Collins[1] and Anand Pandyan[1]

15.1 Problem and background

In addition to the use of electrical simulation (ES) in spasticity management, this chapter introduces the pathophysiology of spasticity, as it is not covered in Chapter 2. Neuromuscular Physiology. It also presents an overview of the measurement and management of spasticity to put the use of ES in context and emphasise the complexity of the topic.

Spasticity is associated with upper motor neuron (UMN) syndromes such as after a stroke or traumatic brain injury (TBI), in multiple sclerosis (MS) and post spinal cord injury (SCI). Spasticity is a complex problem and can be affected by movement as well as affecting movement. Spasticity is managed as part of the cluster of UMN impairments using a range of interventions including ES. As will be discussed in this chapter ES is most often used in conjunction with other interventions to treat spasticity as part of the UMN syndrome. Although the term spasticity is commonly used there is a lack of consensus around a precise definition of spasticity. An original definition proposed by Lance[*] suggests that spasticity is one component of UMN syndrome that can result from an increased stretch reflex, resulting from increased motor-neuro excitability [1]. A review of the literature suggests that the term spasticity is often used when one or many of the following clinical conditions are noted: spastic dystonia (a muscle in a constant state of contraction leading to a change in posture in a state of rest), stretch induced activation of in a relaxed muscle, spasms (transient but continuous muscle contraction), clonus (transient and intermittent rhythmic muscle contraction), abnormal movement patterns and impaired co-contraction between agonist and antagonist muscles [2]. There are two updated definitions of spasticity that try to capture this complex clinical presentation [3,4]. Spasticity is often treated and managed through rehabilitation. There are a range of interventions currently used to treat spasticity that have varying evidence to support their effectiveness. Example treatments include

[1]Faculty of Health and Social Sciences, Bournemouth University, UK
[*]Lance defined Spasticity as a motor disorder related to velocity-dependent increase in tonic stretch reflexes associated with exaggerated tendon from increased excitability of the neurons involved in the stretch reflex [1].

exercise, stretching, posture management, medications and ES. Often interventions are not used in isolation, interventions are used to manage and treat a cluster of UMN impairments. This chapter will focus on spasticity and interventions specifically focusing on the use of ES.

The primary cause for spasticity is a disruption of the descending inhibitory tracts. However, spasticity is rarely seen in patients with a pure cortico-spinal tract lesion and is commonly associated with disruptions in the indirect descending tracts, primarily the reticulospinal tract (Figure 15.1) [3]. The incidence and presentation of spasticity varies between different neurological conditions. The incidence may also depend on the method of assessment used as many clinical scales measure stiffness rather than spasticity. Estimates cited below are based on assessment of stiffness (as opposed to direct measures of spasticity), e.g. using Modified Ashworth Scale (MAS). In people with MS 60–80% of people have

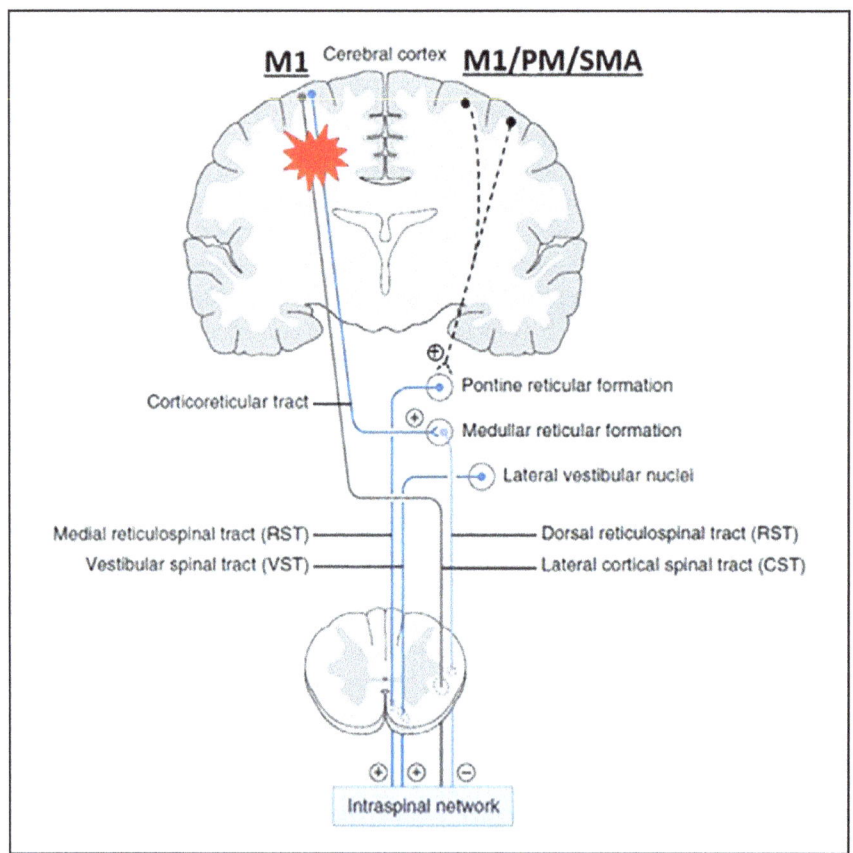

Figure 15.1   Connections between the cerebral cortex and descending tracts and how an injury (such as a stroke) in the M1 area of the cerebral cortex can contribute to spasticity. Adapted from Li et al. [3].

spasticity [5]. In people with stroke, 25.3% of people have spasticity, this increases to 39.5% if the individual has paresis as a post-stroke impairment, and around 9.4% have severe spasticity as defined by a MAS of ≥ 3 [6]. In people with SCI the incidence of spasticity at community discharge is around 65%, with 20%–27% reporting spasticity impacting on function. Individuals with American Spinal Injury Association Impairment Scale (ASIA) grade C had the greatest prevalence of spasticity and associated functional limitations [7]. It is possible these estimates are underreporting the prevalence and incidence of spasticity [3].

Because it is one component of the UMN syndrome, spasticity usually presents with a range of other motor impairments, i.e. weakness and limb deformities/contractures, and sensory impairments. Together, these impairments may impact on a person's ability to do their everyday activities such as washing and dressing as well as impact on their mobility such as rising from a chair and walking. Spasticity is treated holistically as part of a group of interconnecting impairments. If one were to use the World Health Organization's International Classification of Functioning, Disability and Health (ICF) [8], see Figure 15.2, in addition to treating spasticity, it is necessary to manage the limb deformities and weakness, whilst simultaneously training the ability to carry out activities of daily living (e.g. difficulty getting dressed, getting in/out of bed, sit-stand, walking). For example, a clinician might use FES to help to activate the leg muscles during treadmill training to facilitate gait after stroke. FES can contribute to both improving muscle function, strength and gait, as well as reducing spasticity [9].

Pandyan *et al.* [10] have explored spasticity's impact on individuals living with spasticity within the framework of the ICF. The work demonstrates an algorithm suggesting how to manage various aspects of patient reported problems associated with spasticity – spasms, pain, stiffness and function. This work has been used as a basis for summarising the associated impairments with spasticity, how we measure them, and the management (both focal and global management), this is summarised in Table 15.1.

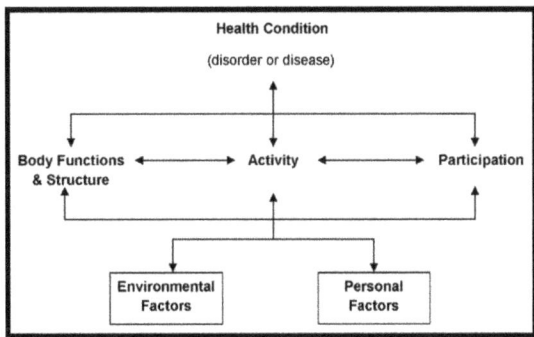

Figure 15.2   World Health Organization's International Classification of Functioning and Disability and Health [8]

Table 15.1 Impairment, measurement and treatment associated with spasticity. Adapted from [10].

ICF domain	Impairment	Measurement	Focal intervention	Global intervention
Body structure and function	Pain	Numerical pain rating scale Visual Analogue Scale	• Positioning • Cryotherapy • ES (i.e. TENS)	• Pharmacology (oral medication to reduce pain)
	Spasm	Spasm Frequency Scale	• Pharmacology targeted intramuscular injections to reduce spasticity and spasm, e.g. Botulinum toxin. • ES (i.e. TENS, FES) • Cryotherapy	• Pharmacology oral medication to reduce spasticity and spasm and pain. • Intrathecal pumps to reduce spasticity and spasm (Baclofen pump)
	Stiffness	• Modified Ashworth Scale • Range of motion assessment • Muscle length assessment	• Positioning • Stretching • Strengthening • ES • Botulinum Toxin • Cryotherapy	• Pharmacology • Oral medication to reduce pain and spasticity. • Intrathecal pump (Baclofen) • Positioning sleep system • General exercise and physical activity
	Contracture		• Positioning • Stretching • Pharmacology botulinum toxin • Posture management and sleep systems. • Exercise and movement (prevent progression) • Surgical intervention (e.g. tendon lengthening) • ES	• Pharmacology to reduce spasticity. • Intrathecal pump, e.g. Baclofen pump • Postural management and sleep systems
	Weakness		• Strengthening, e.g. progressive resistance strengthening	• Aerobic exercise • Functional training and mobility

(Continues)

		Outcome measures	Interventions	
		• Medical Research Council Muscle Grading Scale • Functional assessment • Standardised outcome measures	• Active range of motion • ES (i.e. FES)	• Self-care • Physical activity • Sport
	Balance	• Standardised outcome measures, e.g. single leg stand, BERG balance test	• Targeted static and dynamic balance activities • Progressive balance challenges • Progressive change to vestibular system	• Functional mobility • Physical activity • Strengthening
Activity	Functional mobility	• Standardised outcome measures, e.g. Barthel, Functional Independence Measures	• Progressive functional training of the impaired mobility, e.g. transfers, gait • Task-specific training • Strengthening of weakened muscles contributing to limited mobility • ES (i.e. FES)	• Physical activity • Aerobic exercise • Balance re-education
	Self-care	• Standardised outcome measure, e.g. Barthel Index, Functional Independence Measure	• Strengthening of weakness muscles contributing to impaired task • Task-specific training • Progressive functional training of impaired tasks, e.g. dressing, bathing • ES (i.e. FES)	• Physical activity • Aerobic exercise • Balance re-education
	Fatigue	• Standardised outcome measures, e.g. Fatigue Assessment Scale • Disease-specific measure, e.g. Modified Fatigue Impact Scale	• Pharmacology • Cognitive Behavioural Therapy • ES (i.e. FES)	• Graded exercise • Physical activity • Aerobic exercise • Pacing • Mindfulness/relaxation/meditation
Participation	Quality of Life	• EQ-5D-5L (Euroqual) • Disease-specific quality of life measure, e.g. Stroke Impact Scale	• Targeted treatment for anxiety and depressions • Cognitive behavioural therapy	• Support groups – disease/condition specific. • Physical activity • Carer support • Mindfulness/relaxation/meditation

The decision to treat spasticity, including using ES will depend on its presentation and concomitant impairments. For example, if a person reports difficulty using the upper limb for activities of daily living, interventions may include improving upper limb function, often by targeting impairments, such as strengthening, and task-specific practice combined with ES. If a person reports difficulty with mobility or gait, interventions may include FES, not only to improve function, but also to reduce calf spasticity secondary to a more normal gait pattern. FES may also be combined with cycling (especially in SCI) and in so doing may reduce spasticity in other muscle groups such as the quadriceps. Most interventions explored within this chapter, including ES, are appropriate for individuals that have spasticity and some level of voluntary control.

Spasticity can negatively impact on the individual's function, activities of daily living, self-care, mobility, and bowel and bladder function, psychological aspects and quality of life. For example, upper limb spasticity can inhibit an individual's ability to undertake self-care and hygiene. Whereas lower limb spasticity can impact on an individual's ability to walk and negotiate stairs. Spasticity can contribute to decreased or limited joint/limb movement and reduced mobility, thereby putting the individual at risk of developing a contracture and skin breakdown [11,12]. Spasticity should be managed and treated when it interferes with activity, self-care or hygiene [5]. Spasticity that is poorly managed can lead to changes in the biomechanical properties of the soft tissues and present as stiffness and lead to contractures and is often and inappropriately equated with non-neural spasticity. However, not all spasticity has negative consequences and in some cases, spasticity may help to facilitate activity such as transfers and gait, especially in children with cerebral palsy.

15.1.1   *Pathophysiology of spasticity and contractures*

The pathophysiological manifestation of spasticity results from neurological disruptions (i.e. neural components), however, the secondary consequences of immobility and spasticity can lead to concomitant change in the biomechanical properties of muscles and associated soft tissues (i.e. contractures) and the force production capacity (weakness and fatigue). Therefore, in practical terms one is often treating two or more clinical conditions concurrently, i.e. spasticity, contractures and/or weakness.

Damage to the UMN can alter the communication between the brain and the spinal cord, this is summarised below [2,13–17] and in Figure 15.3.

1. Disinhibition of the spinal reflexes including the withdrawal reflex and the stretch reflex due to changes in membrane potential and action potential threshold.
2. A natural response to weakness. Weakness occurs due to impaired descending input to the muscle leading to an impaired ability to voluntarily activate the muscle. Inability to contract a muscle can contribute to joint activity limitations. The body's natural response process to this reduced activity is to upregulate the excitability of the motor neuron to increase the output from the motor neuron and these are discussed below.

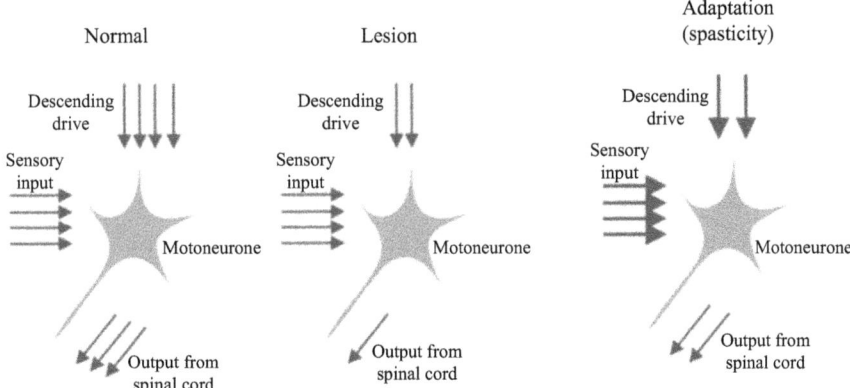

Figure 15.3   Adaptive changes in the cortical and spinal networks that contribute to spasticity from the work of Nielsen [18]

(a)   Altered spinal modulation (reciprocal inhibition – the activation of ago-nist muscles and relaxation or inhibition of antagonist muscles) and changes in supra spinal modulation. Altered spinal modulation occurs due to loss of descending motoneuron input (caused by UMN lesion) where the input is insufficient to maintain the normal output of the cell. To account for this change in input there is upregulation of membrane channels to increase the excitability of the membrane and decrease the threshold for action potential generation. Signals are received from sur-viving motor tract fibres, collateral sprouting and new synaptic site formation.

(b)   Changes in the properties of the motor neuron. With an UMN lesion there is loss of the excitatory drive to the motor neuron due to disruption of the descending tracts (reticulospinal and brain stem tracts). The motor neuron thus adapts to the reduced neural drive through changes to sodium and calcium channels that regulate excitability and action potential generation reducing the threshold for action potential generation.

(c)   Reduced post activation depression in people with spasticity.

(d)   Plasticity within the spinal cord can result from one or all the above

15.1.2   Muscle mechanical properties and spasticity

There are changes to the mechanical properties of spastic muscles that are not capable of normal movement. For example, atrophy of type II muscle fibres, structural changes in type I fibres [4] and a decrease in sarcomere numbers and increase in connective tissue and fat cells within the muscle [4,13,19]. Collagen cross bridges develop within tendons leading to an increase in stiffness. The col-lagen cross bridges occur because of normal tendon loading forces being lost in patients with weakness. These soft tissue changes can lead to shorter muscles, increased resistance to stretch and changes in joint position [4,13]. Moreover, the changes to the mechanical properties of the muscle can lead to additional stiffness and difficulty using the limb/joint which can lead to reduced use and the exacerbation of a contracture further compromising skin integrity and hygiene problems in those with severe deformities.

15.1.3   Measurement of spasticity

Spasticity measurement can be undertaken in three ways; electrophysiological measures, biomechanical measures and clinical measures [4,19]. Table 15.3 summarises the measures used in spasticity assessment. Despite the existence of a range of measures, there have been challenges to the measurement of spasticity due to the lack of consensus around a definition, measurement tools and lack of a core outcome set for measurement consistency across studies and in clinical practice [16]. The lack of a valid, reliable and commonly agreed outcome measures contributes to challenges in comparing the effect of interventions for spasticity, and accurately identifying if the intervention was successful in reducing spasticity.

Clinicians rarely measure spasticity directly (i.e. using a neurophysiological measure such as the H-Reflex or EMG that are discussed below). Neither do they use a controlled measure of stiffness which may provide indirect measures of spasticity under certain circumstances (i.e. a biomechanical measure). The primary reason being that these biomechanical measures are not easily implementable in clinical practice. Many clinical centres (and much of the research) use clinical measures such as a variation of the Ashworth Scale or the Tardieu scale to assess spasticity, despite these measures being invalid. However, if one were to start from a patient perspective it would make logical sense to use measures that are appropriate to the clinical problems described by the patient such as limitations in mobility, strength or self-care (i.e. Barthel Index, Functional Independence measure). Starting from a patient perspective and their reported clinical problems aligns with the biopsychosocial approach and holistic person-centred care. Changes in spasticity can indirectly be measured through changes in function. Table 15.3 describe the various outcome measures used to assess spasticity.

Electrophysiological measurement assesses or quantifies the electrical responses of the nerves/muscles to an externally imposed stretch or an electrical stimulus such as the H-Reflex or F-Wave [4]. A summary of the electrophysiological measures are shown in Table 15.3. If you want to know more about these reflexes, please refer to the work by Voerman and colleagues [20] and Malhotra and colleagues [16]. Electrophysiological measures may provide a valid measure of aspects of spasticity, but the reliability remains a challenge. These measures are also not practical in clinical practice due to the training required to conduct the assessment and the equipment needed [4]. Thus, clinicians do not routinely use them.

[20] Biomechanical measures assess the properties of muscles, limbs and joints in response to movement [9]. Traditionally stiffness (the extent to which the joint opposes the movement) can be an indirect measure of spasticity under certain

circumstances, i.e. when there are no contractures or biomechanical changes in the muscle and associated soft tissue structures. Stiffness is measured through the observation/measurement of how the limb responds during an externally imposed passive movement [2,4]. Biomechanical measures can be combined with ES to provide additional quantitative data around what is happening in the muscle, e.g. if it is trust spasticity or stiffness. For a summary of the biomechanical measures please see Table 15.2 and for a deeper dive in the different biomechanical measurement please see the paper by Wood and colleagues [21].

The challenges of biomechanical measurement are limb shape and size, it is measuring both neural and non-neural aspects of spasticity, it is difficult to separate these two factors in measurement as well as the need for an external rig to undertake the measurement these methods are not applicable to clinical practice [21].

Clinical scales assess a range of aspects of spasticity, assessing stiffness as an indirect measure of spasticity, measuring clinical presentations such as spasms and clonus, or the impact of spasticity on function, see Table 15.3 for a summary of measures. There are many clinical or observational scales available, with the most common clinical measure being the Modified Ashworth Scale (MAS) and Modified Tardieu Scale (MTS) [22]. The following section will discuss the MAS and Modified Tardieu Scale in detail, details of other scales used can be found in Table 15.3.

Table 15.2   Summary of the biomechanical measures of spasticity [21]

Category	Definition	Measurement	Practical considerations
Biomechanical	Assess the behaviour of muscles, limbs and joints in response to movement. Explore and measure stiffness which may be an indirect measure of spasticity	Manual instrumented – Stretching the limb passively and measuring stiffness by quantifying both the displacement and the force/moment required to stretch the joint. Controlled displacement – controlled joint rotation during which the opposing moment/ stiffness is measured. Controlled Torque – external torque placed on the limb – resultant joint angles are measured. Gravitational Method – lift limb into extension and release and observe the swing and angular joint motion.	These methods can be combined with EMG for more accurate and reliability findings. The assessments measure both neural and non-neural aspects of spasticity and it is difficult to separate the two.

Table 15.3 Summary of the clinical scales to measure spasticity

Category	Definition	Measurement	Practical considerations
Clinical scales	Clinical outcome measure used to assess spasticity which is completed by the health care practitioner	Spasm Frequency Scale-measures the frequency of spasms associated with spasticity in one day. Modified Ashworth scale and Ashworth Scale (MAS/AS) measure limb resistance to passive movement. The Tardieu Scale and Modified Tardieu Scales (TS and MTS) measure limb stiffness at varying velocities. Patient reported Impact of Spasticity measure (PRISM) spasticity impacts on quality of life Modified Composite Spasticity Scale measures tendon jerk and muscle resistance during passive movement	Easy to administer. The MAS/AS and TS/MTS are not valid or reliable this can be due to the experience of the rater and the limb/joint assessed.

The original Ashworth Scale was developed by Ashworth in 1964. Researchers Bohannon and Smith added an additional measurement 1+ to improve the sensitivity of the scale which is now the MAS [22], please see for complete details of the MAS and its measurement. The MAS is widely used in research and practice for its ease of use, and it does not require any specialist equipment or training thus it is accessible to clinicians. However, the MAS has varied reliability depending on the joint being measured and the clinical population, e.g. stroke, SCI, MS with claims the elbow is the most reliable joint [23–25]. The MAS does not incorporate the velocity component of spasticity [22,26]. Another clinical scale commonly used is the Tardieu Scale (TS) and Modified Tardieu Scale (MTS), which assesses resistance to passive movement at both slow and fast velocities. The TS and MTS utilise the same grading scale, however the MTS has additional modifications to standardise the use of the measure such as limb placement and limb alignment, describes the MTS. In Tardieu's preliminary work he identified that the stretch reflex was triggered at a specific speed, and that this speed would be different for different people thus incorporating different speeds of assessment [21]. The MTS incorporates an angle of catch and the speed at which the catch was obtained to describe the spasticity. The MTS has similar challenges around reliability as the MAS [25], demonstrating poor to good reliability with wide limits of agreement suggesting measurement variability and lack of precision [26,27].

The Spasm Frequency Scale is used to assess the frequency of spasms throughout the day. The scoring ranges from 0, no spasms, to 4 which is ten or more spasm per day or continuous contraction. The Spasm Frequency Scale has demonstrated good intra-rater and interrater reliability in people with SCI [28].

Often clinicians will also use functional outcome measures to evaluate if and how spasticity impacts on the person's ability to complete everyday activities or undertake mobility/gait. There are a range of different functional assessments that can be used g from generic assessments to disease-specific assessments each with their own advantages and disadvantages. Another important aspect to consider is how spasticity impacts on a person's quality of life or health related quality of life. This can be explored through patient reported outcome measures of health-related quality of life of which there are generic measures as well as disease-specific measures.

A weakness of the clinical scales is that they are not able to distinguish between spasticity and contractures and they may not be sufficiently sensitive to measure change in spasticity in response to an intervention. The evidence of the reliability of the MAS and MTS lacked reporting of the associated confidence intervals, which is problematic as the variability of the measurement is not known. In these circumstances, estimating the minimal clinical important difference (MCID), i.e. the minimal change needed in an outcome measure to be meaningful to a patient, is not possible.

Advantages of the clinical scales are ease of use, cost and accessibility in practice. Development of a core outcome set would facilitate standardisation of measurement for both research and clinical practice and facilitate the comparison between research studies for spasticity management and best practice outcomes. We recommend that clinicians use outcome measures and measure change in aspects that are important to the patient and that relate to the patients specific and individual problems, e.g. strength, mobility, self-care. Assessing these areas within the framework of the ICF (Figure 15.2 and Table 15.1) will ensure person centred and meaningful care.

15.1.4 Spasticity management

There are a range of interventions for spasticity management, and inconclusive evidence to support their efficacy. The aim of treating spasticity is aligned with the ICF and the goals of reducing pain, improving function and participation and pre-venting secondary complications such as skin breakdown and contractures. Interventions can be classified into rehabilitation, pharmacological and surgical. There are challenges in understanding the effectiveness of the different interven-tions, including ES, due to the complexity of the UMN syndrome, absence of reliable measures and lack of a clear understanding of the underlying mechanisms.

15.1.5 Rehabilitation interventions

Rehabilitation interventions often address a range of problems associated with the UMN syndrome as well as spasticity per se, such as contractures, muscle weakness, impaired motor control, balance, posture and gait. Because they are not used in isolation it is difficult to know which mechanism is contributing to change in function or spasticity. For example, ES could be used in conjunction with strength training or functional training to improve the muscle response (activation), functional activity ability, as well as spasticity. Spasticity can be affected by abnormal movement patterns and abnormal motor control. Improvement in motor control and reduction of abnormal movement patterns may therefore contribute to a reduction in spasticity. This is evidenced below in the literature in which, for example, FES can contribute to improved gait parameters as well as a reduction in spasticity. Cause, effect and interaction are unclear and understanding is hampered by paucity of effective outcome measures.

Rehabilitation interventions can reduce spasticity for a brief period but long-term spasticity may remain unchanged [16]. Short-term reduction in spasticity however provide a window to improve function.

Additional interventions for spasticity are pharmacological interventions and invasive surgical interventions, these are often used in conjunction with ES or rehabilitation interventions. For additional information regarding pharmacological and surgical interventions please refer to the work by Chang and colleagues [29], Lindsey and colleagues [30], Pandyan and colleagues [2], Winston and colleagues [31], Hashemi and colleagues [32] and Kakodkar and colleagues [33].

In summary, interventions for spasticity are not used in isolation, for example, using pharmacologic interventions or ES in conjunction with rehabilitation interventions such as exercise or prolonged stretching. When treating spasticity, the clinician needs to approach the person with spasticity holistically considering the ICF, the impacts of spasticity on the person's function and participation, if they are taking medication how this might impact on their engagement in rehabilitation, and the measurement tools available to assess spasticity and function. Considering these aspects and how spasticity is specifically impacting the person will help the clinician to identify which combination of interventions to use and how to measure progress. In an ideal world we would use a combination of interventions such as BoNTA combined with ES and an orthotic such as a contracture correction device (CCD). However, this does not happen in clinical practice outside of specialist spasticity clinics.

15.2   Review of the evidence for ES to manage spasticity

There are many terms used to describe ES, which are defined in the Introduction to the book. The following section will use the term ES to refer to the therapeutic use of ES, e.g. mobilising the joint through full range of motion; the term FES will refer to ES used during a functional activity, e.g. walking or reaching; and the term TENS to refer to stimulation at a sensory level that does not cause a motor response, e.g. to manage pain.

ES can modulate spasticity in several ways. Firstly, through stimulation of the non-spastic antagonist, which can inhibit the spastic muscle via the reciprocal

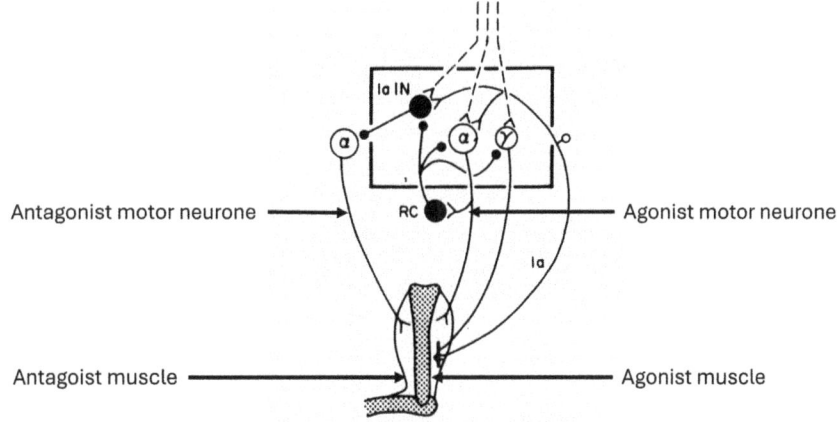

Figure 15.4   Possible di-synaptic reciprocal inhibitory pathway

inhibition pathway. Reciprocal inhibition is a normal spinal process that improves efficiency of muscle activity by 'switching off' the antagonist when the agonist is active. Causing a muscle contraction by ES can have the same inhibitory effect on the spastic muscle. Figure 15.4 illustrates how ES to the non-spastic triceps brachii muscle can inhibit the spastic biceps brachii. There are numerous other examples of how this can be used clinically, for example, ES to the ankle dorsiflexors during walking using a drop-foot stimulator can inhibit the spastic calf muscle and stimulation of wrist extensors, facilitate reciprocal inhibition of spastic wrist flexors. ES along with a muscle contraction enhances the impact of reciprocal inhibition [34], and although the effects on the spastic muscle are likely to be short lived [14], they may result in a more normal movement pattern, which if learnt can lead to positive neuroplastic changes, rather than abnormal neuroplastic changes and increased spasticity.

Activation of the agonistic muscle leads to feedback via the Ia afferent neuron. The Ia afferent generates an excitatory signal via a mono-synaptic connection to the agonistic muscle and simultaneously provides and inhibitory signal via a di-synaptic (IaIN) connection to the antagonistic muscle. It is believed that ES of the agonistic muscle can lead to a reduction in spasticity in the antagonistic muscle via this di-synaptic pathway. Second, ES can be applied to a spastic muscle (at a sensory level or higher to achieve a muscle contraction). This approach potentially reduces spasticity and stiffness through autogenic inhibition which is a reduction in the excitability of a contracting or stretched muscle [35]. Autogenic inhibition can be through reduction of an overactive alpha motor neuron or post activation depression decreasing the likelihood of another depolarisation. Again, the effects on the muscle and spasticity are likely to be short lived [35].

Thirdly, ES can reduce pain and may promote positive neuroplastic changes and changes to function and spasticity. There are different theories on the mechanism of action for TENS in pain reduction. For example, TENS activates afferent fibres to the cortex which in turn activates inhibitory systems thereby decreasing pain. Another example is high frequency TENS (>50 Hz) that can increase endorphins in the cerebrospinal fluid [36].

There is some evidence that ES can reduce muscle spasms (in cerebral palsy) as measured with the Spasm Frequency Scale [37]) and in people with SCI as measured with the Spinal Cord Assessment Tool for Spastic Reflexes [38]. This is thought to be through a reduction of spinal excitability due to accommodation of the ES.

People with spasticity also have a range of other UMN impairments which interact and impact on movement control. Therefore, there is an interaction between UMN impairments such that, if ES of weaker muscles reduces antagonist spasticity, this may allow for a more normal movement pattern to be use for a period linking back to reciprocal inhibition. In this way ES may impact positively on spasticity and neuroplasticity [9]. Finally, the long-term effects of rehabilitation interventions are largely unknown, including the long-term impact of ES. Whilst evidence suggests that treatment effects of existing protocols are short lived, it is possible that more intensive or longer lasting treatment may have different effects. This is likely to be the case when rehabilitation has a positive influence over movement performance and functional tasks, but to what extent changes are due to natural recovery and what to do with ES or other interventions remains unknown.

The following section discusses the current evidence for using ES as a stand-alone intervention or in combination with other rehabilitation interventions for people with SCI, stroke and MS. Varied stimulation protocols and settings were used throughout the research. For a summary of the stimulation protocols used in the literature please refer to Table 15.4.

15.2.1   *ES and spasticity in SCI*

SCI results from damage to the spinal cord (usually around 2-3 segments) including the long descending (efferent) tracts that take motor information from the CNS to the periphery and ascending (afferent) tracks bring sensory information from the periphery to the CNS. SCI can be a complete injury in which there is loss of nerve signal communication from the level of injury distally, or incomplete injury in which some motor and sensory signals are transmitted leading to some loss of motor and sensory function. There is some evidence that the level of injury and type of injury may be associated with spasticity. For example, around 93% of individuals with an ASIA (American Spinal Injury Association Impairment Scale) A cervical injury and 78% of ASIA B-D cervical injuries have spasticity [39]. The ASIA scale is used to describe and classify SCIs based on the type of injury and associated impairments. The area or level of the spinal cord which is injured impacts on the presentation and impairments. SCI involving the cervical spinal cord present with weakness and sensory changes of the upper limbs, trunk and lower limb, respiration can also be affected. SCI involving the thoracic spinal segments can present with weakness and sensory changes to the trunk and lower

Table 15.4   ES settings used in the literature

Type of stimulation	Stimulator settings	Frequency	Muscles of stimulation	Population	Study/references
FES	35 Hz pulse width 280 μs in constant mode	20–30 minute 12 Weeks FES during Gait	Peroneal nerve and anterior tibialis motor point	Stroke	Sabut et al. (2011)
	35 Hz Pulse width 350 μs Intensity to a motor response of acceptable personal tolerance	15–20 minutes Three times week three weeks FES during gait	Quadriceps, hamstrings, peroneals and plantar flexors muscle groups.	Stroke	Hakakzadeh et al. (2021)
	Biphasic 40 Hz pulse width between 30 and 350 μs amplitude up to 100mA	Wear for most of the day. 12 weeks	Peroneal nerve	Stroke	Johnson et al. (2004)
	50 Hz Pulse width 200 μs Delivered for six seconds (ramp up 1 second, ramp down two seconds) 12 second burst Intensity to elicit finger extension without pain	30 minutes a day Five days a week Four weeks	extensor digitorum communis	Stroke	Lee et al. (2018)
	33 Hz Intensity up to patient tolerance	30–45 minutes Four times per week Two weeks FES with cycling	Quadriceps, hamstrings and gluteals	SCI	Ralston et al. (2013)
	Biphasic rectangular pulses Frequency of 35 Hz Pulse width 300 μs Ramp up time – three seconds Hold duration five seconds Ramp down time two seconds Rest time ten seconds	One session of 30 minutes	Quadriceps and adductor muscle groups Plantar flexors (motor points)	Spinal cord injury	Sivaramakrishnan et al. (2018)

(Continues)

Table 15.4 ES settings used in the literature (Continued)

Type of stimulation	Stimulator settings	Frequency	Muscles of stimulation	Population	Study/references
	50 Hz Pulse width 200 μs Intensity to participant tolerance	30 minutes Three times a week Four weeks	quadriceps, hamstrings and gluteal muscles	Multiple sclerosis	Backus *et al.* (2020)
	Constant current 30 Hz Pulse width 300 μs Intensity to induce muscle contraction	Six sessions over two weeks FES with cycling	Quadriceps hamstrings	Multiple sclerosis	Szecsi *et al.* (2009)
Surface NMES	Russian current: 50 Hz Phase duration 200 μs; On time: ten seconds, Off time: 50 seconds Intensity to muscle contraction Faradic current: 100 Hz Pulse duration 100 μs Pulse interval 900 μs Surge duration: four seconds Rest: six seconds	Ten minutes Five days a week Six weeks	Common peroneal nerve and anterior tibialis (motor point)	Stroke	Ganesh *et al.* (2014)
	Biphasic symmetrical rectangular waves 20Hz Pulse duration 200 μs Intensity to maximum tolerable contraction	20 minutes Six days per week One month	Gastrocnemius Achilles tendon junction	Stroke	Chen *et al.* (2005)
	40 Hz Pulse width 300 μs On time 15 seconds Off time 15 seconds Six second ramp up and six second ramp down Intensity to maximum range of wrist and finger	30 minutes 2–3 times a day Five days a week Six weeks	Wrist and finger extensors – common extensor digitorum	Stroke	Malhotra *et al.* (2013)

(Continues)

	Parameters	Duration	Target	Condition	Reference
	50 Hz Pulse duration 200 μs One second ramp up Two seconds ramp down 12 seconds burst Intensity to elicit finger extension	30 minutes Five days a week Four weeks	Finger extensors	Stroke	Lee et al. (2018)
	Three protocols All had: Pulse duration 250 μs Duty cycle 40 seconds on and ten second off Intensity slightly above motor threshold Varied frequencies 50 Hz; 100 Hz; and 200 Hz	One 30-minute session	Anodal electrode on the tibialis anterior muscle belly and cathode on the deep peroneal nerve	Stroke	Koyama et al. (2015)
	Muscle contraction protocol (agonist and antagonist): 30 Hz Pulse width 300 μs Burst duration four seconds Ramp up one second Off time four seconds Intensity 300% of motor threshold Sensory protocol: 30 Hz Pulse width 100 μs Burst duration four seconds Ramp up one second Off time four seconds Intensity at 80% motor threshold	All protocols delivered for 45 minutes (each protocol one session)	Tibialis anterior (agonist) Or triceps surae (antagonist)	Spinal cord injury	Van der Salm et al. (2006)
TENS	Square pulses 100 Hz Pulse duration 50 ms Intensity 2–3× sensory threshold	30 minutes vs 60 minutes	Common peroneal nerve and tibialis anterior (motor point)	Stroke	Laddha et al. (2015)
	100 Hz Pulse width 200 μs Intensity 2× sensory threshold	20 minutes 5 days a week 6 weeks	Peroneal nerve	Stroke	Jung et al. (2017)

(Continues)

Table 15.4 ES settings used in the literature (Continued)

Type of stimulation	Stimulator settings	Frequency	Muscles of stimulation	Population	Study/references
	100Hz Pulse width 125 μs Continuous mode Below motor threshold but at a comfortable sensation	60 minutes or eight hours Two weeks	Quadriceps	Multiple sclerosis	Miller *et al.* (2007)
	100 Hz Pulse width 300 μs (Applied to the spinal cord level of the muscle with spasticity)	20 minutes Daily Four weeks	Gastrocsoleus	Multiple sclerosis	Armuthu *et al.* (2003)
	Symmetric biphasic rectangular waves 100 Hz Pulse duration 200 μs Intensity 2× sensory threshold	50 minutes Five days a week Three weeks	Bilateral common peroneal nerves	Spinal cord injury	Oo *et al.* (2015)
	Biphasic square wave pulses Frequency 100 Hz Pulse duration 300 μs Intensity increased until further intensity cause discomfort or muscle contraction	One session of 30 minutes	Quadriceps and adductor muscle groups Plantar flexors (motor points)	Spinal cord injury	Sivaramakrishnan *et al.* (2018)
	Sensory protocol: 30 Hz Pulse width 100 μs Burst duration four seconds Ramp up one second Off time four seconds Intensity at 80% motor threshold	45 minutes (one session)	Tibialis anterior (agonist) Or triceps surae (antagonist)	Spinal cord injury	Van der Salm *et al.* (2006)
	100 Hz	15 days	Tibial nerve	Spinal cord injury	Aydn *et al.* (2005)

limbs. SCI in the lumbar region (below T12) are associated with weakness and sensory changes to the lower limbs. Injury to associated lower motor neurons (LMN) can impair or destroy the reflex arc as well as communication between the UMN and LMN. This can result in paralysis to the lower limbs as well as loss of bowel and bladder function.

15.2.1.1   ES used in conjunction with other interventions

Much of the research in SCI is around using ES in conjunction with other therapeutic interventions. There is inconclusive evidence of the efficacy of ES as an intervention for spasticity in people with SCI. The evidence of using ES with other interventions has been summarised in three systematic reviews. The first systematic review of ES alongside physiotherapeutic interventions for spasticity in people with SCI, demonstrated through a meta-analysis, that there was no significant difference in spasticity following ES [40]. However, three studies demonstrated a decrease in spasticity post ES as measured by various clinical measures. For example, a crossover randomised controlled trial (RCT) by Ralston [41] compared FES during cycling for 2 weeks versus a control phase of no FES or cycling with one week washout. The experimental FES interventions demonstrated a decrease in spasticity as measured with the MAS and Patient Reported Impact of Spasticity Measure (PRISM) [41]. Alternatively, Alashram and colleagues (2023) completed a systematic review of FES combined with cycling post SCI (without meta-analysis) [42]. The authors identified that six of the nine studies demonstrated a significant decrease in spasticity as measured with a range of clinical scales, e.g. MAS post FES and cycling. The simulation parameters ranged from frequencies between 20 and 50 Hz, pulse width between 250 and 00 μs, at a range of amplitudes to generate a muscle contraction [42]. Massey and colleagues conducted a systematic review and meta-analysis of five RCTs and quantitative synthesis of 17 non-RCT studies exploring ES (FES, TENS, transcutaneous spinal cord stimulation) to reduce spasticity in people with SCI [43]. The meta-analysis of three studies using the MAS demonstrated that ES was effective at reducing spasticity compared to the control group. However, there was not a statistically significant difference (n=2) between the ES and control groups when spasticity was measured with the pendulum test. Quantitative synthesis demonstrated that 22/29 studies demonstrated an improvement in spasticity following ES, of which 13 studies demonstrated significant improvements in spasticity [43].

The use of TENS alongside physical therapy in people with SCI has been explored. Oo and colleagues [44] conducted an RCT exploring TENS (60-minute sessions) (n=8) prior to physical therapy compared to physical therapy alone for the control group (n=8). Immediately following the first session and after the final session there was a significant reduction in spasticity as measured with the composite spasticity score, as well as a significant reduction in ankle clonus after the final session [44]. Similarly, Aydn [45] explored TENS (n=11) versus oral baclofen (n=10) demonstrating that TENS contributed to a significant decrease in spasticity (as measured by the MAS), a decrease in the H-reflex maximum amplitude and an improvement in function. The outcomes of the TENS group were similar to the group that received Baclofen without the unwanted side effects [45]. This high-lights the benefits of TENS versus oral medication.

Researchers have explored the effects of one session of stimulation on spasticity. Sivaramakrishnan and colleagues [46] completed a randomised crossover study with ten participants exploring TENS versus FES, separated by 24 hours between sessions. The results identified there was a significant reduction in spasticity as measured with the MAS in the hip adductors and knee extensors for up to four-hour post stimulation (FES and TENS) with TENS having a greater decrease in spasticity. There was not a significant decrease in plantar flexor spasticity [46]. The authors also explored the length of time post intervention for which spasticity remained reduced demonstrating that following FES there was a spasticity reduction for up to four hours, following TENS there was a reduction up to 24 hours [46].

Researchers have explored the impact of stimulating different muscles using motor vs sensory stimulation and the impact on reducing spasticity [35]. Van der Salm and colleagues (2006) [47] undertook a placebo-controlled study with repeated measurement with ten participants with SCI exploring three stimulation protocols – agonist ES, antagonist ES and a sensory stimulation to reduce triceps surae spasticity. The placebo intervention utilised electrodes and a stimulation unit but the unit was not durned on. Stimulation of the agonist muscle resulted in a significant decrease in triceps surae spasticity immediately post treatment which was associated with a significant reduction in reflex sensitivity.

15.2.1.2   ES and spasticity in stroke
Stroke is injury/ischemia to the brain due to lack of blood flow (loss of oxygen) to brain tissue resulting from a thrombus (ischaemic stroke), or rupture of a blood vessel (haemorrhagic stroke) leading to bleeding in the brain.

There is a lack of long-term follow up in rehabilitation studies, including studies of ES. Often the final assessment is close to when the intervention ends. Researchers have identified key time points for assessment of sensorimotor function in response to interventions to be at three months, six months and 12 months post stroke [48]. However, when considering the longer-term impacts of ES in people after stroke, there is evidence that there is a change in spasticity at the end of the treatment period which lessens with time (reduction in effect one hour post intervention and no reduction 24 hours after ending the ES) [49]. Long -term follow up can be challenging due to morbidity following stroke, individuals moving into supported care facilities, and the shorter-term mechanism of action of rehabilitation interventions. As identified with the research SCI, the stimulation parameters and dose of stimulation in people with brain injury is variable between studies. ES to reduce spasticity has been used both as a stand-alone intervention as well as combined with other pharmacological and therapeutic interventions. Spasticity is not treated in isolation but as part of a group of impairments. These will be discussed separately below.

15.2.1.3   ES as a stand-alone intervention
The prime aim of many of the studies using ES is not always to reduce spasticity, but to improve function, strength or motor control. Assessment of spasticity is

therefore a secondary outcome measure, as spasticity reduction can contribute to improved function and motor control.

Chen and colleagues [50] conducted a trial with 24 participants; 12 in the experimental group receiving ES to the gastrocnemius and 12 in the control group which received sham stimulation. The experimental group received ES for 20 minutes, six days per week, for 1 month (four weeks) [50]. The findings demonstrated 8–12 participants had a decrease in ankle plantar flexor spasticity as measured by the MAS, however significance was not explored. There was a significant decrease in the Hmax and Mmax ratio (determined by the maximum H reflex divided by the maximum M wave and representing reflex excitability) and significant increase in the H-reflex latency post ES. These findings demonstrate changes to neurophysiological mechanisms that contribute to spasticity, such that there was a reduction in reflex excitability which can contribute to reduced spasticity [50]. Alongside a reduction in spasticity the participants who received ES had a significant decrease in the time to complete the 10 m walk test demonstrating the link between spasticity and function.

Yuzer and colleagues (2017) undertook a controlled trial (n=30 participants) to determine the effect of FES on wrist spasticity after stroke compared to a usual care control group [51]. FES was FES was applied for 30 minutes, 5 days a week for 20 sessions. Post intervention 80% of participants had an improvement in MAS from baseline however significance was not explored [51].

Koyama explored TENS applied to the anterior tibialis muscles and deep personal nerve in 20 people with chronic stroke for one session of 30 minutes at varying TENS frequencies (50, 100 and 200 Hz) [52]. Electrophysiological measures were explored including the H reflex, reciprocal inhibition and D1 (alpha motor neuron) inhibition. None of the protocols impacted on reciprocal inhibition, however D1 alpha motor neuron inhibition improved significantly following 200 Hz TENS. These findings suggest that 200 Hz TENS contributed to an increase in presynaptic inhibition of alpha motor neurons in the soleus muscle which may be useful in reducing spasticity [52].

15.2.1.4   ES used in conjunction with other interventions

Rehabilitation interventions for UMN impairments including spasticity are not used in isolation, but as part of a holistic approach. As previously mentioned, spasticity would be treated as part of a cluster or UMN impairments with a range of interventions. The following section discusses using ES in conjunction with other interventions such as gait training, BoNTA injections, task-specific practice and usual care as would be done in clinical practice. The main goal of the studies discussed below is not always to reduce spasticity but to improve function, motor control and quality of life, with spasticity reduction a secondary outcome measure.

15.2.1.5   FES and gait training

FES has been used in conjunction with gait training. For example, a case series (n=6 participants) explored FES to the quadriceps, hamstrings, peroneii and plantar flexors muscle groups with gait treadmill training for three weeks (ten minutes week one, increased to 15–20 minutes week 2–3). The results demonstrated in no significant change in ankle plantar flexor or knee extensor muscle spasticity as measured by the MAS [53]. However, significant changes in active ROM and gait speed were present following the FES gait training intervention [53]. Alternatively, Sabut and colleagues undertook a prospective interventional study with 51 participants. The experimental group (n=27) received FES with gait training for 12 weeks (20–30-minute treatment sessions) compared to the control group receiving convention rehabilitation (n=24) [9]. The findings demonstrate a significant reduction in plantar flexor spasticity (37.5% decrease in FES group versus 21.2% decrease in the control group) [9]. The difference in the outcomes between the two studies may be the length of intervention three weeks versus 12 weeks, the electrode placement, the muscles of investigation.

15.2.1.6   Stimulation and BoNTA

Research supporting the use of ES in conjunction with botulinum toxin type A (BoNTA) is mixed. Johnson and colleagues [54] undertook an RCT(n = 18) exploring BoNTA injections to the medial and lateral heads of the gastrocnemius and the tibialis posterior along with FES for foot drop (n = 10) compared to usual care physiotherapy (n = 8) [54]. The findings demonstrated a downwards trend in spasticity of the gastrocnemius and quadriceps femoris muscles as well as improvement in walking speed, neither was statistically significant [54]. Lee and colleagues (2018) undertook a pilot study with 15 people with chronic stroke exploring the use of BoNTA and ES for the upper limb [55]. The results demonstrate a significant reduction in wrist flexor spasticity as well as improvements in hand function and ROM [55]. Picelli and colleagues (2021) [56] completed a systematic review exploring the combination of BoNTA and ES. Their findings demonstrate that of the nine studies included, three found no improvement in spasticity or function combining ES with BoNTA; however, the other six studies demonstrated improved spasticity or function when combining BoNTA with ES versus the control treatment [56]. The mechanism of action of combining ES with BoNTA is thought to be due to the ES 'boosting' the effects of BoNTA through enhancing the uptake process at the presynaptic cholinergic nerve terminals [56].

15.2.1.7   Stimulation and task-specific practice

Jung and colleagues (2017) completed a RCT of 40 people after stroke to explore TENS combined with sit-stand training to improve balance and function versus sham stimulation [57]. The findings demonstrated the TENS group had a significantly greater decrease in spasticity as measured with the composite spasticity score compared to the control group alongside significant improvements in postural sway and hip extension [57].

Laddha and colleagues completed a three arm RCT (n = 3) exploring TENS with task-oriented exercises to reduce ankle plantar flexor spasticity [58]. This study comprised three groups group 1 – task-oriented training, group 2 – task-oriented training with 30 minutes of TENS and group 3 – task-oriented training with 60 minutes of TENS; spasticity was measured using the Modified Composite

Spasticity Scale [32,58]. All three groups demonstrated a mean reduction in ankle spasticity, between group comparisons show the greatest mean change in spasticity was greater in group 3 (60 minutes of TENS), followed by group 2 (30 minutes of TENS) and group 1 no TENS. The reduction in group 3 revelated a significant minimal clinical dateable difference which was determined by [58]. the clinical importance (standardised effect size) through 'dividing the mean difference un outcome between the intervention and control groups by the combined standard deviation of the two groups' [58].

Ganesh and colleagues 2017 [59] conducted a RCT (n=83) to explore two types of stimulation current on plantar flexor spasticity ES versus Russian current. Both stimulation protocols were alongside. task-oriented practice [59]. The results were that both Faradic and Russian currents resulted in a significant decrease in MAS for ankle plantar flexors (soleus and gastrocnemius) with small to medium minimal detectable clinical difference noted favouring Faradic current [59].

15.2.2 Stimulation with routine therapy after stroke

ES has been used in conjunction with routine rehabilitation with varying outcomes. Malhotra and colleagues (2013) undertook secondary data analysis of a RCT (n = 90). ES to the wrist and finger extensors for 30 minutes, twice a day for five days a week, for six weeks in people with acute stroke [14]. The findings demonstrated there was no change in wrist flexor stiffness and abnormal muscle activity measured via EMG, in other words the stimulation protocol had no impact on spasticity [14].

A pilot RCT (n = 19) by Almutairi and colleagues explored ES to the dorsi-flexors (n=10) for 30 minutes three times a week for four weeks with usual care versus usual care with sham stimulation (n=9) for people after stroke to reduce spasticity [60]. The results demonstrated a significant improvement in spasticity as measured with the MAS as well as significant improvements in the 10 m walk test and Functional Ambulation category in the ES group [60].

A systematic review and meta-analysis included 12 RCTs of ES combined with rehabilitation interventions demonstrated a significant improvement in spasticity [61]. The authors also identified that five RCTs that applied NMES of the leg muscles demonstrated a significant improvement in spasticity, whereas 6 RCTs applied ES to the wrist and 4 RCTs applied ES to the elbow with no significant improvement in spasticity [61].

A systematic review of TENS to reduce spasticity was conducted by Mahmood and colleagues (2018) including 15 studies in the narrative synthesis and seven studies in the meta-analysis [62]. Synthesis of the evidence identified that TENS parameters varied between studies, 14/15 studies used 99 Hz or 100 Hz frequencies, one study used 50 Hz frequency. Nine of the studies used a TENS intensity between two and three times the sensory threshold; while three studies used intensities up to 60 mA, one used subsensory threshold, and two used a bearable pain threshold. Most studies used TENS in square pulses with a pulse width of 200 or 100 μs. Placement of electrodes varied between placing the electrode on the muscle with spasticity, the muscle antagonist to the muscle with spasticity, along the nerve or acupressure points. Of the various electrode placements, electrodes placed over the acupressure points were not effective in reducing spasticity. Interventions ranged between 2 and 12 weeks in most studies. The number of TENS sessions ranged from a single session per week to seven sessions per week. The duration of TENS ranged from less than 20 minutes to 60 minutes. Overall, the meta-analysis along with the narrative synthesis identified that TENS caused a significant reduction in spasticity [62]. Mahmood's findings are supported by a review of systematic reviews identifying moderate quality evidence that TENS and NMES as an adjunct to physiotherapy can help to reduce spasticity in people with chronic stroke [63]. The work by Mahmood highlights the variation in stimulation parameters used, length of intervention and electrode placement, despite these differences overall TENS had a positive impact on spasticity. The most common frequency used was around 100 Hz, electrode placement along the muscle belly or nerve path was effective, and there were no negative or adverse effects on spasticity from using TENS.

FES is being combined with new technology such as computer brain interface interventions. For example, 51 people with stroke and upper limb impairment underwent three months of brain compute interface training (wrist extension tasks) with electroencephalogram (EEG) and FES for the wrist extensors. The outcomes of the study included a significant improvement in spasticity measured with the MAS and a significant improvement in upper limb function (Fugl-Meyer) [64].

Future work is needed to identify the most effective stimulation protocol(s) and develop a common stimulation protocol to ease comparison between studies and help clinicians to use ES in the most effective way.

15.2.3   ES and spasticity in MS

MS is a progressive neurological condition arising from demyelination of the central nervous system. There are three main types of MS (1) primary progressive – progressive decline in function; relapsing remitting – (2) period of exacerbation and recovery which overtime result in a decline in function; and secondary progressive – (3) when relapsing remitting becomes a progressive decline [2]. Impairments associated with MS are related to the area of the CNS that is demyelinated, example impairments include weakness, sensory changes and spasticity which can impact on the person's everyday activities and quality of life.

ES to reduce spasticity has been used both as a stand-alone intervention as well as combined with other pharmacological and therapeutic interventions, please see appendix one for a summary of stimulation settings.

15.2.4   Stimulation as a stand-alone intervention

Miller and colleagues [65] have explored the use of TENS at the quadriceps muscle at different doses eight hours versus 60 minutes per day to reduce spasticity in a crossover study design (n=32) [33]. There was a small non-significant reduction in the Global Spasticity Scale (GSS) following 60 minutes of TENS, and larger but non-significant reduction in GSS following eight hours of TENS [65]. Furthermore, the group that received 8 hours of TENS had a significant reduction in muscle spasms and pain [65].

Armutlu and colleagues explored the use of spinal TENS for plantar flexor spasticity in ten people with primary and secondary progressive MS [66]. The results demonstrated a significant reduction in spasticity as measured by myoe-lectrical activity (reduction in muscle amplitude via EMG) and the MAS [66].

The difference in findings between in these two studies could be the type of stimulation Armutlu and colleagues utilised spinal stimulation identifying a sig-nificant difference in spasticity, whereas Miller used peripheral stimulation without a significant change in spasticity. More research is needed to explore the impact of spinal stimulation versus peripheral stimulation and the combination of both in the reduction of spasticity and impact on function.

The use of stimulation to reduce spasticity has been compared to pharmaco-logical interventions such as Baclofen. Shaygannejad and colleagues completed a RCT (n = 52) with one group receiving TENS and the other group receiving Baclofen to reduce spasticity as measured with the MAS [67]. Following four weeks, both demonstrated a statistically significant reduction in spasticity. The TENS group also had a statistically significant reduction in MAS compared to the Baclofen group; the Baclofen group reported side effects (e.g. dizziness, drowsi-ness and gastrointestinal) [67]. This research suggests that TENS was more effec-tive in reducing spasticity in this group of people with MS without the unwanted side effects of taking Baclofen.

15.2.5   Stimulation in conjunction with other therapeutic interventions

15.2.5.1   FES and cycling

Combining FES with cycling is a common intervention for people with MS that have higher levels of impairment and lower extremity weakness as it can contribute to functional and cardiovascular improvements both of which are important health benefits [68]. For example Szecsi and colleagues (2009), explored FES and cycling alongside routine physiotherapy over six sessions in two weeks in 12 people with MS [69]. The findings identified a significant reduction in spasticity pre/post the FES cycling sessions measured using the MAS, but no significant long term (first to last training day) reduction in spasticity [69].

Backus and colleagues (2020) completed a pilot study (n=14) exploring FES with cycling in people with MS who were non-ambulatory, training three times a week for four weeks [70]. The findings demonstrate no significant difference in lower limb spasticity. However, the participants did demonstrate improvements in pain, the social subscale of the SF-36 and were able to cycle against increased resistance over the four weeks [70].

A systematic review identified nine papers exploring FES and cycling in people with MS of which five investigated the impact of FES and cycling on spasticity. Of these five papers, three demonstrated a non-significant reduction in spasticity, two demonstrated a significant reduction in spasticity [68]. The studies in the systematic review employed stimulation at lower frequencies between 33 and 45 Hz, treatment were sessions between once and three times a week, with interventions ranging from 2 to 24 weeks. As in the research in people with stroke and SCI, the optimal stimulation protocol needs to be developed and shared as best practice. This will contribute to improved clinical decision making for clinicians and facilitate comparison between studies.

15.3   Clinical considerations for ES

There are many clinical considerations when employing an ES protocol. First, is electrode placement on the agonist muscle, antagonist muscle or along the nerve pathway. Most of the research discussed in this chapter utilised electrode placement along the agonist (hyperactive) muscle or the nerve pathway. Stimulating the hyperactive muscle can contribute to reduced activation through autogenic inhibition of Renshaw cells. Stimulating the antagonist muscle reduces activation through linking neural pathways stimulating the non-spastic muscle helps to decrease stiffness in the spastic muscle. The latter stimulating the antagonist muscles reduces spasticity through reciprocal inhibition where there is agonist alpha motor neuron inhibition due to a contraction of the antagonist muscle. Stimulating the muscles to achieve a muscle contraction may also contribute positively to the biomechanical properties of the muscle leading to reduced spasticity.

   Although the evidence presented is inconclusive as to the efficacy of ES in reducing muscle spasticity, ES is associated with minimal adverse events or reactions and there are none of the systematic side effects that are often seen with oral medications to reduce spasticity. The evidence demonstrates that even when spasticity is not significantly improved, there are other benefits for the person with spasticity, such as improved pain or function which can contribute to improved overall quality of life.

   A range of stimulation settings were used across studies which is a weakness of the current evidence base. Moving forward, identifying the most effective stimulation protocols can could? lead into practical guidelines for research and clinicians to ease clinical decision making. Three case studies are detailed below, these do not represent best practice, but the stimulation parameters most used that resulted in a decrease in spasticity.

15.4   Future directions of engineering and scientific research and development, clinical applications and use

There are many different directions for scientific research and development of ES for clinical use. The complexity around UMN lesions, spasticity and rehabilitation interventions continues to be an area of interest for researchers. Unpicking the relationship between impairments and their impact on function, as well as the mechanism of action of how ES and other rehabilitation interventions impact on spasticity is needed. Neuroscience research is continuingly developing and changing area, future work developing a better understanding of how ES might impact on neuroplasticity and recovery is needed.

The development of sensitive and specific outcome measures for spasticity that can be applied in clinical practice and research is needed. The ICF and a holistic approach to spasticity management should be considered when developing spasticity assessment tools. Additionally, a core outcome set for spasticity measurement in research would standardise outcome measurement, ease comparisons between studies, improve our understanding of both the short-term and longer-term effects of ES, all of which would contribute to improved clinical decision making around spasticity management.

New and developing ways of using ES to manage spasticity are emerging in the research. Examples are combining surface spinal activated ES with peripheral activated ES and using ES in combination with computer brain interface interventions [71] demonstrating benefits for managing spasticity.

Implantable spinal cord stimulation devices may be beneficial in the for people with complete SCI to reduce spasticity, more research is needed [72]. Other implantable devices are an implantable wireless neurostimulator devices such as the radio frequency microstimulator (RFM) are being explored. The RFM is a microstimulator that is implanted near a nerve or motor point through a minimally invasive procedure which can then deliver ES. Following a 12-week intervention using RFM people with stroke demonstrated an improvement in upper limb function and spasticity [73].

A developing subtype of ES called balanced-charged kilohertz (kHz) frequency alternating current has the potential to block verve conduction and thus may be useful in reducing spasticity; however more research is needed to understand the mechanisms of action of kHz stimulation [74].

There are new and exciting developments in technology such as artificial intelligence (AI) and we do not yet know how this will be used in neuroscience and in understanding ES or spasticity.

The following case studies are simulated patients based on data from clinical experience or research studies.

15.5   Case studies

15.5.1   Case study example 1

15.5.1.1   Background

A person with a posterior circulation stroke of the posterior communicating artery one month ago. The person presents with left lower limb weakness rated 4/5 on the Oxford Grading Scale, left upper limb weakness rated 3/5 on the Oxford Grading Scale throughout their upper limb, and decreased balance. They have decreased fine motor control in their hand with difficulty releasing objects, object manipulation and doing fine selective movements of the fingers. The person has spasticity in their left elbow and hand without limb deformity, however the hand often rests in flexion but can be opened and stretched into extension. Muscle isolation was not possible. Using the MAS to assess spasticity they are rated a 2/4 with resistance felt through the second half of the available range of motion. The spasticity and weakness are impacting on their everyday activities such as manipulating a fork, knife and toothbrush, difficulty with opening cupboards deformity, however the hand often rests in flexion but can be opened and stretched into extension. Muscle isolation was not possible. Using the MAS to assess spasticity they are rated a 2/4 with resistance felt through the second half of the available range of motion. The spasticity and weakness are impacting on their everyday activities such as manipulating a fork, knife and toothbrush, difficulty with opening cupboards and doors with their left hand, doing up buttons and opening containers.

15.5.1.2   Clinical indication
Upper limb spasticity and weakness that is impacting on their ability to do everyday activities such as dressing and eating.

15.5.1.3   Outcome measures

- MAS – left elbow and hand 2/4
- Action Research Arm Test – 37/57 demonstrating moderate ability
- Bartel Index – 60/100
- Stroke Impact Scale – 65/100

15.5.1.4   Interventions
The intervention and management plan would include patient education on exercise, safety, mobility, self-care and fatigue management. Task-specific practice would include standing, transfer and gait activities as well as activities around ADLs such as reach, grasp and object manipulation, transfer of objects, objects of different sizes, weights and textures, and fine motor activities. Exercise/strengthening exercises would incorporate muscles of the upper limb, lower limb and trunk and would be progressed as needed. Aerobic exercise would be encouraged such as stationary cycling, treadmill training (body weight supported treadmill training) or arm ergometer. ES specifically TENS would be used prior to rehabilitation sessions.

15.5.1.5   Electrical stimulation

- Electrode placement: Dorsal aspect of forearm along wrist extensor muscles
- Type of stimulation: TENS
- Stimulation parameters:
  - Frequency 100 Hz
  - Pulse width 200 μs
  - Intensity – two times the sensory threshold

- Stimulation protocol
  - ○ 45 minutes before rehabilitation
  - ○ Five times a week
  - ○ Eight weeks

15.5.1.6   Expected treatment effects

Following the intervention, given the stimulation protocol used we would expect a reduction in pain with the effects being carried over for a longer duration after stimulation ends. There would also be a reduction in spasticity in the shorter term which may or may not be reflected in the MAS score (due to reliability and sensitivity of the measure). Furthermore, we would expect an improvement in arm function related to task-specific practice and neuroplasticity (due to practice and antidromic stimulation of the central nervous system) which would be reflected in the ARAT score (45/57). As the person's strength (due to ES improving contractility) and function progress we would also expect an improvement in their functional ability assessed with the Bartel Index [75] and quality of life reflected in the Stroke Impact Scale (75/100).

15.5.2   Case study example 2

15.5.2.1   Background

A person with relapsing remitting MS for the past 20 years presents with spasticity in bilateral lower extremities which is in a flexor muscle pattern (muscle isolation was not possible), using the MAS the person is between a 2/4 and 3/4 with resistance through a little over half the range of motion. Additionally, the person has bilateral weakness measured with the Oxford Grading Scale- hips 3/5, knees 3/5, ankle dorsiflexion 2/5 and plantarflexion 2/4. They have oedema at bilateral feet and ankles due to muscle weakness and prolonged sitting. The person has difficulty getting their ankle to neutral, lacking around 10° to get to neutral, this is due to both spasticity and muscle tightness. The ankle can be stretched passively to neutral but not beyond. They have decreased knee extension lacking around 15° of knee extension because of tight flexors and weakness. Furthermore, the person wears bilateral ankle foot orthoses (AFO) due to dorsiflexor weakness and foot drop. They use a wheelchair for community mobility and a wheeled Zimmer frame in their home. They have a bath chair for washing and dressing and a perching stool in their kitchen. The person is interested in improving their strength activity tolerance, be able to walk longer distances within a break and to reduce leg stiffness due to spasticity.

15.5.2.2   Clinical indication

Individual with MS and lower limb spasticity, weakness and oedema that is impacting on their mobility

15.5.2.3   Outcome measures

- MAS – bilateral ankles 3/4 with clonus, bilateral knees 2/4, bilateral hips 2/4.
- Timed up and Go – (used with wheeled Zimmer frame and bilateral AFO's) 35 seconds
- Modified Fatigue Impact Scale – 40/84

15.5.2.4   Interventions

The interventions and management plan would include patient education on safety, pressure relief, skin integrity, exercise, positioning, stretching, functional mobility, self-care and fatigue management. Strengthening exercises for the trunk and lower limbs would be given as well as for the upper limbs to maintain their strength and contribute to improved activity tolerance. Stretching for knee extension and ankle dorsiflexion would be completed. Functional training and task-specific training such as sit-stand, transfers, gait, indoor/outdoor mobility would be incorporated with FES and cycling to improve strength, range of motion, spasticity and oedema. Balance training incorporating both static and dynamic activities.

15.5.2.5   Electrical stimulation

- Electrode placement: Quadriceps, hamstrings and gluteus muscles.
- Type of stimulation: FES during cycling
- Stimulation parameters:
    - Frequency: 35 Hz,
    - Pulse width: 250 μs
    - Intensity: to muscle contraction

- Stimulation protocol: 30 minutes three times a week for ten weeks

15.5.2.6   Expected treatment effects

After completing the intervention, we would expect to see an improvement in muscle strength, and overall endurance/aerobic capacity which would be reflected in improvement in the Modified Fatigue Impact Scale (32/100) and Timed up and Go (30 seconds). There is a possibility that muscles spasms may reduce spasticity may remain unchanged within the MAS but the person would have greater ease undertaking daily activities.

15.5.3   *Case study example 3*

15.5.3.1   Background

A 20-year-old presents with an incomplete SCI at T11 and T 12, ASIA C classification following a motor bike accident around ten months ago. They have lower extremity weakness 2/5 throughout the lower extremities and. They also have diminished sensation to light touch in bilateral lower extremities with bowel and bladder incontinence. They are experiencing spasticity and muscle spasms in their bilateral lower extremities which is impacting on function such as transfers, posture, positioning and quality of life. They use a wheelchair for mobility and want to work towards more standing/ambulation. The spasticity is in the pattern (3/4 MAS) (muscle isolation was not possible) with clonus at the ankle with resistance through more than half the available range of motion, there is no limb deformity. The person experiences muscle spasms rated a 3/4 on the Spasm Frequency Scale the spasms can impact on the person's ability to transfer and are disruptive to their everyday activities.

15.5.3.2   Clinical indication

Lower limb spasticity and weakness that is impacting on their function and quality of life

15.5.3.3   Outcome measures

- MAS – 3/4 at the hip, knee and ankle
- Spasm Frequency Scale 3/4
- Trunk Control Test – 48/100
- SF-12-40

15.5.3.4   Interventions

The intervention and management plan would include patient education on safety, skin integrity, pressure relief, transfers, exercise, stretching, mobility and self-care. Exercises for strengthening the lower limbs would be started in gravity eliminated positioning and progressed as needed. Strengthening exercises for the trunk and upper limbs would also be prescribed to maintain and progress muscle strength to aid in transfers and mobility. Functional training and task-specific training around transfers, mobility and self-care. Stretching exercises to maintain muscle length. Trunk control and balance interventions to progress trunk control to facilitate easier mobility and transfers. Aerobic exercise using an arm ergometer or assisted cycling would be recommended. ES would be used such as TENS as an adjunct to the rehabilitation treatment.

15.5.3.5   Electrical stimulation

- Electrode placement: Anterior tibialis
- Type of stimulation: TENS
- Stimulation parameters:
  o Frequency: 100 Hz
  o Pulse width: 200 μs
  o Intensity at twice the sensory threshold

- Stimulation protocol: 20 minutes three times a week for ten weeks

15.5.3.6   Expected treatment effects

Following ten weeks of rehabilitation we would expect there to be a reduction in the frequency of muscle spasms as measured by the Spasm Frequency Scale, pain reduction and a reduction in spasticity in the shorter term as measured with the MAS which may or may demonstrate a change (due to reliability and sensitivity). Strengthening and functional training would contribute to an improvement on the Trust Control Test (60/100) and overall muscle strength. An improvement in strength, spasticity, spasm frequency and function would contribute to an improvement in quality of life as measured with the SF-12 (50/100). However, if this person was depending on their low-level spasticity to help with activity there is a potential risk that ES may hinder their rehabilitation.

15.6   Practical considerations

Suputtitada and colleagues developed the 'Best practice Guidelines for the Management of Patient with Post-Stroke Spasticity' (2024) [75]. Within their guidelines the use of TENS to reduce post-stroke spasticity is rated as grade A evidence supporting TENS use in spasticity reduction. The National Clinical Guidelines for Stroke (2023) spasticity recommendations include incorporating positioning, passive movement and active movement, pain control, intramuscular botulinum toxin, oral medications such as Baclofen and intrathecal Baclofen pumps [76]. Following a botulinum toxin injection individuals should be considered for ES to maintain range of movement as an adjunct but should not be treated with ES on its own. The timing of ES was explored in a systematic review identifying ES to be most effective when started rapidly after the BoNTA injection such as the same day or next day [56]. However, the guidelines recommend the use of ES as an intervention for other UMN impairments such as muscle weakness and motor control.

The National Institute for Health and Care Excellence (NICE) guideline for MS was updated in June 2022 [77]. For spasticity management in people with MS the guidelines recommend pharmacologic intervention such as Baclofen, referral to a specialist team and THC-CBD spray. However, FES is recommended for foot drop.

The Best Practice Guidelines for Spine Injury (2022) recommend the following interventions for the management of spasticity: physiotherapy such as strengthening, stretching, postural management, and orthoses, pharmacologic interventions such as Baclofen, and intrathecal Baclofen through a Baclofen pump [78].

There are additional practical considerations to weight up when using ES in people with UMN lesions. First, is the ability of an individual to don and doff the electrodes and set up the stimulator. The individual may require support to use the simulation, this should be considered when considering the use of ES. Individuals with stroke, MS and SCI may also have upper limb impairments which may contribute to hand and arm weakness and impaired dexterity to use the stimulator. In addition to upper limb weakness, individuals with changes to cognition may need support to use ES in donning/doffing the electrodes, using/setting up the stimulator and reminders of when to use the stimulation. Changes to sensation is another impairment often seen in UMN lesions and may impact on the use of ES. For example, stimulation at a sensory level such as TENS may be challenging in people with sensory changes as they may not be able to identify the sensory threshold. Finally, this chapter discussed ES within the context of different UMN conditions, however it is possible to utilise a protocol in one condition for people with a different condition.

Clinicians should use their clinical reasoning, the available evidence and the individuality of their patient to identify if ES would be beneficial to their patient to reduce spasticity.

15.7   Suggestions for future research

Development of studies with robust designs and longer term follow up will help to identify any long-term effects of stimulation on spasticity. Research is needed to identify the most effective stimulation protocols for spasticity in people with neurological disorders. The most effective protocols can be developed into best practice recommendations which would inform clinical decision making for practice. The protocols need to identify the dose of the intervention such as length of stimulation session, frequency of stimulation session, length of intervention, as well as the stimulator settings. The protocols may vary depending on the target muscle and the population such as SCI, stroke or MS and at what point in the disease progression or the rehabilitation pathway the person is at. There needs to be the development of valid, reliable, and sensitive clinically practical outcome measures to assess spasticity to distinguish between spasticity and mechanical stiffness; this will help clinicians and researchers to better understand the impact interventions have on spasticity. Finally, to ease comparisons between studies a core outcome set for spasticity would be beneficial to ensure consistent outcome measure use.

References

[1]   Lance J. *Spasticity Disordered Motor Control*. In: Feldman RJ, and Young RR (eds). Chicago: Year Book; 1980. pp. 485–94.

[2]   Pandyan AD, Hermens HJ, and Conway BA. *Neurological Rehabilitation Spasticity and Contractures in Clinical Practice and Research*. Boca Raton: Taylor and Francis Group; 2018.

[3]   Li S, Francisco GE, and Rymer WZ. A new definition of poststroke spasticity and the interference of spasticity with motor recovery from acute to chronic stages. *Neurorehabil Neural Repair*. 2021;35(7):601–10.

[4]   Burridge JH, Wood DE, Hermens HJ, *et al*. Theoretical and methodological considerations in the measurement of spasticity. *Disabil Rehabil*. 2005; 27:69–80.

[5]   Amatya B, Khan F, La Mantia L, Demetrios M, and Wade DT. Non pharmacological interventions for spasticity in multiple sclerosis. *Cochrane Database Syst Rev*. 2013;2013:1–40.

[6]   Zeng H, Chen J, Guo Y, and Tan S. Prevalence and risk factors for spasticity after stroke: a systematic review and meta-analysis. *Front Neurol*. 2021; 11:1–12.

[7]   Holtz KA, Lipson R, Noonan VK, Kwon BK, and Mills PB. Prevalence and effect of problematic spasticity after traumatic spinal cord injury. *Arch Phys Med Rehabil*. 2017;98(6):1132–8.

[8]   World Health Organization. *International Classification of Functioning, Disability and Health*. ICF; 2001.

[9]   Sabut SK, Sikdar C, Kumar R, and Mahadevappa M. Functional electrical stimulation of dorsiflexor muscle: effects on dorsiflexor strength, plantarflexor

spasticity, and motor recovery in stroke patients. *NeuroRehabilitation*. 2011;29 (4):393–400.

[10]   Pandyan A, Stewart C, Farmer S, and Bromwich W. Assistive technology in the management of spasticity. In 2013.

[11]   Malhotra S, Pandyan AD, Rosewilliam S, Roffe C, and Hermens H. Spasticity and contractures at the wrist after stroke: time course of development and their association with functional recovery of the upper limb. *Clin Rehabil*. 2011;25(2):184–91.

[12]   Tariq H, Collins K, Tait D, Dunn J, Altaf S, and Porter S. Factors associated with joint contractures in adults: a systematic review with narrative synthesis. *Disabil Rehabil*. 2023;45:1755–72.

[13]   Kuo CL, and Hu GC. Post-stroke spasticity: a review of epidemiology, pathophysiology, and treatments. *Int J Gerontol*. 2018;12:280–4.

[14]   Malhotra S, Rosewilliam S, Hermens H, Roffe C, Jones P, and Pandyan AD. A randomized controlled trial of surface neuromuscular electrical stimulation applied early after acute stroke: effects on wrist pain, spasticity and contractures. *Clin Rehabil*. 2013;27(7):579–90.

[15]   Nielsen JB, Christensen MS, Farmer SF, and Lorentzen J. Spastic movement disorder: should we forget hyperexcitable stretch reflexes and start talking about inappropriate prediction of sensory consequences of movement? *Exp Brain Res*. 2020;238(7–8):1627–36.

[16]   Malhotra S, Pandyan AD, Day CR, Jones PW, and Hermens H. Spasticity, an impairment that is poorly defined and poorly measured. *Clin Rehabil*. 2009; 23(7):651–8.

[17]   Motta-Oishi AAP, Magalhães FH, and Mícolis de Azevedo F. Neuromuscular electrical stimulation for stroke rehabilitation: is spinal plasticity a possible mechanism associated with diminished spasticity? *Med Hypotheses*. 2013;81(5):784–8.

[18]   Nielsen JB, Willerslev-Olsen M, and Lorentzen J. Pathophysiology of spasticity. In: Pandyan AD, Hermens HJ, and Conway BA (eds) (eds.) . *Pathophysiology of Spasticity in Neurological Rehabilitation: Spasticity and Contractures in Clinical Practice and Research*. Boca Raton: Taylor and Francis Group; 2018. pp. 32–44.

[19]   Kheder A, and Nair KPS. Spasticity: pathophysiology, evaluation and management. *Pract Neurol*. 2012;12:289–98.

[20]   Voerman GE, Gregorič M, and Hermens HJ. Neurophysiological methods for the assessment of spasticity: the Hoffman reflex, the tendon reflex, and the stretch reflex. *Disabil Rehabil*. 2005;27:33–68.

[21]   Wood DE, Burridge JH, van Wijck FM, *et al.* Biomechanical approaches applied to the lower and upper limb for the measurement of spasticity: a systematic review of the literature. *Disabil Rehabil*. 2005;27:19–33.

[22]   Bohannon RW, and Smith MB. Interrater reliability of a modified Ashworth scale of muscle spasticity. *Physical Therapy*. 1987;67(2):206–207.

[23] Zurawski E, Behm K, Dunlap C, *et al.* Interrater reliability of the modified Ashworth scale with standardized movement speeds: a pilot study. *Physiother Can.* 2019;71(4):348–54.

[24] Meseguer-Henarejos AB, Sánchez-Meca J, López-Pina JA, and Carles-Hernández R. Inter- and intra-rater reliability of the modified Ashworth scale: a systematic review and meta-analysis. *Eur J Phys Rehabil Med.* 2018; 54(4):576–590.

[25] Li F, Wu Y, and Li X. Test-retest reliability and inter-rater reliability of the modified Tardieu scale and the modified Ashworth scale in hemiplegic patients with stroke. *Eur J Phys Rehabil Med.* 2014;50(1):9–15.

[26] Ben-Shabat E, Palit M, Fini NA, Brooks CT, Winter A, and Holland AE. Intra-and interrater reliability of the modified tardieu scale for the assessment of lower limb spasticity in adults with neurologic injuries. *Arch Phys Med Rehabil.* 2013;94(12):2494–501.

[27] Naghdi S, Ansari NN, Ghorbani-rad S, Senobari M, and Sahraian MA. Intra-rater reliability of the modified Tardieu scale in patients with multiple sclerosis. *Neurol Sci.* 2017;38(1):93–9.

[28] Baunsgaard CB, Nissen UV, Christensen KB, and Biering-Sørensen F. Modified Ashworth scale and spasm frequency score in spinal cord injury: reliability and correlation. *Spinal Cord.* 2016;54(9):702–8.

[29] Chang E, Ghosh N, Yanni D, Lee S, Alexandru D, and Mozaffar T. A review of spasticity treatments: pharmacological and interventional approaches. *Crit Rev Phys Rehabil Med.* 2013;25(1–2):11–22.

[30] Lindsay C, Kouzouna A, Simcox C, and Pandyan AD. Pharmacological interventions other than botulinum toxin for spasticity after stroke. *Cochrane Database Syst Rev.* 2016;2016:1–52.

[31] Winston P, Mills PB, Reebye R, and Vincent D. Cryoneurotomy as a per-cutaneous mini-invasive therapy for the treatment of the spastic limb: case presentation, review of the literature, and proposed approach for use. *Arch Rehabil Res Clin Transl.* 2019;1(3–4):100030.

[32] Hashemi M, Sturbois-Nachef N, Keenan MA, and Winston P. Surgical approaches to upper limb spasticity in adult patients: a literature review. *Front Rehabil Sci.* 2021;2:1–12.

[33] Kakodkar P, Fallah A, and Tu A. Systematic review on use and efficacy of selective dorsal rhizotomy (SDR) for the management of spasticity in non-pediatric patients. *Child's Nervous System.* 2021;37:1837–1847. 30 October 2024. Available from: https://doi.org/10.1007/s00381-021-05167-y.

[34] Takahashi Y, Fujiwara T, Yamaguchi T, *et al.* Voluntary contraction enhances spinal reciprocal inhibition induced by patterned electrical stimulation in patients with stroke. *Restor Neurol Neurosci.* 2018;36(1):99–105.

[35] Dewald JPA, Given JI, and Rymer WZ. Long-lasting reductions of spasticity induced by skin electrical stimulation. *IEEE Trans Rehabil Eng.* 1996; 4:231–242.

[36]   Vance CGT, Dailey DL, Chimenti RL, Van Gorp BJ, Crofford LJ, and Sluka KA. Using TENS for pain control: update on the state of the evidence. *Medicina (Lithuania)*.2022;58:2–27.

[37]   Yiğitoğlu P. Effectiveness of electrical stimulation after administration of botulinum toxin in children with spastic diplegic cerebral palsy: a prospective, randomized clinical study. *Turk J Phys Med Rehabil*. 2019;65 (1):16–23.

[38]   Carty A, McCormack K, Coughlan GF, Crowe L, and Caulfield B. Alterations in body composition and spasticity following subtetanic neuromuscular electrical stimulation training in spinal cord injury. *J Rehabil Res Dev*. 2013; 50(2):193.

[39]   Adams MM, and Hicks AL. Spasticity after spinal cord injury. *Spinal Cord*. 2005;43:577–86.

[40]   Thomaz SR, Cipriano G, Formiga MF, *et al.* Effect of electrical stimulation on muscle atrophy and spasticity in patients with spinal cord injury – a systematic review with meta-analysis. *Spinal Cord*. 2019;57:258–66.

[41]   Ralston KE, Harvey LA, Batty J, *et al.* Functional electrical stimulation cycling has no clear effect on urine output, lower limb swelling, and spasticity in people with spinal cord injury: a randomised cross-over trial. *J Physiother*. 2013;59(4):237–43.

[42]   Alashram AR, Annino G, and Mercuri NB. Changes in spasticity following functional electrical stimulation cycling in patients with spinal cord injury: a systematic review. *J Spinal Cord Med*. 2022;45:10–23.

[43]   Massey S, Vanhoestenberghe A, and Duffell L. Neurophysiological and clinical outcome measures of the impact of electrical stimulation on spasticity in spinal cord injury: systematic review and meta-analysis. *Front Rehabil Sci*. 2022;3:1058663.

[44]   Oo WM. Efficacy of addition of transcutaneous electrical nerve stimulation to standardized physical therapy in subacute spinal spasticity: a randomized controlled trial. *Arch Phys Med Rehabil*. 2014;95(11):2013–20.

[45]   Aydin G, Tomruk S, Keleş I, Demir SÖ, and Orkun S. Transcutaneous electrical nerve stimulation versus baclofen in spasticity: clinical and electrophysiologic comparison. *Am J Phys Med Rehabil*. 2005;84(8):584–92.

[46]   Sivaramakrishnan A, Solomon JM, and Manikandan N. Comparison of transcutaneous electrical nerve stimulation (TENS) and functional electrical stimulation (FES) for spasticity in spinal cord injury - A pilot randomized cross-over trial. *J Spinal Cord Med*. 2018;41(4):397–406.

[47]   Van Der Salm A, Veltink PH, IJzerman MJ, Groothuis-Oudshoorn KC, Nene AV, and Hermens HJ. Comparison of electric stimulation methods for reduction of triceps surae spasticity in spinal cord injury. *Arch Phys Med Rehabil*. 2006;87(2):222–8.

[48]   Kwakkel G, Lannin NA, Borschmann K, *et al.* Standardized measurement of sensorimotor recovery in stroke trials: consensus-based core recommendations from the stroke recovery and rehabilitation roundtable. *Int J Stroke*. 2017;12(5):451–61.

[49] Pandyan A, Granat M, and Stott D. Effects of electrical stimulation on flexion contractures in the hemiplegic wrist. *Clin Rehabil.* 1997;11(2):123–30.

[50] Chen SC, Chen YL, Chen CJ, Lai CH, Chiang WH, and Chen WL. Effects of surface electrical stimulation on the muscle-tendon junction of spastic gastrocnemius in stroke patients. *Disabil Rehabil.* 2005;27(3):105–10.

[51] Nakipoğlu Yuzer GF, Köse Dönmez B, and Özgirgin N. A randomized controlled study: effectiveness of functional electrical stimulation on wrist and finger flexor spasticity in hemiplegia. *J Stroke Cerebrovasc Dis.* 2017;26 (7):1467–71.

[52] Koyama S, Tanabe S, Takeda K, Sakurai H, and Kanada Y. Modulation of spinal inhibitory reflexes depends on the frequency of transcutaneous electrical nerve stimulation in spastic stroke survivors. *Somatosens Mot Res.* 2016;33(1):8–15.

[53] Hakakzadeh A, Shariat A, Honarpishe R, *et al.* Concurrent impact of bilateral multiple joint functional electrical stimulation and treadmill walking on gait and spasticity in post-stroke survivors: a pilot study. *Physiother Theory Pract.* 2021;37(12):1368–76.

[54] Johnson CA, Burridge JH, Strike PW, Wood DE, and Swain ID. The effect of combined use of botulinum toxin type A and functional electric stimulation in the treatment of spastic drop foot after stroke: a preliminary investigation. *Arch Phys Med Rehabil.* 2004;85(6):902–9.

[55] Lee JM, Gracies JM, Park SB, Lee KH, Lee JY, and Shin JH. Botulinum toxin injections and electrical stimulation for spastic paresis improve active hand function following stroke. *Toxins (Basel).* 2018;10(11):426–437.

[56] Picelli A, Filippetti M, Sandrini G, *et al.* Electrical stimulation of injected muscles to boost botulinum toxin effect on spasticity: rationale, systematic review and state of the art. *Toxins.* 2021;13:291–296.

[57] Jung KS, In TS, and Cho HY. Effects of sit-to-stand training combined with transcutaneous electrical stimulation on spasticity, muscle strength and balance ability in patients with stroke: a randomized controlled study. *Gait Posture.* 2017;54:183–7.

[58] Laddha D, Ganesh GS, Pattnaik M, Mohanty P, and Mishra C. Effect of transcutaneous electrical nerve stimulation on plantar flexor muscle spasticity and walking speed in stroke patients. *Physiother Res Int.* 2016;21 (4):247–56.

[59] Ganesh GS, Kumari R, Pattnaik M, *et al.* Effectiveness of Faradic and Russian currents on plantar flexor muscle spasticity, ankle motor recovery, and functional gait in stroke patients. *Physiother Res Int.* 2018;23(2):1–8.

[60] Almutairi SM, Khalil ME, Almutairi N, *et al.* Effects of neuromuscular electrical stimulation on spasticity and walking performance among individuals with chronic stroke: a pilot randomized clinical trial. *Healthcare.* 2023; 11(3137):1–10.

[61] Stein C, Fritsch CG, Robinson C, Sbruzzi G, and Plentz RDM. Effects of electrical stimulation in spastic muscles after stroke: systematic review and

meta-analysis of randomized controlled trials. *Stroke; J Cereb Circ.* 2015;46 (8):2197–205.

[62] Mahmood A, Veluswamy SK, Hombali A, Mullick ANM, and Solomon JM. Effect of transcutaneous electrical nerve stimulation on spasticity in adults with stroke: a systematic review and meta-analysis. *Arch Phys Med Rehabil.* 2019;100:751–68.

[63] Brusola G, Garcia E, Albosta M, Daly A, Kafes K, and Furtado M. Effectiveness of physical therapy interventions on post-stroke spasticity: an umbrella review. *NeuroRehabilitation.* 2023;52:349–63.

[64] Sebastián-Romagosa M, Cho W, Ortner R, *et al.* Brain computer interface treatment for motor rehabilitation of upper extremity of stroke patients—a feasibility study. *Front Neurosci.* 2020;14:1–12.

[65] Miller L, Mattison P, Paul L, and Wood L. The effects of transcutaneous electrical nerve stimulation (TENS) on spasticity in multiple sclerosis. *Mult Scler.* 2007;13(4):527–33.

[66] Armutlu K, Meriç A, Kirdi N, Yakut E, and Karabudak R. The effect of transcutaneous electrical nerve stimulation on spasticity in multiple sclerosis patients: a pilot study. *Neurorehabil Neural Repair.* 2003;17(2):79–82.

[67] Shaygannejad V, Janghorbani M, Vaezi A, Haghighi S, Golabchi K, and Heshmatipour M. Comparison of the effect of baclofen and transcutaneous electrical nerve stimulation for the treatment of spasticity in multiple sclerosis. *Neurological Research.* 35(6):636–641.

[68] Scally JB, Baker JS, Rankin J, Renfrew L, and Sculthorpe N. Evaluating functional electrical stimulation (FES) cycling on cardiovascular, musculoskeletal and functional outcomes in adults with multiple sclerosis and mobility impairment: a systematic review. *Mult Scler Relat Dis.* 2020; 37:101485.

[69] Szecsi J, Schlick C, Schiller M, Pöllmann W, Koenig N, and Straube A. Functional electrical stimulation-assisted cycling of patients with multiple sclerosis: biomechanical and functional outcome - a pilot study. *J Rehabil Med.* 2009;41(8):674–80.

[70] Backus D, Burdett B, Hawkins L, Manella C, McCully KK, and Sweatman M. Outcomes after functional electrical stimulation cycle training in individuals with multiple sclerosis who are nonambulatory: a pilot study. *Int J MS Care.* 2017;19(3):113–21.

[71] Sieghartsleitner S, Sebastián-Romagosa M, Cho W, *et al.* Upper extremity training followed by lower extremity training with a brain-computer interface rehabilitation system. *Front Neurosci [Internet].* 2024;18:1346607.

[72] Nagel SJ, Wilson S, Johnson MD, *et al.* Spinal cord stimulation for spasticity: historical approaches, current status, and future directions. *Neuromodulation: Technol Neural Interface.* 2017;20(4):307–21.

[73] Merrill DR. Review of electrical stimulation in cerebral palsy and recommendations for future directions. *Dev Med Child Neurol.* 2009;51(S4):154–65.

[74] Kilgore KL, and Bhadra N. Reversible nerve conduction block using kilo-hertz frequency alternating current. *Neuromodulation: Technol Neural Interface*. 2014;17(3):242–55.

[75] Suputtitada A, Chatromyen S, Chen CPC, and Simpson DM. Best practice guidelines for the management of patients with post-stroke spasticity: a modified scoping review. *Toxins*. 2024;16:98–117.

[76] National Clinical Guideline for Stroke for the UK and Ireland. London; Intercollegiate Stroke Working Party; 2023 May 4 [accessed 14 February 2024].

[77] National Institute for Health and Care Excellence (NICE). *Multiple sclerosis in adults: management (NG220)*. 2022. Available at https://www.nice.org.uk/guidance/ng220 [Accessed 16 February 2024].

[78] Trauma Quality Programs American College of Surgeons. Best Practice Guidelines Spine Injury; 2022. Available at https://www.facs.org/media/k45gikqv/spine_injury_guidelines.pdf [Accessed 16 February 2024].

Chapter 16

Paediatric applications

Kirsten Hart[1], Matt White[2], Chris Smith[3] and Kristin Girshin[4]

16.1 Defining the problem

This chapter will describe the practical application of surface electrical stimulation (ES) in the management of children with congenital and acquired long-term conditions that are attributed to impairment of the central and peripheral nervous systems. The chapter will describe the unique differences in physiology of neuromuscular function of the child and provide the information required to improve accessibility of ES delivery to the child.

A sound understanding of the basics to ES is essential to successful clinical application but also confidence in the language required to improve acceptance of the device. It is hoped this chapter provides the basic foundations from which healthcare professionals can build and develop their clinical knowledge and skills in the use of this modality with children.

The scope of the population covered in this chapter will include developmental ages up to pre-adolescents (i.e. <12 years), application of ES beyond this age group should be considered in line with the management of adults. Whilst there are similarities to use with adults, there are key and noteworthy differences to application in this pre-adolescent group that will be described.

Additional challenges when considering application of ES to this younger population include:

- The heterogeneity of the population (physically, cognitively and developmentally) which limits standardisation of ES treatment regimes.
- Diversity of size and shape of each child, which limits availability of hardware available to the clinician, especially for the under 5-year-olds or the particularly small child.

[1]The National Spinal Injury Centre, Stoke Mandeville Hospital, Buckinghamshire Hospitals NHS Trust, Aylesbury, Bucks, UK
[2]Restorative Therapies Inc., Warwick, Warwickshire, UK
[3]Case Management Works Ltd., Colchester, Essex, UK
[4]In Focus Pediatric and Neuro Physical Therapy Centre, Upland, California, USA

- The impact of family dynamics and child lifestyle which can have both an extremely positive, but also challenging impact to the use of any therapy regime, including ES.

Aspects of the available evidence, related to ES use within a child population, is described in Section 16.2 of this chapter, use of available hardware in Section 16.3 and suggested approaches to clinical application in Section 16.4.

This first section includes the impact of clinical implications to smaller muscle and limb size, the difference in muscle fibre composition in the prepubescent child, differences in muscle activation, the impact of growth and development as well as the impact of force, length and abnormal muscle contraction, all of which impact on the practical application and delivery of ES as a therapy.

16.1.1 Paediatric muscle

Whilst it may be commonly believed that 'muscle is muscle', there are key differences of pre-adolescent muscle. These differences are taken into consideration when selecting an exercise intervention with ES. These areas were described by the authors (Hart and White) in the 2020 Association of Paediatric Chartered Physiotherapy (APCP) Electrical Stimulation Guidance Document [1].

16.1.1.1 Muscle fibre composition

This section describes muscle fibre composition and effect on ES use. Further information regarding muscle fibre type and muscle properties can be found in Chapter 2. Table 16.1 also provides a summary of muscle fibre type and function.

In a 2012 review [2], a range of factors were described that result in child-adult muscle activation differences. One such factor noted is that pre-adolescent children have a greater proportion of Type I muscle fibres resulting in lower relative maximal strength compared to adults. A greater proportion of Type I muscle fibres also results in increased muscle fatigue resistance, and faster recovery rates following therapeutic intervention/exercise. The clinician may therefore find the quality of muscle contraction produced by ES much reduced in a younger child (not just down

Table 16.1   Summary of muscle fibre type and impact

Muscle properties	Type I	Type 2a	Type 2b
Fibres per motor neuron	Few	Many	Many
Motor neuron size	Small	Large	Large
Speed of contraction	Slow	Fast	Fast
Force of contraction	Low	High	High
Size	Smaller	Large	Large
Fatigue resistance	Fatigue resistant	Less resistant	Easily fatigued
Mitochondrial density	High	Lower	Low
Myoglobin content	High	Lower	Low
Aerobic capacity	High	Medium	Low
Anaerobic capacity	Low	High	Very high

to smaller muscle groups) as there is less Type 2 fibre to produce strength. The effect of power output using ES will also be lower, e.g., power output when using an ES leg ergometer, will be much lower until the child approaches puberty, when muscle composition (and thus power output) will start to mirror that of the adult.

16.1.1.2 Metabolic profile

Pre-pubertal children are also noted to fatigue less when compared to untrained adults, and further are metabolically comparable to well-trained adult endurance athletes [3]. This may be attributed to pre-pubertal children's greater reliance upon oxidative energy sources [4,5], adding support to that demonstrating the noted child-adult difference in muscle fibre type [2]. Birat *et al.* [3] suggested the finding could not be explained by the fact that children performed more (especially aerobic) physical activity since the boys and men in their study had to meet the same strict inclusion criteria and their relative VO2max were not significantly different.

16.1.1.3 Motor unit activation

As pre-pubertal children have a greater proportion of Type I muscle fibres, how they recruit muscle fibres to produce a muscle contraction varies in comparison to adult muscle.

Differences in distribution of muscle fibre types are suggested to impact exercise and training [5]. With less ability to recruit Type 2 muscle fibres, low repetition, high resistance training regimes are less effective for the pre-pubescent child. Instead, it is recommended that pre-pubertal muscle favour high repetition, medium resistance activities which translates well to ES use during therapy.

A summary of the possible factors that explain the child/adult muscle differences were summarised by Dotan *et al.* [2] and can be seen in Table 16.2.

16.1.2 Impact of cerebral palsy on muscle physiology

In addition to the differences in paediatric muscle composition, neurodisability adds further changes to the structure and function of muscle tissue whilst the child grows. Various authors have reported morphological differences in the muscle fibres of children with cerebral palsy (CP) [6], including:

- Increased incidence of muscle fibre atrophy
- Increased intramuscular fat and connective tissue
- Increased percentage of histochemically identified Type 1 muscle fibres

16.1.2.1 Impact of muscular force and muscle length

Several studies have indicated that Individuals with CP lack voluntary force production. Specifically, within extensors but also generally [7]. This is further compounded by the simultaneous activation of a muscle and its antagonist [8] and increased ankle stiffness.

Children with hypertonic (spastic) CP often develop muscle shortening and then joint contractures. Ultrasonographic measurements in individuals with unilateral CP demonstrated muscle volume differences in gastrocnemius across sides

Table 16.2 Evaluation of factors that may explain observed differences between children and adults in muscular performance and metabolic response to exercise

Observed differences (dimensionally normalised relative to adults)		Factors						
		Muscle anatomy		Muscle metabolism	Muscle function			
		↓ Relative Muscle Size	↓ Type-II Muscle-Fibre Composition (*)	↓ Anaerobic, ↑ Oxidaive Profile	↑ Agonist-Antagonist Co-Activation (**)	↓ Intra-Muscular Synchronisation	↓ Volitional Muscle Activation	↓ Type-II Motor-Unit Activation
Acute Exercise	↓ Maximal Isometric Strength	√	√√	—	—	√	√√	√√
	↓ Short-Term Power	√	√√	√√	√√	√√	√√	√√
	↓ Force Kinetics	—	√√	—	√√	√√	—	√√
	↓ Force-Velocity Relationships	—	√√	—	√√	√√	—	√√
	↓ Q30 & Mean Power Frequency	—	√√	—	—	√√	—	√√
	↑ Muscle Endurance	—	√√	√√	—	√	√	√√
	↑ Recovery	—	—	—	—	√	√√	√√
Response to Resistance Training (non-hypertrophic)		—	—	—	√	√	√√	√√
Metabolic Responses	↓ Peak [La]	√	√√	√√	—	—	√	√√
	↑ Lactate Threshold in Blood & Muscle	—	√√	√√	—	—	—	√√
	↑ Intra-Cellular Pi/PCr Threshold	—	√√	√√	—	—	√	√

↑ Intra-Cellular PCr-Recovery Kinetics	—	√√	—	√	√√
↑ Fat, ↓ CHO Utilisation	—	√√	—	—	√√
↑ VO_2 Kinetics	—	√√	—	—	√√

Data from Dotan et al. (2012) [2].

Key to symbols:

'*' Assuming the possibility that the type-1 ber composition of prepubertal children is higher than that of adults by as much as 10–15%.

'**' Differences in coactivation may account for part of the observed strength child-adult difference in multi-joint and dynamic movements, but apparently, not in single-joint, isometric contractions.

'↓' Lower in children

'↑' Higher in children

'—' Marginal or no effect

'√' Can provide partial explanation

'√√' Can provide full explanation

of more than 50% to that of typical age matched peers [9]. This deficit would indicate that ES or indeed any intervention which had the potential to increase muscle volume would be of benefit with CP.

There has also been some uncertainty in the literature regarding muscle fascicle length, with some studies reporting shorter lengths [10] and others reporting no difference to age matched children [11]. In current literature it is unclear what the impact this may have on the effectiveness or practical application of ES on this population as a result.

Whilst contractures in CP are often thought to be thixotropic (i.e. stiff but reducing during movement) because of muscle overactivation, increased intrinsic passive stiffness of the tissue may also contribute to the overall impact. Although most studies report a decrease in muscle fibre size [12], It has not been observed that the shape of individual muscle fibres change as drastically as some other muscular disorders.

Muscle weakness has long been recognised as a major feature of CP. Contrary to previously held beliefs (that muscles in children with CP are incapable of adapting), Moreau *et al.* [13] describe muscle to be plastic and strength training interventions, in children with CP, impact to a similar magnitude to that observed in healthy adult muscle. A recent systematic review of muscle strength training in CP by Merino-Andrés *et al.* [14] concluded that strength training had positive functional effects on muscle strength, (however it should be noted none of the strength programmes reviewed in this study included ES). It is fair to expect however that repeated and prolonged periods of ES, with the aim to impact muscle strength, would also have a positive effect on muscle fibre size over time, as seen with progressive resisted strengthening [15]. One study reported increased fibre-size variability in patients with CP, and also found that patients with CP were more likely to have a strong predominance of one fibre type over the other (either Type 1 or Type 2). These differences were often greater than 40%, whereas typically developing children of all ages showed no predominance [16].

Even so, variation of muscle fibre type in individuals with CP does exist due to the nature of the condition, and further studies are needed to understand the impact of muscle fibre type predominance. However, the proportion of each fibre type is not likely to cause dramatic functional impairment from a physiologic perspective, but rather reflects an altered use of the muscle. It may be that ES is an intervention to support more typical use, but more research is needed.

Global trends in surgical interventions have changed dramatically in the past twenty years with orthopaedic multilevel surgery procedures, such as those pioneered by Gage [17], making way for more neurology led spasticity management in the form of Botulinum Toxin [18] and selective dorsal rhizotomy (SDR) [19]. Although research is limited [20], there is promise in the combination of ES with activity-based training.

As discussed previously, populations of children with CP are heterogenous; so, research can be challenging. However, the potential impact of ES at addressing dysfunctional muscle use, or lack of use of muscle, is clear. The physiological differences within the specifics of CP muscle do not preclude the use of ES and they arguably provide a method to improve these differences to values closer to age matched peers.

16.2 An overview of the available research

The use of ES with children is not new [21,22]. However, due to the heterogeneity of childhood, paediatric studies can lack statistical power and clinical utility remains a topic of debate in some areas of application. There is a distinct lack of clinical guidelines for the use of ES in paediatric patients. Notably, the use of ES is absent from NICE guidance documents for the management of paediatric neuro-logical conditions, except for a generic reference, in a Urinary incontinence NICE guideline, to the use of ES to assist in pelvic floor control in patients with stress incontinence but the potential to actively engage pelvic floor muscles [23]. However, this document is not specific to paediatric patients and instead considers patients across the lifespan with urinary incontinence and neurological disease Paediatric professional bodies may support members with advice regarding ES application in children [1], but no standardised advice is available.

The most comprehensive review and current best level of evidence of the use of ES in children (up to 18 years of age) was conducted by Bosques *et al.* [24]. This broad-focus review explored the application of 'therapeutic electrical stimulation', that being any application of ES in a therapeutic capacity, whether targeting a muscular contraction or activating the sensory system. In total 37 articles published between 2003 and 2013 were reviewed. Of note was the wide range in the level of evidence reviewed, from level 1b, on the Oxford Scale for evidence-based medi-cine, (defined as 'Individual RCT (with narrow Confidence Interval)') to 4 ('Case-series (and poor-quality cohort and case-control studies)') [25].

The authors noted that most studies reviewed recruited participants with CP (77.78%), however their analysis of data stratified the evidence by impairment (Table 16.3), thereby generating greater applicability of their findings to general paediatric rehabilitation and clinical practice. In this section we will use the impairment stratifications identified by Bosques *et al.* [24], summarise their

Table 16.3   Impairment stratification used by Bosque et al. [24]

Muscle properties	Type I	Type 2a	Type 2b
Fibres per motor neuron	Few	Many	Many
Motor neuron size	Small	Large	Large
Speed of contraction	Slow	Fast	Fast
Force of contraction	Low	High	High
Size	Smaller	Large	Large
Fatigue resistance	Fatigue resistant	Less resistant	Easily fatigued
Mitochondrial density	High	Lower	Low
Myoglobin content	High	Lower	Low
Aerobic capacity	High	Medium	Low

conclusions and attempt to outline how the evidence generated since 2013 has further informed clinical practice and identified different applications of ES in paediatric patients.

16.2.1   Existing impairment stratifications

16.2.1.1   Muscle weakness

Addressing muscle weakness is one of the primary factors in the decision making of clinicians to apply ES. Bosques *et al.* [24] concluded that the studies in their review indicated that therapeutic ES has the potential to increase muscle strength in patients with CP and spinal cord injury (SCI). However, what was unclear from their analysis was the clinical significance of reported changes in muscle volume, strength and posture. Furthermore, they noted that some studies reported effects being temporary and failed to persist beyond the termination of treatment and concluded that it is unknown whether a longer treatment duration, or continuous treatment, is required to maintain the effects. Future studies (within a paediatric population) may benefit from selecting outcomes related to meaningful participation in functional activities, rather than the impact on muscular structure and performance.

In a more recent scoping review, Yan and Vassar [26] reviewed literature pertaining to a wide range of paediatric conditions, including stroke, SCI, myelo-meningocele, scoliosis, obstetric brachial plexus injury, genetic neuromuscular disease and other neuromuscular conditions. They reported varied results, with some studies reporting statistically significant improvements in muscle strength in the stimulated muscle groups, and others finding no difference to conventional or sham treatments, or controls. It should be noted that positive results were reported in those conditions predicated by central nervous system dysfunction (stroke, SCI and myelomeningocele). Conditions with peripheral dysfunction were supportive of ES interventions (with the exception of obstetric brachial plexus injury) where two small RCTs found greater improvements in muscle strength compared to conventional treatments.

Unfortunately, due to the heterogeneity in therapeutic approach, and often a disconnection between therapeutic stimulation protocols and established muscle strength training protocols, no optimal dosage of ES for muscle strengthening for use in a paediatric population has been identified in the literature. Accordingly, the clinician is recommended to judge the efficacy of their treatment provision by the quality of muscle contraction achieved by their individualised parameter selection. The authors would recommend to aim to progressively increase the range of motion and the level of opposing resistance applied against the movement generated by the target muscle group.

16.2.1.2   Spasticity and tonicity

The management of tone and spasticity is another major clinical indication for the application of ES in paediatric patients. Bosques *et al.* [24] concluded from the evidence they reviewed, the use of neuromuscular electrical stimulation (NMES – stimulation applied with the goal of eliciting a muscle contraction) may be useful in temporarily decreasing spasticity in patients with both upper and/or lower limb involvement, specifying that it must be applied for at least 30 minutes per day between three and five times per week. Furthermore, when NMES is applied to the antagonist muscle following Botulinum toxin, they concluded that the addition of NMES may accelerate the chemical denervation of the agonist muscle, and that a combined treatment of Botulinum toxin and NMES has been shown to potentially improve function and motor control where Botulinum toxin alone has not. They also noted that when used immediately after Botulinum toxin injection, NMES may increase uptake. This has been supported with combined treatment of Botulinum toxin with antagonist ES was more effective than either intervention alone [27]. However, there is evidence in animal studies that Botulinum toxin not only affects the targeted muscle tissue but also has deleterious distal effects [28], so caution is advised to the clinician combining ES with Botulinum toxin interventions to be aware of the risk of greater circulation of Botulinum toxin to surrounding tissues leading to chemical denervation in non-target muscles which may lead to deleterious effects on tone, spasticity and function.

In a 2022 scoping review of neuromuscular ES interventions for CP, Greve *et al.* [29] cites seven studies that investigated the use of NMES or TENS (transcutaneous electrical nerve stimulation – a method of pain relief that involves the use of a mild electrical current.) for spasticity reduction. In all studies stimulation was applied to the antagonist muscle group, and selected groups included hip adductors, quadriceps, hamstrings, gastrocnemius and tibialis anterior. Of these seven trials, five reported significant benefit, notably these five were higher quality randomised controlled trial (RCT) or prospective controlled trials. However, once again wide-ranging stimulation parameters and protocols leave optimal application procedures uncertain.

Adding some weight to the application of ES as part of a combined approach, Liu *et al.* [30] conducted a RCT of 32 children aged 2–12 years old. They investigated the effects of transcranial pulsed current stimulation (tPCS) used concurrently with TENS (for 30 minutes) and 30 minutes of physiotherapy, compared to 30 minutes of physiotherapy alone. Non-invasive brain stimulation (NIBS), of which tPCS is one specific approach, involves the transmission of an electromagnetic pulse through the skull to target specific areas of the brain. Further discussion of NIBS is outside the scope of this work. Liu *et al.* [30] selected the combination of tPCS and TENS to investigate the hypothesis that the application of different modes of stimulation directed toward the skull, spine and lower limbs would be effective in reducing lower limb spasticity in patients with severe CP (graded III and IV on the Gross Motor Function Classification System). Both groups underwent five treatment sessions per week for twelve weeks. At the end of the intervention period, they reported significant improvements in Modified Ashworth Scale and Modified Tardieu Scale scores in the lower limbs in the intervention group but declines in all measures from baseline in the control group. However, significant limitations of the study, including failure to control for the additional 30 minutes per session in the intervention group, and no comparison to either non-invasive neuromodulation intervention in isolation, limits the impact of the results.

Over recent years, a new ES full body suit, applying whole body stimulation via multiple built-in electrodes has gained popularity for the management of dystonia and spasticity by applying low voltage currents at or around the sensory threshold, to the antagonist of the dystonic muscle, for 60 minutes a day on alternate days. This approach is intended to utilise the principle of reciprocal inhibition by activating afferent nerve fibres in the opposing muscle to that which requires therapeutic intervention, which activate inhibitory Ia interneurons in the spinal cord and reduces the excitability of the target muscle [31]. Details about the device can be sought through the NICE MedTech Innovation Briefing [32]. As the therapy involves sitting at rest whilst the stimulation is applied, it can be delivered at home or alongside/before other therapeutic interventions. However, despite positive anecdotal findings, experimental trials in paediatric patients with CP have failed to demonstrate significant effects of using the ES suit on dystonia or spasticity within the paediatric population [33].

ES interventions are largely well supported in the literature for the management of tone and spasticity in paediatric patients, however, heterogenous therapeutic approaches prevent conclusions to be drawn over optimal dosage. Despite this, combining ES interventions with other therapeutic interventions, appears to be the most promising means of optimising tone and spasticity management.

16.2.1.3   Contracture management/prevention of deformity

Linked to the management of tone and spasticity, ES can be applied in clinical practice to aid in contracture management. Commonly, ES is applied to the antagonist muscle to provide an 'active stretch' to the contracted or at-risk tissues. Bosques *et al.* [24] concluded that ES can be applied to children with upper motor neuron (UMN) lesions that commonly lead to posterior chain limitations. Additionally, they reported that from the studies they considered, infants with congenital torticollis may benefit from including ES with traditional therapies. Later research into children with CP has failed to provide greater clarity. Investigations into the effects of both motor and sensory ES interventions on passive range of motion of the knee and ankle reported no significant difference to standard interventions [34]. In a small pilot study of children with scoliosis secondary to severe CP, Ko *et al.* [35] found that highly intensive ES applied to the convex side of the curvature (twice a day for one hour over three months) resulted in significant improvements to the severity of the scoliosis and improved sitting balance at both one and three months.

It is reasonable to consider the ES is as effective as many other tools (such as stretching and splinting) used to address muscle contractures and prevent emerging deformities, however, all non-surgical approaches reported offer small increments to range of motion and are therefore of questionable clinical significance. Therefore, ES, or any other non-surgical approach to contracture management, should be applied prophylactically where possible and used as part of a diverse multidisciplinary therapeutic approach.

16.2.1.4   Upper limb dysfunction

Due to its effect upon muscle function and spasticity, ES interventions may also be applied with the goal of enhancing upper limb function. Bosque *et al.* [24] concluded that motor ES has been shown to result in statistically significant improvements in upper limb function outcome measure such as the 9-hole peg test, Peabody Developmental Scale and Jebsen Hand Test when applied for a minimum of 20 minutes, five days per week over a period of two to six weeks. They also reported that in those with hemiplegic CP, ES applied alongside constraint-induced movement therapy (CIMT) yielded favourable results, and infants suffering from Neonatal Brachial Plexus Palsy may experience improvements to their upper limb muscle activation, in spite of their lower motor neuron injury. As will be discussed later, when treating paediatric patients, their parents/guardians/carers are pivotal to any intervention, especially those involving long term adherence. Musselman *et al.* [36] conducted a case study on a two-year-old child with a perinatal stroke where the ES intervention was delivered at home by the child's mother, supported remotely by a clinician. They reported high levels of adherence and lent support to the feasibility of delivering an eight-week ES protocol to a two-year-old. In addition, significant improvements in the Melbourne Assessment 2's (MA2) range of motion, accuracy and fluency subscales were reported. Later, in a qualitative study, Swaffield *et al.* [37] investigated mothers' perceptions of functional ES interventions for upper limb dysfunction in their children (aged between 3 and 6 years) with hemiparesis caused by perinatal stroke. Researchers found that mothers perceived that their children benefitted physically, functionally and psychologically from the ES intervention.

In addition to motor ES protocols for upper limb function, sensory protocols have also been employed and combined with therapeutic exercise to enhance hand function in 29 children with hemiplegic CP [38]. When compared to a control group receiving traditional physical therapy, those who also received 30 minutes of TENS therapy (pulse duration 250 μs and frequency 100 Hz) to the wrist extensors completed the Jebsen Taylor Hand Function Test 48% faster and improved ABILHAND-Kids questionnaire scores by 23%.

ES interventions have been demonstrated to improve upper limb function in paediatric patients with stroke and CP and are viewed positively by parent's who, outside of a research setting, are the primary facilitator of ES interventions for their children.

16.2.1.5   Gait deficits

Gait dysfunction is common in paediatric patients with neurological disability. Applications of ES for gait enhancement need to be applied in a coordinated pattern at the correct phase of gait. In the case of crouch gait this can pose a problem due to the pathological gait pattern failing to allow for triggering with the commonly applied heel switch or angle detection methods. However, despite this Bosque *et al.* [24] was able to conclude from the research considered in this review that children with spastic hemi and diplegic CP, presenting with poor knee control, decreased cadence and crouched gait may benefit from FES (functional electrical stimulation – ES applied to support a function) with or without Botulinum toxin (see also above) relating to combining ES and Botulinum toxin) when provided for 30–60 minutes per day on five-six days per week.

Rather than considering 'functional' applications of ES (i.e. timed during the gait cycle), the influence of motor ES applications on gait function has been widely explored in children with CP. In a systematic review and meta-analysis, Chen *et al.* [39] identified 14 RCTs including a total of 421 children with CP and identified that when compared with physical therapy, motor ES treatment resulted in a greater improvement in walking speed (standardised mean difference = 0.29; 95% confidence interval = 0.02–0.57; Figure 16.1) and the standing (dimension D), walking, running and jumping (dimension E) dimensions of the Gross Motor Function Measure (GMFM) (standardised mean difference = 1.24; 95% confidence interval = 0.64–1.83; Figure 16.2).

Shideler *et al.* [40] reported in 2020 that they are working toward a hybrid exoskeleton, combining ES with robotic assisted motion to address crouch gait in children with CP. In their case report, they stated that they were able to provide a precisely timed contraction of the vastus lateralis and rectus femoris during stance using a custom, digitally controlled ES system, supporting immediate improvement

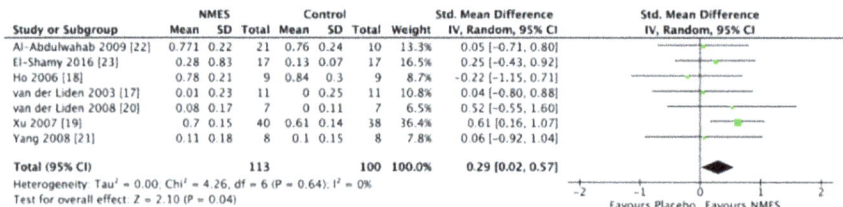

Study or Subgroup	NMES Mean	SD	Total	Control Mean	SD	Total	Weight	Std. Mean Difference IV, Random, 95% CI
Al-Abdulwahab 2009 [22]	0.771	0.22	21	0.76	0.24	10	13.3%	0.05 [-0.71, 0.80]
El-Shamy 2016 [23]	0.28	0.83	17	0.13	0.07	17	16.5%	0.25 [-0.43, 0.92]
Ho 2006 [18]	0.78	0.21	9	0.84	0.3	9	8.7%	-0.22 [-1.15, 0.71]
van der Liden 2003 [17]	0.01	0.23	11	0	0.25	11	10.8%	0.04 [-0.80, 0.88]
van der Liden 2008 [20]	0.08	0.17	7	0	0.11	7	6.5%	0.52 [-0.55, 1.60]
Xu 2007 [19]	0.7	0.15	40	0.61	0.14	38	36.4%	0.61 [0.16, 1.07]
Yang 2008 [21]	0.11	0.18	8	0.1	0.15	8	7.8%	0.06 [-0.92, 1.04]
Total (95% CI)			113			100	100.0%	0.29 [0.02, 0.57]

Heterogeneity: Tau² = 0.00; Chi² = 4.26, df = 6 (P = 0.64); I² = 0%
Test for overall effect: Z = 2.10 (P = 0.04)

Favours Placebo Favours NMES

Figure 16.1   The effect of motor ES on walking speed compared with conventional therapy in children with cerebral palsy [39]

Study or Subgroup	NMES Mean	SD	Total	Control Mean	SD	Total	Weight	Std. Mean Difference IV, Random, 95% CI
Chan 2004 [24]	4.25	14.27	6	1.37	7.28	6	9.6%	0.23 [-0.90, 1.37]
Duymaz 2018 [30]	14.8	7.89	20	5.6	4.32	20	12.3%	1.42 [0.72, 2.12]
Gao 2010 [28]	30.5	1.35	12	21.3	2.15	12	6.6%	4.95 [3.22, 6.68]
Jiang 2008 [26]	10.85	5.24	14	6.71	3.97	14	11.8%	0.86 [0.08, 1.64]
Kerr 2006 [25]	2.9	9.51	18	-0.16	8.79	22	12.7%	0.33 [-0.30, 0.96]
Li 2013 [29]	14.28	6.53	14	7.18	5.04	11	11.3%	1.16 [0.29, 2.02]
Xu 2007 [19]	24.6	9.73	40	15.9	9.28	38	13.6%	0.91 [0.44, 1.37]
Yang 2008 [21]	2.69	1.61	8	1.58	3.2	8	10.5%	0.41 [-0.58, 1.41]
Zhang 2009 [27]	16.48	5.76	21	4.14	4.22	18	11.5%	2.37 [1.53, 3.20]
Total (95% CI)			153			149	100.0%	1.24 [0.64, 1.83]

Heterogeneity: Tau² = 0.62; Chi² = 39.34, df = 8 (P < 0.00001); I² = 80%
Test for overall effect: Z = 4.07 (P < 0.0001)

Favours Placebo Favours NMES

Figure 16.2   The effect of motor ES on gross motor function dimensions D and E compared with conventional therapy in children with cerebral palsy [39]

of the midstance peak knee extension. Their work to integrate ES with a hybrid exoskeleton continues.

16.2.1.6   Foot drop

Foot drop is a prevalent condition in paediatric neurological conditions, resulting in complications including, falls, injuries to the ankle and increased effort of walking [41]. Foot drop is generally managed in the same manner as in adults (see also chapter 7 for information on managing adult foot drop), using FES neuro-prostheses that deliver stimulation to the muscles of the anterolateral leg to generate dorsiflexion. Children may however benefit from use of smaller lighter ES devices that allow them more freedom of movement or quicker set up, i.e. the use of wireless devices or of tilt switches rather than footswitches. Candidates for such devices are ambulatory children with UMN lesions, most commonly diplegic and hemiplegic CP [24] and incomplete SCI. In these populations, ES has been shown to have both an orthotic effect upon gait, with improved initial contact angles and dorsiflexion range during swing [42] and is tolerated for at least six hours' use per day. In addition, there are therapeutic effects on gastrocnemius spasticity, community mobility and balance skills [43]. However, ES interventions for foot drop do not perform significantly differently to ankle-foot-orthoses, and as tolerance can impede success, patient selection for ES correction of foot drop is critical [42].

16.2.1.7   Poor trunk control

Impairments to postural control require careful intervention throughout childhood to optimise spinal and peripheral alignment into adulthood. ES can be used alongside bracing, seating and sleeping systems to achieve this. Applying ES to the trunk for 30 minutes per day, five times per week for at least four weeks has been demonstrated to be effective in improving sitting balance in children with CP [24].

Taping is an alternative, non-invasive means of managing impairments to trunk control and thus investigations have been conducted comparing the effects of taping and ES interventions. Karabay *et al.* [44] conducted a randomised controlled study of 75 children with CP with a control group of neurodevelopment therapy (NDT) only, and two experiment groups of NDT with taping and NDT with motor ES. All subjects received four sessions of 75 minutes NDT for four weeks with/without their randomised intervention. They reported that both of the intervention groups resulted in greater improvements in kyphosis angle (assessed by radiograph) and seated dimension of the GMFM. Furthermore, they concluded that the improvements in the motor ES group were superior to the taping group in both measures. A later RCT by Elbasan *et al.* [45] went on to demonstrate that, in a slightly smaller sample of 45 children with CP, the combination of motor ES, taping and NDT was superior to motor ES and NDT, and NDT alone for enhancing postural control. This was measured by the seated section of GMFM and the Seated Postural Control (SPC) Measure [46]. However, the efficacy of ES interventions can be enhanced when used in combination with balance training and kinesio taping [45]. Surface spinal stimulation has also been found to achieve upright sitting in a small feasibility trial of eight children with SCI [47].

Long term use of ES for postural control is poorly tolerated in the authors' experience, however short-term applications in combination with active postural re-education offers a viable means of enhancing active control in those with sensorimotor and proprioceptive deficits.

16.2.1.8   Deconditioning

It is widely accepted that paediatric patients with neurological dysfunction have lower levels of exercise engagement and impaired cardiorespiratory and cardiometabolic function when compared to their able-bodied peers. Motor-targeted ES interventions have been shown to assist in addressing this deconditioning in children with decreased lower extremity function and ambulatory capacity [24].

Further research has explored comparisons of ES induced/assisted exercise in patients with CP and the consensus of findings suggests that there is no superiority of ES interventions to active exercise (such as adapted cycling and goal-based exercise) in those with sufficient capacity [48,49], but that the addition of motor ES to active exercise may raise the intensity of the activity [50].

In the experience of the authors, the use of ES interventions to address deconditioning may provide greater benefit in those patients with lower functional capacities due to a greater impairment of volitional activity. Furthermore, individual choice and motivations of the patient are highly influential in the effectiveness of such ES applications.

16.2.1.9   Disuse osteopenia/osteoporosis

Within the paediatric population, evidence is sparse as to the impact of ES application upon disuse osteopenia and osteoporosis. Bosque *et al.* [24] concluded from evaluation of two independent studies that the engagement in FES ergometry may be beneficial in paediatric patients with SCI who are at risk of disuse osteopenia or osteoporosis. Since this publication, no studies have further investigated the effect of ES applications on bone health in the paediatric population. Anecdotally, in patients capable of weight bearing, load and impact-based interventions, with and without ES are utilised as a means for enhancing bone health and reducing the risk or incidence of disuse osteopenia/osteoporosis. In patient's incapable of volitional load bearing, ES interventions may offer potential solutions but must be applied consistently and over the long term. Further information on the use of ES and osteopenia/osteoporosis can be found in the literature pertaining to adult care.

16.2.1.10   Dysphagia

ES has been widely explored within adult patients with dysphagia but less so with paediatric patients. In conducting their 2016 review, Bosque *et al.* [25] reported on a single research trial utilising a specific stimulator and concluded that those with acquired dysphagia may be better candidates than those with primary dysphagia. Since this publication, further research has been conducted into the use of ES for the management of oropharyngeal dysphagia in children, but in a 2023 review citing seven peer reviewed journal articles, Gosa and Marshall [51] concluded that there is very little empirical evidence to support the application of ES interventions for either acquired or primary dysphagia in paediatric patients (aged 1–18 years).

16.2.2 Updated areas of ES application

The section above, on 'Existing Impairment Stratifications' incorporated and expanded upon the work of Bosque *et al.* [24] within the defined categories identified by the authors and presented in Table 16.3. Since the publication of Bosque *et al.*'s study [24], a further decade of research and clinical application has explored the use of ES within the paediatric population, resulting in several additional areas of application that are available to the paediatric therapist.

16.2.2.1 Urinary tract dysfunction

Despite its omission from the Bosque review [24], the use of ES for the management of urinary tract dysfunction has been investigated and practised for more than two decades. A 2011 systematic review [52] evaluated 17 studies where ES had been applied to either the sacrum (surface or implanted), posterior tibial percutaneous, anogenital, endoanal or intravesical nerves for the treatment of lower urinary tract dysfunctions (including overactive bladder (OAB), dysfunctional voiding and underactive bladder) in children not related to congenital abnormalities or neurological disease. They found that, like much ES research in children, neither the sample populations, the duration and frequency of stimulation sessions, or the stimulation parameter selections were homogenous, and rates of resolution of OAB, urgency and daytime incontinence varied widely (36–86%). They did, however, identify that transcutaneous sacral stimulation appeared to be superior to sham therapy. In later work, Quintiliano *et al.* [53] conducted a RCT to compare the efficacy of transcutaneous parasacral ES with oxybutynin for the treatment of OAB in 28 (final analysis on 26 following two withdrawals) children without signs of neurological disease or urinary tract anatomical problems. They reported that not only was parasacral ES as effective as oxybutynin in treating OAB, it had no discernible side effects and was more effective than oxybutynin in improving constipation. Oxybutynin however was more effective in improving voiding frequency [53]. Whilst promising, the authors noted that these results lacked power to determine superiority of ES over oxybutynin as power calculations for patient recruitment were based upon ES being noninferior. They also noted the lack of blinding as a limitation of the study. Further, larger scale studies would be required to rigorously investigate the superiority of transcutaneous parasacral ES over the use of oxybutynin in this population.

16.2.2.2 Gastric dysfunction

Increasing attention has been focused on the use of ES in treating gastric disorders in children, particularly gastroparesis and slow transit constipation. Slow transit constipation is characterised by slow proximal colonic transit and is recognised as an underdiagnosed condition with potentially severe consequences including the need for colectomy [54]. The primary modes of application have either applied an interferential current, with electrodes placed anteriorly on the abdomen and posteriorly to the erector spinae to produce an area of interference targeted at the colon, or TENS with current applied using two electrodes on the abdomen and two on the back. Both treatment methods have been demonstrated to improve between one and four outcome measures including frequency of defecation, soiling, Bristol Stool Scale, radionuclear transit studies and quality of life. However, in their systematic review, Lu *et al.* [54] state that despite providing some support for the use of ES applications to treat slow transit constipation, the heterogeneity and weak study design of most of the six trials they included, limits the weight of this support. However, subsequent RCTs with larger participant numbers lend further support to the use of ES for slow transit constipation. Sharifi-Rad *et al.* [55] conducted an RCT on 90 Iranian children aged 5–13 years with half receiving interferential stimulation with pelvic floor exercises and the other half sham stimulation with pelvic floor exercises. They reported significantly greater treatment success in the intervention group (88.4% compared to 43.2%, p>0.003), with lower median constipation scores [56]: 4 vs 8, p<0.000) and normalised stool form in 75.6% of cases compared to 45.5% of controls (p<0.01). Recently, a smaller, randomised crossover design study of 18 children with CP [57], reported that laxatives with TENS resulted in increased defecation frequency, wetter stools and reduced colonic transit time, compared to laxatives alone.

Invasive ES, delivered via implantation of a neurostimulator to the abdomen, has been employed to treat gastroparesis with positive effects beyond 12-months follow up (mean follow up 3.5 years), with significant improvements in individual symptoms of gastroparesis and total symptom score [58].

16.2.2.3   Intensive care unit acquired weakness (ICUAW)

During admission to intensive care units (ICU), critically ill patients can acquire neuropathies and/or myopathies, referred to as ICU acquired weakness (ICUAW), for which there are currently only few preventative and no therapeutic interventions known to address the short- and long-term consequences [59]. One preventative method is the use of ES. To explore the efficacy of ES in preventing ICUAW, Magalhães *et al.* [60] conducted a scoping review and found significant heterogeneity in the current modulation and duration of sessions, but most applications within the paediatric ICU (PICU) were administered to patients with neurological disorders, and generally favoured higher intensity, shorter durations applications seeking muscle contractions. They were however able to conclude that the use of ES was significantly more effective that sham therapy in outcomes of spasticity, bone mineral density, disability, gait and cardiorespiratory performance. ES promises to be a potentially effective approach for the prevention of ICUAW but further trials, considering homogenous populations and consistent therapeutic parameters will clarify the efficacy in applying ES to critically ill children in the PICU. For further information on the use of ES in the management of adult critical care refer to Chapter 13.

16.2.2.4 Emergence agitation

Emergence agitation occurs in 10-50% of children emerging from anaesthesia [61] and involves a variety of symptoms including restlessness, disorientation, excitation, non-purposeful movement, inconsolability, thrashing and incoherence [62]. A potential means of mitigating this phenomenon was investigated by Hijikata *et al.* [63] who applied ES to the 'Heart 7' acupuncture point bilaterally during surgery (Heart 7 is located at the wrist crease, on the radial side of the flexor carpi ulnaris tendon, between the ulna and the pisiform bones). They conducted a double blind, RCT of 120 patients aged 18–96 months, comparing the application of ES with a sham-ES control. In the intervention group stimulation was started at the time of the first incision and terminated at the end of the surgery. They reported a significantly lower incidence of emergence agitation in the intervention group (31.7%) compared to the control (56.7%). The same research group later completed a similar trial but only applied ES to the 'Heart 7' unilaterally [64]. In this trial they reported no significant differences to the control group. These trials suggest that the application of low risk, non-invasive ES during surgery may offer benefit to a significant number of young children in reducing the likelihood of experiencing emergence agitation, but only when ES is applied bilaterally.

16.2.2.5 Summary of paediatric clinical application

Despite the weight of research studies focused upon condition specific variables and the structural differences previously noted in the skeletal muscles of children with CP (as described in the section above on 'Impact of muscular force and muscle length', ES interventions can and have been demonstrated to be effective within the literature across a range of, predominantly (but not exclusively) upper motor neuron affected conditions. Therefore, commonalities in the stimulation approach apply across different diagnoses when there is a common goal of ES application. Thus, in the authors' opinion, decision making to apply ES should be considered on an *impairment*, as opposed to a *condition*, basis.

16.3 An overview of the current state of the art

Clinical application of ES in paediatric patients, like in adults, remains a gross, relatively broad stroke approach. Whilst sound anatomical knowledge forms the basis of electrode placement, a trial-and-error approach remains necessary to elicit the desired muscular response. Motor point testing devices can aid in the accuracy of electrode placement locations, but these are not widely used within clinical practice.

When applying single, double or multiple channel ES, hardware selection mirrors that of adults. Devices with greater capability to manipulate stimulation parameters will provide greater flexibility to the therapist and a more effective stimulation application for the paediatric patient. This is especially relevant for devices which allow reduced pulse widths to be used for greater specificity when targeting contractions in small muscle groups (for example the finger extensors of a very young child). In the selection of a suitable stimulator, consideration of the child and/or parent's ability to comprehend and work with multiple channels is required. In the same way, electrode selection needs to be appropriate for the size of the target stimulation area, to ensure accurate delivery of stimulation to the target muscle groups and not to other local groups which may affect/impair the success of the treatment.

When applying ES for the enhancement of function, the electrodes are often housed within an orthosis and in most cases are available in an 'extra small', or paediatric size, making them more appropriate for the paediatric patient. However, suitability will often depend upon the girth of the child's limb to ensure sufficient contact of the electrodes against the skin and the correct location of the electrodes for the target effect. Sometime, despite a reduced/paediatric size of the ES orthotic, the relative size of the ES device (compared to the child's size), remains cumbersome, which may impact function (Figure 16.3).

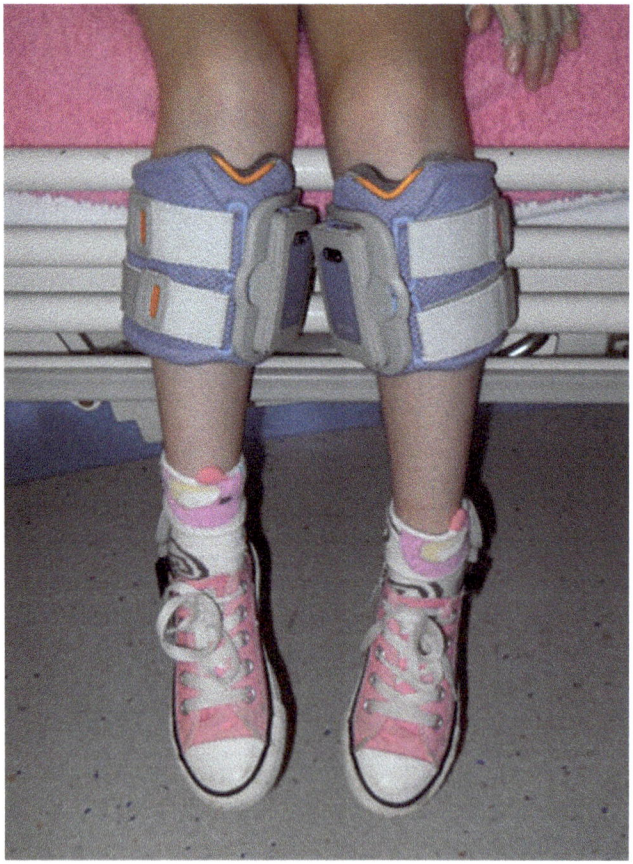

Figure 16.3   Foot drop stimulation.
'Foot drop' devices may appear relatively more cumbersome for the young child compared to the adult

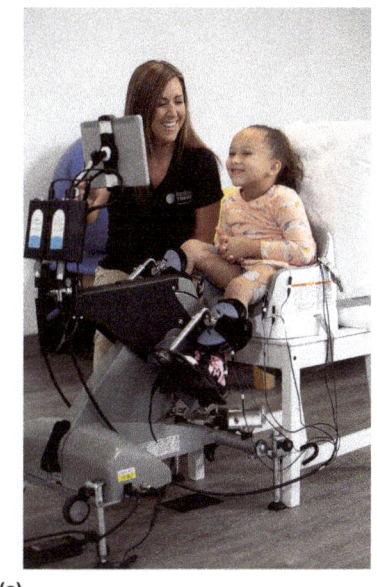
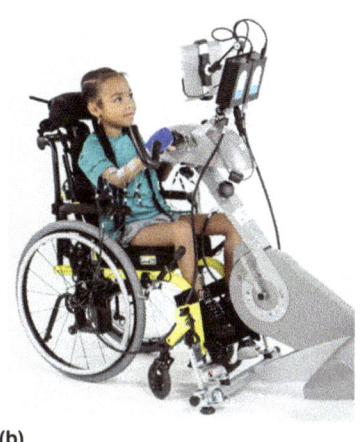

(a)   (b)

Figure 16.4   FES arm and leg bike ergometry.
A – Paediatric versions of FES leg bike ergometry. B – Paediatric arm ergometry.
By kind permissions of Restorative Therapies, leading global manufacturer of FES
Integrated Therapy Systems

FES ergometry has been discussed in the sections above on 'Spasticity and Tonicity' and 'Disuse osteopenia/osteoporosis' and manufacturers of such devices have sought to make them accessible to younger children by designing ergometers with smaller footplates and crank radius, as well as redesigning the interfaces to enable children to be able to see the visual, real-time outputs of their session (Figure 16.4). Additionally, the multichannel sensory stimulation suit discussed in the section above on 'Spasticity and Tonicity' has also been produced in a wide range of sizes to enable accurate alignment of the electrodes to the targeted regions of the body for therapeutic application (Figure 16.5). However, in both aforementioned cases, the devices may at times remain too large to be usable for the very young child (and is often related to leg length), especially those with neurological conditions from birth impacting the rate of growth compared to their peers.

In summary, many commercially available ES units are appropriate for application on even very young paediatric patients (in some cases weeks old). For those devices which require a more bespoke fitting to enable enhanced function, there remains work to be done in the development of appropriately sized, easy to use devices for paediatric patients.

(a)   (b)

Figure 16.5   Child stimulation suits.
Paediatric and child versions of the Mollii Suit, by kind permissions of Otto Bock
Healthcare PLC

16.4   Practical clinical considerations for the use of electrical stimulation with children

Applications pertinent to the younger (pre-adolescent) child will be explored in this section. This will include use to support developing function/habilitation, restorative approaches and the specific practical techniques undertaken for successful application.

16.4.1   Principles of clinical application

ES can be used to provide both motor and sensory input to the child during functional or static tasks via the delivery of stimulation to unused or weakened muscle(s).

Age is not a contraindication to use of ES as a modality, Justice *et al.* [65] cited consistent benefits of ES in neonatal brachial plexus palsy in children as young as three weeks old without detrimental effect. However, for the young and cautious child, who perhaps has been over exposed (through necessity) to a medical model of examination and assessment, the mere sight of a clinician dressed in a uniform holding yet another examination/treatment device, may result in an anxious and uncooperative child, therefore care and a considered approach is required.

Accordingly, a sensitive and playful approach to treatment is not only necessary but essential if the application of ES is to be successful and useful. This may translate into the use of age-appropriate language to describe each aspect of ES (and how ES may feel) as well as distraction techniques and play to make the therapy fun and more appealing.

The clinician should also expect that the progression of use may likewise be slower – not only because the child may not tolerate an explorative approach (for the more cautious child it would not be unusual to use the equipment switched off or at a non-therapeutic level, in order to gain compliance, due to fear or due to the effect of an unusual sensation) but also as a consequence of a growing and developing nervous system and musculoskeletal system. Thus, the clinician benefits from a clinically reasoned and rationalised ES approach prior to application, and objectively analysing, recording and progressing effect over time.

16.4.2 Habilitation

Habilitation therapy aims to support children in acquiring functional motor skills they have not yet developmentally mastered. The child may present with weak non-functional movement below the level of impairment. ES is a useful modality to support the required specific patterned training to develop learning [66]. Active participation in the activity and a high number of repetitions are required to achieve a change in skill [67] which ES therapies are able to provide. In the presence of regular, repeated excitation, nerve cells learn to associate with one another – Hebbs rule [68]; 'Cells that fire together, wire together'. For example, ES leg ergometry/cycling can be a useful modality prior to the acquisition of ambulation. Whilst carried out in a static sitting position it allows the child to see their legs moving and attempt to join in. Over time (and with developing function) this activity can be translated to a standard fixed wheel trike to duplicate the activity and effort carried out with more volitional ability. Similarly, this bilateral patterned activity may later translate to upright stepping/mobility. Alternatively, stepping can be facilitated with ES in a de-weighted upright position allowing the child to see their leg stepping and allowing them to join in with the effort/activity. Over time support provided by ES can be reduced or removed and the activity trailed with the child's own volitional control once the child understands what is expected and is motivated to perform the activity. Over time repetitions and duration can be increased as well as the complexity of the movement. For considered areas of ES application through development see Table 16.4.

16.4.3 Restorative approaches

Neuroplasticity is similar to habilitation but refers to structural and functional neuronal change, brought about by training, function and experience [69]. Neuronal connections increase rapidly in the first two years of life [69] and continue through growth and development. Thereafter, for the child with neurological impairment, the period of growth is a useful period to influence and enhance this neuronal change. The child may present with significant loss below the level of impairment, of which the clinician hopes to exploit the plastic potential for change. ES can be a useful tool to enhance the experience of movement and subsequent neuronal

Table 16.4 Consideration of electrical stimulation application through stages of development

Developmental stages	New born	Infant – siting	Infant – crawling	Toddler	Pre-school/school age
Examples of developmental areas of need	Positioning	Sitting balance Trunk control Reaching Bimanual Integration	Mobility Pulls to stand – start to initiate upright weight bearing. Upper limb function Short attention span	Upright Mobility Weight bearing Upper limb function – simple ADLs	Upright mobility Weight bearing
ES considerations for treatment with ES	What can they move? What can they feel? What is asymmetrical? Will they be a candidate for ES/do they respond to ES?	What can they move? What can they feel? Do they respond to ES? Establishing routine use of ES over areas of paralyses to enhance muscle strength and foster function	Assess ability to engage/duration of treatment. Break into parts ES and limb movements. Bringing together into FES bike use to develop patterned functional activity ready for mobility	Combining gait with ES in simplest form Standing with ES Activities with ES, i.e. kicking ball in sitting with ES or reaching for toy with ES	Gait with ES – multi channel or FES Fitness Function Daily routines (school/home/peers)

adaptation. The principles 'use it or lose it' and 'use it and grow it' are based on the principles of neuronal plasticity and are highly relevant to the neurologically impaired child [69].

This type of ES may be used for single or multiple muscles with less than grade 3 muscle strength and can be carried out at any age. ES is applied bilaterally and involves stimulation of both agonist and antagonist muscle groups in a coordinated activity. Similarly recent advances in restorative therapies for children using ES include non-invasive spinal neuromodulation which has been demonstrated in initial trials to produce statistically significant improvement in function [70].

As with other types of ES, both conventional stimulation and neuromodulation therapies are most effective when applied alongside volitional effort and ideally, used as part of an active exercise or function. The chosen task undertaken should be as practical and engaging as possible for the child and should be carried out as regularly as possible, ideally up to five times per week with high repetitions of task-specific activity where possible. The use of novel activities and those of interest to the child also provides an enriched experience – required to further enhance cortical learning.

16.4.4   Application of electrical stimulation

The instruction below will consider the needs of the child, the family, the clinician, the device and how maximal effectiveness of all leads to the most effective clinical application for the child. This will be addressed according to what to consider prior to application, during application and after application.

16.4.4.1   Consideration prior to application

1. Following a full assessment of the child, consider the child's needs/impairment and how ES may be indicated or may enhance treatment, e.g. consider physical benefits to be gained via ES and the expected outcomes to treatment. Also consider pertinent contraindications or cautions and risks to treatment, including any allergies, the child may present with (see the section above on precautions and contraindications to the use of ES).

   (a) Thought should be paid to the following areas of need:
   - Normal growth & development / age.
     - Consider the developmental age of the child and the activity to be performed (and whether this is age and neurodevelopmentally appropriate) and the necessary muscle(s) to be selected. See Table 16.4 for consideration of ES application through stages of development)
   - The Impact of the impairment / aims of treatment.
     - What movement can ES support to enhance normal function or to support loss or weakness?
   - Activity & participation.
     - Consider what task, game or activity will suit both the developmental age of the child and the chosen area of ES and/or impairment.

- Family and child routines.
  - ○ If the plan is for continuation of the treatment at home, consider if this will fit in with all the other areas of need the child and the family are already committed to at home, or could it replace something else? Support the parent/carer develop an understanding of therapeutic goals so they can develop their own home-based activities to fit around family life.
- The multiple professionals (and perspectives) surrounding the child.
  - ○ Children very often will have many more professionals involved in their care (compared to their adult counterparts). Consider whether all professionals working with the child are on board with the effect and plan for use of ES or will the parent/child be unsupported by professionals who do not agree or are not comfortable or competent with provision of the stimulation technique.
- Gather sufficient background knowledge and understanding of ES to be able to explain the rationale for use, to enable the necessary and usual consents to treatment (of the parent/caregiver but also the child (as age appropriate). Use of explanation, leaflets, websites, photos, videos may all be useful.
  - ○ Explain the procedure in child-friendly language, i.e. use of the 'tickle box' or the 'buzzy bee box'.
  - ○ Explain what the therapy entails and what you hope to achieve from use – if the parent/caregiver is onside, interested and enthusiastic about the use then generally so will the child. It is useful to ask the parent to remain calm and relaxed during initial treatment to enable the child to do similar.
  - ○ Provide parents/caregivers with appropriate training and education on the procedure, including electrode placement and stimulation techniques, so they can continue the therapy at home if necessary. Consider competence and competencies. Ensure they understand the importance of adhering to safety guidelines and consulting with you/their healthcare professionals for ongoing support

2. Demonstrate the ES device on yourself/your colleague or on the child's parent/caregiver so the child knows what to expect.
   - Use the parent/caregiver if you (and they) are confident they can maintain a relaxed, fun and interested demeanour otherwise demonstrate it on yourself (nothing will put a child off use of ES more than watching their parent/caregiver jump and squeal when the amplitude is turned up).

3. Consider appropriate clinical outcome measures to support the efficacy of your therapy (e.g. Oxford Scale, Visual Analogue Scale, 10 m walk test) as well as perhaps considering use of videos and photographs.

4. Consider which muscle(s) are to be stimulated according to the activity selected, and developmental age.

5. Select the appropriate ES device that will match your expected outcomes. Familiarise yourself (including appropriate training) with the device operation (via user manual/device manufacturer representative) – reduce the extent of experimental parameter selection on the child.

6. Select appropriate electrodes. Proper electrode usage and placement is crucial for effective and safe ES.

   • Electrode size: When choosing electrodes for the child, take into account their smaller body size, and the size and location of the muscles to be stimulated.

      o Electrodes that are too large may cause unwanted overflow to surrounding tissues.

      o Electrodes that are too small may be uncomfortable. This may result in difficulty reaching a high enough intensity to achieve the desired outcome.

      o Consider potential allergies when choosing electrode type.

   • Consider using electrodes with adhesive properties that are suitable for sensitive skin. Rubber electrodes (with use of contact gel) may also be considered. It is important to ensure that the electrodes are properly positioned and secure to avoid any discomfort or skin irritation.

   • If you expect the child may be anxious about the electrodes being applied to them, start by encouraging the child to stick stickers on themselves or yourself and build up to use with electrodes (without being connected to the stimulation device). Self-adhesive electrodes can also be trimmed into fun shapes to make them more appealing (caution not to cut the core of the electrode or make the electrode too small). Colourful/character stickers can be applied to the electrodes to make them more appealing or allow the child to draw on their electrodes prior to application (Figure 16.6).

   • Consider timely replacement of the electrodes (in accordance with the electrode manufactures advice) to enhance the impact of therapeutic dose but also to negate the risk of skin insult/burns. Adhesive electrodes should not be used once the contact adhesive has dried out and is no longer sticky.

7. Choose the appropriate therapy parameters to achieve the desired outcome (frequency, pulse duration, duty cycle, duration of treatment, repetitions of the activity)

   • Parameters are chosen considering the age, strength and general health of the child.

   • The frequency and duration of the stimulation should be tailored to the child's specific needs and condition.

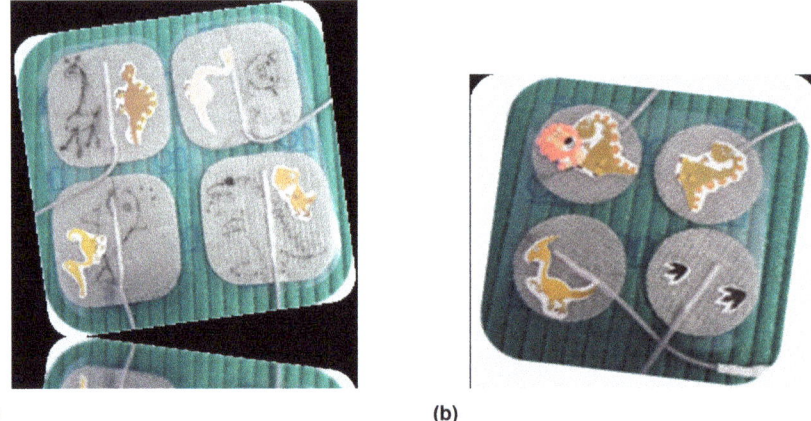

(a)   (b)

Figure 16.6   Child friendly electrode application.
Use of stickers and drawings on electrodes can help improve acceptance.

- Start with lower intensity levels (amplitude and pulse duration/width) and gradually increase based on the child's tolerance and muscle response. A wider pulse duration/width may allow you to use less amplitude if sensation is a limiting factor.

16.4.4.2   Consideration during application

Paediatric ES can be a daunting experience for children, so it is essential to create a positive and engaging environment and involve the child in the process as much as possible. Use distractions or play-based activities to keep the child relaxed and focused during all parts of the session.

1. Explain what the device is going to do – use lay/child terminology.
   - Consider the terminology used to describe the therapy (e.g. describing the sensation of stimulation as 'buzzy bees' rather than stimulation, describe the electrodes as 'stickers' rather than electrodes and the device as the 'tickle box' rather than ES)
2. Prepare the child/young person:
   - The child should be encouraged (as age appropriate) to handle the device and help attach the wires into the electrodes. Compliance and tolerance can be improved by allowing the child to connect the wires and ready the device themselves (with support) (Figure 16.7).
   - For the very anxious child, build confidence using the device in stages, e.g. confidence with stickers, then electrodes then with device connected (but not switched on).
     - Prepare the skin in the target area (skin should be clean, remove excessive hair if necessary – clip hair don't shave).
     - Place the area of the body to be treated in an appropriate position.

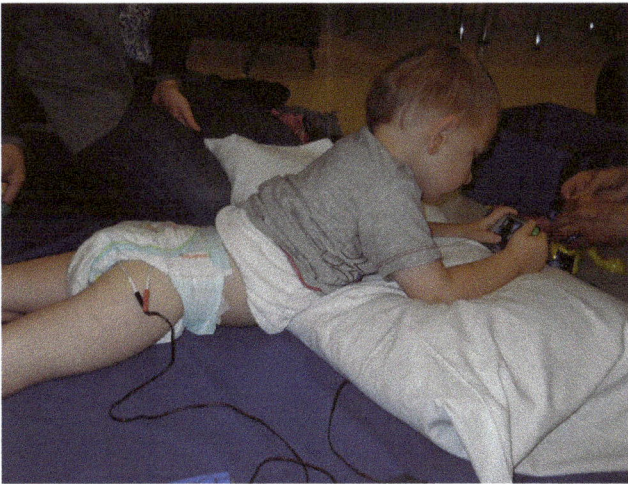

Figure 16.7   Allowing the child to support control of their own NMES unit. Involving the child in the use of electrical stimulation devices. Allowing them to connect wires or control amplitude can improve acceptance and compliance.

      o   For muscular applications, consider placing the limb/patient in a task specific position, or a neutral, anatomical/mid-range muscle posture.

3. Place the electrodes over the motor point (if possible) or muscle belly of the muscle(s) selected for ES application.
   - For the very young child the muscle belly is not always easy to locate, in this instance start at a mid-point, similar to where you would expect the anatomical alignment to be.
4. Advise the child and parent/caregiver the treatment is commencing and (having previously chosen and set your parameters) gradually increase the amplitude until the desired functional/muscular effect is produced – tolerance and response may vary from session to session.
   - It may be useful to 'fish' with the electrode during treatment to find the maximal position for stimulation, pen electrodes may also be useful to consider when muscle groups are small.
5. The child should be (as developmentally appropriate) an active participant in the chosen ES supported task and encouraged, as amplitude rises, to initiate or join in with any movement elicited. Alternatively, dependent on age and ability, the task may benefit from either clinician assistance of the chosen movement (so the child can learn/feel the movement requested) or the child's may be encouraged to follow their own instincts to move/play/function.
6. Consider incorporating play and/or distraction to aid tolerance and maximise therapeutic dose.

- The child should be granted the freedom to move freely during treatment and make mistakes along the way to learn the chosen task. Use of a trigger switch may be useful to support ES elicited activity for children too young to follow instruction. This may also result in choosing smaller, more portable devices that can be clipped to clothing (to not impede movement) or consider active movement in a stationary position by use of a treadmill for larger ES devices.
- Vary practice to incorporate a variety of games, tasks, activities and toys. Consider how it could be incorporated into their wider therapy programme.
- When utilising ES to determine sensation level or awareness (in the very young child), it may be useful that the child is distracted watching a screen or playing with a parent/caregiver. This is a useful strategy for the child who may not be able to describe feeling otherwise, more often when distracted with play (if they do have an awareness of sensation) they will turn to the limb having ES applied as the amplitude is raised gently. For those with little or no sensation the child will continue to be distracted by play. A full or altered level of sensation can be a limiter to use of ES as a treatment modality (if the child perceives the sensation as discomfort) whilst they do not have the comprehension to understand impact or effect.

7. The clinician should be mindful and aware of signs of fatigue in the child and discontinue treatment accordingly – endurance improving with time/use. Once put off from a session of uncomfortable ES delivery it can be a slow journey back to compliance thereafter. Over enthusiastic parents/carers and therapists alike should be dissuaded from the assumption that 'more is better'.

8. Constant monitoring and observation during the ES session is crucial to ensure the child's safety and comfort. Keep a close eye on the child's responses, such that any signs of discomfort, pain or adverse reactions are noted. If any issues arise, stop ES and reconsider treatment parameters, and affect.

9. Be aware of the possibility of seldom (but nonetheless reported) adverse responses including skin irritation, dizziness, abnormal vital signs, epileptic episodes. Should they occur, stop ES and resolve before re-starting the stimulation program.

16.4.4.3   Considerations after application

1. When the session has finished, turn down the amplitude gradually.
2. Unplug connections carefully as directed by user guides/representatives.
3. Remove electrodes carefully, two handed and in line with any hairs. DO NOT REMOVE ELECTRODES FROM THE PATIENT WHILST STIMULATION IS BEING DELIVERED
4. Check skin – in many applications an erythema is expected underneath the site of the electrode. If it disappears within 60–120 min that is a normal response. If

it continues for 24 hours or more it is considered 'electrically induced dermatitis' and stimulation over that site must discontinue until the skin recovers.
5. If appropriate, stick the electrodes back onto the correct side of their protective backing (a water mist applied to the backing can prolong lifetime of the electrodes) and store in airtight pack/container.
6. Evaluate the therapeutic effects, record and plan follow up.

16.5   Precautions and contraindications

Whilst paediatric ES can be beneficial in many cases, there are certain precautions to consider outlined below (Table 16.5). Contraindications align with those applicable to adult use – the reader is directed to these chapters for further information which discuss adult contraindications.

Table 16.5   Precautions and contraindications

Implanted medical devices:	Children with implanted medical devices, such as pacemakers, defibrillators or deep brain stimulators, may have specific precautions regarding electrical stimulation. It is essential to consult with the child's medical lead to determine whether electrical stimulation is safe or could interfere with the functioning of the implanted device.
Age/cognition of child	Age or cognition is not a contraindication to the use of electrical stimulation, but a different approach may be required with the very young child due to cognition, understanding and communication
Attention span	The impact of attention span will depend on the intended use of the device, if a functional activity is intended and the child loses interest quickly, keep sessions short and build gradually. Use of a favourite toy, game or task may help maintain their attention progressively.
Compliance/motivation	Understanding the impact and intention of the desired strategy may help improve motivation. Consider the reason for noncompliance and how parameter selection may impact this, e.g. if the treatment is uncomfortable or too challenging (or not challenging enough)
Adipose	Higher stimulation intensities (pulse duration or amplitude) are required to produce muscle contractions with increased adipose tissue as the stimulation has to travel further/produces increased resistance to elicit the desired effect.
Anxiety	Start conservatively to ensure comfort is maintained. Stimulation parameters can always be increased gradually, at future sessions, to ensure optimal muscle contraction is eventually achieved.
Seizures	Children with a history of epilepsy or seizure disorders require careful consideration when using electrical stimulation.

(Continues)

Table 16.5   (Continued)

	Seizures can potentially be triggered by electrical stimulation, especially if the parameters (i.e. frequency, intensity) are not appropriately adjusted. Therefore, consultation with the child's medical lead to assess the safety of electrical stimulation use would be recommended.
Tumours	Patients with a cancerous tumour in the area of the electrical stimulation should not be treated with electrical stimulation as increased local blood flow may increase tumour growth. However electrical stimulation applied on an area of the body remote from the tumour may be acceptable after medical clearance. Seek advice from the patient's medical team.
Understanding/awareness of parent/carer	Information provided to the parent/caregiver will improve the understanding and acceptance of electrical stimulation treatment for the child/young person. The safe and low level of electrical application, the need to apply the intervention regularly for successful implementation, how the intervention 'feels' and what the intended goal of the intervention is, are all areas that are useful to discuss and describe. More generally with all treatment modalities if the parent champions a treatment, it will be endured and used more effectively by the child compared to disengaged parental support. Consider if further information regards the impact of treatment may help to gain the confidence and interest of the parent/caregiver. Provide education so the parent can deliver the treatment without clinical support which may help adoption of the routine as less exercise like and more akin to a daily health routine like teeth brushing.
Volitional activity	Electrical stimulation should be used in combination with available volitional activity to improve strength and sensory-motor control. However, if the intensity (amplitude and pulse duration) is excessive this will impede the volitional ability of the child/young person.
Hair	Excessive hair will prevent adhesive electrodes from sticking and increase impedance. Hair should be clipped, NOT shaved, as this will increase skin irritation. Securing water-based electrodes may be a preference.
Limb length	For functional antigravity movements limb length will often dictate the intensity required to perform the activity. Longer limbs require greater force to elicit movement.
Muscle bulk	Increased muscle bulk will require higher intensities (pulse duration or amplitude) to recruit maximum muscle fibres. Reduced muscle bulk will require lower intensities to prevent overflow and co-contraction. As such the treatment parameters for a toddler and a pre-teen will vary.
Range of motion	Ensure there is sufficient range of motion in the joints affected by the muscle contraction desired, e.g. sufficient ankle range of motion when using electrical stimulation for foot drop

Table 16.5 *(Continued)*

	stimulation. However, the use of electrical stimulation to improve joint range can also be a useful goal of intervention.
Sensitivity	Some patients may have sensation or the ability to feel the effects of the electrical current flow. In this instance electrical stimulation must be comfortable if treatment is to be successful and to be repeated.
Sensory over responsiveness	Some children with neurodisabilities may have sensory processing difficulties, including hypersensitivity or sensory over responsiveness. Electrical stimulation can be overwhelming and uncomfortable for these children, potentially leading to adverse reactions. A thorough evaluation of sensory profiles is important before implementing electrical stimulation.
Skin surface	Make sure the area where the electrodes will be placed is clean and hydrated (moisturised/oily skin will prevent the electrodes adhering to the skin). Electrical stimulation requires proper electrode placement and good skin integrity. If a child has compromised skin, such as severe dermatitis, burns or open sores, electrical stimulation should be avoided in those areas.
Active infections	Electrical stimulation should not be applied directly to areas with active infections or open wounds. Infections can be worsened, and it may impede the healing process.
Swelling/oedema	Higher stimulation intensities (pulse duration or amplitude) are required to produce muscle contractions in the presence of swelling/oedema.
Time	As with any muscle training, a weak muscle will fatigue more quickly. Short sessions may be required initially (i.e. 10 min or less), building to longer sessions. Some devices have pre-set programmes with pre-set parameters and session times. When considering treatment time, the therapist should consider how many repetitions are achieved in each treatment session and how this compares to more conventional muscle strengthening to evoke an increased in strength or quality of muscle movement

Adapted from the work of Hart/White [1].

16.6 The future face of paediatric electrical stimulation

In this chapter, we discuss the application of ES in paediatrics, and highlight its use in improving muscle function, promoting motor recovery, offering neural prosthetic control and improving activities of daily living for various paediatric patient populations. ES plays a vital role in children with upper motor neuron involvement (e.g. CP or brain injury), lower motor neuron pathology (e.g. brachial plexus injury), mixed upper and lower motor neuron disorders (e.g. paediatric SCI and myelomeningocele or spina bifida) and youngsters with skeletal maturation disorders, including idiopathic spine deformity. The future of ES in paediatric physical therapy care includes:

- Tailored treatment with appropriate diagnostic reasoning and assessment of motor and sensory nerve innervation.
- Standardisation of care in ES treatments including dosage, electrode location and integrated use of ES with other interventions
- Collaboration in ongoing research to advance evidence-based practices.

16.6.1   Tailored treatment

Standardised assessment and diagnostic reasoning will justify the inclusion of ES in treatment plans for children. This knowledge will help to construct a framework for treatment. For example, in children with SCIs, it's important to recognise that not all patients with the same level of injury have identical innervation status below the zone of SCI. A child's medical diagnosis does not directly correspond to treatment outcome and further assessment of current or prior muscle denervation in medical conditions such as SCIs and spina bifida, is vital for tailoring treatment. Electromyographic testing can guide reasonable goal setting and, when there is potential for recovery, it will help set the timeline for physical intervention to improve muscle performance and function. In future practice the physical therapist will identify specific questions to be answered by clinical electromyography and evoked potential studies or, in the case of CP or brain injury (when one or more muscles may be out-of-phase), by kinesiological electromyography with intramuscular EMG recording.

Physical therapists are a proactive member of the paediatric medical team. Therapists provide realistic short and long-term expectations, interpret clinical assessments, including clinical and/or kinesiological EMG; and collaborate on decisions regarding injections of chemical denervation agents and/or surgical procedures. Physical therapists are leaders in the consideration of the use of more complex and expensive hybrid orthotics and ES systems, and implanted systems.

16.6.2   Standardisation of care

The future of physical therapy treatment includes standardised consideration of ES in the plan of care for children, as for adults, with the theme of improving function in the developing child. Clinical guidelines from professional physical therapy associations provide recommendations of care that guide treatment. For the physical therapist, providing ES in the child's plan of care includes multiple goals and considers the application of non-invasive proximal, peripheral, transcranial stimulation or referral for implanted ES. When ES proves essential to maintaining goals, the physical therapist will recommend the child for neural prosthetic consideration. Families are supported and educated on the use of ES in daily home routine as appropriate.

16.6.3   Collaboration

Continued collaboration between medical providers, researchers and technology developers are crucial for realising the full application of ES in improving the lives of children. Continued collaboration with researchers will help to understand ES's impact on neuroplasticity to promote adaptive changes, long-term effects and mechanisms involved in motor changes. Research on the use of ES in paediatrics encompasses the location of electrode placement (peripheral vs. proximal vs. implantable), comfortable parameter settings and dosage, and best therapeutic practices (interventions) to incorporate with ES to functional outcomes. Research is funded and designed to appropriately classify participants on nerve innervation not medical diagnosis and provides reproducible clinical use.

16.6.4   The future is here

Looking forward, ES will be a standard of care and routinely included in treatment options for common paediatric disorders. Physical Therapists are responsible to gather evaluation findings to determine if children are candidates for ES and provide appropriately dosed intervention based on documented evidence of improvements. Easily accessible, wearable, multichannel stimulators with programmable pulse duration, pulse repetition rate, ramp, duty cycle, amplitude modulation and switch triggered devices are available for clinical and personal use. Continued clinician support to integrate ES into standard of care practice will continue to advance the use of ES. Physical therapists can educate families and children to understand reasonable expectations and the outcomes of treatment. Ongoing clinical research will provide evidence to progress treatment and integrate into best practices in paediatric healthcare. The future holds the promise of early childhood disability recognition, determination of the underlying causes of impairments and excellence in health care to improve the life of our youngest citizens with paediatric physical therapy care.

References

[1]   Association of Paediatric Chartered Physiotherapists. Guidance for paediatric physiotherapist working with surface electrical stimulation in neurorehabilitation (2020). From: http://apcp.csp.org.uk/publications/guidance-paediatric-physiotherapists-working-surface-electrical-stimulation (Accessed 03/02/2024).

[2]   Dotan R, Mitchell C, Cohen R, Klentrou P, Gabriel D, and Falk B. Child—adult differences in muscle activation—a review. *Pediatric Exercise Science*. 2012;24(1):2–21.

[3]   Birat A, Bourdier P, Piponnier E, *et al.* Metabolic and fatigue profiles are comparable between prepubertal children and well-trained adult endurance athletes. *Frontiers in Physiology*. 2018;9:387.

[4]   Tonson A, Ratel S, Le Fur Y, Vilmen C, Cozzone PJ, and Bendahan D. Muscle energetics changes throughout maturation: a quantitative 31P-MRS analysis. *Journal of Applied Physiology*. 2010;109(6):1769–78.

[5]   Faigenbaum AD, Westcott WL, Loud RL, and Long C. The effects of different resistance training protocols on muscular strength and endurance development in children. *Pediatrics*. 1999;104(1):e5–e5.

[6]   Stackhouse SK, Binder-Macleod SA, and Lee SC. Voluntary muscle activation, contractile properties, and fatigability in children with and without cerebral palsy. *Muscle & Nerve: Official Journal of the American Association of Electrodiagnostic Medicine*. 2005;31(5):594–601.

[7]   Tammik K, Matlep M, Ereline J, Gapeyeva H, and Pääsuke M. Quadriceps femoris muscle voluntary force and relaxation capacity in children with spastic diplegic cerebral palsy. *Pediatric Exercise Science*. 2008;20(1):18–28.

[8]   Damiano DL, Martellotta TL, Sullivan DJ, Granata KP, and Abel MF. Muscle force production and functional performance in spastic cerebral palsy: relationship of cocontraction. *Archives of Physical Medicine and Rehabilitation*. 2000;81(7):895–900.

[9]   Malaiya R, McNee AE, Fry NR, Eve LC, Gough M, and Shortland AP. The morphology of the medial gastrocnemius in typically developing children and children with spastic hemiplegic cerebral palsy. *Journal of Electromyography and Kinesiology*. 2007;17(6):657–63.

[10]   Mohagheghi AA, Khan T, Meadows TH, Giannikas K, Baltzopoulos V, and Maganaris CN. In vivo gastrocnemius muscle fascicle length in children with and without diplegic cerebral palsy. *Developmental Medicine & Child Neurology*. 2008;50(1):44–50.

[11]   Barber LE, Hastings-Ison TA, Baker R, Barrett RO, and Lichtwark G. Medial gastrocnemius muscle volume and fascicle length in children aged 2 to 5 years with cerebral palsy. *Developmental Medicine & Child Neurology*. 2011;53(6):543–8.

[12]   Marbini A, Ferrari A, Cioni G, Bellanova MF, Fusco C, and Gemignani F. Immunohistochemical study of muscle biopsy in children with cerebral palsy. *Brain and Development*. 2002;24(2):63–6.

[13]   Moreau NG, and Lieber RL. Effects of voluntary exercise on muscle structure and function in cerebral palsy. *Developmental Medicine & Child Neurology*. 2022;64(6):700–8.

[14]   Merino-Andrés J, Garcia de Mateos-Lopez A, Damiano DL, and Sánchez-Sierra A. Effect of muscle strength training in children and adolescents with spastic cerebral palsy: a systematic review and meta-analysis. *Clinical Rehabilitation*. 2022;36(1):4–14.

[15]   McNee AE, Gough M, Morrissey MC, and Shortland AP. Increases in muscle volume after plantarflexor strength training in children with spastic cerebral palsy. *Developmental Medicine & Child Neurology*. 2009;51(6):429–35.

[16]   Rose J, Haskell WL, Gamble JG, Hamilton RL, Brown DA, and Rinsky L. Muscle pathology and clinical measures of disability in children with cerebral palsy. *Journal of Orthopaedic Research*. 1994;12(6):758–68.

[17]   Gage JR, Schwartz MH, Koop SE, Novacheck TF, eds. *The Identification and Treatment of Gait Problems in Cerebral Palsy*. New York: John Wiley & Sons; 2009.

[18]   Multani I, Manji J, Hastings-Ison T, Khot A, and Graham K. Botulinum toxin in the management of children with cerebral palsy. *Pediatric Drugs*. 2019;21(4):261–81.

[19] Park TS, Dobbs MB, and Cho J. Evidence supporting selective dorsal rhizotomy for treatment of spastic cerebral palsy. *Cureus*. 2018;10(10):e3466.

[20] Steinbok P, Reiner A, and Kestle JR. Therapeutic electrical stimulation following selective posterior rhizotomy in children with spastic diplegia cerebral palsy: a randomized clinical trial. *Developmental Medicine & Child Neurology*. 1997;39(8):515–20.

[21] Carmick J. Clinical use of neuromuscular electrical stimulation for children with cerebral palsy, part 1: lower extremity. *Physical Therapy*. 1993;73 (8):505–13.

[22] Reed B. The physiology of neuromuscular electrical stimulation. *Pediatric Physical Therapy*. 1997;9(3):96–102.

[23] National Institute for Health and Care Excellence. Urinary incontinence in neurological disease: assessment and management. *Clinical Guideline [CG148]*. 2012.

[24] Bosques G, Martin R, McGee L, and Sadowsky C. Does therapeutic electrical stimulation improve function in children with disabilities? A comprehensive literature review. *Journal of Pediatric Rehabilitation Medicine*. 2016;9(2):83–99.

[25] Oxford Centre for Evidence-based Medicine. *Levels of evidence (March 2009)*. From: https://www.cebm.ox.ac.uk/resources/levels-of-evidence/oxford-centre-for-evidence-based-medicine-levels-of-evidence-march-2009 (Accessed 15/01/2024).

[26] Yan D, and Vassar R. Neuromuscular electrical stimulation for motor recovery in pediatric neurological conditions: a scoping review. *Developmental Medicine & Child Neurology*. 2021;63(12):1394–401.

[27] Elnaggar RK, Alqahtani BA, and Elbanna MF. Functional outcomes of botulinum neurotoxin-A injection followed by reciprocal electrical stimulation in children with cerebral palsy: a randomized controlled trial. *Restorative Neurology and Neuroscience*. 2020;38(6):431–41.

[28] Frick CG, Fink H, Blobner M, and Martyn J. A single injection of botulinum toxin decreases the margin of safety of neurotransmission at local and distant sites. *Anesthesia and Analgesia*. 2012;114(1):102.

[29] Greve KR, Joseph CF, Berry BE, Schadl K, and Rose J. Neuromuscular electrical stimulation to augment lower limb exercise and mobility in individuals with spastic cerebral palsy: a scoping review. *Frontiers in Physiology*. 2022;13:951899.

[30] Liu Z, Dong S, Zhong S, *et al.* The effect of combined transcranial pulsed current stimulation and transcutaneous electrical nerve stimulation on lower limb spasticity in children with spastic cerebral palsy: a randomized and controlled clinical study. *BMC Pediatrics*. 2021;21(1):1–7.

[31] Pennati GV, Bergling H, Carment L, Borg J, Lindberg PG, and Palmcrantz S. Effects of 60 min electrostimulation with the EXOPULSE mollii suit on objective signs of spasticity. *Frontiers in Neurology*. 2021;12:706610.

[32] NICE Medtech Innovation Briefing. *Mollii suit for spasticity, [MIB100]*, Published: 28 March 2017. From: www.nice.org.uk/guidance/mib100. Accessed 11 November 2023.

[33]   Arkkukangas M, Hedberg Graff J, and Denison E. Evaluation of the electrodress Mollii® to affect spasticity and motor function in children with cerebral palsy: seven experimental single-case studies with an ABAB design. *Cogent Engineering*. 2022;9(1):2064587.

[34]   Svane C, Nielsen JB, and Lorentzen J. Nonsurgical treatment options for muscle contractures in individuals with neurologic disorders: a systematic review with meta-analysis. *Archives of Rehabilitation Research and Clinical Translation*. 2021;3(1):100104.

[35]   Ko EJ, Sung IY, Yun GJ, Kang JA, Kim J, and Kim GE. Effects of lateral electrical surface stimulation on scoliosis in children with severe cerebral palsy: a pilot study. *Disability and Rehabilitation*. 2018;40(2):192–8.

[36]   Musselman KE, Manns P, Dawe J, Delgado R, and Yang JF. The feasibility of functional electrical stimulation to improve upper extremity function in a two-year-old child with perinatal stroke: a case report. *Physical & Occupational Therapy in Pediatrics*. 2018;38(1):97–112.

[37]   Swaffield E, Yang JF, Manns P, Chan K, and Musselman KE. Parents' perceptions of functional electrical stimulation as an upper limb intervention for young children with hemiparesis: qualitative interviews with mothers. *BMC Pediatrics*. 2022;22(1):346.

[38]   Alhusaini AA, Fallatah S, Melam GR, and Buragadda S. Efficacy of transcutaneous electrical nerve stimulation combined with therapeutic exercise on hand function in children with hemiplegic cerebral palsy. *Somatosensory & Motor Research*. 2019;36(1):49–55.

[39]   Chen YH, Wang HY, Liao CD, Liou TH, Escorpizo R, and Chen HC. Effectiveness of neuromuscular electrical stimulation in improving mobility in children with cerebral palsy: a systematic review and meta-analysis of randomized controlled trials. *Clinical Rehabilitation*. 2023;37(1):3–16.

[40]   Shideler BL, Bulea TC, Chen J, Stanley CJ, Gravunder AJ, and Damiano DL. Toward a hybrid exoskeleton for crouch gait in children with cerebral palsy: neuromuscular electrical stimulation for improved knee extension. *Journal of Neuroengineering and Rehabilitation*. 2020;17(1):1–4.

[41]   Singleton C, Jones H, and Maycock L. Functional electrical stimulation (FES) for children and young people with cerebral palsy. *Paediatrics and Child Health*. 2019;29(11):498–502.

[42]   Moll I, Marcellis RG, Fleuren SM, *et al.* Functional electrical stimulation during walking in children with unilateral spastic cerebral palsy: a randomized cross-over trial. *Developmental Medicine & Child Neurology*. 2023;66 (5):598–609.

[43]   Pool D, Valentine J, Bear N, Donnelly CJ, Elliott C, and Stannage K. The orthotic and therapeutic effects following daily community applied functional electrical stimulation in children with unilateral spastic cerebral palsy: a randomised controlled trial. *BMC Pediatrics*. 2015;15(1):1–0.

[44]   Karabay İ, Doğan A, Ekiz T, Köseoğlu BF, and Ersöz M. Training postural control and sitting in children with cerebral palsy: kinesio taping vs.

neuromuscular electrical stimulation. *Complementary Therapies in Clinical Practice.* 2016;24:67–72.

[45] Elbasan B, Akaya KU, Akyuz M, and Oskay D. Effects of neuromuscular electrical stimulation and Kinesio Taping applications in children with cerebral palsy on postural control and sitting balance. *Journal of Back and Musculoskeletal Rehabilitation.* 2018;31(1):49–55.

[46] Fife S, Roxborough LA, Story M, and Field D. Reliability of a measure to assess outcomes of adaptive seating in children with neuromotor disabilities. *Canadian Journal of Rehabilitation.* 1993;7:11–11.

[47] Keller A, Singh G, Sommerfeld JH, *et al.* Noninvasive spinal stimulation safely enables upright posture in children with spinal cord injury. *Nature Communications.* 2021;12(1):5850.

[48] Armstrong EL, Boyd RN, Horan SA, Kentish MJ, Ware RS, and Carty CP. Functional electrical stimulation cycling, goal-directed training, and adapted cycling for children with cerebral palsy: a randomized controlled trial. *Developmental Medicine & Child Neurology.* 2020;62(12):1406–13.

[49] Özen N, Unlu E, Karaahmet OZ, Gurcay E, Gundogdu I, and Umay E. Effectiveness of functional electrical stimulation-cycling treatment in children with cerebral palsy. *Malawi Medical Journal.* 2021;33(3):144–52.

[50] Sansare A, Harrington AT, Wright H, *et al.* Aerobic responses to FES-assisted and volitional cycling in children with cerebral palsy. *Sensors.* 2021; 21(22):7590.

[51] Gosa MM, and Marshall J. Effect of device-assisted therapy in children with oropharyngeal dysphagia: a rapid review. *Perspectives of the ASHA Special Interest Groups.* 2023;9(1);215–227.

[52] Barroso Jr U, Tourinho R, Lordêlo P, Hoebeke P, and Chase J. Electrical stimulation for lower urinary tract dysfunction in children: a systematic review of the literature. *Neurourology and Urodynamics.* 2011;30(8):1429–36.

[53] Quintiliano F, Veiga ML, Moraes M, *et al.* Transcutaneous parasacral electrical stimulation vs oxybutynin for the treatment of overactive bladder in children: a randomized clinical trial. *The Journal of Urology.* 2015;193 (5S):1749–53.

[54] Lu ML, He J, and Lu S. Electrical stimulation therapy for slow transit constipation in children: a systematic review. *International Journal of Colorectal Disease.* 2015;30:697–702.

[55] Sharifi-Rad L, Ladi-Seyedian SS, Manouchehri N, *et al.* Effects of interferential electrical stimulation plus pelvic floor muscles exercises on functional constipation in children: a randomized clinical trial. *Official Journal of the American College of Gastroenterology/ ACG.* 2018;113(2):295–302.

[56] Fichtner-Feigl S, Sailer M, Höcht B, and Thiede A. Development of a new scoring system for the evaluation of incontinence and constipation in children. *Coloproctology.* 2003;25(1):10–15.

[57] Soumena RZ, Darma A, Ranuh RG, Gunawan PI, Sumitro KR, and Sudarmo SM. Efficacy of transcutaneous electrical nerve stimulation on slow transit

constipation in children with cerebral palsy. *Bali Medical Journal*. 2023;12 (1):192–6.

[58]   Islam S, McLaughlin J, Pierson J, Jolley C, Kedar A, and Abell T. Long-term outcomes of gastric electrical stimulation in children with gastroparesis. *Journal of Pediatric Surgery*. 2016;51(1):67–71.

[59]   Vanhorebeek I, Latronico N, and Van den Berghe G. ICU-acquired weakness. *Intensive Care Medicine*. 2020;46(4):637–53.

[60]   Magalhães P, Figueirêdo BB, Vasconcelos A, de Andrade ÉM, Dornelas de Andrade A, and Reinaux C. Is transcutaneous electrical muscle stimulation an alternative for preventing acquired muscle weakness in the pediatric intensive care unit? A scoping review. *Pediatric Pulmonology*. 2019;54(8):1108–16.

[61]   Cole JW, Murray DJ, McAllister JD, and Hirshberg GE. Emergence behaviour in children: defining the incidence of excitement and agitation following anaesthesia. *Pediatric Anesthesia*. 2002;12(5):442–7.

[62]   Lee SJ, and Sung TY. Emergence agitation: current knowledge and unresolved questions. *Korean Journal of Anesthesiology*. 2020;73(6):471–85.

[63]   Hijikata T, Mihara T, Nakamura N, Miwa T, Ka K, and Goto T. Electrical stimulation of the heart 7 acupuncture site for preventing emergence agitation in children. *European Journal of Anaesthesiology*. 2016;33(7):535–42.

[64]   Nakamura N, Mihara T, Hijikata T, Goto T, and Ka K. Unilateral electrical stimulation of the heart 7 acupuncture point to prevent emergence agitation in children: a prospective, double-blinded, randomized clinical trial. *PloS One*. 2018;13(10):e0204533.

[65]   Justice D, Awori J, Carlson S, Chang KW, and Yang LJ. Use of neuromuscular electrical stimulation in the treatment of neonatal brachial plexus palsy: a literature review. *The Open Journal of Occupational Therapy*. 2018;6(3):10.

[66]   Sadowsky CL, Becker D, Bosques G, *et al.* Rehabilitation in transverse myelitis. *CONTINUUM: Lifelong Learning in Neurology*. 2011;17(4):816–30.

[67]   Charman RA. Chapter 5: Motor learning, In: Trew M, and Everett T. (eds) *Human Movement: An Introductory Text*. (No Title). Churchill Livingstone, London, 1997.

[68]   Hebb DO. The organization of behavior: a neuropsychological theory. *Psychology Press*. London, 2005.

[69]   Mundkur N. Neuroplasticity in children. *The Indian Journal of Pediatrics*. 2005;72:855–7.

[70]   Hastings S, Zhong H, Feinstein R, *et al.* A pilot study combining noninvasive spinal neuromodulation and activity-based neurorehabilitation therapy in children with cerebral palsy. *Nature Communications*. 2022;13(1):5660.

Index

www.ingramcontent.com/pod-product-compliance
Ingram Content Group UK Ltd.
Pitfield, Milton Keynes, MK11 3LW, UK
UKHW020742140325
456129UK00001B/1

9 781839 538766